T0310306

# DRYING PHENOMENA

# DRYING PHENOMENA
## THEORY AND APPLICATIONS

**İbrahim Dinçer**
**and**
**Calin Zamfirescu**
*University of Ontario Institute of Technology, Oshawa, ON, Canada*

*Library of Congress Cataloging-in-Publication Data*

Dinçer, İbrahim, 1964– author.
Drying phenomena : theory and applications / İbrahim Dinçer and Calin Zamfirescu.
   pages  cm
Includes bibliographical references and index.
ISBN 978-1-119-97586-1 (cloth)
1. Drying.  I. Zamfirescu, Calin, author.  II. Title.
TP363.D48 2016
664′.0284–dc23

                                                2015025655

A catalogue record for this book is available from the British Library.

Set in 10/12pt Times by SPi Global, Pondicherry, India
Printed and bound in Singapore by Markono Print Media Pte Ltd


1   2016

# Contents

# Preface

Drying, as an energy-intensive process, plays a major role in various sectors, ranging from food industry to wood industry, and affects economies worldwide. Drying applications consume a noticeable part of the world's produced energy and require a careful attention from microlevel to macrolevel applications to make them more efficient, more cost effective, and more environmentally benign. Bringing all these dimensions into the designs, analyses, and assessments of drying systems for various practical applications is of paramount significance.

This book offers a unique coverage of the conventional and novel drying systems and applications while keeping a focus on the fundamentals of drying phenomena. It includes recent research and contributions in sustainable drying systems and integration with renewable energy. The book is expected to serve the drying technology specialists by providing comprehensive tools for system design, analysis, assessment, and improvement. This is essentially a research-oriented textbook with comprehensive coverage of the main concepts and drying systems designs. It includes practical features in a usable format for the design, analysis, multicriteria assessment, and improvement of drying processes and systems which are often not included in other solely academic textbooks. Due to an extensive coverage, practicing engineers, researchers, and graduate students in mainstream engineering fields of mechanical and chemical engineering can find useful information in this book.

The book consists of 11 chapters which amalgamate drying technology aspects starting from basic phenomena to advanced applications, by considering energy, exergy, efficiency, environment, economy, and sustainability issues. The first chapter covers in broad manner introductory topics of thermodynamics, energy, exergy, and transient heat transfer and mass transfer, so as to furnish the reader with sufficient background information necessary for the rest of the book.

Chapter 2 covers the basics of drying, introducing the drying phases and the related phenomena of heat and moisture transfer. The moist materials are characterized and classified (e.g., hydroscopic, nonhygroscopic, capillary, etc.) in relation with the mechanisms of moisture diffusion and associated phenomena such as shrinkage. Introduction to diffusion modeling through porous media and moist solids is provided.

Chapter 3 comprehensively classifies and describes drying devices systems. Two- and three-dimensional explanatory sketches are presented to facilitate the systems explanation. The most relevant processes occurring in drying systems and devices are presented for natural and forced drying.

Chapter 4 introduced the energy and exergy analyses for drying processes and systems. There are only few studies in the literature that treat the exergy analysis of drying processes and system; most of the published research limit to energy analyses only. Therefore, this chapter aims to fill this gap and provides a comprehensive method for irreversibility analysis of drying using exergy as a true method to identify the potentials for system improvement. Performance assessment of drying systems based on energy and exergy efficiency is explained in detail. Some relevant drying systems are analyzed in detail such as direct combustion dryers, fluidized bed dryers, and heat pump dryers.

Chapter 5 focuses on analytical methods for heat and moisture transfer. The solutions for moisture transfer in basic geometries such as infinite slab, infinite cylinder, and sphere are given. Parameters such as drying coefficient and lag factor which are essential for analytical modeling of the processes are introduced. The chapter also teaches about the analytical expressions for drying time of object with regular and irregular geometry and the so-called shape factors for drying time. One important aspect is represented by determination of moisture transfer diffusivity and moisture transfer coefficient in drying operation. A comprehensive method to determine these parameters based on the experimental drying curve is introduced. Also, the chapter allocates sufficient space to the analytical formulation and treatment of the process of simultaneous heat and moisture transfer. In this respect, the Luikov equations and other formulations for simultaneous heat and moisture transfer are presented and the impact of sorption–desorption isotherms is explained. A summary of drying curve equations and models is given.

Numerical heat and moisture transfer is treated extensively in Chapter 6. Finite difference schemes and three types of weighted residual numerical methods (finite element, finite volume, and boundary element) are introduced in sufficient detail. The subsequent part of the chapter is structured in three sections corresponding to one-, two-, and three-dimensional numerical analysis of heat and moisture transfer covering Cartesian, cylindrical, polar, and spherical coordinate systems. The influence of external flow field on heat and moisture transfer inside the moist material is also discussed.

Drying parameters and correlations are presented in Chapter 7. Selected correlations are introduced for quick, firsthand calculation of essential drying parameters such as drying time, moisture diffusivity, moisture transfer coefficient, binary diffusion coefficient, drying coefficient, and lag factor. An interesting and useful graphical method for moisture transfer parameters determination in drying processes is given.

Chapter 8 introduces the exergoeconomic and exergoenvironmental analyses for drying processes and systems. Here, the economic value of exergy is emphasized together with its role in economic analysis and environmental impact assessment of drying technologies. Two exergoeconomic methods and their application to drying are presented, namely, the energy–cost–exergy–mass and the specific exergy cost methods. The use of exergy and exergy destruction for environmental impact assessment of drying systems is explained.

Chapter 9 concentrates on optimization of drying processes and system. Optimization is crucial for the design of better systems with improved efficiency, effectiveness, more economically attractive and sustainable, and having a reduced environmental impact. It is important to formulate technical, economic, and environmental objective functions, and this aspect is extensively explained in the chapter. Single-objective and multiobjective optimizations are discussed.

Chapter 10 is about sustainability and environmental impact assessment of drying systems. Here, sustainability as a multidimensional parameter is defined and the most important sustainability indicators are introduced. An exergy-based sustainability assessment method is proposed which accounts for energy, environment, and sustainable development. Various aspects are discussed such as reference environment models and environmental impacts and the role of exergy destruction-based assessment of environmental impact of drying systems. A case study is treated comprehensively regarding the life cycle exergo-sustainability assessment of a heat pump dryer.

Some selected novel drying systems and applications are presented in Chapter 11 based on a literature review. The use of superheated steam as drying medium appears very promising and consists of a novel development trend on drying technology. Chemical heat pump-assisted dryers emerged as a technology push. Very impressive developments in spray drying are reported to cover drying and production of nanoparticles and microcapsules. These emerging technologies are relevant in medicine for nanotherapeutics, in pharmaceutical industry for drug delivery, and in food industry for foodstuff encapsulation. Other emerging technologies and applications such as ultrasonic drying and membrane-assisted air conditioning are reviewed.

The book comprises a large number of numerical examples and case studies, which provide the reader with a substantial learning experience in analysis, assessment, and design of practical applications. Included at the end of each chapter is the list of references which provides the truly curious reader with additional information on the topics yet not fully covered in the text.

We hope that this book brings a new dimension to drying technology teaching and learning, promoting up-to-date practices and methods and helping the community implement better solutions for a better, more sustainable future.

We acknowledge the assistance provided by Dr. Rasim Ovali for drawing various illustrations of the book.

We also acknowledge the support provided by the Natural Sciences and Engineering Research Council of Canada and Turkish Academy of Sciences.

Last but not least, we warmly thank our wives, Gulsen Dincer and Iuliana Zamfirescu, and our children Meliha, Miray, Ibrahim Eren, Zeynep, and Ibrahim Emir Dincer and Ioana and Cosmin Zamfirescu. They have been a great source of support and motivation, and their patience and understanding throughout this book have been most appreciated.

İbrahim Dinçer and Calin Zamfirescu
Oshawa, September 2015

# Nomenclature

| | |
|---|---|
| $a$ | empirical constant |
| $a$ | acceleration, m/s$^2$ |
| $a$ | general parameter |
| $a$ | thermal diffusivity, m$^2$/s |
| $a$ | regression coefficient |
| $a_1, a_2$ | constants |
| $a_w$ | water activity |
| $A$ | area (general; or area normal to the flow of heat or mass), m$^2$ |
| $A$ | discretization parameter |
| $A$ | factor in Eq. (7.8) |
| $\mathcal{A}$ | discretization matrix |
| AC | annual consumption |
| AI | annual income, $ |
| $A_n$ | factor in Eq. (5.10) |
| AP | annual production, units |
| $Ar$ | Archimedes number |
| AR | aspect ratio |
| ASI | aggregated sustainability index |
| $b$ | general parameter |
| $b$ | regression coefficient |
| $\mathit{b}$ | numerical scheme parameter |
| $B$ | driving force |
| $B$ | discretization parameter |
| $Bi$ | Biot number |
| $Bi_m$ | Biot number for moisture transfer |
| $B_n$ | factor in Eq. (5.10) |
| $c$ | speed of light in vacuum, m/s |
| $C$ | specific heat, J/kg K |
| $C$ | coefficients for numerical schemes |

| $C$ | molar concentration, mol/l |
|---|---|
| $C$ | cost, $ |
| $\dot{C}$ | cost rate, $/h |
| CEF | consumed energy fraction |
| $C_{ex}$ | exergy price, $ |
| CExF | consumed exergy fraction |
| CIEx | exergy based capital investment effectiveness |
| $C_m$ | moisture (or mass) concentration, kg/m$^3$ |
| COP | coefficient of performance |
| $C_p$ | specific heat, J/kg K |
| CP | capital productivity |
| CRF | capital recovery factor |
| CSF | capital salvage factor |
| $C_v$ | specific heat at constant volume, kJ/kg K |
| $d$ | diameter, m |
| $d$ | constant |
| $D$ | diffusion coefficient, m$^2$/s |
| $D$ | moisture diffusivity, m$^2$/s |
| $D_c$ | binary diffusion coefficient for water vapor in air, m$^2$/s |
| DDTOF | dimensionless drying time objective function |
| DE | drying effectiveness |
| $D_{eff}$ | effective diffusion coefficient, m$^2$/s |
| DEI | dryer emission indicator |
| $D_h$ | hydraulic diameter, m |
| $Di$ | Dincer number |
| $Di_m$ | Dincer number for mass transfer |
| DPV | drying product value |
| DQ | drying quality |
| $D_T$ | Soret coefficient for thermal diffusion, kg/m s K |
| $e$ | specific energy, kJ/kg |
| $e$ | elementary charge, C |
| $e$ | mass specific energy, kJ/kg |
| $E$ | shape factor |
| $E$ | energy, J |
| $\dot{E}$ | energy rate, W |
| EcI | eco-indicator |
| EE | embodied energy, GJ/t |
| EEOF | energy efficiency objective function |
| EF | ecological footprint |
| EI | environmental impact |
| $E_{in}OF$ | energy input objective function |
| EPC | environmental pollution cost, $/kg |
| $EPC_{ex}$ | exergetic environmental pollution cost, $/GJ |
| ex | specific exergy, kJ/kg |
| Ex | exergy amount, kJ |

| $\dot{\mathrm{E}}\mathrm{x}$ | exergy rate, kW |
|---|---|
| ExCI | specific exegetic capital investment |
| ExCDR | construction exergy expenditure to lifecycle exergy destruction ratio |
| ExIE | exergetic investment efficiency |
| ExEOF | exergy efficiency objective function |
| EUR | energy utilization ratio |
| $f$ | friction coefficient |
| $f$ | function |
| $f(r)$ | distribution of pores radius |
| $F$ | force, N |
| $F$ | Faraday constant, C/mol |
| $F$ | function |
| $F$ | radiative forcing, W/m$^2$ |
| $\mathcal{F}$ | dimensionless parameter |
| $F_1, F_2$ | series expansions for shape factors |
| $Fo$ | Fourier number |
| $F_{\mathrm{obj}}$ | objective function |
| $Fo_{\mathrm{m}}$ | Fourier number for mass transfer (dimensionless time) |
| $g$ | gravity constant ($= 9.81$ m/s$^2$) |
| $g$ | specific Gibbs free energy, kJ/kg |
| $G$ | basis weight |
| GC | generated capital, $ |
| GEI | grid emission indicator, g/kW h |
| GF | greenization factor |
| $Gr$ | Grashof number |
| $Gu$ | Gukhman number |
| GWP | global warming potential |
| $Gz$ | Graetz number |
| $h$ | specific enthalpy, kJ/kg |
| $h$ | Planck constant, kJ s |
| $H$ | enthalpy, kJ |
| $h_{\mathrm{m}}$ | moisture transfer coefficient, m/s |
| HR | Hausner ratio |
| HT | halving time |
| $h_{\mathrm{tr}}$ or $h$ | heat transfer coefficient, W/m$^2$ K |
| $i$ | inflation rate |
| $I$ | irradiation, W/m$^2$ |
| $I$ | electric current, A |
| Ind | indicator |
| $I_{\mathrm{v}}$ | luminous intensity, cd |
| $j$ | diffusive mass flux, kg/m$^2$ s |
| $j$ | mass flux, kg/m$^2$ s |
| $J_0$ | zeroth-order $J$ Bessel function |
| $J_1$ | first-order $J$ Bessel function |
| $J_{\mathrm{m}}$ | mass flux, kg/m$^2$ s |

| $\mathcal{J}_m, \mathcal{J}_q$ | boundary intervals |
|---|---|
| $k$ | thermal conductivity, W/m K |
| $k$ | drying rate, $s^{-1}$ |
| $K_{1,2}$ | parameters |
| $k$ | constant; coefficient, or parameter |
| $k_B$ | Boltzmann constant, J/K |
| $k_m$ | mass transfer coefficient, $s^{-1}$ |
| $l$ | (characteristic) length, m |
| $L$ | length, characteristic length or thickness, m |
| L | bed height, m |
| $L_c$ | (characteristic) dimension, m |
| LCC | levelized cost of consumables, \$/unit |
| $LCEI_{ex}$ | Life cycle exergetic emission indicator, g/kW h |
| LCSI | lifecycle sustainability index |
| $Le$ | Lewis number |
| LF | lag factor |
| LHV | lower heating value, MJ/kg |
| LPP | levelized product price \$ |
| LPPOF | levelized product price objective function |
| LT | life cycle time, years |
| $m$ | index |
| $m$ | mass, kg |
| $m$ | mass ratio |
| $\dot{m}$ | mass flow rate, kg/s |
| $\dot{m}''$ | mass flux, $kg/m^2 s$ |
| $m, n, p$ | number of elements (vector) |
| $M$ | molecular weight, kg/kmol |
| $M_a$ | relative molecular mass of air, kg/kmol |
| MEPC | molar environmental pollution cost, \$/kmol |
| $M_v$ | molecular mass of vapor, kg/mol |
| $n$ | index, exponent, number |
| $n$ | empiric exponent |
| $n$ | mole number, kmol |
| $n$ | adiabatic exponent |
| $n$ | system lifetime |
| $n$ | normal to surface |
| $N$ | number of particles |
| $N_A$ | Avogadro's number |
| NH | number of halving times |
| $n_{hour}$ | number of hours of operation, h |
| NI | net income, \$ |
| NSI | normalized sustainability index |
| $Nu$ | Nusselt number |
| $P$ | pressure, kPa |
| $P_a$ | partial pressure of air, Pa |

| | |
|---|---|
| $P_{am}$ | mean of partial pressures of air over the product surface and in drying air, Pa |
| PBP | payback period, years |
| $Pe$ | Péclet number |
| PoI | point of impingement |
| PP | performance parameter |
| $Pr$ | Prandtl number |
| $P_v$ | partial pressure of vapor, Pa |
| $P_{va}$ | partial pressure of vapor in drying air, Pa |
| $P_v^*$ | saturated vapor pressure, Pa |
| PVF | present value factor |
| $P_{vm}$ | mean of partial vapor pressures of vapor over the product surface and in drying air, Pa |
| $P_{vo}$ | vapor pressure over the product surface, Pa |
| PWI | present worth income, $ |
| PWF | present worth factor |
| $q$ | heat rate per unit area, W/m$^2$; flow rate per unit width or depth |
| $\dot{q}$ | heat flux, W/m$^2$ |
| $q$ | heat flux, W/m$^2$ |
| $Q$ | heat flux, J or kJ |
| $Q$ | quantity (amount) |
| $\dot{Q}$ | heat transfer rate, W |
| $\dot{Q}''$ | heat flux per unit of surface, W/m$^2$ |
| QP | quality parameter |
| $r$ | radial coordinate; radius, m |
| $r$ | aerodynamic resistance, m/s |
| $r$ | real discount rate |
| $r$ | latent heat, J/kg |
| $r$ | particle coordinate, m |
| $r$ | distance normal to the flow of heat, m |
| $r$ | mesh parameter |
| $R$ | loss ratio |
| $R$ | radius, radius of a single particle, m |
| $\mathcal{R}$ | universal gas constant, kJ/kg K |
| $Ra$ | Rayleigh number |
| RC | specific resource consumption |
| RD | relative drying |
| $Re$ | Reynolds number |
| RI | relative irreversibility |
| $\mathfrak{R}_n$ | residual function |
| $R_{pai}$ | practical application impact ratio |
| RPC | removal pollution cost |
| $R_{si}$ | sectorial impact ratio |
| $R_{ti}$ | technological impact ratio |
| $R_v$ | gas constant for water vapor, J/kg/K |
| $s$ | specific entropy, kJ/kg |

$\dot{S}$              entropy rate, kW/K
$S$              entropy, kJ/K
$S$              drying coefficient, s$^{-1}$
$S$              surface, m$^2$
$\dot{S}$              entropy rate, W/K
$Sc$              Schmidt number
SE              specific GHG emissions, kg$_{GHG}$/GJ
SEI              sustainability efficiency indicator
$S_g$              gas phase saturation
$Sh$              Sherwood number
SI              exergetic sustainability index
SIOF              sustainability index objective function
SP              span
SPI              sustainable process index
SRW              specific reversible work
SR              shrinkage ratio
$St$              Stanton number
SV              salvage value, $
$t$              time, s
$t$              tortuosity factor
$T$              temperature, K
$\mathcal{T}$              temperature function, K
$t_{05}$              halftime, h
$t_c$              tax credit
TCD              tax credit deduction, $
TExDOF              total exergy destruction objective function
$t_i$              tax on income
TI              taxable income, $
$T_m$              mean temperatures of product surface and drying air, °C
$T_{ma}$              mean absolute temperatures of product surface and drying air, K
$T_o$              surface temperature, K
TOI              tax on income, $
$t_{op}$              operational time, h
TOP              tax on property, $
$t_p$              tax on property
$t_s$              tax on salvage
$u$              specific internal energy, kJ/kg
$u$              velocity in $x$ direction
$u$              displacement, m
$U$              internal energy, kJ
U              flow velocity of drying air, m/s
U              economic utility
$v$              specific volume, m$^3$/kg
$v$              velocity in $y$ direction
$v$              velocity, m/s

| | |
|---|---|
| $V$ | volume, $m^3$ |
| $V$ | velocity, m/s |
| $\dot{V}$ | volumetric flow rate, $m^3$/s |
| $V_0$ | standard ideal gas volume, $m^3$/kmol |
| $u$ | velocity (speed), m/s |
| $w$ | mass specific work, kJ/kg |
| $w$ | weighting factors |
| $W$ | work, kJ |
| $\dot{W}$ | work rate, kW |
| $\mathcal{W}$ | moisture content function, kg/kg dry basis |
| $W$ | moisture content kg water/kg dry material |
| $\bar{W}$ | average moisture content, kg/kg |
| $x$ | quality, kg/kg |
| $x$ | Cartesian coordinate, m |
| $x_s$ | degree of saturation |
| $X_v$ | volumetric moisture content, $m^3/m^3$ |
| $y$ | mole fraction |
| $y$ | Cartesian coordinate, m |
| $y$ | dimensional coordinate, m |
| $Y$ | characteristic dimension (length), spatial dimension, m |
| $z$ | Cartesian coordinate, m |
| $z$ | axial coordinate, thickness, m |
| $Z$ | compressibility factor |

**Greek Letters**

| | |
|---|---|
| $\alpha$ | volume fraction of air |
| $\beta$ | enhancement factor |
| $\beta$ | volume-shrinkage coefficient |
| $\beta$ | length ratio |
| $\gamma$ | parameter |
| $\gamma$ | quality factor |
| $\gamma$ | climate sensitivity factor |
| $\delta$ | thickness, length, coordinate, m |
| $\delta$ | space increment, m |
| $\delta$ | thermal gradient coefficient, $K^{-1}$ |
| $\Delta h_{lv}$ | latent heat of vaporization, J/kg |
| $\Delta t$ | time step, s |
| $\varepsilon$ | void fraction |
| $\varepsilon$ | phase conversion |
| $\varepsilon$ | volumetric fraction of vapor |
| $\zeta$ | dimensionless coordinate |
| $\eta$ | energy efficiency |
| $\eta$ | dynamic viscosity, Pa/s |
| $\eta$ | dimensionless space variable |
| $\theta$ | total specific energy of flowing matter, kJ/kg |

| | |
|---|---|
| $\theta$ | dimensionless temperature |
| $\mu$ | dynamic viscosity, kg/ms |
| $\mu$ | chemical potential, kJ/kg |
| $\mu$ | diffusion resistance factor; root of the transcendental characteristic equation |
| $\mu_1$ | first eigenvalue |
| $\mu_n$ | $n$th eigenvalue |
| $\nu$ | kinematic viscosity, m$^2$/s |
| $\xi_M$ | specific mass capacity (kg mol/kJ) |
| $\xi_T$ | specific temperature coefficient (kg/kg K) |
| $\rho$ | density, kg/m$^3$ |
| $\rho_{dr}$ | bone dry density, kg/m$^3$ |
| $\sigma$ | Stefan–Boltzmann constant, W/m$^2$ K$^4$ |
| $\sigma$ | surface tension, N/m |
| $\sigma$ | standard average |
| $\tau$ | time constant, s |
| $\tau$ | residence time, s |
| $\tau$ | atmospheric lifetime, s |
| $\vartheta_{contact}$ | contact angle |
| $\phi$ | relative humidity |
| $\phi, \Phi$ | dimensionless moisture content |
| $\Phi_s$ | sphericity |
| $\varphi$ | total specific exergy, kJ/kg |
| $\varphi$ | porosity, m$^3$/m$^3$ |
| $\varphi$ | relative humidity |
| $\varphi$ | zenith angle |
| $\varphi$ | trial function |
| $\psi$ | exergy efficiency |
| $\psi$ | test function |
| $\omega$ | humidity ratio |
| $\Omega$ | domain of decision variables |

## Subscripts

| | |
|---|---|
| 0 | reference state |
| 0 | dry material |
| 0.5, 1, ½, ¼, ⅛, 2 | indices |
| 0.5 | half time |
| $\infty$ | bulk |
| a | (dry) air; medium; surroundings |
| act | activation |
| acum | accumulated |
| air | air |
| am | air mixer |
| ap | air penetration process |
| AP | air pollution |
| avg | average |
| b | boundary, dry bulb, bulk |

| | |
|---|---|
| b | fluidized bed |
| bw | bounded moisture |
| c | characteristic, critical, convection |
| c | cyclone |
| cap | capital |
| ch | chemical |
| CIE | capital investment effectiveness |
| cmp | compressor |
| comb | combustor |
| cond | condenser |
| conc | concentration |
| CO | carbon monoxide |
| cons | consumed |
| csteel | carbon steel |
| cv | control volume |
| cyl | cylinder |
| d | destroyed, dew point, drying |
| da | drying air |
| dissip | dissipation |
| dr | dryer |
| deliv | delivered |
| e | equilibrium |
| Eef | effective effusion |
| ef | effective |
| en | energetic |
| ex | exergy, exergetic |
| evap | evaporator |
| f | fluid; final; flow; force; formation, fuel |
| fa | fan |
| fc | feeder/conveyor |
| fg | liquid–vapor equilibrium |
| fi | filter |
| g | gas, global, generation |
| gen | generated |
| gt | gas turbine generator |
| H | high-temperature |
| ha | humid air |
| hp | heat pump |
| $i, j, k$ | indices |
| i, in | initial |
| in | input |
| int | internal |
| k | conduction |
| ke | kinetic energy |
| l | liquid, lateral |
| lam | laminar |

| lc | lifecycle |
|---|---|
| liq | liquid |
| loss, lost | lost |
| lv | liquid–vapor |
| L | low-temperature |
| m | mass, environment, material, moisture, moist material, market |
| m | monolayer |
| ma | material-to-air (binary coefficient) |
| mat | materials |
| mf | minimum fluidization |
| mm | moist material |
| mr | moisture removal |
| $n$ | normal direction |
| nf | nonflow |
| oc | other cost |
| occ | other cost creation |
| o&m | operation and maintenance |
| opt | optimum |
| out | output |
| p | particle |
| p, prod | product |
| pe | potential energy |
| ph | physical |
| pr | pollutant removal |
| pw | pollutant waste |
| $Q$ | heat |
| r | reduced |
| r | refrigerant |
| r | removed moisture |
| $R$ | radius |
| rec | recovered |
| ref | reference |
| rev | reversible |
| rf | recirculation flap |
| s | surface; solid, saturation, dry solid surface |
| sat | saturation |
| sc | supplementary combustor |
| sep | separator |
| shape | shape |
| slab | slab |
| sph | sphere |
| ssteel | stainless steel |
| surface | surface |
| sys | system |
| tot | total |

| | |
|---|---|
| tr | heat transfer |
| turb | turbulent |
| tv | throttling valve |
| used | utilized or used |
| v | vapor |
| w | wet bulb, water, wind, moisture, vapor |
| wb | wet bulb |
| wm | wet material |
| $x$ | $x$ direction |
| $y$ | $y$ direction |

## Superscripts

| | |
|---|---|
| $(\bar{\ })$ | average value |
| $''$ | saturation condition |
| 0 | reference state with respect to dry air |
| ch | chemical |
| $q$ | discretized time index |
| $Q$ | heat |

# 1

# Fundamental Aspects

## 1.1 Introduction

Drying is a key industrial process of great practical importance in chemical and pharmaceutical industries, agriculture and food processing, pulp and paper, wood and minerals processing, solid fuels preparation (e.g., biomass or coal drying). It consists of a mass transfer process aimed at removing a solvent – in general water (or moisture) – from a solid, liquid, or a semi-solid (a highly viscous liquid). Thence, drying is a thermally driven separation process and typically occurs by evaporation of the solvent (the moisture) or by sublimation or by a super-critical process that avoids solid–liquid boundary, or by reverse osmosis.

The process of drying is recognized as one of the most energy intensive process among the separation technologies. For example, according to Mujumdar (2006) drying energy sector in North America is just responsible of ~15% energy consumption. It requires a source of heat and sometimes it necessitates maintaining deep vacuum for effective moisture removal. Drying can be also applied in some cases followed by heat addition and moisture removal by sublimation. In addition, the method can be integrated with other types of separation technologies such as centrifugal draining which require the application of strong centrifugal forces.

In the drying sector, it is aimed to make drying processes more efficient, more cost effective, more environmentally benign, and more sustainable. Thus, this requires optimization methods to be applied to these processes. Furthermore, there is large panoply of materials spanning from thick slabs to nano-powders which require specific methods of drying.

Understanding drying processes requires the application of analysis methods from thermodynamics, heat, mass and momentum transfer, psychometrics, porous media, materials science, and sometimes chemical kinetics altogether. Some specific processes that occur during drying and must be considered in the analysis are crystallization and allotropic transition or shrinkage, texture change, porosity change, and fracture. Drying is a transient process; therefore, changing

*Drying Phenomena: Theory and Applications*, First Edition. İbrahim Dinçer and Calin Zamfirescu.

of moisture removal rate must be accounted for. Depending on the drying material (e.g., sample size, porosity, tortuosity) and drying conditions, drying can extend from milliseconds to couple of months.

In this chapter, fundamental aspects on drying are reviewed. First some general physical notions and basic properties and quantities are presented. Thermodynamics fundamentals are introduced. The name "thermodynamics" is meaningful: it came from the Greek words "therme," which means "heat," and "dynamis," which means power. Therefore, "thermodynamics" suggests a science of conversion of thermal energy into mechanical energy. In various sources, *thermodynamics* is defined as the science of energy and entropy. In this book, we define *thermodynamics* as the science of energy and exergy. This definition is more correct since both energy and exergy are given in the same units and can be used for performance assessment through energy and exergy efficiencies. Furthermore, this makes both first-law and second-law of thermodynamics very significant concepts as energy comes from the first law and exergy from the second law. These two laws essentially govern thermodynamic systems.

Thermodynamics differentiates two categories of energies: (i) organized (such as mechanical, electrical or electromagnetic, photonic, gravitational) and (ii) disorganized (which is thermal energy or "heat"). According to the second law of thermodynamics (SLT), which will be detailed further in this chapter, the organized forms of energy can be completely converted in any other forms of energy. However, thermal energy cannot be fully converted in organized forms of energy due to the intrinsic irreversibilities. In separation technologies – such as drying – Gibbs free energy is an important parameter that quantifies the required work (organized energy) to drive the process. Irreversibilities within the system can be determined according to the SLT. In this chapter, exergy analysis is expanded as a method to quantify the irreversibilities which helps in design improvement. The advantage of exergy method springs from the fact that it relates the thermodynamic analysis to the reference environment.

Exergy analysis is useful in identifying the causes, locations, and magnitudes of process inefficiencies and irreversibility. The exergy is a quantitative assessment of energy usefulness or quality. Exergy analysis acknowledges that, although energy cannot be created or destroyed, it can be degraded in quality, eventually reaching a state in which it is in complete equilibrium with the surroundings and hence of no further use for performing tasks (i.e., doing work).

Besides thermodynamics, a good amount of this chapter is dedicated to heat and mass transfer with emphasis on moisture diffusion in steady state and transient regime. In addition porous media are analyzed and characterized based on their porosity, tortuosity, and other parameters which affect moisture transfer. Psychometrics and most air modeling though energy and exergy methods is analyzed in detail.

## 1.2   Fundamental Properties and Quantities

In this section several thermodynamics properties and other physical quantities are covered to provide adequate preparation for the study of drying processes, systems and applications. Adoption of a *system of units* is an important step in the analyses. There are two main systems of units: the *International System of Units*, which is normally referred to as SI units, and the *English System of Units* (sometimes referred as *Imperial*). The SI units are used most widely throughout the world, although the English System is traditional in the United States. In this

**Table 1.1**   The fundamental quantities of the International System of Units

| Quantity | Definition | Measurement unit |
|---|---|---|
| Length ($l$) | Geometric distance between two points in space | m – meter; the length of the path traveled by light in vacuum during a time interval of 1/299,792,458 of a second |
| Mass ($m$) | Quantitative measurement of inertia that is the resistance to acceleration | kg – kilogram; the weight of the International Prototype of the kilogram made in platinum–iridium alloy |
| Time ($t$) | A measurable period that measures the progress of observable or non-observable events | s – second; duration of 9,192,631,770 periods of the radiation corresponding to the transition between the two hyperfine levels of the ground state of the cesium 133 atom |
| Electric current ($I$) | Rate of flow of electric charge | A – ampere; constant current which, if maintained in two straight parallel conductors of infinite length, of negligible circular cross-section, and placed 1 m apart in vacuum, would produce between these conductors a force equal to $2 \times 10^{-7}$ N/m of length |
| Thermodynamic temperature ($T$) | A measure of kinetic energy stored in a substance which has a minimum of zero (no kinetic "internal" energy) | K – Kelvin; the fraction 1/273.16 of the thermodynamic temperature of the triple point of water (having the isotopic composition defined exactly by the following amount of substance ratios: 0.00015576 mol of $^2$H per mole of $^1$H, 0.0003799 mol of $^{17}$O per mole of $^{16}$O, and 0.0020052 mol of $^{18}$O per mole of $^{16}$O) |
| Amount of substance ($n$) | The number of unambiguously specified entities of a substance such as electrons, atoms, molecules, and so on | mol – mole; the amount of substance of a system which contains as many elementary entities as there are atoms in 0.012 kg of carbon 12 |
| Luminous intensity ($I_v$) | Luminous flux of a light source per direction and solid angle, for a standard model of human eye sensitivity | cd – candela; the luminous intensity, in a given direction, of a source that emits monochromatic radiation of frequency 540 THz and that has a radiant intensity in that direction of 1/683 W/sr |

*Source*: ISU (2006).

book, SI units are primarily employed. However, relevant unit conversions and relationships between the International and English unit systems for fundamental properties and quantities are listed in Appendix A.

The International System of Units comprises seven basic quantities which are shown in Table 1.1. They include length, mass, time, electric current, thermodynamic temperature, amount of substance, and luminous intensity. All other quantities and units beside the fundamental ones are denoted as "derived quantities/units" and can be determined based on physical laws from the fundamental ones. Derived quantities such as specific volume and pressure are very important in thermodynamics.

Table 1.2 gives definitions of some quantities and notions which are very relevant in thermodynamics. The molecular mass of a substance represents the ratio between mass and the

**Table 1.2** Some quantities relevant in thermodynamics

| Quantity | Definition | Measurement unit |
|---|---|---|
| Molecular mass ($M$) | Ratio between mass and the amount of substance. $M = m/n$ | kg/kmol – kilogram per kilomole; the mass of one thousand moles of a substance |
| Specific volume ($v$) | Represents the volume of unit mass. $v = V/m$ | $m^3$/kg – cubic meter per kilogram; the volume a mass of 1 kg of a substance or assembly of substances |
| Velocity ($v$) | Rate of position change with respect to a reference system: $v = dl/dt$ | m/s – meter per second; a change of position of 1 m occurred during the time interval of 1 s |
| Acceleration ($a$) | Rate of velocity change: $a = dv/dt$ | $m/s^2$ – meter per squared second; a change of velocity of 1 m/s occurred during a time interval of 1 s |
| Force ($F$) | A measure of action capable to accelerate a mass: $F = ma$ | N – Newton; the force which accelerate a body of 1 kg with 1 $m/s^2$ |
| Pressure ($P$) | Represents the force exerted by unit of area: $P = F/A$ | Pa – Pascal; represents the force of 1 N (Newton) exerted on 1 $m^2$ surface |

| Notion and definition relationship | Explanatory sketch |
|---|---|
| Average velocity cross sectional velocity, $v$<br><br>$v \equiv \dfrac{1}{A}\displaystyle\int_A v_L dA$ (m/s)<br><br>Volumetric flow rate, $\dot{V}$<br>$\dot{V} \equiv vA$ (m³/s)<br>Mass flow rate, $\dot{m}$<br>$\dot{m} \equiv \dfrac{\dot{V}}{v}$ (kg/m³)<br>where $v$ is specific volume | 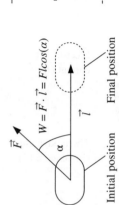 |
| Mechanical work<br>$W \equiv Fl\cos(\alpha)$<br>Gravity work<br>$W = mgz$<br>where $g = 9.81$ m/s² is the gravitational acceleration | 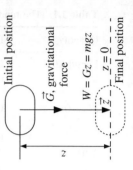 |

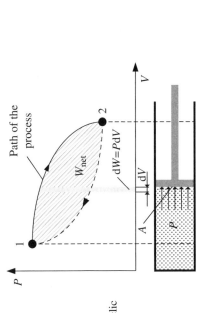

Boundary work

$$dW_b = F\,dl = PA\,dl \equiv P\,dV$$

$$W_b \equiv \int_1^2 P\,dV$$

where

$dl$ = infinitesimal boundary displacement

$A$ = moving boundary area

Note: Work is the area below the path of process; if the process is cyclic then the network is the area enclosed by the cycle path

Flow work:

$$dW_f = d(F\,l) = d(PA\,l) \equiv d(PV)$$

$$W_f \equiv \int d(PV) = PV$$

Specific flow work:

$$dw_f = d(Pv)$$

where $v$ = specific volume

Specific flow work rate:

$$d\dot{w}_f = \dot{m}\,d(Pv)$$

where $\dot{m}$ = mass flow rate

amount of substance. The ratio between volume and the mass of a material represents its specific volume, whereas the reciprocal of this is the density. The velocity is a vector that expresses the rate of position change of a material point. Acceleration is the rate of velocity change that expresses the change of movement due to action of forces. Pressure represents the force exerted per unit of area. The SI unit for pressure is Pa (Pascal), where $1\,Pa = 1\,N/m^2$, whereas in the English System it is pounds force per square foot, $lbf/ft^2$.

The atmosphere that surrounds the earth can be considered a reservoir of low-pressure air. Its weight exerts a pressure which varies with temperature, humidity, and altitude. Atmospheric pressure also varies from time to time at a single location, because of the movement of weather patterns. The standard value of the *atmospheric pressure* (or the pressure of standard atmosphere) is 101,325 Pa or 760 mmHg. Atmospheric pressure is often measured with an instrument called barometer; thence, the name of *barometric pressure*. While these changes in barometric pressure are usually less than 12.5 mm of mercury, they need to be taken into account when precise measurements are required.

**Gauge Pressure**. The *gauge pressure* is any pressure for which the base for measurement is atmospheric pressure expressed as kPa (gauge). Atmospheric pressure serves as a reference level for other types of pressure measurements, for example, gauge pressure. As shown in Figure 1.1, the gauge pressure is either positive or negative, depending on its level above or below atmospheric level. At the level of atmospheric pressure, the gauge pressure becomes zero.

**Absolute Pressure**. A different reference level is utilized to obtain a value for absolute pressure. The absolute pressure can be any pressure for which the base for measurement is

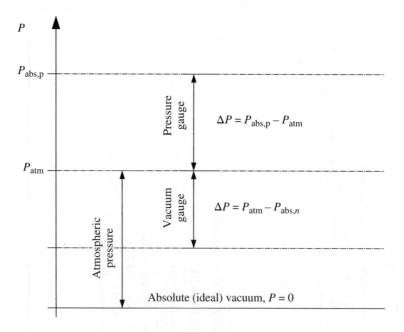

**Figure 1.1**   Illustration of pressures for measurement

a complete vacuum, and is expressed in kPa (absolute). Absolute pressure is composed of the sum of the gauge pressure (positive or negative) and the atmospheric pressure as follows:

$$\text{Pressure (gauge)} + \text{atmospheric pressure} = \text{Pressure (absolute)} \qquad (1.1)$$

For example, to obtain the absolute pressure, we simply add the value of atmospheric pressure. The absolute pressure is the most common one used in thermodynamic calculations, despite having the pressure difference between the absolute pressure and the atmospheric pressure existing in the gauge being what is read by most pressure gauges and indicators.

**Vacuum**. A vacuum is a pressure lower than atmospheric and occurs only in closed systems, except in outer space. It is also called *negative gauge pressure*. In engineering, vacuum is usually divided into four levels: (i) low vacuum representing pressures above 133 Pa (~1 Torr) absolute, (ii) medium vacuum varying between 1 and $10^{-3}$ (0.1333 to 133 Pa) absolute, (iii) high vacuum ranging between $10^{-3}$ and $10^{-6}$ Torr absolute (1 Torr = 133.3 Pa), and (iv) very high vacuum representing absolute pressure below $10^{-6}$ Torr. The *ideal vacuum* is characterized by the lack of any form of matter.

**Universe**. The *Universe* represents the cosmic system comprising matter and energy. Furthermore, *matter* is defined as particles with rest mass which includes *quarks* and *leptons* which are able to combine and form *protons* and *neutrons*. *Chemical elements* are matter formed by combinations of protons, neutrons, and electrons, whereas electrons are a class of leptons.

**Thermodynamic System**. By definition a *thermodynamic system* is a part of the universe, delimited by a real or imaginary boundary that separates the system from the rest of the universe, whereas the rest of the universe is denoted as the *surroundings*. If a thermodynamic system exchange energy but not matter with its surroundings it is said to be a *closed thermodynamic system* or *control mass*; see the representation from Figure 1.2a. A closed thermodynamic system which does not exchange energy in any form with its surroundings is denoted as *isolated system*. On the other hand an *open system* or *control volume* is a thermodynamic system which – as represented in Figure 1.2b can interact with its surroundings both by mass and energy transfer. Figure 1.3 shows an isolated thermodynamic system.

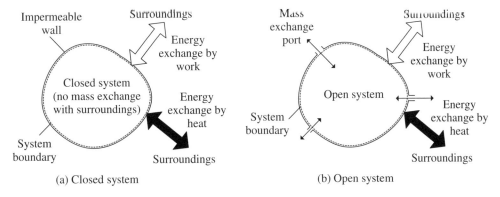

**Figure 1.2**   Illustrating the concept of thermodynamic system

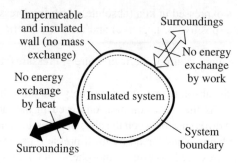

**Figure 1.3**    Representation of an isolated thermodynamic system

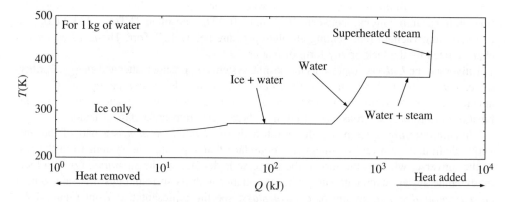

**Figure 1.4**    Representation of phase diagram of water

The interaction between a thermodynamic system and its surrounding take place by inter-mediate of *forces* or *force fields*. Several sources define that the physics notion of *field* repre-sents a region of universe which is affected by a force. Consequently, there is possible that no matter exist in a region of space where a field manifest. However, the contrary is not true, namely, fields exist in any region of the universe (viz. any thermodynamic system), even where matter (such as particles with rest mass) is inexistent.

Bulk matter or *substance* is formed by groups of atoms, molecules, and clusters of them. There are four *forms of aggregation* of substances, denoted also as phases or states, namely – *solid*, *liquid*, *gas*, and *plasma*. Each of the properties of a substance in a given state has only one definite value, regardless of how the substance reaches the state. Temperature and specific volume rep-resent a set of thermodynamic properties that define completely the thermodynamic state and the state of aggregation of a substance. The thermodynamic state of a system can be modified via various interactions, among which heat transfer is one. Heat can be added or removed from a system. When sufficient heat is added or removed at a certain condition, most substances undergo a state change. For pure substances the temperature remains constant until the state change is complete. This can be from solid to liquid, liquid to vapor, or vice versa.

Figure 1.4 shows the phase diagram of water, using the Engineering Equation Solver (EES, 2013) which is a typical example of temperature non-variation during latent heat exchange as in

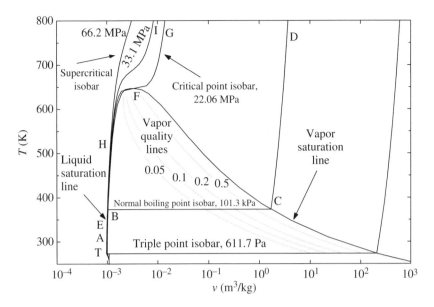

**Figure 1.5** *T–v* diagram for pure water

melting and boiling. Ice reaches its melting point at 273.15 K. During the melting process, an ice-water mixture is formed. Due to phase change, the temperature remains constant (see figure), although heat is continuously added. At the end of the melting process all water is in liquid state of aggregation. Water is further heated and its temperature increases until it reaches the boiling point at 373.15 K. Additional heating produces boiling which evolves at constant temperature, while a water + steam mixture is formed. The boiling process is completed when all liquid water is transformed in steam. Further heating leads to temperature increase and generation of superheated steam.

A representation of solid, liquid, and vapor phases of water is exhibited also on a temperature–volume ((*T–v*) diagram in Figure 1.5 using the EES). Similar diagrams can be constructed for other substances. In this diagram "T" is the triple point of water. The triple point represents that thermodynamic state where solid, liquid, and vapor can coexist. Triple point of water occurs at 273.16 K, 6.117 mbar and specific volume is 1.091 dm³/kg for ice, 1 dm³/kg for liquid water, and 206 m³/kg for vapor. Below the triple point isobar there is no liquid phase. A sublimation or de-sublimation process occurs which represents phase transition between solid and vapor.

Between triple point isobar and critical point isobar three phases do exist: solid, liquid, and vapor. In addition there are defined thermodynamic regions of subcooled liquid, two-phase, and superheated vapor. Subcooled liquid region exist between critical isobar and liquid saturation line (see the figure). The two-phase region is delimited by liquid saturation line at the left, vapor saturation line at the right, and triple point isobar at the bottom. Superheated vapor exist above the vapor saturation line and below the critical isobar. At temperatures higher than the temperature of critical point and above the critical isobar there is a thermodynamic region denoted as "*supercritical fluid region*" where the substance is neither liquid nor gas, but has some common properties with gases and with liquid; supercritical fluids will be discussed in detail in other chapters of the book.

The specific volume along the boiling pallier can be expressed based on *vapor quality*, which is defined based $v, v', v''$, the specific volumes of mixture, saturated liquid, and saturated vapor, respectively, according to

$$x = \frac{v - v'}{v'' - v'} \tag{1.2}$$

On the diagram from Figure 1.5 the constant vapor quality lines and state points A to I are indicated. These state points are representative for various processes as follows:

- A–B–C–D: Constant pressure process.
- A–B: Represents the process where water is heated from the initial temperature to the saturation temperature (liquid) at constant pressure. At point B, the water is a fully saturated liquid with a quality $x = 0$, but no water vapor has formed.
- B–C: Represents a constant-temperature vaporization process in which there is only phase change from a saturated liquid to a saturated vapor. As this process proceeds, the vapor quality varies from 0% to 100%. Within this zone, the water is a mixture of liquid and vapor. At point C we have a completely saturated vapor and the quality is 100%.
- C–D: Represents the constant-pressure process in which the saturated water vapor is superheated with increasing temperature.
- E–F–G: Represents a non-constant-temperature vaporization process. In this constant-pressure heating process, Point F is called the critical point where the saturated-liquid and saturated-vapor states are identical. The thermodynamic properties at this point are called critical thermodynamic properties, for example, critical temperature, critical pressure, and critical specific volume.
- H–I: Represents a constant-pressure heating process in which there is no change from one phase to another (only one is present). However, there is a continuous change in density during this process.

The pressure versus specific volume diagram of water is presented in Figure 1.6, by using the EES software. In this plot are indicated the saturation lines where liquid and vapor reach the saturation temperature at a given pressure. Observe that the specific volume of saturated vapors is 1000 times higher for normal boiling point isotherm. The normal boiling point isotherm corresponds to a temperature of 373.15 K. Many other simple substances have qualitatively similar behavior as shown in the $P$–$v$ diagram for water.

The *pressure versus temperature diagram* is also an important tool which shows phase transitions of any substance. In Figure 1.7 it is presented the $P$–$T$ diagram of water by using the EES software. There are four regions delimited in the diagrams: solid, vapor, liquid, and supercritical fluid. The phase transition lines are sublimation, solidification, boiling, critical isotherm, and critical isobar; the last two lines are represented only for supercritical region (at pressure and temperature higher than critical).

Beside pressure, volume and temperature there are other important parameters that describe a thermodynamic state; these parameters are denoted as *state functions*. According to the thermodynamic definition a *state function* is a property of a system that depends only on current state parameters. When a change occurs, the state function is not influenced by the process in

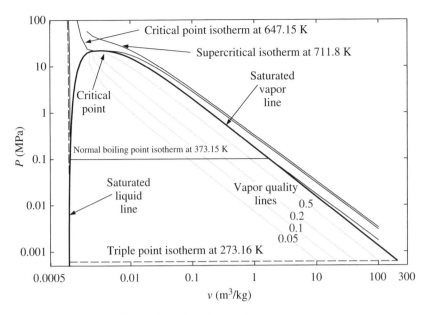

**Figure 1.6**  *P–v* diagram for pure water

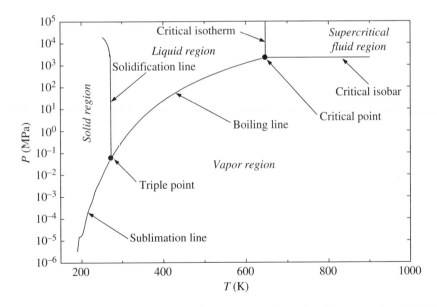

**Figure 1.7**  Pressure versus temperature diagram of water (data from Haynes and Lide (2012))

which the transformation is performed. Internal energy is a quantity which plays the role of a state function.

*Internal energy* represents a summation of many microscopic forms of energy including vibrational, chemical, electrical, magnetic, surface, and thermal. Internal energy is an extensive quantity. By definition, an extensive thermodynamic quantity depends on the amount of matter. When the internal energy $U$ is divided by the mass of the control volume $m$, then an intensive property is obtained $u = U/m$ specific internal energy. Intensive properties are independent on the amount of matter (example of intensive properties are pressure, temperature, specific volume, etc.).

For many thermodynamic processes in closed systems, the only significant energy changes are of internal energy, and the significant work done by the system in the absence of friction is the work of pressure–volume expansion, such as in a piston–cylinder mechanism. Other importance energy change is that occurring due to liquid–vapor phase change. The vapor quality changes during phase change processes such as boiling, condensation, absorption, or desorption. Thence, the specific internal energy of the two-phase liquid–vapor mixture changes according to the change $\Delta x$ in vapor quality. In general, the specific internal energy of a mixture of liquid and vapor at equilibrium can be written as follows:

$$u = (1-x)u' + xu'' \tag{1.3}$$

where $u'$ is the specific internal energy of the saturated liquid whereas $u''$ is the specific internal energy of saturated vapor. *Enthalpy* is another state function. Specific enthalpy, usually expressed in kilojoule per kilogram is defined based on internal energy and pressure and specific volume. According to its definition the specific enthalpy is given by

$$h = u + Pv \tag{1.4}$$

*Entropy* is another important state function defined by ratio of the heat added to a substance to the absolute temperature at which it was added, and is a measure of the molecular disorder of a substance at a given state. Entropy quantifies the molecular random motion within a thermodynamic system and is related to the *thermodynamic probability* ($p$) of possible microscopic states as indicated by Boltzmann equation $S = k_B \ln p$. Entropy is an extensive property whereas specific enthalpy ($s$) is an intensive property.

It is known that all substances can "hold" a certain amount of heat; that property is their thermal capacity. When a liquid is heated, the temperature of the liquid rises to the boiling point. This is the highest temperature that the liquid can reach at the measured pressure. The heat absorbed by the liquid in raising the temperature to the boiling point is called *sensible heat*. The thermodynamic quantity known as *specific heat* is a parameter that can quantify the state change of a system which performs a process with sensible heat exchange. Specific heat is defined based on internal energy or enthalpy, depending on the nature of the thermodynamic process. The specific heat at constant volume is equal to the change in internal energy with temperature at constant volume as defined in the following:

$$C_v \equiv \left( \frac{\partial u}{\partial T} \right)_v \tag{1.5}$$

The *specific heat at constant pressure* $C_p$ which represents the amount of heat required to increase the temperature of a system evolving at constant pressure with 1 K. The specific heat is the change of enthalpy with temperature at constant pressure defined according to

$$C_p \equiv \left(\frac{\partial h}{\partial T}\right)_P \tag{1.6}$$

The heat required for converting liquid to vapor at the same temperature and pressure is called *latent heat*. This is the change in enthalpy during a state change (the amount of heat absorbed or rejected at constant temperature at any pressure, or the difference in enthalpies of a pure condensable fluid between its dry saturated state and its saturated liquid state at the same pressure).

Fusion is associated with the melting and freezing of a material. For most pure substances there is a specific melting/freezing temperature, relatively independent of the pressure. For example, ice begins to melt at $0\,°C$. The amount of heat required to melt 1 kg of ice at $0\,°C$ to 1 kg of water at $0\,°C$ is called the latent heat of fusion of water, and equals 334.92 kJ/kg. The removal of the same amount of heat from 1 kg of water at $0\,°C$ changes it back to ice.

A number of fundamental physical constants are very relevant for thermodynamics; examples are the universal gas constant, Boltzmann constant, Faraday constant, elementary electric charge. In addition, some standard parameters such as standard atmospheric pressure and temperature, standard molar volume, solar constant are very important for thermodynamic analysis. Table 1.3 presents a fundamental physical constants and standard parameters. In the table there are indicated the constant name, its symbol, the value and units, and a brief definition.

## 1.3   Ideal Gas and Real Gas

Ideal gas theory is very important for analysis of drying processes because in most of the situations moisture content is extracted in form of water vapor which behaves as an ideal gas. An ideal gas can be described in terms of three parameters: the volume that it occupies, the pressure that it exerts, and its temperature. The practical advantage of treating real gases as ideal is that a simple equation of state with only one constant can be applied in the following form:

$$PV = mRT \quad \text{or} \quad Pv = RT \quad \text{or} \quad Pv = \mathcal{R}T \tag{1.7}$$

where $P$ is the pressure in Pa, $V$ is the gas volume in $m^3$, $m$ is mass of gas in kg, $T$ is gas temperature in K, $R$ is known as gas constant and is given in J/kg K, $v$ is mass specific volume in $m^3$/kg, $v$ is molar specific volume in $m^3$/kmol, $\mathcal{R}$ is the *universal gas constant* of 8.134 J/mol K. Observe that gas constant is specific to each particular gas and depends on the universal gas constant and the molecular mass of the gas according to

$$R = \frac{\mathcal{R}}{M} \tag{1.8}$$

Equation (1.7) is named "the thermal equation of state" of the ideal gas because it expresses the relationship between pressure, specific volume and temperature. It is possible to express the

**Table 1.3**  Fundamental constants and standard parameters

| Constant/ parameter | Value | Definition |
| --- | --- | --- |
| Speed of light in vacuum | $c = 299{,}792{,}458 \text{ m/s}$ | Maximum speed at which matter and information can be transported in the known cosmos |
| Elementary charge | $e = 1.60218 \times 10^{-19} \text{ C}$ | Electrical charge carried by a single proton |
| Faraday's constant | $F = 96{,}485 \text{C/mol}$ | Electric charge of 1 mol of electrons |
| Gravitational acceleration | $g = 9.80665 \text{ m/s}^2$ | Standard gravitational acceleration represents the gravitational force ($G$) per unit of mass. $g = G/m$ |
| Planck's constant | $h = 6.626 \times 10^{-37} \text{ kJ s}$ | Indicates the magnitude of energy of a quanta (particle) which expresses the proportionality between frequency of a photon and its energy according to $E = h \cdot \nu$ |
| Boltzmann constant | $k_B = 1.3806 \times 10^{-23} \text{ J/K}.$ $k_B = \mathcal{R}/N_A$ | Represents a measure of kinetic energy of one molecule of ideal gas |
| Number of Avogadro | $N_A = 6.023 \times 10^{26} \text{ molecules/kmol}$ | Ratio of constituent entities of a bulk substance to the amount of substance. $N_A = N/n$ |
| Standard atmospheric pressure | $P_0 = 101.325 \text{ kPa}$ | Pressure of the terrestrial atmosphere at the level of sea in standard conditions |
| Universal gas constant | $\mathcal{R} = 8.314 \text{J/molK}$ $\mathcal{R} = Pv/T$ | Represents a measure of kinetic energy of one mole of an ideal gas at molecular level |
| Stefan–Boltzmann constant | $\sigma = 5.670373 \times 10^{-8} \text{ W/m}^2 \text{ K}^4$ | Constant in Stefan–Boltzmann law expressing the proportionality between fourth power of temperature and black body's emissive power |
| Standard temperature | $T_0 = 298.15 \text{K}$ | Standard room temperature ($25\,°\text{C}$) |
| Standard ideal gas volume | $V_0 = 24.466 \text{ m}^3/\text{kmol}$ | Volume of 1 mol of ideal gas at $P_0 = 101.325 \text{ kPa}$, $T_0 = 298.15 \text{ K}$ |

ideal gas equation in terms of internal energy, specific volume and temperature. In this case the equation of state is called – caloric equation of state. In particular, for ideal gas only, the internal energy depends on temperature only. The caloric equation of state for a monoatomic ideal gas is $u = 1.5\,\mathcal{R}\,T$, where $u$ is the molar specific internal energy. Since $h = u + Pv$ it results that the enthalpy of monoatomic ideal gas is given by $h = 2.5\,\mathcal{R}\,T$. Combining the enthalpy and internal energy expressions in specific terms, the known Robert Meyer equation for ideal gas can be derived:

$$C_p \left[ \frac{\text{J}}{\text{kg K}} \right] = C_v + R \tag{1.9}$$

It can be remarked that for ideal gas the internal energy is a function of temperature only. Therefore, specific heat for ideal gas is $C_v = 1.5\,R$ and $C_p = 2.5\,R$. Thence, the ratio of specific heat at constant pressure and constant volume known as the *adiabatic exponent*, namely:

$$\gamma = \frac{C_p}{C_v}$$

has the following values for ideal gas – monoatomic gas 1.4 and $5/3 = 1.67$ for diatomic gas.

There are some special cases if one of $P$, $v$, and $T$ is constant. At a fixed temperature, the volume of a given quantity of ideal gas varies inversely with the pressure exerted on it (in some books this is called Boyle's law), describing compression as

$$P_1 V_1 = P_2 V_2 \tag{1.10}$$

where the subscripts refer to the initial and final states.

Equation (1.10) is employed by analysts in a variety of situations: when selecting an air compressor, for calculating the consumption of compressed air in reciprocating air cylinders, and for determining the length of time required for storing air. If the process is at constant pressure or at constant volume, then Charles' law applies:

$$\frac{V_1}{T_1} = \frac{V_2}{T_2} \text{ and } \frac{P_1}{T_1} = \frac{P_2}{T_2} \tag{1.11}$$

If the number of moles of ideal gas does not change in an enclosed volume, then the combined ideal equation of state is

$$\frac{P_1 V_1}{T_1} = \frac{P_2 V_2}{T_2} \tag{1.12}$$

If there is no heat exchange with the exterior ($dq = 0$), then the process is called adiabatic. Also, if a process is neither adiabatic nor isothermal, it could be modeled as polytropic. Table 1.4 gives the principal features of simple processes for ideal gas. Figure 1.8 shows representation in $P$–$v$ diagram for four simple processes with ideal air, modeled as ideal gas, by using the EES software.

**Table 1.4** Simple thermodynamic processes and corresponding equations for ideal gas model

| Process | Definition | Equation | Work expression |
|---|---|---|---|
| Isothermal | $T = \text{const.}$ | $P_1 v_1 = P_2 v_2$ | $w_{1-2} = P_1 v_2 \ln(v_2/v_1)$ |
| Isochoric | $v = \text{const.}$ | $\dfrac{P_1}{T_1} = \dfrac{P_2}{T_2}$ | $w_{1-2} = 0$ |
| Isobaric | $P = \text{const.}$ | $\dfrac{v_1}{T_1} = \dfrac{v_2}{T_2}$ | $w_{1-2} = P(v_2 - v_1)$ |
| Polytropic | $Pv^n = \text{const.}$ | $\dfrac{P_2}{P_1} = \left(\dfrac{V_1}{V_2}\right)^n = \left(\dfrac{T_2}{T_1}\right)^{n/(n-1)}$ | $w_{1-2} = \dfrac{1}{n-1}(P_2 v_2 - P_1 v_1)$ |
| General | $P, v, T$ vary constant mass | $\dfrac{P_1 v_1}{T_1} = \dfrac{P_2 v_2}{T_2}$ | $w_{1-2} = \displaystyle\int_1^2 P\,dv$ |

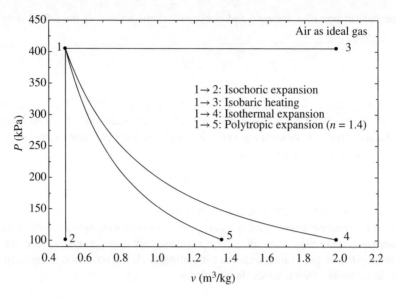

**Figure 1.8** Ideal gas processes represented in *P–v* diagram

The entropy change of an ideal gas with constant specific heats is given by the following equations, depending on the type of process (at constant pressure or at constant volume):

$$s_2 - s_1 = C_{v0} \ln\left(\frac{T_2}{T_1}\right) + R \ln\left(\frac{v_2}{v_1}\right) \text{ and } s_2 - s_1 = C_{v0} \ln\left(\frac{T_2}{T_1}\right) + R \ln\left(\frac{P_2}{P_1}\right) \qquad (1.13)$$

For the entire range of states, the ideal gas model may be found unsatisfactory. Therefore, the *compressibility factor* (*Z*) is introduced to measure the deviation of a real substance from the ideal-gas equation of state. The compressibility factor is defined by the following relation:

$$Z \equiv \frac{Pv}{RT} \qquad (1.14)$$

where specific volume is expressed on mass basis.

The order of magnitude is about 0.2 for many fluids. For accurate thermodynamic calculations compressibility charts can be used, which express compressibility factor as a function of pressure and temperature. In this way, an equation of state is obtained based on compressibility factor is the following:

$$Pv = ZRT$$

where the compressibility factor is a function of pressure and temperature.

According to the so-called *principle of corresponding states* compressibility factor has a quantitative similarity for all gases when it is plotted against reduced pressure and reduced temperature. The reduced pressure is defined by the actual pressure divided by the pressure of the critical point

$$P_r = \frac{P}{P_c}$$

where subscript c refers to critical properties and subscript r to reduced properties. Analogously, the reduced temperature is defined by

$$T_r = \frac{T}{T_c}$$

The compressibility charts showing the dependence of compressibility factor on reduced pressure and temperature can be obtained from accurate $P, v, T$ data for fluids. These data are obtained primarily based on measurements. Accurate equations of state exist for many fluids; these equations are normally fitted to the experimental data to maximize the prediction accuracy. A generalized compressibility chart $Z = f(P_r, T_r)$ is presented in Figure 1.9. As seen in the figure, at all temperatures $Z$ tends to 1 as $P_r$ tends to 0. This means that the behavior of the actual gas closely approaches ideal gas behavior, as the pressure approaches zero.

In the literature, there are also several equations of state for accurately representing the $P$–$v$–$T$ behavior of a gas over the entire superheated vapor region, for example, the Benedict–Webb–Rubin equation, the van der Waals equation, and the Redlich and Kwong equation. However, some of these equations of state are complicated, due to the number of empirical constants they contain, and are more conveniently used with computer software to obtain results. The most basic

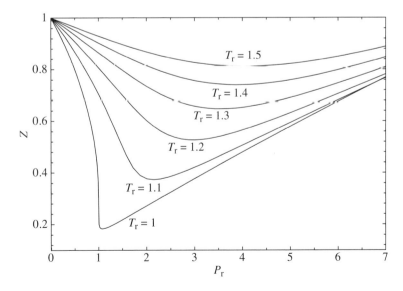

**Figure 1.9** Generalized compressibility chart averaged for water, oxygen, nitrogen, carbon dioxide, carbon monoxide, methane, ethane, propane, n-butane, iso-pentane, cyclohexane, n-heptane

**Table 1.5**   Description of the van der Waals equation of state

| Item | Equation |
|------|----------|
| Reduced pressure, temperature, and specific volume | $P_r = \dfrac{P}{P_c}; T_r = \dfrac{T}{T_c}; v_r = \dfrac{v}{v_c}$ |
| Reduced internal energy | $u_r = \dfrac{u}{(P_c\, v_c)}$ |
| Thermal equation of state | $P_r = 8\,T_r/(3\,v_r - 1) - 3/v_r^2$ |
| Caloric equation of state | $u_r = 4\,\mathcal{R}\,T_r$ |

**Table 1.6**   Comparison of Dalton and Amagat models

| Definition | Dalton model | Amagat model |
|------------|--------------|--------------|
| Assumptions | $T$ and $V$ are constant<br>$P_{tot} = P_1 + P_2 + \cdots + P_N$ | $T$ and $P$ are constant<br>$V_{tot} = V_1 + V_2 + \cdots + V_N$ |
| Equations for the components | $P_i V = n_i\,\mathcal{R}\,T$ | $P V_i = n_i\,\mathcal{R}\,T$ |
| Equation for the mixture | $P_{tot} V = \left(\sum n\right)\mathcal{R}\,T$ | $P V_{tot} = \left(\sum n\right)\mathcal{R}\,T$ |

**Table 1.7**   Relevant parameters of ideal gas mixtures

| Parameter | Equation |
|-----------|----------|
| Total mass of a mixture of $N$ components | $m_{tot} = \sum m_i$ |
| Total number of moles of a mixture of $N$ components | $n_{tot} = \sum n_i$ |
| Mass fraction for each component | $c_i = m_i/m_{tot}$ |
| Mole fraction for each component | $y_i = \dfrac{n_i}{n_{tot}} = \left(\dfrac{P_i}{P_{tot}}\right)_{\text{Dalton model}} = \left(\dfrac{V_i}{V_{tot}}\right)_{\text{Amagat model}}$ |
| Molecular weight of the mixture | $M_{mix} = \dfrac{m_{tot}}{n_{tot}} = \dfrac{\sum (n_i\, M_i)}{n_{tot}} = \sum (y_i\, M_i)$ |
| Internal energy of the mixture | $U_{mix} = \sum (n_i\, U_i)$ |
| Enthalpy of the mixture | $H_{mix} = \sum (n_i\, H_i)$ |
| Entropy of the mixture | $S_{mix} = \sum (n_i\, S_i)$ |
| Entropy difference for the mixture | $S_2 - S_1 = -\mathcal{R} \sum (n_i \ln y_i)$ |

equation of state is that of *van der Waals* which is capable to predict the vapor and liquid saturation line and a qualitatively correct fluid behavior in the vicinity of the critical point. This equation is described in Table 1.5.

In many practical situation mixtures of real gases can be approximated as mixtures of ideal gases. There are two ideal gas models for gas mixtures: the Dalton model and Amagat model.

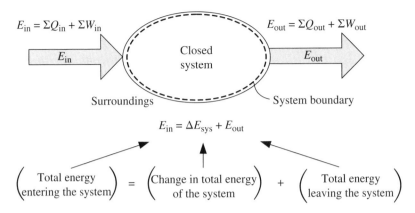

**Figure 1.10**   Illustrating the first law of thermodynamics written for a closed system

For both models it is assumed that each gas is unaffected by the presence of other gases. The Dalton model assumes that the mixture is at constant temperature and volume whereas the Amagat volume considers the case when temperature and pressure are constant. Table 1.6 gives a comparison between models of Dalton and Amagat for ideal gas mixtures. The equations relating the thermodynamic parameters of the component gases with the parameters of the mixture are given in Table 1.7.

## 1.4   The Laws of Thermodynamics

There are three laws of thermodynamics. The *zeroth law of thermodynamics* is a statement about thermodynamic equilibrium expressed as follows: "if two thermodynamic systems are in thermal equilibrium with a third, they are also in thermal equilibrium with each other." A system at internal equilibrium has a uniform pressure, temperature and chemical potential throughout its volume.

Note that two thermodynamic systems are said to be in *thermal equilibrium* if they cannot exchange heat, or in other words, they have the same temperature. Two thermodynamic systems are in *mechanical equilibrium* if cannot exchange energy in form of work. Two thermodynamic systems are in *chemical equilibrium* if they do not change their chemical composition. An insulated thermodynamic system is said to be in *thermodynamic equilibrium* when no mass, heat, work, chemical energy, and so on are not exchanged between any parts within the system.

The *First Law of Thermodynamics* (FLT) postulates the energy conservation principle: "energy can be neither created nor destroyed." The FLT can be phrased as "you can't get something from nothing." If one denotes $E$ the energy (in kJ) and $\Delta E_{sys}$ the change of energy of the system, then the FLT for a closed system undergoing any kind of process is written in the manner illustrated in Figure 1.10. There are two mathematical forms for FLT namely on an amount basis or on rate basis. In addition the FLT can be expressed in a mass specific basis. These mathematical formulations are indicated by

$$E_{in} = E_{out} + \Delta E_{sys}, \text{ on amount basis} \tag{1.15a}$$

$$\dot{E}_{in} = \dot{E}_{out} + dE_{sys}/dt, \text{ on rate basis} \tag{1.15b}$$

$$e_{in} = e_{out} + \Delta e_{sys}, \text{ on mass specific basis, } e = E/m \tag{1.15c}$$

Energy can be transferred to or from a thermodynamic system in three basic forms, namely as work, heat, and through energy associated with mass crossing the system boundary. In classical thermodynamics there is, however, a sign convention for work and heat transfer which is the following:

- The heat is positive when it is given to the system, that is $Q = \sum Q_{in} - \sum Q_{out}$ is positive when there is net heat provided to the system.
- The net useful work, $W = \sum W_{out} - \sum W_{in}$ is positive when work is generated by the system.

Using the sign convention, the FLT for closed systems becomes:

$$Q - W = \Delta E_{sys} = m(e_2 - e_1) \tag{1.16}$$

where $e$ is the specific total energy of the system comprising internal energy, kinetic energy, and potential energy, and expressed as follows:

$$e = u + \frac{1}{2}v^2 + gz \tag{1.17}$$

The FLT can be expressed in differential form on a mass basis as follows:

$$de = dq - dw = dq - P\,dv$$

If it is assumed that there is no kinetic and potential energy change, the FLT for closed system becomes

$$du = dq - P\,dv \tag{1.18}$$

If the system is a control volume, then the energy term will comprises the additional term of flow work. In this case the total specific energy of a flowing matter is

$$\theta = u + Pv + 0.5v^2 + gz = h + 0.5v^2 + gz \tag{1.19}$$

Using enthalpy formulation, the FLT for a control volume that has neither velocity nor elevation becomes

$$dh = dq + v\,dP \tag{1.20}$$

The FLT for control volume, using the sign convention for heat and work is formulated mathematically, in rate form, in the following way:

$$\dot{Q} + \sum_{\text{in}} \dot{m}\,\theta = \dot{W} + \sum_{\text{out}} \dot{m}\,\theta + \frac{\mathrm{d}(me)}{\mathrm{d}t} \tag{1.21}$$

Because $u = u(T,v)$ and $h = h(T,P)$ the following two relationships can be obtained from FLT:

$$u = u(T,v) \rightarrow \mathrm{d}u = \left(\frac{\partial u}{\partial T}\right)_v \mathrm{d}T + \left(\frac{\partial u}{\partial v}\right)_T \mathrm{d}v = C_v\,\mathrm{d}T - P\,\mathrm{d}v$$

$$h = h(T,P) \rightarrow \mathrm{d}h = \left(\frac{\partial h}{\partial T}\right)_P \mathrm{d}T + \left(\frac{\partial h}{\partial P}\right)_T \mathrm{d}P = C_p\,\mathrm{d}T + v\,\mathrm{d}P$$

From the foregoing two expressions the pressure and specific volume can be obtained from the specific internal energy as follows:

$$P \equiv -\left(\frac{\partial u}{\partial u}\right)_T \tag{1.22a}$$

$$v \equiv \left(\frac{\partial h}{\partial P}\right)_T \tag{1.22b}$$

Regarding the *SLT* this provides a mean to predict the direction of any processes in time, to establish conditions of equilibrium, to determine the maximum attainable performance of machines and processes, to assess quantitatively the irreversibilities and determine their magnitude for the purpose of identifying ways of improvement of processes. The SLT is related to the concepts of reversibility and irreversibility. One says that a thermodynamic process is reversible if during a transformation both the thermodynamic system and its surroundings can be returned to their initial states. Reversible processes are of three kinds as follows:

- Externally reversible: with no associated irreversibilities outside the system boundary.
- Internally reversible: with no irreversibilities within the boundary of the system during the process.
- Totally reversible: with no irreversibilities within the system and surroundings.

There are two classical statements of SLT which say that heat cannot be completely converted into work although the opposite is possible:

- The Kelvin–Planck statement: It is impossible to construct a device, operating in a cycle (e.g., heat engine), that accomplishes only the extraction of heat energy from some source and its complete conversion to work. This simply shows the impossibility of having a heat engine with a thermal efficiency of 100%.
- The Clausius statement: It is impossible to construct a device, operating in a cycle (e.g., refrigerator and heat pump), that transfers heat from the low-temperature side (cooler) to the high-temperature side (hotter). The Clausius inequality provides a mathematical statement of the SLT namely

$$\oint \frac{dQ}{T} \leq 0 \tag{1.23}$$

where the circular integral indicates that the process must be cyclical.

At the limit when the inequality becomes zero then the processes are reversible (ideal situation). A useful mathematical artifice is to attribute to the integral from Eq. (1.23) a new physical quantity, denoted as the "Generated entropy". Any real process must have generated entropy positive; the following cases may thus occur: (i) $S_{gen} > 0$, real, irreversible process; (ii) $S_{gen} = 0$, ideal, reversible process; (iii) $S_{gen} < 0$, impossible process. Generated entropy of a system during a process is a superposition of entropy change of the thermodynamic system and the entropy change of the surroundings. This will define entropy generated by the system $S_{gen}$

$$S_{gen} = -\oint \frac{dQ}{T} = \Delta S_{sys} + \Delta S_{surr} \geq 0 \rightarrow \Delta S_{sys} > \Delta S_{surr} = \left(\frac{Q}{T}\right)_{surr} \tag{1.24}$$

Since for a reversible process $S_{gen} = 0$ it results that entropy change of the system is the opposite of the entropy change of the surroundings

$$\Delta S_{rev} = -\Delta S_{surr} = \left(\frac{Q}{T}\right)_{rev}$$

Although the change in entropy of the system and its surroundings may individually increase, decrease or remain constant, the total entropy change (the sum of entropy change of the system and the surroundings or the total entropy generation) cannot be less than zero for any process. Note that entropy change along a process 1–2 results from the integration of the following equation:

$$dQ = T\,dS \tag{1.25}$$

and hence

$$S_{1-2} = \int_1^2 \frac{dQ}{T}$$

The SLT is a useful tool in predicting the limits of a system to produce work while generating irreversibilities to various imperfections of energy conversion or transport processes. The most fundamental device with cyclical operation with which thermodynamic operates is the heat engine; other important device is the heat pump. These devices operate between a heat source and a heat sink. A heat sink represents a thermal reservoir capable of absorbing heat from other systems. A heat source represents a thermal reservoir capable of providing thermal energy to other systems. A heat engine operates cyclically by transferring heat from a heat source to a heat sink. While receiving more heat from the source ($Q_H$) and rejecting less to the sink ($Q_C$), a heat engine can generate work ($W$). As stated by the FLT energy is conserved, thus $Q_H = Q_C + W$.

A typical "black box" representation of a heat engine is presented in Figure 1.11a. According to the SLT the work generated must be strictly smaller than the heat input, $W < Q_H$. The thermal

(a)                                                      (b)

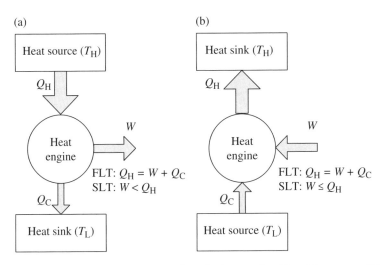

**Figure 1.11**   Conceptual representation of a heat engine (a) and heat pump (b)

efficiency of a heat engine – also known as energy efficiency – is defined as the network generated by the total heat input. Using notations from Figure 1.11a, energy efficiency of a heat engine is expressed (by definition) with

$$\eta = \frac{W}{Q_H} = 1 - \frac{Q_C}{Q_H} \tag{1.26}$$

If a thermodynamic cycle operates as refrigerator or heat pump, then its performance can be assessed by the *Coefficient of Performance (COP),* defined as useful heat generated per work consumed. As observed in Figure 1.11b, energy balance equation for a heat pump is written as $Q_C + W = Q_H$, according to SLT $Q_H \geq W$ (this means that work can be integrally converted in heat). Based on its definition, the *COP* is

$$COP = \frac{Q_H}{W} = \frac{Q_H}{Q_C + W} \tag{1.27}$$

The Carnot cycle is a fundamental model in thermodynamics representing a heat engine (or heat pump) that operates between a heat source and a heat sink, both of them being at constant temperature. This cycle is a conceptual (theoretical) cycle and was proposed by Sadi Carnot in 1824. The cycle comprises fully reversible processes, namely two adiabatic and two isothermal processes. The efficiency of Carnot cycle is independent on working fluid which performs cyclically the processes. Based on the definition of Carnot cycle, it results to $s_2 = s_3$ and $s_4 = s_1$ (for heat engine). The heat transferred at source and sink are $Q_H = T_H(s_3 - s_4) = T_H(s_2 - s_1)$ and $Q_L = T_L(s_2 - s_1)$. Therefore, the energy efficiency of a reversible Carnot heat engine is defined as

$$\eta = 1 - \frac{T_L}{T_H} \tag{1.28}$$

and the COP of the reversible Carnot heat pump becomes

$$COP = \frac{T_H}{(T_H - T_L)} \tag{1.29a}$$

and the COP of the reversible Carnot refrigerator becomes

$$COP = \frac{T_L}{(T_H - T_L)} \tag{1.29b}$$

using the temperature scale $(Q_H/Q_L)_{rev} = (T_H/T_L)$.

In summary, the given Carnot efficiency and Carnot COPs are useful criteria to assess practical heat engines, refrigerators, heat pumps, or other energy conversion systems with respect to the idealized case of reversible devices. Accordingly, energy efficiency ($\eta$) and COP of a reversible thermodynamic cycle (Carnot) is the highest possible and any actual (irreversible) cycle has smaller efficiency ($\eta_{rev} > \eta_{irrev}$) and ($COP_{rev} > COP_{irrev}$).

## 1.5 Thermodynamic Analysis Through Energy and Exergy

### 1.5.1 Exergy

Exergy represents the maximum work which can be produced by a thermodynamic system when it comes into equilibrium with its surroundings environment. This statement assumes that at an initial state there is a thermodynamic system which is not in equilibrium with the environment. In addition it is assumed that – at least potentially – mechanisms of energy (and mass) transfer between the system and the environment must exist – such that eventually the system can evolve such equilibrium condition will eventually occur. The system must at least exchange work with the environment. Another remark is that exergy definition assumes the existence of a reference environment. The system under the analysis will interact only with that environment. Exergy analysis is a method appertaining to engineering thermodynamics and can be used to determine the alleviation of manmade and natural systems from the ideal case. Here, by ideal system one understands a reversible system.

In many practical problems, the reference environment is assumed to be the earth atmosphere, characterized by its average temperature and pressure; often standard pressure and temperature as listed in Table 1.3 are used for reference environment: $P_0 = 101.325 \, kPa$, $T_0 = 298.15 \, K$. In some class of the problems when reacting systems are present, the chemical potential of the reference environment must be specified. In such cases, thermodynamic equilibrium will refer to all possible interaction; one can say that a system is in thermodynamic equilibrium with the environment if it has the same temperature with it (thermal equilibrium), the same pressure with it (mechanical equilibrium) and the same chemical potential with it. Therefore, exergy includes at least two components, one is thermo-mechanical and one is chemical.

Exergy cannot be conserved. Any real process destroys exergy as, similarly, generates entropy. Exergy is destroyed and entropy is generated due to irreversibilities. According to Dincer and Rosen (2012) the exergy of a closed (non-flow) thermodynamic system comprises four terms, namely physical (or thermo-mechanical), chemical, kinetic, and potential. In brief, total exergy of a non-flow system is

$$Ex_{nf} = Ex_{ph} + Ex_{ch} + Ex_{ke} + Ex_{pe} \qquad (1.30)$$

The exergy of a flowing stream of matter $Ex_f$ represents the sum of the non-flow exergy and the exergy associated with the flow work of the stream $(P-P_0)\,V$, therefore

$$Ex_f = Ex_{nf} + (P-P_0)\,V \qquad (1.31)$$

The physical exergy for a non-flow system is defined by

$$Ex_{ph} = (U-U_0) + P_0\,(V-V_0) - T_0\,(S-S_0) \qquad (1.32)$$

where $U$ is internal energy, $V$ volume, and $S$ entropy of closed system which is in non-equilibrium with the environment, $T_0$ is the reference temperature of the surroundings environment, and index 0 refer to the values of the parameters when the system is in thermo-mechanical equilibrium with the environment.

The kinetic and potential exergies of the system equals with the kinetic and potential energies respectively, which are given by known formulas, namely:

$$Ex_{ke} = \frac{1}{2}m\,v^2$$

$$Ex_{pe} = m\,g\,(z-z_0)$$

where $m$ is the system mass, and $v$ is its (macroscopic) velocity, $z$ is the system elevation and $z_0$ is a reference elevation of the environment (e.g., ground level).

Consider a system which is in thermo-mechanical equilibrium with the reference environment (it has the same temperature and pressure as the environment – $T_0, P_0$), but it is not in chemical equilibrium with the reference environment because it has other chemical composition. *Chemical exergy* represents the maximum work that can be extracted during a process when the system composition changes to that of the environment. There are two main components of chemical exergy: (i) exergy due to chemical reaction, (ii) exergy due to concentration difference. When a chemical compound is allowed to interact with the environment, chemical reactions may occur, involving unstable species. Eventually, more stable species are formed and further reaction is not possible.

If a substance is not present in the atmosphere then the reference for zero chemical exergy is the most stable state of that substance in seawater. There are developed tables of chemical exergy of elements in past literature data. A recent source for tabulated data of standard chemical exergy of elements is Rivero and Grafias (2006). Table 1.8 tabulates the chemical exergies of some of the most encountered chemical elements in industrial processes. Standard chemical exergy of elements is useful for calculation of chemical exergy of chemical compounds provided that their Gibbs energy of formation is known.

Moreover, if system compounds have other concentration or other phase as that corresponding to the environment then various processes such as dilution or concentration may occur until

**Table 1.8** Standard chemical exergy of some elements

| Element | B | C | Ca | Cl$_2$ | Cu | F$_2$ | Fe | H$_2$ | I$_2$ | K | Mg |
|---------|-----|--------|-------|-------|-------|-------|-------|--------|-------|-------|-------|
| ex$^{ch}$ (kJ/mol) | 628.1 | 410.27 | 729.1 | 123.7 | 132.6 | 505.8 | 374.3 | 236.12 | 175.7 | 336.7 | 626.9 |
| | Mo | N$_2$ | Na | Ni | O$_2$ | P | Pb | Pt | Pu | Si | Ti |
| ex$^{ch}$ (kJ/mol) | 731.3 | 0.67 | 336.7 | 242.6 | 3.92 | 861.3 | 249.2 | 141.2 | 1100 | 855.0 | 907.2 |

*Source*: Rivero and Garfias (2006).

**Table 1.9** Components molar fraction and standard chemical exergy for terrestrial atmosphere

| Component | N$_2$ | O$_2$ | H$_2$O | Ar | CO$_2$ | Ne | He | Kr |
|-----------|-------|-------|--------|-------|--------|--------|---------|----------|
| $C$ (mol %) | 75.67 | 20.34 | 3.03 | 0.92 | 0.03 | 0.0018 | 0.00052 | 0.000076 |
| ex$^{ch}$ (kJ/mol) | 0.69 | 3.95 | 8.67 | 11.62 | 20.11 | 27.10 | 30.16 | 34.93 |

there is no difference in concentration between system components and the environment. The standard composition of terrestrial atmosphere is given in Table 1.9 together with the chemical exergies of the components; these are very important in calculations for chemical exergy due to concentration difference between the system and the environment.

The chemical exergy depends on the difference between chemical potential of system components ($\sum n_i \mu_i^0$) being in thermo-mechanical equilibrium but not in chemical equilibrium with it ($\mu_i^0$), and the chemical potential of system components ($\mu_i^{00}$) when they are brought in chemical equilibrium with the environment, ($\sum n_i \mu_i^{00}$). Therefore, the chemical exergy of the system is defined as

$$\text{Ex}_{ch} = \sum n_i \left( \mu_i^0 - \mu_i^{00} \right) \tag{1.33}$$

Let us analyze the chemical exergy due to concentration difference between the system and the surroundings environment. Let us assume the thermodynamic system at state 1 in non-equilibrium with the environment. If mass transfer is permitted with the environment a dilution process occurs until the moment when the system components are fully diluted and there is no concentration gradient; this state is denoted with 2. The maximum work extractable from process 1–2 represents the exergy due to concentration difference and is given by

$$\Delta \text{Ex}_{conc}^{ch} = \text{Ex}_1 - \text{Ex}_2 = (U_1 - U_2) + (P_1 V_1 - P_2 V_2) - T_0(S_1 - S_2) = T_0(S_2 - S_1) \tag{1.34}$$

Here, one accounts that the process of diffusion is isothermal and one assumes that the gases involved are ideal gas $U_2 = U_1$ and $P_2 V_2 = P_1 V_1$. Furthermore, according to the FLT $T \, dS = dU + P \, dV$; therefore, for an isothermal process of ideal gas for which $dU = 0$ and $d(PV) = 0$, one has $T \, dS = d(PV) - v \, dP$, or $T \, dS = -v \, dP$. Consequently, the chemical exergy due to difference in concentration of the gas component i having molar fraction $y_i$ is given as follows:

$$\text{Ex}_{conc,i}^{ch} = -\mathcal{R} T_0 \ln(y_i) \tag{1.35}$$

The notion of *Gibbs free energy* is introduced as a *state function* defined by $g = h - Ts$; this function can be used to determine the maximum work related to chemical processes. Rivero and Grafias (2006) give a general equation for chemical exergy calculation of any chemical compound can be derived in a similar manner as illustrated earlier for water chemical exergy. In order to determine the chemical exergy of a compound it is required to know its standard Gibbs free energy of formation, $\Delta^f g^0$. Then, using $\Delta^f g^0$ and the standard exergy of the elements, the following formula must be used to determine the chemical exergy of the compound as

$$\mathrm{ex}^{\mathrm{ch}} = \Delta^f g^0 + \sum_{\mathrm{element}} \left( \nu \, \mathrm{ex}^{\mathrm{ch}} \right)_{\mathrm{element}} \tag{1.36}$$

where $\nu$ is the stoichiometric factor representing the number of moles of element per 1 mol of chemical compound.

An example of chemical exergy calculation is given here for methane. The formation reaction of methane is $C + 2H_2 \rightarrow CH_4$, while the standard formation Gibbs energy $\Delta^f g^0 = -50.53 \, \mathrm{kJ/mol}$. The standard chemical exergy of hydrogen is $236.12 \, \mathrm{kJ/mol}$ while that of carbon is (graphite) is $410.27 \, \mathrm{kJ/mol}$. Therefore, the chemical exergy of methane is computed with Eq. (1.36) as follows: $\mathrm{ex}^{\mathrm{ch}}_{\mathrm{CH_4}} = -50.53 + 410.27 + 2 \times 236.12 = 832 \, \mathrm{kJ/mol}$.

## 1.5.2 Balance Equations

Thermodynamic analysis is generally based on four types of balance equations which will be presented here in detail. These are: mass balance equation, energy balance equation, entropy balance equation (EnBE), and exergy balance equation. Thermodynamic analysis using balance equations is documented in detail in Dincer and Rosen (2012). Here a brief introduction on this method is presented.

### 1.5.2.1 Mass Balance Equation

The effect of mass addition or extraction on the energy balance of control volume is proportional with the "*mass flow rate*" defined as the amount of mass flowing through a cross section of a flow stream per unit of time. For a control volume – according to the "*conservation of mass principle*" – the net mass transferred to the system is equal to the net change in mass within the system plus the net mass leaving the system.

Assume that a number of streams with total mass flow rate $\sum \dot{m}_{\mathrm{in}}$ enter the system while a number of streams of total mass flow rate $\sum \dot{m}_{\mathrm{out}}$ leave the system. Consequently, the mass of the control volume will change with differential amount $dm_{\mathrm{cv}}$. The mass balance equation for a general control volume (see Figure 1.12) can be written for non-steady state system as in Eq. (1.37a) or for a steady state system as in Eq. (1.37b):

$$\mathrm{MBE} : \sum \dot{m}_{\mathrm{in}} = \sum \dot{m}_{\mathrm{out}} + \frac{dm_{\mathrm{cv}}}{dt} \tag{1.37a}$$

$$\mathrm{MBE}_{\mathrm{steady\ flow}} : \sum \dot{m}_{\mathrm{in}} = \sum \dot{m}_{\mathrm{out}} \tag{1.37b}$$

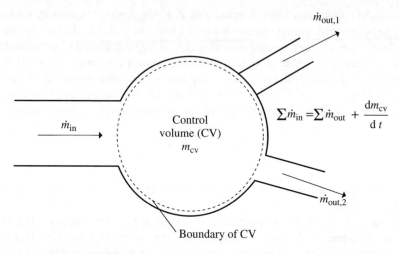

**Figure 1.12**   Illustrative sketch for mass balance equation

### 1.5.2.2   Energy Balance Equation

The energy balance equation is an expression of the FLT with a sign convention relaxed. Therefore the variation of system energy between states 1 and 2 is

$$\Delta E_{\text{sys}} = m\Delta e_{\text{sys}} = m\left[\left(u_2 + \frac{1}{2}v_2^2 + gz_2\right) - \left(u_1 + \frac{1}{2}v_1^2 + gz_1\right)\right] \tag{1.38}$$

For a closed system the energy balance equation is written with the help of the total specific energy of a not-flowing thermodynamic system $e = u + 0.5v^2 + gz$, namely:

$$\text{EBE}_{\text{Closed System}} : \sum \dot{q}_{\text{in}} + \sum \dot{w}_{\text{in}} = \sum \dot{q}_{\text{out}} + \sum \dot{w}_{\text{out}} + \frac{d\dot{e}}{dt} \tag{1.39}$$

The energy balance equation for control volumes must account for the existence of flow work and boundary work and for the rate of change of total energy [d($me$)/d$t$]; thence it can be formulated as follows:

$$\text{EBE}_{\text{Open System}} : \sum_{\text{in}} \dot{m}\theta + \sum \dot{Q}_{\text{in}} + \sum \dot{W}_{\text{in}} = \sum_{\text{out}} \dot{m}\theta + \sum \dot{Q}_{\text{out}} + \sum \dot{W}_{\text{out}} + \left[\frac{d(me)}{dt}\right]_{\text{sys}} \tag{1.40}$$

where $\theta$ is the total *energy of a flowing matter* which represents the sum of internal energy, flow work, kinetic energy, and potential energy defined by

$$\theta = u + Pv + \frac{1}{2}v^2 + gz = h + \frac{1}{2}v^2 + gz \tag{1.41}$$

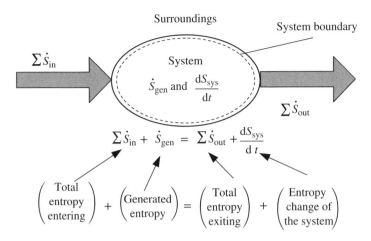

**Figure 1.13**  Explanatory sketch for the entropy balance equation – a statement of SLT

In a steady flow system mass flow rate, pressure, temperature, and so on, do not change in time, thence the integration of the following equations $d(mh) = (\dot{m}h)dt$, $dQ = \dot{Q}\,dt$, and $dW = \dot{W}\,dt$ between initial state 1 and a latter state 2 of the open system is straightforward. In steady flow regime, the EBE can be written in rate form:

$$\text{EBE}_{\text{Steady Flow}} : \dot{m}_1 e_1 + \dot{Q}_{\text{in}} + \dot{W}_{\text{in}} + \sum_{\text{in}} (\dot{m}h) = \dot{m}_2 e_2 + \dot{Q}_{\text{out}} + \dot{W}_{\text{out}} + \sum_{\text{out}} (\dot{m}h) \qquad (1.42)$$

### 1.5.2.3   Entropy Balance Equation

The SLT can be expressed in form of an *EnBE* that states for a thermodynamic system that entropy input plus generated entropy is equal to entropy output plus change of entropy within the system. In other words, the EnBE postulates that the entropy change of a thermodynamic system is equal to entropy generate within the system plus the net entropy transferred to the system across its boundary (i.e., the entropy entering minus the entropy leaving). Entropy can be transferred outside of the system as heat, but it cannot be transferred as work. Figure 1.13 illustrates schematically the EnBE, which is written mathematically according to

$$\text{EnBE} : \sum \dot{S}_{\text{in}} + \dot{S}_{\text{gen}} = \sum \dot{S}_{\text{out}} + \frac{dS_{\text{sys}}}{dt} \qquad (1.43)$$

The entropy transferred across the system boundary or along a process 1–2 is $S_{1-2} = \displaystyle\int_1^2 \frac{dQ}{T}$.

The general EnBE takes special form for closed systems. For a closed system, there is no mass transfer at the system boundary. Therefore, entropy can be transferred only by heat. If the closed system is also adiabatic, then there is neither entropy transfer due to mass nor due to heat transfer, henceforth $\dot{S}_{\text{sys}} = \dot{S}_{\text{gen}}$.

If the closed system is not adiabatic, then the EnBE becomes

$$\text{EnBE}_{\text{closed system}} : \sum_{\text{in}} \left( \int \frac{\mathrm{d}\dot{Q}}{T} \right) + \dot{S}_{\text{gen}} = \frac{\mathrm{d}S_{\text{sys}}}{\mathrm{d}t} + \sum_{\text{out}} \left( \int \frac{\mathrm{d}\dot{Q}}{T} \right) \tag{1.44}$$

The EnBE for an open system (control volume, cv) has the following expression in rate form:

$$\text{EnBE}_{\text{cv}} : \sum_{\text{in}} \left( \int \frac{\mathrm{d}\dot{Q}}{T} \right) + \sum_{\text{in}} \dot{m}s + \dot{S}_{\text{gen}} = \frac{\mathrm{d}S_{\text{CV}}}{\mathrm{d}t} + \sum_{\text{out}} \left( \int \frac{\mathrm{d}\dot{Q}}{T} \right) + \sum_{\text{out}} \dot{m}s \tag{1.45}$$

The EnBE for a steady flow through a control volume must account for the fact that there are no temporal variations of parameters; thence the mass enclosed in the control volume and specific entropy of the control volume remains constant in time; consequently

$$\text{EnBE}_{\text{steady state}} : \sum_{\text{in}} \left( \int \frac{\mathrm{d}\dot{Q}}{T} \right) + \sum_{\text{in}} \dot{m}s + \dot{S}_{\text{gen}} = \sum_{\text{out}} \left( \int \frac{\mathrm{d}\dot{Q}}{T} \right) + \sum_{\text{out}} \dot{m}s \tag{1.46}$$

In the case when $\dot{m}_{\text{out}} = \dot{m}_{\text{in}} = \dot{m}$ applies, the EnBE simplifies to

$$\text{EnBE} : \sum_{\text{in}} \left( \int \frac{\mathrm{d}\dot{Q}}{T} \right) + m(s_{\text{in}} - s_{\text{out}}) + \dot{S}_{\text{gen}} = \sum_{\text{out}} \left( \int \frac{\mathrm{d}\dot{Q}}{T} \right)$$

In the case that the process is adiabatic, there is no heat transfer across the system boundary, therefore, the EnBE simplifies to EnBE : $m s_{\text{in}} + \dot{S}_{\text{gen}} = m s_{\text{out}}$. The generated entropy is the sum of entropy change of the system and of its surroundings. There are three relevant cases that can be assumed at heat transfer across system boundary for determination entropy generation. Consider a thermodynamic system which has a diabatic boundary. As illustrated in Table 1.10, the EnBE for this system is given by the difference between $Q/T_0$ and $Q/T_{\text{sys}}$. It is assumed in this case that there is no wall with finite thickness at the system boundary. Therefore, temperature profile has a sharp change. A more accurate assumption is to assume the existence of a wall at the boundary. In this case there will be a variation of temperature across the wall. In Case 2 represented in Table 1.10 the entropy generation has to be calculated by integration accounting of temperature profile. In the third case, in addition to a wall, one considers the existence of boundary layers at the inner and outer sides of the wall. Therefore, the entropy generation will be the highest in assumption Case 3.

### 1.5.2.4 Exergy Balance Equation

The exergy balance equation introduces the term *exergy destroyed* which represents the maximum work potential that cannot be recovered for useful purpose due to irreversibilities. For a reversible system, there is no exergy destruction since all work generated by the system can be made useful. The exergy destruction and entropy generation are related by the expression

**Table 1.10** Illustration of the effects of wall assumptions considered at entropy associated with heat transfer

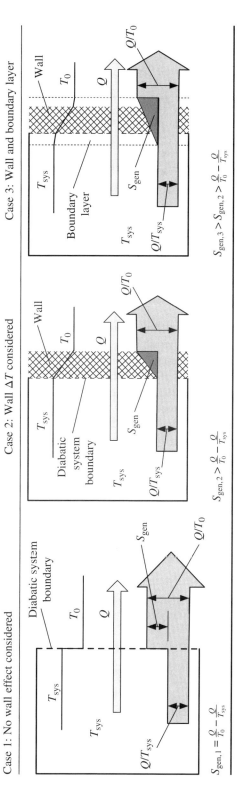

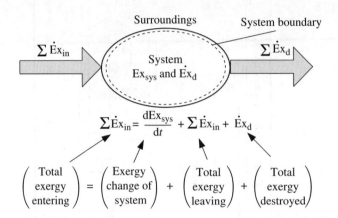

**Figure 1.14**   Explanatory sketch for the exergy balance equation

$Ex_d = T_0 \Delta S_{gen}$, where $T_0$ is the reference temperature. If $Ex_d > 0$ then the process is irreversible; if $Ex_d = 0$ then the process is reversible; if $Ex_d < 0$ the process is impossible.

The total exergy entering a thermodynamic system must be balanced by the total exergy leaving the system plus the change of exergy content of the system plus the exergy destruction. Figure 1.14 shows an explanatory sketch for the exergy balance equation. Exergy can be transferred to or from a system by three means: work, heat, and mass. Therefore, the exergy balance equation can be expressed generally in rate form as

$$\text{ExBE}: \sum_{in} \left[ \dot{W} + m\varphi + \left(1 - \frac{T_0}{T}\right)\dot{Q} \right] = \frac{d\text{Ex}}{dt} + \sum_{out} \left[ \dot{W} + m\varphi + \left(1 - \frac{T_0}{T}\right)\dot{Q} \right] - P_0 \frac{dV_{cv}}{dt} + \dot{Ex}_d$$

$$(1.47)$$

where the *total specific exergy* is defined with $\varphi = (h - h_0) + T_0(s - s_0) + \frac{1}{2}v^2 + g(z - z_0) + ex^{ch}$.

Exergy transfer between the system and surrounding can be done by work, mass transfer, heat transfer. The exergy due to work transfer ($Ex^W$) is by definition equal to the work: $Ex^W = W$. However, if the system impinges against a moving boundary then the exergy must be diminished accordingly, thence $Ex^W = W - P_0(V - V_0)$. The exergy associated to mass transfer ($Ex^m$) is $Ex^m = m\varphi$. The exergy due to heat transfer can be expressed based on Carnot factor according to

$$Ex^Q = \int_{\substack{system \\ boundary}} \left(1 - \frac{T_0}{T}\right) dQ$$

For a thermodynamic system at steady state the ExBE simplifies to

$$\text{ExBE}_{\substack{steady \\ state}} : \sum_{in} \left[ \dot{W} + \dot{m}\varphi + \left(1 - \frac{T_0}{T}\right)\dot{Q} \right] = \sum_{out} \left[ \dot{W} + \dot{m}\varphi + \left(1 - \frac{T_0}{T}\right)\dot{Q} \right] + \dot{Ex}_d \qquad (1.48)$$

If Eq. (1.48) denote $\dot{W} = \sum \dot{W}_{out} - \sum \dot{W}_{in}$ and $\dot{Q} = \sum \dot{Q}_{in} - \sum \dot{Q}_{out}$ and assume that there is no exergy destroyed. Thence, the reversible work can be obtained as follows:

$$\dot{W}_{rev} = \dot{m}(\varphi_1 - \varphi_2) + \sum \left(1 - \frac{T_0}{T}\right) \dot{Q} \tag{1.49}$$

### 1.5.2.5 Formulations for System Efficiency

The term efficiency originates mainly from thermodynamics when the attempt of assessing the heat conversion into work lead to its initial formulation as the "work generated per total heat energy input." However, efficiency as assessment criterion can be applied widely for any systems and processes. A general efficiency expression of a system – as a measure of its performance and effectiveness – is represented by the ratio of useful output per required input. Here it is recognized an efficiency criterion based on FLT, also called *energy efficiency*. If the system is an energy system then its input and output must be forms of energy. Therefore, for an energy system, the energy efficiency is written as

$$\eta = \frac{\dot{E}_{deliv}}{\dot{E}_{cons}} = 1 - \frac{\dot{E}_{loss}}{\dot{E}_{cons}} \tag{1.50}$$

Any source of energy stream is characterized by an associated exergy. By analogy with energy efficiency, the exergy efficiency is defined as the ratio between exergy associated to the useful output and the exergy associated to the consumed input, namely:

$$\psi = \frac{\dot{E}x_{deliv}}{\dot{E}x_{cons}} = 1 - \frac{\dot{E}x_d}{\dot{E}x_{cons}} \tag{1.51}$$

In Table 1.11, the efficiency formulations for main devices used in process engineering are given. Turbine is the first device analyzed in the table. A high enthalpy flow enters the turbine; work is produced, and a lower enthalpy flow exits the turbine. The turbine efficiency quantifies various losses such as the isentropic losses, the heat loses from turbine shell, the friction losses. Isentropic efficiency is one of the most used assessment parameter for turbines. Isentropic efficiency ($\eta_s$) is defined by the ratio of actual power generated to the power generated during an isentropic expansion. For an isentropic expansion there is no entropy generation; the turbine operation is reversible. Therefore, isentropic efficiency is a relative measure of alleviation from thermodynamic ideality.

The expansion process 1–2 is the actual process, while the process 1–2 is the reversible process (isentropic). Exergy efficiency of a turbine is defined as the ratio of generated power and rate of exergy consumed. A compressor is a device used to increase the pressure of a fluid under the expense of work consumption. Compressors are typically assessed by the isentropic efficiency which, for the case of compressors, is the ratio of isentropic work and actual work.

Pumps are organs used to increase the pressure of liquids on the expense of work input. The liquid is incompressible and therefore, the power required for pumping the liquid for a reversible process in $\dot{W}_s = \dot{m}\,v(P_2 - P_1)$. Hydraulic turbines are devices that generate work from

**Table 1.11** Energy and exergy efficiency of some important devices for power generation

| Device | Equation |
|---|---|
| 1. Turbine<br>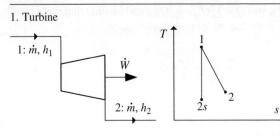 | Balance equations:<br>MBE: $\dot{m}_1 = \dot{m}_2 = \dot{m}$<br>EBE: $\dot{m}_1 h_1 = \dot{W} \rightarrow + \dot{m}_2 h_2$<br>EnBE: $\dot{m}_1 s_1 + \dot{S}_{gen} = \dot{m}_2 s_2$<br>ExBE: $\dot{m}[(h_1 - h_2) - T_0(s_1 - s_2)] = \dot{W} + \dot{E}x_d$<br>Efficiency equations:<br><br>$$\eta = \frac{\dot{W}}{\dot{W}_s} = \frac{\dot{m}(h_1 - h_2)}{\dot{m}(h_1 - h_{2s})}$$<br><br>$$\psi = \frac{\dot{W}}{\dot{E}x_{cons}} = \frac{\dot{W}}{Ex_1 - Ex_2}$$<br><br>$$= \frac{\dot{m}(h_1 - h_2)}{\dot{m}[h_1 - h_2 - T_0(s_1 - s_2)]}$$ |
| 2. Compressor<br>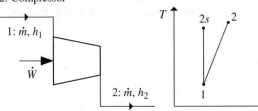 | Balance equations:<br>MBE: $\dot{m}_1 = \dot{m}_2 = \dot{m}$<br>EBE: $\dot{m}_1 h_1 + \dot{W} = \dot{m}_2 h_2$<br>EnBE: $\dot{m}_1 s_1 + \dot{S}_{gen} = \dot{m}_2 s_2$<br>ExBE: $\dot{W} = \dot{m}[h_2 - h_1 - T_0(s_2 - s_1)] + \dot{E}x_d$<br>Efficiency equations:<br><br>$$\eta = \dot{W}_s \dot{W} = \frac{\dot{m}(h_{2s} - h_1)}{\dot{m}(h_2 - h_1)}$$<br><br>$$\psi = \frac{\dot{W}_{rev}}{\dot{E}x_{cons}} = \frac{\dot{m}(h_{2s} - h_1)}{\dot{m}[h_1 - h_2 - T_0(s_1 - s_2)]}$$ |
| 3. Pump<br>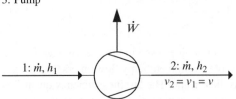 | Balance equations:<br>MBE: $\dot{m}_1 = \dot{m}_2 = \dot{m}$<br>EBE: $\dot{m}_1 h_1 + \dot{W} = \dot{m}_2 h_2$<br>EnBE: $\dot{m}_1 s_1 + \dot{S}_{gen} = \dot{m}_2 s_2$<br>ExBE: $\dot{W} = \dot{m}[h_2 - h_1 - T_0(s_2 - s_1)] + \dot{E}x_d$<br>Efficiency equations:<br><br>$$\eta = \frac{\dot{W}_s}{\dot{W}_{cons}} = \frac{\dot{m}v(P_2 - P_1)}{\dot{m}(h_2 - h_1)}$$<br><br>$$\psi = \frac{\dot{W}_{rev}}{\dot{E}x_{cons}} = \frac{\dot{m}[h_2 - h_1 - T_0(s_2 - s_1)]}{\dot{m}(h_2 - h_1)}$$ |
| 4. Hydraulic turbine<br>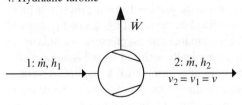 | Balance equations:<br>MBE: $\dot{m}_1 = \dot{m}_2 = \dot{m}$<br>EBE: $\dot{m}_1 h_1 = \dot{W} + \dot{m}_2 h_2$<br>EnBE: $\dot{m}_1 s_1 + \dot{S}_{gen} = \dot{m}_2 s_2$<br>ExBE: $\dot{m}[(h_1 - h_2) - T_0(s_1 - s_2)] = \dot{W} + \dot{E}x_d$<br>Efficiency equations:<br><br>$$\eta = \frac{\dot{W}}{\dot{W}_s} = \frac{\dot{m}(h_1 - h_2)}{\dot{m}v(P_1 - P_2)}$$<br><br>$$\psi = \frac{\dot{W}}{\dot{E}x_{cons}} = \frac{\dot{m}(h_1 - h_2)}{\dot{m}[h_1 - h_2 - T_0(s_1 - s_2)]}$$ |

**Table 1.11**  (*continued*)

| Device | Equation |
|---|---|

**5. Diffuser and nozzle**

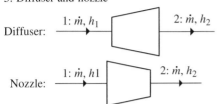

Diffuser:  1: $\dot{m}$, $h_1$    2: $\dot{m}$, $h_2$

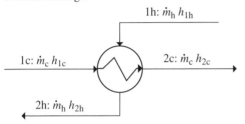

Nozzle:  1: $\dot{m}$, $h1$    2: $\dot{m}$, $h_2$

Balance equation:
MBE: $\dot{m}_1 = \dot{m}_2 = \dot{m}$
EBE: $\dot{m}_1 h_1 = \dot{m}_2 h_2$
EnBE: $\dot{m}_1 s_1 + \dot{S}_{gen} = \dot{m}_2 s_2$
ExBE: $\dot{m}[(h_1 - h_2) - T_0(s_1 - s_2)] = \dot{E}x_d$
Efficiency equations:

$$\eta = \frac{h_1 - h_2}{h_1 - h_{2s}} \text{ and } \psi = 0$$

**6. Heat exchanger**

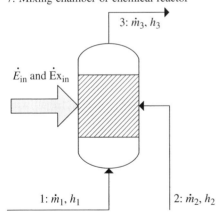

1h: $\dot{m}_h$ $h_{1h}$

1c: $\dot{m}_c$ $h_{1c}$    2c: $\dot{m}_c$ $h_{2c}$

2h: $\dot{m}_h$ $h_{2h}$

Balance equations:
MBE: $\dot{m}_{1h} = \dot{m}_{2h} = \dot{m}_h$ and $\dot{m}_{1c} = \dot{m}_{2c} = \dot{m}_c$
EBE: $\dot{m}_h(h_{1h} - h_{2h}) = \dot{m}_c(h_{2c} - h_{1c})$
EnBE: $\dot{S}_{gen} = \dot{m}_c(h_{2c} - h_{1c}) + \dot{m}_h(h_{2h} - h_{1h})$
ExBE: $\dot{m}_h[(h_{1h} - h_{2h}) - T_0(s_{1h} - s_{2h})]$
$\qquad + \dot{m}_c[(h_{1c} - h_{2c}) - T_0(s_{1c} - s_{2c})] = \dot{E}x_d$

Efficiency equations:

$$\varepsilon = \frac{\dot{Q}_{cold}}{\dot{Q}_{max}} = \frac{(\dot{m}C_p \Delta T)_{cold}}{(\dot{m}C_p)_{min}(T_{1h} - T_{2h})}$$

$$\eta = \frac{\dot{Q}_{cold}}{\dot{Q}_{hot}}; \psi = \frac{(\dot{E}x_{2c} - \dot{E}x_{1c})_{cold}}{(\dot{E}x_{1h} - \dot{E}x_{2h})_{hot}}$$

$$\psi = \frac{\dot{m}_{cold}[h_{2c} - h_{1c} - T_0(s_{2c} - s_{1c})]}{\dot{m}_{hot}[h_{2h} - h_{1h} - T_0(s_{2h} - s_{1h})]}$$

**7. Mixing chamber or chemical reactor**

3: $\dot{m}_3$, $h_3$

$\dot{E}_{in}$ and $\dot{E}x_{in}$

1: $\dot{m}_1$, $h_1$    2: $\dot{m}_2$, $h_2$

Balance equations:
MBE: $\dot{m}_1 + \dot{m}_2 = \dot{m}_3$
EBE: $\dot{m}_1 h_1 + \dot{m}_2 h_2 + \dot{E}_{in} = \dot{m}_3 h_3$;
$\qquad$ assume adiabatic
EnBE: $\dot{m}_1 s_1 + \dot{m}_2 s_2 + \dot{S}_{gen} - \dot{m}_3 s_3$,
$\qquad$ assume adiabatic
ExBE: $\dot{m}_1 ex_1 + \dot{m}_2 ex_2 + \dot{E}x_{in}$
$\qquad = \dot{m}_3 ex_3 + \dot{E}x_d$
Efficiency equations:
$$\eta = \frac{\dot{m}_3 h_3}{\dot{m}_1 h_1 + \dot{m}_2 h_2 + \dot{E}_{in}}$$
$$\psi = \frac{\dot{m}_3 ex_3}{\dot{m}_1 ex_1 + \dot{m}_2 ex_2 + \dot{E}x_{in}}$$

(*continued overleaf*)

**Table 1.11**  (*continued*)

| Device | Equation |
|---|---|
| 8. Separation device (e.g., dryer) | Balance equations: |

Balance equations:

MBE: $\dot{m}_1 = \dot{m}_2 + \dot{m}_3$

EBE: $\dot{m}_1 h_1 + \dot{Q}_{in} + \dot{E}_{in} = \dot{m}_2 h_2 + \dot{m}_3 h_3$

EnBE: $\dot{m}_1 s_1 + \dfrac{\dot{Q}_{in}}{T_{in}} + \dot{S}_{gen} = \dot{m}_2 s_2 + \dot{m}_3 s_3$

ExBE: $\dot{m}_1 ex_1 + \dot{Q}_{in}\left(1 - \dfrac{T_0}{T_{in}}\right) + \dot{Ex}_{in}$

$= \dot{m}_2 ex_2 + \dot{m}_3 ex_3 + \dot{Ex}_d$

Efficiency equations:

$$\eta = \frac{\dot{m}_2 h_2 + \dot{m}_3 h_3}{\dot{m}_1 h_1 + \dot{Q}_{in} + \dot{E}_{in}}$$

$$\psi = \frac{\dot{m}_2 ex_2 + \dot{m}_3 ex_3}{\dot{m}_1 ex_1 + \dot{Q}_{in}\left(1 - \frac{T_0}{T_{in}}\right) + \dot{Ex}_{in}}$$

Stream labels: 3: $\dot{m}_3, h_3$ ; $\dot{E}_{in}, \dot{Ex}_{in}$ and $\dot{Q}_{in}$ ; 1: $\dot{m}_1, h_1$ ; 2: $\dot{m}_2, h_2$

potential energy of a liquid. Nozzles and diffusers are adiabatic devices used to accelerate or decelerate a fluid, respectively. The exergy efficiency of nozzle is nil because they produce no work although the expanded flow has work potential. A heat exchanger is a device which facilitates heat transfer between two fluids without mixing. It is known that heat exchangers are assessed by their effectiveness parameter which represents the ration between the actual amount of heat transfer and the maximum amount of heat possible to transfer.

For a heat exchanger the exergy source is derived from the hot fluid which during the process reduces its exergy. The exergy of the cold fluid represents the delivered exergy as useful product of a heat exchanger. Regarding the energy exchange, ideally if there are no energy losses, all energy from the hot fluid is transferred to the cold fluid. However, some loses are unavoidable in practical systems; therefore one can define an energy efficiency of heat exchangers as ratio between energy delivered and energy consumed. Regarding the exergy efficiency of a heat exchanger, this is given by exergy retrieved from the cold fluid divided by exergy provided by the hot fluid.

Many practical devices are used to mix streams. Mixing chambers accept multiple stream inputs and have one single output. Mixers can be isothermal or one can mix a hot fluid with a cold fluid and so on. A combustion chamber or a reaction chamber can be modeled from thermodynamic point of view as a mixing device, whereas mixing is accompanied by chemical reaction. Very similar with mixers there are stream separators (e.g., a dryer). In this case, an input stream is separated in two (or more) different output streams.

## 1.6  Psychometrics

Psychometrics is the science that studies the properties of moist air and is of great significance in many industrial sectors, and is very relevant in drying technology. Many processes of heating, cooling, drying, humidification require psychometrics analysis. Actually, psychometry

**Table 1.12**  Important notions and properties for psychometrics

| Notion | Definition |
|---|---|
| Dry air | Purified atmospheric air without moisture (water vapors are removed) |
| Moist air | A binary mixture of dry air and water vapor, whereas the mass fraction of water vapor can vary from nearly 0 to 20 g $H_2O$/kg dry air depending on temperature and pressure |
| Saturated air | Moist air with water vapor at saturation condition at specified temperature and pressure |
| Dew point temperature, $T_d$ | Temperature at which water vapor condense when moist air is cooled at constant pressure |
| Dry bulb temperature, $T_b$ | Actual temperature of moist air |
| Wet bulb temperature, $T_w$ | Temperature corresponding to saturation at the actual partial pressure of water in moist air |
| Adiabatic saturation | An adiabatic process of humidification of moist air by adding water vapor until relative humidity becomes 100% |
| Adiabatic saturation temperature, $T_s$ | Temperature obtained during an adiabatic saturation process when relative humidity reached 100% at a given pressure |
| Vapor pressure, $P_v$ | Partial pressure of water vapor in moist air: $P_v = P - P_a$, where $P_a$ is the partial pressure of dry air |
| Relative humidity, $\phi$ | Mole fraction of water vapor in moist air divided by mole fraction in a saturated air at the same temperature and pressure. Because air is ideal gas, mole fractions equal to partial pressures by total pressure. Therefore: $\phi = P_v/P_s$, where $P_s$ is saturation pressure of water at same temperature |
| Humidity ratio, $\omega$ | Mass ratio of water vapor to the amount of dry air contained in a given volume of moist air. By definition, $\omega = m_v/m_a = 0.622 P_v/P_a$. Note also that $\omega P_a = 0.622 \phi P_s$ |
| Degree of saturation, $x_s$ | Actual humidity ratio divided by humidity ratio of a saturated air at same temperature and pressure |

also plays a crucial role in food preservation, especially in cold storage. In order to prevent the spoilage and maintain the quality of perishable products during storage, a proper arrangement of the storage conditions in terms of temperature and relative humidity is extremely important in this regard. Furthermore, the storage conditions are different for each food commodity and should be implemented accordingly.

Table 1.12 gives the descriptions of the most important notions and properties for psychometrics. One of the most important parameter is the dew point temperature. This is the temperature of moist air saturated at the same pressure and with the same humidity ratio as that of the given sample of moist air. The process defining dew point temperature is shown in the temperature-entropy diagram of water from Figure 1.15. If dry air is represented by the thermodynamic state 1 in the diagram, then point 2 represents a thermodynamic state having the dew point temperature. In state 2 the moisture starts to condense at the same pressure corresponding to the partial pressure of water vapor; this is the same as the pressure in state 1.

Another important parameter is the relative humidity. This is defined as the ratio of the mole fraction of water vapor in the mixture to the mole fraction of water vapor in a saturated mixture

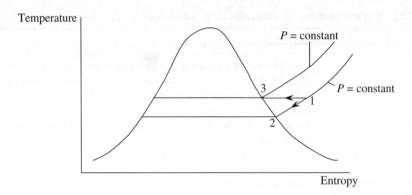

**Figure 1.15**   Diagram illustrating the concepts of dew point and relative humidity

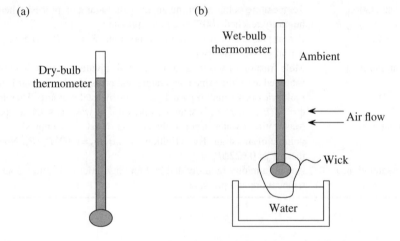

**Figure 1.16**   Schematic representation of (a) dry-bulb and (b) wet-bulb thermometers

at the same temperature and pressure, based on the mole fraction equation since water vapor is considered to be an ideal gas:

$$\phi = \frac{P_v}{P_s} = \frac{P_1}{P_3} = \frac{\rho_v}{\rho_s} = \frac{v_s}{v_v} \tag{1.52}$$

where $P_v$ and $P_s$ are the partial and saturation pressure of vapor, respectively. According to the diagram from Figure 1.15, the pressure of the water vapor at the same temperature as humid air corresponds to state 3.

The use of both a dry-bulb thermometer and a wet-bulb thermometer is very old practice to measure the specific humidity of moist air. This technique is illustrated in Figure 1.16. The dry-bulb temperature is the temperature measured by a dry-bulb thermometer directly. The bulb of

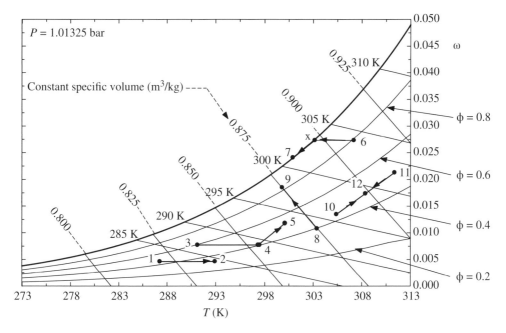

**Figure 1.17**   The Mollier diagram of moist air

the wet-bulb thermometer is covered with a wick which is already saturated with water. When the wick is subjected to an air flow, some of the water in the wick gets evaporated into the surrounding air, therefore resulting in a temperature drop in the thermometer. This final temperature is dependent on the moisture content of the air. It is important to mention that in the past there was a convention that the wicks are boiled in distilled water first and allowed to dry before using them in wet-bulb temperature measurements. Current technology allows for direct determination of the relative humidity using capacitive meter which are constructed with a dielectric material which is sensitive to the moisture content of air.

Five elementary psychometric processes are analyzed next: sensible heating (or cooling), heating with humidification, cooling with dehumidification, evaporative cooling, and adiabatic mixing. These processes are represented in the Mollier diagram, as obtained by using the EES software, as shown in Figure 1.17. Sensible heating 1–2 is the simplest psychometric process. The reverse process 2–1 represents sensible cooling and has a similar thermodynamic treatment. In sensible heating process, heat is added to moist air which produces an increase of air temperature. During this process the mass of air and mass of water vapor does not change. Therefore the humidity ratio remains constant. The dry bulb temperature of air changes, consequently, the relative humidity must change.

The process 3–4–5 in Figure 1.17 represents a heating with humidification of moist air. In this process, humidity in form of saturated water vapor is added to the air stream simultaneously with a process of heat addition. From thermodynamic point of view, this process can be subdivided in two sub-processes 3–4 and 4–5. The process 3–4 is a process of sensible heating evolving at constant humidity ratio. The subsequent process, 4–5, water vapor are added and the humidity ratio increases.

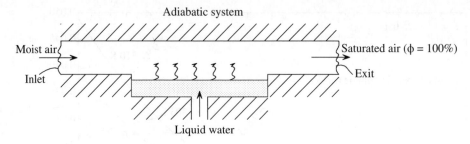

**Figure 1.18**   Representation of the adiabatic saturation process

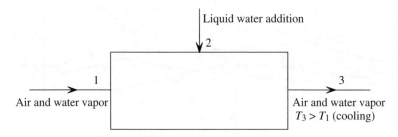

**Figure 1.19**   Illustration of a latent cooling process of humid air

The psychometric process 6–x–7 is cooling with dehumidification. In this process air is cooled until wet bulb temperature by putting it in contact with a colder surface with maintains certain heat transfer rate from the air stream. After moisture condenses, air temperature drops and the condensate can be separated gravitationally.

The adiabatic mixing of two streams of humid air is indicated in Figure 1.17 with the line 10–11–12. In this adiabatic process the resulting stream carries the total enthalpy of the two mixing streams. Another process of particular interest is the adiabatic saturation of moist air.

During adiabatic saturation air and moist air with a relative humidity less than 100% is subjected to liquid water addition. Some of the water evaporates into the mixture and makes it saturated, referring to the 100% relative humidity. In this respect, the temperature of the mixture exiting the system is identified as the *adiabatic saturation temperature* and the process is called the *adiabatic saturation process*. The process is shown in Figure 1.18.

The humid air is considered an ideal gas mixture of water vapor and dry air. In terms of balance equations for the mixing process of water vapor and dry air, there are two important aspects to deal with: the mass balance equation (i.e., the continuity equation) and the energy balance equation (i.e., the FLT). These can be written for both closed and open systems. Let's consider a cooling process, with negligible kinetic and potential energies and no work involved, that has two inputs and one output as illustrated in Figure 1.19. The general mass and energy balance equations may be written as follows:

- The mass balance equation for air is

$$\dot{m}_{a,1} = \dot{m}_{a,3} = \dot{m}_a$$

- The mass balance equation for vapor is

$$\dot{m}_{v,1} + \dot{m}_{l,2} = \dot{m}_{v,3}$$

- The energy balance equation for the mixing process is

$$\dot{Q}_i + \dot{m}_a h_{a,1} + \dot{m}_{v,1} h_{v,1} + \dot{m}_{l,2} h_{l,2} = \dot{m}_a h_{a,3} + \dot{m}_{v,3} h_{v,3}$$

- Rearranged equation in terms of the humidity ratio $\omega = \dot{m}_v / \dot{m}_a$ gives

$$\frac{\dot{Q}_i}{\dot{m}_a} + h_{a,1} + \omega_1 h_{v,1} + (\omega_1 - \omega_2) h_{l,2} = h_{a,3} + \omega_3 h_{v,3}$$

where $\omega_2 = \omega_3$ since there is no more water addition or removal between 2 and 3.

Understanding the dynamics of moisture and air will provide a solid foundation for understanding the principles of cooling and air-conditioning systems. Figure 1.20 shows several processes on the psychometric chart. Figure 1.20a exhibits cooling and heating processes and therefore an example of an increase and decrease in dry-bulb temperature. In these processes, only a change in sensible heat is encountered. There is no latent heat involved due to the constant humidity ratio of the air. Figure 1.20b is an example of a dehumidification process at the-constant dry-bulb temperature with decreasing humidity ratio. A very common example is given in Figure 1.20c which includes both cooling and dehumidification, resulting in a decrease of both the dry-bulb and wet-bulb temperatures, as well as the humidity ratio. Figure 1.20d exhibits a process of adiabatic humidification at the constant wet-bulb temperature (adiabatic case), for instance spray type humidification. If it is done by heated water, it will result in "heat and moisture addiction case." Figure 1.20e displays a chemical dehumidification process as the water vapor is absorbed or adsorbed from the air by using a hydroscopic material. It is isolated because of the constant enthalpy as the humidity ratio decreases. The last one (Figure 1.20f) represents a mixing process of two streams of air (i.e., one at state 1 and other at state 2), and their mixture reaches state 3.

Energy and exergy analysis of basic psychometric processes is introduced next. The considered basic processes are represented schematically in Table 1.13 where the balance equations are given. The total exergy of humid air may be expressed in function of specific heat as indicted in the past work of Wepfer et al. (1979)

$$\frac{ex}{T_0\left(C_{p,a} + \omega C_{p,v}\right)} = \left(\frac{T}{T_0} - 1 - \ln\frac{T}{T_0}\right) + \frac{(1+\tilde{\omega})R_a \ln\frac{P}{P_0}}{C_{p,a} + \omega C_{p,v}} + \frac{R_a\left[(1+\tilde{\omega})\ln\frac{1+\tilde{\omega}_0}{1+\omega} + \tilde{\omega}\ln\frac{\tilde{\omega}}{\tilde{\omega}_0}\right]}{C_{p,a} + \omega C_{p,v}}$$

$$(1.53)$$

where $C_{p,a}$, $C_{p,v}$ are the specific heat of dry air and water vapor, respectively, $R_a$ is the gas constant for air, and $\tilde{\omega} = 1.608\omega$.

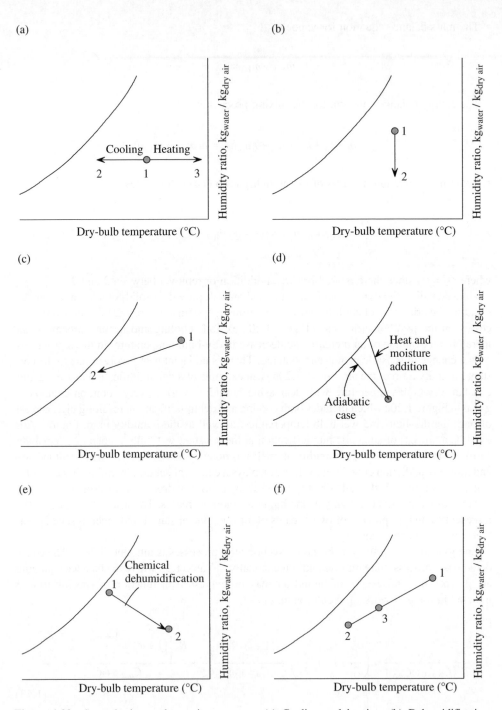

**Figure 1.20**  Some basic psychometric processes: (a) Cooling and heating. (b) Dehumidification.
(c) Cooling and dehumidification. (d) Adiabatic humidification. (e) Chemical dehumidification.
(f) Mixture of two moist air flows

**Table 1.13** The balance equations of basic psychometric processes

| Process and diagram | Balance equation |
|---|---|
| *Process 1–2*: Sensible heating<br>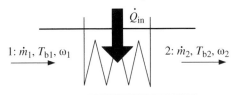 | MBE dry air: $\dot{m}_1 = \dot{m}_2$<br>MBE water vapor: $\dot{m}_{w1} = \dot{m}_{w2}$<br>EBE: $\dot{Q}_{in} + \dot{m}_1 h_1 = \dot{m}_2 h_2$<br>EnBE: $\dfrac{\dot{Q}_{in}}{T} + \dot{m}_1 s_1 + \dot{S}_{gen} = \dot{m}_2 s_2$<br>ExBE: $\dot{Q}_{in}\left(1 - \dfrac{T_0}{T}\right) + \dot{m}_1 ex_1 = \dot{m}_2 ex_2 + \dot{E}x_d$<br>where $ex = (h - h_0) - T_0(s - s_0)$ |
| *Process 3–4–5*: Heating with humidification<br>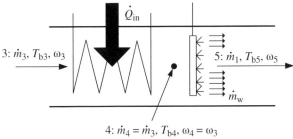 | MBE dry air: $\dot{m}_3 = \dot{m}_4 = \dot{m}_5$<br>MBE for water vapor: $\dot{m}_{w3} + \dot{m}_{w4} = \dot{m}_{w5}$<br>where $\dot{m}_{w4} = \dot{m}_w$ (see diagram)<br>EBE: $\dot{Q}_{in} + \dot{m}_3 h_3 + \dot{m}_w h_w = \dot{m}_5 h_5$<br>EnBE: $\dfrac{\dot{Q}_{in}}{T} + \dot{m}_3 s_3 + \dot{m}_w s_w + \dot{S}_{gen} = \dot{m}_5 s_5$<br>ExBE:<br>$\dot{Q}_{in}\left(1 - \dfrac{T_0}{T}\right) + \dot{m}_3 ex_3 + \dot{m}_w ex_w = \dot{m}_5 ex_5 + \dot{E}x_d$<br>where $ex = (h - h_0) - T_0(s - s_0)$ |
| *Process 6–x–7*: Cooling with dehumidification<br>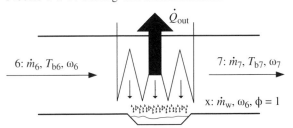 | MBE dry air: $\dot{m}_6 = \dot{m}_7$<br>MBE water vapor: $\dot{m}_6 \omega_6 = \dot{m}_7 \omega_7 + \dot{m}_w$<br>EBE: $\dot{m}_6 h_6 = \dot{Q}_{out} + \dot{m}_w h_w + \dot{m}_7 h_7$<br>EnBE: $\dot{m}_6 s_6 + \dot{S}_{gen} = \dfrac{\dot{Q}_{out}}{T} + \dot{m}_w s_w + \dot{m}_7 s_7$<br>ExBE:<br>$\dot{m}_6 ex_6 = \dot{Q}_{in}\left(1 - \dfrac{T_0}{T}\right) + \dot{m}_w ex_w + \dot{m}_7 ex_7 + \dot{E}x_d$<br>where $ex = (h - h_0) - T_0(s - s_0)$ |
| Process 8–9: Evaporative cooling<br>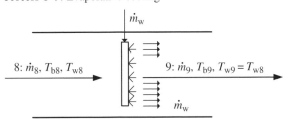 | MBE dry air: $\dot{m}_8 = \dot{m}_9$<br>MBE water vapor: $\dot{m}_8 \omega_8 + \dot{m}_w = \dot{m}_9 \omega_9$<br>EBE: $\dot{m}_8 h_8 + \dot{m}_w h_w = \dot{m}_9 h_9$<br>EnBE: $\dot{m}_8 s_8 + \dot{m}_w h_w + \dot{S}_{gen} = \dot{m}_9 s_9$<br>ExBE: $\dot{m}_8 ex_8 + \dot{m}_w ex_w = \dot{m}_9 s_9 + \dot{E}x_d$<br>where $ex = (h - h_0) - T_0(s - s_0)$ |
| *Process 10–11–12*: Adiabatic mixing<br>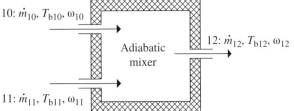 | MBE dry air: $\dot{m}_{10} + \dot{m}_{11} = \dot{m}_{12}$<br>MBE water: $\dot{m}_{10} \omega_{10} + \dot{m}_{11} \omega_{11} = \dot{m}_{12} \omega_{12}$<br>EBE: $\dot{m}_{10} h_{10} + \dot{m}_{11} h_{11} = \dot{m}_{12} h_{12}$<br>EnBE: $\dot{m}_{10} s_{10} + \dot{m}_{11} s_{11} + \dot{S}_{gen} = \dot{m}_{12} s_{12}$<br>ExBE: $\dot{m}_{10} ex_{10} + \dot{m}_{11} ex_{11} = \dot{m}_{12} ex_{12} + \dot{E}x_d$<br>where $ex = (h - h_0) - T_0(s - s_0)$ |

*Note*: States numbers for processes correspond with processes as given in Figure 1.17.

For the sensible heat addition process 1–2, the useful effect can be considered as the heat addition amounting $(\dot{m}_2 h_2 - \dot{m}_1 h_1)$, which, according to EBE is the same as $\dot{Q}_{in}$. Therefore, the energy efficiency of sensible heating of air is given by the following equation:

$$\eta_{1-2} = \frac{\dot{m}_2 h_2 - \dot{m}_1 h_1}{\dot{Q}_{in}}$$

Notice that if the system is well insulated, then the ideal margin of energy efficiency is 100%. Another view on efficiency of sensible heat addition is as follows: useful output is represented by the total enthalpy of the heated stream $\dot{m}_2 h_2$ while the consumed energy is the sum of total enthalpy of the input stream and the heat added to the system, $\dot{Q}_{in} + \dot{m}_1 h_1$. Based on this thinking, energy efficiency for process 1–2 is given by

$$\eta_{1-2} = \frac{\dot{m}_2 h_2}{\dot{Q}_{in} + \dot{m}_1 h_1}$$

The exergy efficiency formulations for the sensible heating process of moist air are in correspondence with the foregoing two definition variants for energy efficiency, namely

$$\psi_{1-2} = \frac{\dot{m}_2 ex_2 - \dot{m}_1 ex_1}{\dot{Q}_{in}(1 - T_0/T)}$$

corresponding to the first variant of energy efficiency, and

$$\psi_{1-2} = \frac{\dot{m}_2 ex_2}{\dot{Q}_{in}(1 - T_0/T) + \dot{m}_1 ex_1}$$

corresponding to the second variant of energy efficiency definition

$$\text{Energy efficiency}: \eta_{3-4-5} = \frac{\dot{m}_5 h_5 - \dot{m}_3 h_3}{\dot{Q}_{in} + \dot{m}_w h_w}$$

$$\text{Exergy efficiency}: \psi_{3-4-5} = \frac{\dot{m}_5 ex_5 - \dot{m}_3 ex_3}{\dot{Q}_{in}(1 - T_0/T) + \dot{m}_w ex_w}$$

whereas the energy and exergy efficiency of the same process when the total enthalpy of heated and humidified air is considered the system output and the total enthalpy of stream in state 3 is an input. Thence, the efficiency becomes

$$\text{Energy efficiency}: \eta_{3-4-5} = \frac{\dot{m}_5 h_5}{\dot{Q}_{in} + \dot{m}_w h_w + \dot{m}_3 h_3}$$

$$\text{Exergy efficiency}: \psi_{3-4-5} = \frac{\dot{m}_5 ex_5}{\dot{Q}_{in} + \dot{m}_w ex_w + \dot{m}_3 ex_3}$$

The useful in a cooling and dehumidification process must be equal to heat output plus the total enthalpy of air at the exit. The energy consumed may be considered to be the total enthalpy of the air stream at the input. Therefore, one has

$$\text{Energy efficiency}: \eta_{6-x-7} = \frac{\dot{Q}_{out} + \dot{m}_7 h_7}{\dot{m}_6 h_6}$$

$$\text{Exergy efficiency}: \psi_{6-x-7} = \frac{\dot{Q}_{out}(1 - T_0/T) + \dot{m}_7 ex_7}{\dot{m}_6 ex_6}$$

The evaporative cooling process is based on water spraying through a sprinklers system to produce cooling. Thus the humidity ratio and the relative humidity increase both. The useful output can considered the total enthalpy of the output stream whereas the consumed energy is a summation of the total enthalpy of input stream and total enthalpy of injected water vapors. Based on this definition, the energy and exergy efficiencies of the evaporative cooling process are as follows:

$$\text{Energy efficiency}: \eta_{8-9} = \frac{\dot{m}_9 h_9}{\dot{m}_8 h_8 + \dot{m}_w h_w}$$

$$\text{Exergy efficiency}: \psi_{8-9} = \frac{\dot{m}_9 ex_9}{\dot{m}_8 ex_8 + \dot{m}_w ex_w}$$

Thus, it appears logically to consider that the total enthalpy of stream in state 12 is the output whereas the total enthalpy sum of the input streams $\dot{m}_{10} h_{10} + \dot{m}_{11} h_{11}$ represents the energy input. The energy efficiency is

$$\eta_{10-11-12} = \frac{\dot{m}_{12} h_{12}}{\dot{m}_{10} h_{10} + \dot{m}_{11} h_{11}} \equiv 1$$

Since the process is adiabatic according to energy balance equation it results that the total enthalpy in the output stream is equal to the total enthalpy sum for input streams. Therefore, energy efficiency of this process, provided that there are no heat losses, must be equal to unity (ideal case). Regarding the exergy efficiency, this is defined by

$$\psi_{10-11-12} = \frac{\dot{m}_{12} ex_{12}}{\dot{m}_{10} ex_{10} + \dot{m}_{11} ex_{11}}$$

which must be always smaller than 1 due to the inherent exergy destruction.

## 1.7   Heat Transfer

### 1.7.1   General Aspects

Drying is essentially a process of heat and mass transfer. The heating and cooling of a material, the evaporation of water vapors, and the removal of heat liberated by a chemical reaction are common examples of processes that involve heat transfer or heat and mass transfer. It is of great importance for drying technologists and engineers to understand the physical phenomena and practical aspects of heat and mass transfer, along with knowledge of the basic laws, governing equations, and related boundary conditions.

There must be a driving force for a process of heat and mass transfer to occur. This driving force is in general a gradient of concentration or a gradient of partial pressure or a gradient of temperature. For example, consider that when a long slab of food product is subjected to heating on the left side, the heat flows from the left-hand side to the right-hand side, which is colder. It is said that heat tends to flow from a point of high temperature to a point of low temperature, owing to the temperature difference driving force.

This section focuses on heat transfer only. Many of the generalized relationships used in heat transfer calculations have been determined by means of dimensional analysis and empirical considerations. It has been found that certain standard dimensionless groups appear repeatedly in the final equations. Reynolds number is one important dimensionless number which is essential for heat transfer through forced convection. In any convection process there are two significant parameters such as velocity and viscosity. If the flow occurs at a relatively low velocity and/or with a highly viscous fluid, resulting in a fluid flow in an orderly manner without fluctuations, the flow is referred to as laminar. As the flow velocity increases and the viscosity of fluid decreases, the fluctuations will take place gradually, referring to a *transition state* which is dependent on the fluid viscosity, the flow velocity, and the geometric details.

In this regard, the Reynolds number represents the characteristics of the flow conditions relative to the transition state. As the flow conditions deviate more from the transition state, a more chaotic flow field, that is, turbulent flow occurs. It is obvious that increasing Reynolds number increases the chaotic nature of the turbulence. Turbulent flow is therefore defined as a characteristic representative of the irregularities in the flow field. The Reynolds number is defined by

$$Re = \frac{Vd}{\nu} = \frac{\rho Vd}{\mu} \tag{1.54}$$

The Reynolds number indicates the ratio of inertia force to viscous force. At high Reynolds numbers the inertia forces predominate, resulting in turbulent flow, while at low Reynolds numbers the viscous forces become dominant which makes the flow laminar. In a circular duct, the flow is laminar when $Re$ is less than 2100 and turbulent when $Re$ is greater than 4000.

In Eq. (1.54) the parameter $\rho$ is the density of the fluid while $V$ is fluid velocity and $d$ is the characteristic length of the flow which is related to the thickness of boundary layer. In general, $d$ is taken as the hydraulic diameter of the flow defined as $d = 4A/p$, where $A$ is the cross sectional area and $p$ is the duct diameter. In approximate terms this is the diameter of a disc with the same area to that of the cross sectional area of the flow.

The parameter $\mu$ is the *dynamic viscosity* which is the ratio of a shear stress to a fluid strain (velocity gradient) and $\nu$ is the *kinematic viscosity* which is defined as the ratio of dynamic viscosity to density. In gases, the viscosity increases with increasing temperature, resulting in a greater molecular activity and momentum transfer. However, in liquids, molecular cohesion between molecules considerably affects the viscosity, and the viscosity decreases with increasing temperature due to the fact that the cohesive forces are reduced by increasing the temperature of the fluid (causing a decrease in shear stress), resulting in an increase of the rate of molecular interchange; therefore, the net result is apparently a reduction in the viscosity.

The coefficient of viscosity of an ideal fluid is zero, meaning that an ideal fluid is inviscid, so that no shear stresses occur in the fluid, despite the fact that shear deformations are finite.

From the viscosity point of view, the types of fluids may be classified into Newtonian and non-Newtonian fluids. Newtonian fluids have a dynamic viscosity dependent upon temperature and pressure and independent of the magnitude of the velocity gradient. The non-Newtonian fluids are very common in practice and have a more complex viscous behavior due to the deviation from the Newtonian behavior. Some example fluids are slurries, polymer solutions, oil paints, toothpaste, sludge, and so on.

Another dimensionless group of major importance is the Prandtl number representing the ratio kinematic viscosity and thermal diffusivity. The kinematic diffusivity is a quantity which expresses momentum transfer, whereas thermal diffusivity is related to heat transfer by combined conduction and convection. Therefore, Prandtl number reflects an analogy between heat and momentum transfer processes. Another diffusivity of interest is the diffusivity of mass, which is quantified with the help of the diffusivity coefficient, generally denoted with the symbol $D$. The ratio between thermal diffusivity coefficient and mass diffusivity coefficient is known as Lewis number. The magnitude of the diffusivity of momentum with respect to the diffusivity of mass is quantified by Schmidt number which represents the ratio between kinematic viscosity and diffusion coefficient.

All three types of diffusivity coefficients mentioned here, namely, the diffusivity of momentum ($v$, viscosity), the diffusivity of heat ($a$, thermal diffusivity coefficient), and the diffusivity of mass coefficient ($D$) are measured in units of $m^2/s$. Table 1.14 gives the most important dimensionless groups used for heat and mass transfer analysis. In the utilization of these groups, care must be taken to use equivalent units so that all the dimensions cancel out.

**Table 1.14** Dimensionless criteria for heat and mass transfer modeling

| Dimensionless number | Symbol | Definition | Application |
| --- | --- | --- | --- |
| Biot | $Bi$ | $hY/k$ | Steady- and unsteady-state conduction |
| Capillary | $Ca$ | $\mu V/\sigma$ | Mass transfer through porous media |
| Darcy | $Da$ | $K/d^2$ | Mass transfer through porous media |
| Fourier | $Fo$ | $at/Y^2$ | Unsteady-state conduction |
| Graetz | $Gz$ | $GY^2C_p/k$ | Laminar convection |
| Grashof | $Gr$ | $g\beta\Delta TY^3/v^2$ | Natural convection |
| Lewis | $Le$ | $a/D$ | Diffusion; mass transfer |
| Rayleigh | $Ra$ | $Gr\,Pr$ | Natural convection |
| Nusselt | $Nu$ | $hY/k_f$ | Natural or forced convection, boiling, or condensation |
| Péclet | $Pe$ | $UY/a = Re \times Pr$ | Forced convection (for small $Pr$) |
| Prandtl | $Pr$ | $C_p\mu/k = v/a$ | Natural or forced convection, boiling, or condensation |
| Reynolds | $Re$ | $VY/v$ | Forced convection |
| Schmidt | $Sc$ | $v/D$ | Diffusion; mass transfer |
| Sherwood | $Sh$ | $KY/D$ | Convective and diffusive mass transfer |
| Stanton | $St$ | $h/\rho UC_p = Nu/Re\,Pr$ | Forced convection |

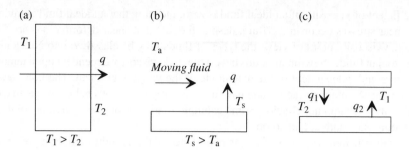

**Figure 1.21** Schematic representations of heat transfer modes: (a) Conduction through a solid. (b) Convection from a surface to a moving fluid. (c) Radiation between two surfaces

## 1.7.2   Heat Transfer Modes

There are three basic heat transfer modes as represented in Figure 1.21, namely conduction, convection and radiation. Typically, heat transfer through solid slabs is purely by conduction as shown in the configuration form (Figure 1.21a). As shown in Figure 1.21b, if a fluid flow over a solid body of different temperature, then the heat transfer mechanism is typically by convection. Also, between the two plates as shown in Figure 1.21c, placed in rarefied air and having a large temperature difference between them the heat transfer is by radiation.

### 1.7.2.1   Conduction Heat Transfer

Conduction is a mode of transfer of heat from one part of a material to another part of the same material, or from one material to another in physical contact with it, without appreciable displacement of the molecules forming the substance. For example, the heat transfer in a flat material subject to heating by contacting it with a hot plate at one of its surface is by conduction.

In solid objects, the conduction of heat is partly due to the impact of adjacent molecules vibrating about their mean positions and partly due to internal radiation. When the solid object is a metal, there are also large numbers of mobile electrons which can easily move through the matter, passing from one atom to another, and they contribute to the redistribution of energy in the metal object. Actually, the contribution of the mobile electrons predominates in metals, which explains the relation that is found to exist between the thermal and electrical conductivity of such materials.

The Fourier's law states that the instantaneous rate of heat flow through an individual homogeneous solid object is directly proportional to the cross-sectional area $A$ (i.e., the area at right angles to the direction of heat flow) and to the temperature difference driving force across the object with respect to the length of the path of the heat flow, $dT/dx$. This is an empirical law based on observation.

Figure 1.22 presents an illustration of Fourier's law of heat conduction. Here, a thin slab object of thickness $dx$ and surface area $A$ has one face at a temperature $T$ and the other at a lower temperature $(T - dT)$ where heat flows from the high-temperature side to the low-temperature

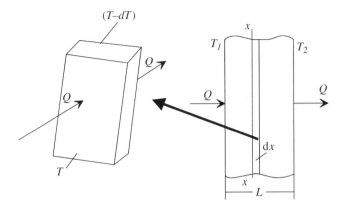

**Figure 1.22**  Schematic illustration of conduction in a slab object

side, with a temperature change in the direction of the heat flow d$T$. Therefore, under Fourier's law the heat transfer equation results to

$$\dot{Q} = -kA\frac{dT}{dx} \tag{1.55}$$

In Fourier equation $k$ is known as *thermal conductivity*, which is defined as the heat flow per unit area per unit time when the temperature decreases by 1° in unit distance. Its units are usually written as W/m K. For most solids, thermal conductivity values are approximately constant over a broad range of temperatures and can be taken as constants. Under this assumption, integrating Eq. (1.55) from $T_1$ to $T_2$ and from 0 to $L$ (the thickness of the slab), the solution becomes

$$\dot{Q} = -k\frac{A}{L}(T_2 - T_1) \tag{1.56}$$

### 1.7.2.2  Convection Heat Transfer

Convection is the heat transfer mode that takes place within a fluid by mixing one portion of the fluid with another. Convection heat transfer may be classified according to the nature of the flow. When the flow is caused by some mechanical or external means such as a fan, a pump, or atmospheric wind, it is called *forced convection*. On the other hand, for *natural (free) convection* the flow is induced by buoyancy forces in the fluid that arise from density variations caused by temperature variations in the fluid. For example, when a hot food product is exposed to the atmosphere, natural convection occurs, whereas in a cold store forced-convection heat transfer takes place between air flow and a food product subject to this flow.

The heat transfer of heat through solid objects is by conduction alone, whereas the heat transfer from a solid surface to a liquid or gas takes place partly by conduction and partly by convection. Whenever there is an appreciable movement of the gas or liquid, the heat transfer by conduction in the gas or liquid becomes negligibly small compared with the heat transfer by convection. However, there is always a thin boundary layer of liquid on a surface, and through this thin film the heat is transferred by conduction. The convection heat transfer occurring within a fluid is due to the combined effects of conduction and bulk fluid motion. Generally the heat that is transferred is the *sensible*, or internal thermal, heat of the fluid. However, there are convection processes for which there is also *latent* heat exchange, which is generally associated with a phase change between the liquid and vapor states of the fluid.

The so-called Newton's law of heat transfer states that the heat transfer from a solid surface to a fluid is proportional to the difference between the surface and fluid temperatures and the surface area. This is a particular nature of the convection heat transfer mode and is defined as

$$\dot{Q} = hA\,(T_s - T_f) \tag{1.57}$$

where $h$ is referred to as the *convection heat transfer coefficient*, measured in W/m$^2$ K.

The heat transfer coefficient encompasses all the effects that influence the convection mode and depends on conditions in the boundary layer, which is affected by factors such as surface geometry, the nature of the fluid motion, and the thermal and physical properties.

A typical problem is the heat transfer across a wall which combines conduction and convection heat transfer modes. Consider the heat transfer from a high-temperature fluid A to a low-temperature fluid B through a wall of thickness $x$ as shown in Figure 1.23. In fluid A the temperature decreases rapidly from $T_A$ to $T_{s1}$ in the region of the wall, and similarly in fluid B from $T_{s2}$ to $T_B$. In most cases the fluid temperature is approximately constant throughout its bulk, apart from a thin film ($\Delta_A$ or $\Delta_B$) near the solid surface bounding the fluid. The heat transfers per unit surface area from fluid A to the wall, conduction heat transfer through the wall and that from the wall to fluid B can be expressed in terms of heat flux density $q = \dot{Q}/A$. For a steady-state heat transfer case, heat flux density is the same everywhere across; thence one must have

$$\rightarrow q = h_A(T_A - T_{s1})$$

$$\rightarrow q = \frac{k_A}{L}(T_{s1} - T_{s2})$$

$$\rightarrow q = h_B(T_{s2} - T_B)$$

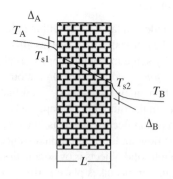

**Figure 1.23**  A wall subject to convection heat transfer from both sides and heat conduction through wall

These equations can be solved for $q$, $T_{s1}$, and $T_{s2}$. The result for heat flux density is

$$q = \frac{T_A - T_B}{1/h_A + L/k + 1/h_B}$$

Furthermore, an overall heat transfer coefficient $H$ can be defined such that $\dot{Q} = H A (T_A - T_B)$, namely:

$$H = \left( h_A^{-1} + \frac{x}{k} + h_B^{-1} \right)^{-1} \tag{1.58}$$

### 1.7.2.3  Radiation Heat Transfer

An object emits radiant energy in all directions unless its temperature is absolute zero. If this energy strikes another object, part of it may be absorbed and part may be reflected. Heat transfer from a hot to a cold object in this manner is known as radiation heat transfer. It is clear that the higher the temperature, the greater is the amount of energy radiated. If, therefore, two objects at different temperatures are placed such that the radiation from each object is intercepted by the other, then the body at the lower temperature will receive more energy than it radiates, and thereby its internal energy will increase; in conjunction with this the internal energy of the object at the higher temperature will decrease.

The fractions of the radiation absorbed, reflected, and transmitted are called the absorptivity $\alpha$, the reflectivity $r$, and the transmittivity $t$, respectively. By definition, $\alpha + r + t = 1$. For most solids and liquids in practical applications, the transmitted radiation is negligible and hence $\alpha + r = 1$. A body which absorbs all radiation is called a *blackbody*. For a blackbody, $\alpha = 1$ and $r = 0$.

The Stefan–Boltzmann law was found experimentally by Stefan, and proved theoretically by Boltzmann. It states that the emissive power of a blackbody is directly proportional to the fourth power of its absolute temperature. The Stefan–Boltzmann law enables calculation of the amount of radiation emitted in all directions and over all wavelengths simply from knowledge of the temperature of the blackbody. This law is given as follows:

$$\dot{E}_b = \sigma T_s^4$$

where $\sigma$ stands for the Stefan–Boltzmann constant, and its value is $5.669 \times 10^{-8}$ W/m$^2$ K$^4$ while $T_s$ stands for the absolute temperature of the surface. The energy emitted by a non-blackbody becomes $\dot{E} = \varepsilon \sigma T_s^4$, where $\varepsilon$ is a sub-unitary factor called *emissivity*. Therefore, the heat flux density exchanged between two parallel plates at temperatures $T_s$ and $T_a$ is given by

$$q = \varepsilon \sigma \left( T_s^4 - T_a^4 \right) \tag{1.59}$$

The heat transfer through radiation mode is highly nonlinear. However, in many situations the linearization of the heat transfer by radiation may be convenient. This can be done by introducing a linearized heat transfer coefficient according to the following equation:

$$h_r = \varepsilon \sigma \left(T_s + T_a\right)\left(T_s^2 + T_a^2\right) \tag{1.60}$$

With the help of Eq. (1.60) the heat transfer rate between to planar bodies of area $A$ which exchange heat by radiation is given by the following linear expression:

$$\dot{Q} = h_r A \left(T_s - T_a\right) \tag{1.61}$$

In the majority of practical situation heat transfer is a combined process of convection and radiation heat exchange. In this case the total heat transfer rate results as a superposition of convective and radiative effects, as follows:

$$\dot{Q}_{tot} = \dot{Q}_{conv} + \dot{Q}_{rad} = \left(h_c + h_r\right) A \left(T_s - T_a\right) \tag{1.62}$$

### 1.7.2.4 Convection Heat Transfer Correlations

Heat transfer coefficient $h$ can be determined based on various correlations; the value of the heat transfer coefficient depends on the flow regime as quantified by Reynolds number and on the fluid type as quantified by other dimensionless groups such as Prandtl number or Grashof number and so on, depending on the case. The scale analysis theory shows that the heat transfer coefficient is related to the boundary thickness which is the main characteristics of flow geometry, affecting the heat transfer. Furthermore, heat transfer coefficient must be proportional with thermal conductivity of the fluid. Nusselt number – given in Table 1.14 represents the proportionality factor between the heat transfer coefficient and the ratio $k/L_c$, where $L_c$ represents the characteristic length (denoted $Y$ in the table). For internal flows in ducts $L_c$ is taken as the hydraulic diameter. Therefore:

$$h = Nu \frac{k}{d} \tag{1.63}$$

Note that the values for Nusselt number for flow in tubes vary from approximately 3.6 when tube wall is isothermal to approximately 4.4 when wall tube is crossed by a uniform heat flux. Some of the most used correlations for Nusselt number are given in Table 1.15. In the table, the correlations for boiling and condensation are not included. This is because, in general, heat transfer correlations for these processes are found in the literature in form of heat transfer coefficients. One of the very important correlations is that given by Gorenflo (1993) for nucleate boiling:

$$\frac{h_{nb}}{h_0} = \left(1.2 \, P_r^{0.27} + 2.5 P_r + \frac{P_r}{1 - P_r}\right)\left(\frac{q}{20,000}\right)^{0.9 - 0.3 P_r} \tag{1.64}$$

**Table 1.15**  Correlations for Nusselt number for various flow configurations

| Type of the flow and configuration | Cases and correlations |
|---|---|
| Laminar slug flow in isothermal tubes | Circle $Nu = 5.8$; Octagon $Nu = 5.5$; Hexagon $Nu = 5.4$; Square $Nu = 4.9$ |
| Laminar slug flow at uniform flux tube wall | Circle $Nu = 8.0$; Octagon $Nu = 7.7$; Hexagon $Nu = 7.5$; Square $Nu = 7.1$ |
| Fully developed laminar flow | Circle $Nu \cong 4.0$; Octagon $Nu \cong 3.8$; Hexagon $Nu \cong 3.7$; Square $Nu \cong 3.3$ |
| Fully developed turbulent duct flow of gases and liquids | Dittus–Boelter:<br>$Nu = 0.023\,Re^{\frac{4}{5}}\,Pr^{\frac{1}{3}}$ for: $2500 \le Re \le 10^6$ and $0.7 \le Pr \le 120$<br>Gnielinski:<br>$Nu = 0.0214(Re^{0.8} - 100)Pr^{0.4}$, for $10^4 \le Re \le 5 \times 10^6$ and $0.5 \le Pr \le 1.5$<br>$Nu = 0.012(Re^{0.87} - 280)Pr^{0.4}$, for $3000 \le Re \le 10^6$ and $1.5 \le Pr \le 500$ |
| Flat plate at uniform temperature | $Nu_Y = 0.332\,Re_Y^{0.5}\,Pr^{0.33}$, where $Y$ is the flow entrance length |
| Staggered tube bundle | $Nu = 0.4\,Re^{0.6}\,Pr^{0.36}$ for $10^3 \le Re \le 2 \times 10^5$, $l_{transversal} > 2\,l_{longitudinal}$ |
| Laminar natural convection, vertical plate | $Nu = \dfrac{4}{3}\left(\dfrac{0.316\,Pr^{\frac{5}{4}}\,Gr}{2.44 + 4.88\,Pr^{0.5} + 4.95\,Pr}\right)^{0.25}$ |
| Laminar natural convection, horizontal plate | $Nu_{T=const.} = 0.394\,Gr^{0.2}\,Pr^{0.25}$ and $Nu_{q=const.} = 0.501\,Gr^{0.2}\,Pr^{0.25}$ |
| Laminar natural convection, vertical (v) horizontal (h) cylinder. Cylinder height ($l$), cylinder diameter ($d$) | $Nu_h = 2 + 0.5(Gr\,Pr)^{0.25}$, $\dfrac{3}{4}Nu_v = \left[\dfrac{7\,Gr\,Pr^2}{5(20 + 21\,Pr)}\right]^{0.25} + \dfrac{272 + 315\,Pr}{747 + 735\,Pr}\left(\dfrac{l}{d}\right)$ |

*Source*: Bejan and Kraus (2003).

In Eq. (1.64) the parameter $h_0$ is a reference heat transfer coefficient for nuclear boiling which corresponds to a reduced pressure $P_r = 0.1$ and a heat flux at the wall of 20 kW/m². The reduced pressure represents the ratio between the actual operating pressure and the critical pressure of the fluid. The values for $h_0$ were determined for many fluids of practical importance and they are tabulated. The range of practical values for $h_0$ is from 2500 up to 25,000 W/m² K.

The heat transfer coefficient for flow boiling (convective boiling) can be determined by the correlation of Steiner and Taborek (1992). This correlation is an asymptotic superposition of nucleate boiling and forced convection heat transfer effects. The correlation is as follows:

$$h_{fb} = \left[(h_{nb}F_{nb})^3 + (hF)^3\right]^{\frac{1}{3}} \tag{1.65}$$

where heat transfer coefficient $h_{nb}$ should be estimated with the correlation of Gorenflo for nucleate boiling (see Eq. (1.64)) whereas $h$ is estimated with the equation of Gnielinski (see Table 1.15) for fully developed turbulent duct flow. The factors $F_{nb}$ and $F$ correspond to nucleate boiling and convective components of heat transfer. The nucleate boiling factor is given by the expression

$$\frac{F_{nb}}{r_u^{0.13}} = \left[ 2.8\,P_r^{0.45} + \left( 3.4 + \frac{1.7}{1-P_r^7} \right) P_r^{3.7} \right] \left( \frac{q}{q_0} \right)^{0.8-0.1\exp(1.75P_r)} \left( \frac{d}{0.01} \right)^{-0.4} \left[ 0.38 + 0.2\ln M + 28 \left( \frac{M}{1000} \right)^2 \right]$$

where $r_u = 0.1 - 18\,\mu m$ is the surface roughness, $d$ is the duct hydraulic diameter, $M$ is the molecular mass of the fluid and $q_0$ is a reference heat flux tabulated for each fluid.

The convective factor form Eq. (1.65) is estimated based on the following equation, which is valid for low up to average flow vapor quality $x$ and for $\rho'/\rho'' = 3.75 - 5000$; where $\rho'$ and $\rho''$ are the densities of saturated liquid and saturated vapor:

$$F = \left[ (1-x)^{1.5} + 1.9x^{0.6} \left( \frac{\rho'}{\rho''} \right)^{0.35} \right]^{1.1}$$

The heat transfer at film condensation tubes has been predicted by Nusselt who developed a well-known correlation. The correlation can be applied for vertical plates ($\zeta = 0.943$) and as well for condensation on horizontal cylinders ($\zeta = 0.729$). The correlation is as follows:

$$h_{cond,\,out} = \zeta \left\{ \left[ \rho' g k'^3 \Delta h_{lv} (\rho' - \rho'') \right] / \left[ \mu' Y (T_{sat} - T_w) \right] \right\}^{0.25} \tag{1.66}$$

where the saturation temperature of the fluid corresponding to the local pressure is noted with $T_{sat}$ whereas the temperature of the colder wall is $T_w$. The characteristic length is $Y$ and this is the height of the vertical plate or the outer diameter of the horizontal cylinder, depending on the case. The superscript ' (prime) indicates a saturated liquid sate whereas the double prime ($''$) represents a saturated vapor state. The latent heat of condensation is denoted with $\Delta h_{lv}$.

The heat transfer at stratified flow condensation in tubes can be estimated by the correlation of Chato (1962) as follows:

$$h_{cond,\,in} = 0.555 \frac{k'}{d} \left[ \frac{\rho' g (\rho' - \rho'') d^3 \Delta h_{lv}}{k' \mu' (T_{sat} - T_w)} \right]^{0.25} \tag{1.67}$$

### 1.7.3   Transient Heat Transfer

In many practical cases, heat transfer which occur during drying processes is in a transient regime. More attention should be paid to the transient conduction heat transfer. Consider a thermodynamic system which is subjected to a heat transfer process at its boundary. The initial temperature of the body is $T_i$. Due to heat addition or removal, the body temperature may increase or decrease in time, depending on the case. The energy balance equation states that the net energy rate entering the body through heat transfer must be the same as the rate of energy increase of the body. Figure 1.24 shows an arbitrary solid of volume $V$ which is subjected to

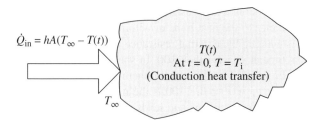

**Figure 1.24** Simple model transient conduction heat transfer

convective heat transfer at its surface of area $A$. The energy balance is written mathematically as follows:

$$h A \left(T_\infty - T\right) \mathrm{d}t = \rho\, V\, C_p\, \mathrm{d}T \Leftrightarrow -\frac{h L_\mathrm{c}}{\rho\, C_p\, L_\mathrm{c}^2}\, \mathrm{d}t = \frac{\mathrm{d}(T - T_\infty)}{T - T_\infty} \tag{1.68}$$

Here, the ratio of body volume and its surface area represents a characteristic length $L_\mathrm{c} = V/A$. The product between the heat transfer coefficient and the characteristic length has the meaning of an equivalent thermal conductivity of the surrounding fluid (gas or liquid) in which the solid body is immersed: $k_\mathrm{f} = h L_\mathrm{c}$. The product $\rho\, C_p$ from Eq. (1.68) can be expressed based on the thermal conductivity of the solid $k_\mathrm{s}$ and the thermal diffusivity of the solid $a_\mathrm{s}$ as follows: $\rho\, C_p = k_\mathrm{s}/a_\mathrm{s}$. Consequently, the RHS of Eq. (1.68) becomes $\mathrm{d}(t/\tau)$ where $\tau$ is denoted as time constant and defined by

$$\tau = \left(\rho\, V\, C_p\right)/(h A) = \left(L_\mathrm{c}^2/a_\mathrm{s}\right)(k_\mathrm{s}/k_\mathrm{f}) = Bi\left(L_\mathrm{c}^2/a_\mathrm{s}\right) \tag{1.69}$$

where the Biot number is introduced which represents the dimensionless ratio between the heat transfer rate by heat convection at the solid surface and the heat transfer rate by conduction. According to this definition and Eq. (1.69), the mathematical expressions for Biot number are as follows:

$$Bi = \frac{h L_\mathrm{c}}{k_\mathrm{s}} = \frac{k_\mathrm{f}}{k_\mathrm{s}} = \frac{h}{k_\mathrm{s}/L_\mathrm{c}} = \frac{a_\mathrm{s}\,\tau}{L_\mathrm{c}^2} \tag{1.70}$$

The heat transfer formulation given by Eq. (1.68) assumes that the temperature within the solid body is uniform, or at least can be reasonably approximated with an averaged value. This type of heat transfer solution is known as *lumped capacity model approximation*. Henceforth, the time constant $\tau$ must be small such that any temperature change at the surface is propagated fast through the body interior. Consequently, smaller is the Biot number, better is the lumped capacity approximation. It is widely accepted that the lumped capacity model gives reasonable approximation for $Bi \leq 0.1$. Provided that $Bi$ number is in the required range and the initial temperature $T_\mathrm{i} = T(t=0)$ is specified, Eqs. (1.68) and (1.69) can be solved to determine the temperature variation within the solid body in time. The analytical solution is as follows:

$$T(t) = T_\mathrm{i} + (T_\mathrm{i} - T_\infty)\exp\left(-\frac{t}{\tau}\right) \tag{1.71}$$

The lumped capacity model is widely used in many practical problems whenever there is a negligible internal resistance to the heat transfer within the product ($Bi \leq 0.1$). In the opposite situation when there is a negligible external resistance to the heat transfer at the surface of the object Biot number is typically larger than 100. In such situations, transient heat conduction through the solid must be solved for with specific methods which involve integration of partial differential equations. The case $0.1 \leq Bi \leq 100$ is the most realistic and common situation for practical transient heat transfer problems and leads to finite internal and external resistances to heat transfer from the product to the surrounding medium. Let us consider this last case together with the following additional assumptions:

- The solids are isotropic and have constant thermophysical properties and heat transfer parameters.
- The initial temperature of the solid is uniform; it is noted with $T_i$.
- The thermophysical properties of the surrounding fluid are constant.
- There is negligible internal heat generation in the solid.
- The heat transfer problem can be approximated as one-dimensional; the coordinate is noted $y$.
- A dimensionless temperature is defined as follows:

$$\theta = \frac{T(y,t) - T_a}{T_i - T_a}$$

The governing differential heat conduction equation in rectangular, cylindrical, and spherical coordinates for infinite slab ($m = 0$, $y = z$), infinite cylinder ($m = 1, y = r$), and spherical body ($m = 2, y = r$) can be written in terms of dimensionless temperature difference in a compact form as follows:

$$\left(\frac{a}{y^m}\right) \frac{\partial}{\partial y} \left[ y^m \left(\frac{\partial \theta}{\partial y}\right) \right] = \frac{\partial \theta}{\partial t} \tag{1.72a}$$

$$\text{infinite slab}: a \frac{\partial^2 \theta}{\partial z^2} = \frac{\partial \theta}{\partial t} \tag{1.72b}$$

$$\text{infinite cylinder}: \frac{a}{r} \frac{\partial}{\partial r} \left[ r \frac{\partial \theta}{\partial r} \right] = \frac{\partial \theta}{\partial t} \tag{1.72c}$$

$$\text{sphere}: \frac{a}{r^2} \frac{\partial}{\partial r} \left[ r^2 \frac{\partial \theta}{\partial r} \right] = \frac{\partial \theta}{\partial t} \tag{1.72d}$$

The following initial and boundary conditions are considered, $\theta(y,0) = 1$, $\partial \theta(0,t)/\partial y = 0$, and $-k[\partial \theta(Y,t)/\partial y] = h\theta(Y,t)$ where $Y = l$ for infinite slab and $R$ for infinite cylinder and for spherical body. The Fourier number can be derived based on Eq. (1.71) as a dimensionless time. Fourier number is important when Biot number is in the range of approximately 0.1–100. According to the definition, Fourier number is expressed by

$$Fo = \frac{a_s t}{L_c^2} \tag{1.73}$$

**Table 1.16**  Mathematical relationships for basic transient heat transfer

| Infinite slab | Infinite cylinder | Spherical body |
|---|---|---|
| $A_1 = 2\sin\mu_1/(\mu_1 + \sin\mu_1\cos\mu_1)$ | $A_1 = 2Bi/\left[(\mu_1^2 + Bi^2) + J_0(\mu_1)\right]$ | $A_1 = 2Bi\sin\mu_1/(\mu_1 - \sin\mu_1\cos\mu_1)$ |
| $\cot(\mu_1) = (1/Bi)\mu_1$ | $J_0(\mu_1)/J_1(\mu_1) = (1/Bi)\mu_1$ | $\cot(\mu_1) = (1 - Bi)/\mu_1$ |
| $\mu_1 = -419.24G^4 +$ $2013.8G^3 - 3615.8G^2$ $+ 2880.3G - 858.94$ | $\mu_1 = -3.4775G^4 +$ $25.285G^3 - 68.43G^2$ $+ 82.468G - 35.638$ | $\mu_1 = -8.3256G^4 +$ $54.842G^3 - 134.01G^2$ $+ 145.83G - 58.124$ |

**Table 1.17**  Analytical solutions for transient heat transfer through the semi-infinite solid

| Boundary and initial conditions | Temperature and surface heat flux solution |
|---|---|
| $T(x>0, t=0) = T_i$ $T(x=0, t) = T_i + \xi t^{0.5n}$ $T(x=\infty, t) = T_i$ | $T(x,t) = T_i + 4\xi\Gamma(1 + 0.5n)\left[i^n\mathrm{erfc}\left(\dfrac{0.5x}{\sqrt{at}}\right)\right]$ $q(x=0,t) = \dfrac{2^{n-1}}{\sqrt{a}} t^{0.5(n-1)} ka\Gamma(1 + 0.5n)\left[i^{n-1}\mathrm{erfc}(0)\right]$ |
| $T(x>0, t=0) = T_i$ $q(x=0, t) = q_0$ $T(x=\infty, t) = T_i$ | $T(x,t) = T_i + 2q_0 \dfrac{\sqrt{at}}{k}\left[i^1\mathrm{erfc}\left(\dfrac{0.5x}{\sqrt{at}}\right)\right]$ |
| $T(x>0, t=0) = T_i$ $-k\dfrac{\partial T(x=0,t)}{\partial x} = h[T_\infty - T(x=0,t)]$ $T(x=\infty, t) = T_i$ | $\dfrac{T(x,t) - T_i}{T_\infty - T_i} = \mathrm{erfc}\left(\dfrac{0.5x}{\sqrt{at}}\right) - \exp\left[\dfrac{hx}{k} + \dfrac{h^2 at}{k^2}\right]\mathrm{erfc}\left[\dfrac{h\sqrt{at}}{k} + \dfrac{0.5x}{\sqrt{at}}\right]$ |

*Source*: Bejan and Kraus (2003).

The solution to the transient heat conduction problem expressed by Eq. (1.73) is given in the form of a series. The solution gives in fact the relationship between the dimensionless temperature $\theta$ and dimensionless time $Fo$. The following equation expresses the transient temperature within the solid body:

$$\theta(y, Fo) = \sum_{n=1}^{\infty} A_n B_n \tag{1.74}$$

From the practical cooling experience, the duration from the beginning until the Fourier number reaches 0.2 is negligible in many cooling applications of products. Therefore, we take into consideration that the Fourier number is greater than 0.2, which means that the period passed between 0 and 0.2 is negligible if compared to the entire period, provided that Eq. (1.74) can be simplified by taking $n = 1$ and ignoring the remaining terms; $\theta = A_1 B_1$. Here the equations for the factor $A_1$ are given in Table 1.16. The factor $B_1 = \exp(-\mu_1^2 Fo)$. The dimensionless center temperature for a solid product subject to cooling can be written as given earlier in terms of a lag factor $(G)$ and heating or cooling coefficient $(C)$ as $\theta = G\exp(-Ct)$; see Lin et al. (1993).

One-dimensional conduction through semi-infinite solid is one of the most fundamental heat transfer problem in transient regime. In many practical problems this type of configuration can be encountered. The heat transfer equation for 1D transient conduction is in this case according to Eq. (1.72b). The solutions of this problem can be determined analytically for three types of boundary conditions: (i) specified surface temperature, (ii) specified surface heat flux, and (iii) surface convection. Table 1.17 gives the analytical solutions for transient heat transfer in the semi-infinite solid. In the solution gamma and complementary error (erfc) functions are used. The notation $i^n$erfc stands for the $n$th integral of the erfc function.

## 1.8 Mass Transfer

Mass transfer analysis and modeling is crucial for design and optimization of unit operations and especially of drying processes and devices. Mass transfer is analogous with heat transfer; the difference is that the transferred quantity is a mass flux instead of a heat flux and the driving force is a gradient of concentration rather than a gradient of temperature. In mass transfer processes chemical species move from higher concentration regions to lower concentration region within a control volume. Heat and mass transfer processes often occur simultaneously.

The process of mass transfer occurs by mass diffusion or mass convection. Mass diffusion is a molecular transport process in which a species diffuses through a material or other chemical species according to the gradient of concentration. Diffusion mass transfer is governed by the well-known Fick's law (which is analogous with Fourier law for heat convection). The law of Fick of diffusion through a surface of area $A$ is written as follows:

$$\dot{m} = -DA\frac{dC_m}{dx} \tag{1.75}$$

Here, the flow is assumed one-dimensional and perpendicular to the considered area. Furthermore, $C_m$ is the mass based concentration of the species which diffuses, expressed in kg/m$^3$ (this is different than density). The equation introduces the quantity $D$ known as diffusion coefficient (m$^2$/s). Fick's law of diffusion gives an expression of mass transfer by diffusion. The problems involving mass diffusion can include generation or of chemical species and in such instances the mass transfer should be better modeled on mole basis. For the processes that do not change chemical species, mass basis modeling can be preferred. Here we introduce the diffusive molar flux ($j_n$, measured in mol/m$^2$ s) defined as the molar flow rate divided to the cross sectional area; similarly the diffusive mass flux ($j_m$, measured in kg/m$^2$ s) is introduced as follows:

$$j_n = \frac{\dot{n}}{A} = -D\frac{dC}{dx} \tag{1.76a}$$

$$j_m = \frac{\dot{m}}{A} = -D\frac{dC_m}{dx} \tag{1.76b}$$

Here, above $C$ is the molar concentration and $C_m$ is mass concentration of species that diffuses. The diffusion coefficient bears the name of *binary coefficient*. For gases, $D$ is strongly temperature dependent increasing its order of magnitude hundreds time when temperature grows 10 times. Diffusion coefficient for ideal gases can be determined theoretically based on kinetic theory which shows that $D$ is proportional to $T^{1.5}$ and inverse proportional with pressure. However, for practical cases, the diffusion coefficient is determined experimentally for pair of substances. For example, diffusion coefficients of water vapor in air is 0.25 mm$^2$/s whereas diffusion coefficient of benzene in oxygen is 0.04 mm$^2$/s. A widely used approximation for the binary diffusion coefficient of water vapor in air is given in Cengel and Boles (2011): $D \cong 0.02T^2P$ ($\mu$m$^2$/s) with $T$ in Kelvin and $P$ in bar.

Diffusion coefficients can be predicted for various cases, such that diffusion through gases, diffusion through liquids or diffusion through liquids. Among those cases, diffusion through solids is very relevant for drying. Diffusion coefficients in solids are lower than in gases and liquids. In some situation when porous solids are involved the diffusion process alleviates from Fick laws, although in most of practical problems Fick law offers a good approximation. In the case when solid is highly porous then the diffusion process is diminished according to a void fraction $\varepsilon$ which is a sub-unitary parameter expressing the ratio between voids volume and the bulk (overall) volume of the porous material. The diffusion coefficient can be diminished even more due to an increased length of the diffusive path. This effect is taken into account by tortuosity $t$ which is typically a parameter higher then unity. Thence, for diffusion in solids an effective diffusion coefficient can be introduced as follows:

$$D_{\text{eff}} = \frac{\varepsilon}{t} D \tag{1.77}$$

The diffusion coefficient can be estimated based on mass and heat transfer analogy or by mass and momentum transfer analogy. When mass and heat transfer analogy is used, the Lewis number must be employed which is defined as shown in Table 1.14 according to the following equation:

$$Le = \frac{a}{D} = \left(\frac{\delta_{\text{th}}}{\delta_{\text{m}}}\right)^n \tag{1.78}$$

where exponent $n \cong 3$ and $\delta$ is the boundary layer thickness for heat diffusion (subscript "th") and for mass diffusion (subscript m). Furthermore, according the mass and momentum analogy, the diffusion coefficient can be related to kinematic viscosity with the help of Schmidt number

$$Sc = \frac{\nu}{D} = \left(\frac{\delta_{\text{d}}}{\delta_{\text{m}}}\right)^n \tag{1.79}$$

Here, $\delta_{\text{d}}$ is the thickness of the dynamic boundary layer; in both Eqs. (1.78) and (1.79) exponent $n \cong 3$ for many practical cases.

Another mechanism of mass transfer occurs by convection of mass. Mass convection involves a combined process of fluid motion and mass diffusion through boundary layer altogether. Thence, in mass convection there is formed a concentration boundary layer across which the concentration of species changes gradually. Assume that the concentration difference across the concentration boundary layer at an interface is denoted with $C_s - C_\infty$. In this case, the mass transfer through convection across the surface $A$ is

$$\dot{m} = h_{\text{m}} A (C_{\text{m,s}} - C_{\text{m},\infty}) \tag{1.80}$$

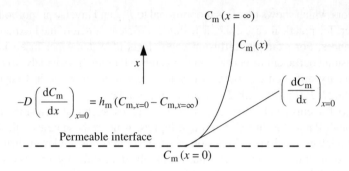

**Figure 1.25**  Mass transfer process at an interface, through a mass transfer boundary layer

In Eq. (1.80) the coefficient of mass transfer $h_m$ is introduced having the units of (m/s). Figure 1.25 shows the mass transfer process at a permeable interface. Flow of a species diffuses through an interface. Further, the species diffuses in a fluid medium due to a concentration gradient. The mass continuity equation at the interface requires that mass transferred by diffusion balances with the mass transferred by convection. Accordingly one has

$$-D\left(\frac{dC_m}{dx}\right) = h_m[C_m(x=0) - C_m(x=\infty)] \tag{1.81}$$

The mass transfer coefficient $h_m$ can be determined based on Sherwood ($Sh$) and Stanton ($St$) numbers defined as indicated in Eq. (1.82). Sherwood number is defined based on a characteristic length $L_c$ (often taken as the thickness of the boundary layer for mass transfer $L_c = \delta_m$). Stanton number is defined based on the velocity of bulk flow.

$$Sh = \frac{h_m L_c}{D} \tag{1.82a}$$

$$St = \frac{h_m}{V_{bulk}} \tag{1.82b}$$

Correlations for mass transfer coefficient are often expressed as a correlation of the type $Sh = f(Re, Sc)$; in this correlation Sherwood number is the analogous of Nusselt number for heat transfer whereas Schmidt number is the equivalent of Prandtl. Table 1.18 gives a number of representative correlations for Sherwood number. Two cases from Table 1.18 are of special interest. First is the so-called Reynolds analogy which treats the hypothetical case when momentum, heat and mass diffusivities are similar: $v = a = D$. In this case the thickness of the velocity boundary layer must be similar to the thickness of boundary layer for mass transfer, thence $Sh = 0.5 f\, Re$.

The Chilton–Colburn analogy predicts the thickness of the mass transfer boundary layer for cases when Schmidt number is different than Prandtl number ($Sc \neq Pr \neq 1$). It is shown that in this case Sherwood number can be expressed with $Sh = 0.5 f\, Re\, Sc^{0.33}$. Note that this

**Table 1.18** Correlations for Sherwood number

| Case description | Sherwood number correlation |
|---|---|
| 1. Reynolds analogy | $Sh = 0.5 f\, Re$, where |
| | $f = 2\tau_w / (\rho V_\infty^2)$, $Sc \cong Pr \cong 1$ |
| 2. Chilton–Colburn analogy | $Sh = 0.5 f\, Re\, Sc^{\frac{1}{3}}$, $0.6 < Sc < 3000$ |
| 3. Natural convection over vertical plate | $Sh = 0.59\, (Gr\, Sc)^{0.25}$, $10^5 < Gr\, Sc < 10^9$ |
| | $Sh = 0.1\, (Gr\, Sc)^{\frac{1}{3}}$, $10^9 < Gr\, Sc < 10^{13}$ |
| 4. Forced convection flow over a plate | $Sh = 0.664\, Re^{0.5} Sc^{\frac{1}{3}}$, $Sc > 0.5, Re < 5 \times 10^5$ |
| | $Sh = 0.037\, Re^{0.8} Sc^{\frac{1}{3}}$, $Sc > 0.5,\ 5 \times 10^5 < Re < 10^7$ |
| 5. Fully developed turbulent flow in smooth circular tubes | $Sh = 0.023\, Re^{0.8} Sc^{0.4}$, $0.7 < Sc < 160, Re > 10^4$ |

*Source*: Cengel and Boles (2011).

analogy can be expressed in an equivalent form as the ratio of heat transfer and mass transfer coefficient, namely

$$\frac{h}{h_m} = \rho\, C_p\, Le^{\frac{2}{3}}$$

Therefore, the diffusive mass flux at the interface is

$$j_m = \frac{h_m}{\rho\, C_p\, Le^{\frac{2}{3}} R} (\rho_s - \rho_\infty) \tag{1.83}$$

The use of Chilton–Colburn analogy is useful to predict the evaporation rate at the surface of a moist (humid) solid or at the surface of a liquid. Furthermore, the heat transfer rate due to evaporation and the surface temperature can be predicted. In this respect, the water vapor concentration results from application of the ideal gas law $P = \rho R T$, where $P$ is the partial pressure. Furthermore, the vapors carry a heat equivalent to the latent heat of evaporation ($\Delta h_{lv}$) at local partial pressure. Consequently, heat and mass transfer at superficial evaporation can be described by the following equations:

$$j_m = \frac{h_m}{\rho\, C_p\, Le^{\frac{2}{3}} R} \left( \frac{P_s}{T_s} - \frac{P_\infty}{T_\infty} \right) \tag{1.84a}$$

$$\dot{Q} = \frac{h_m \Delta h_{lv}}{\rho\, C_p\, Le^{\frac{2}{3}} R} \left( \frac{P_s}{T_s} - \frac{P_\infty}{T_\infty} \right) \tag{1.84b}$$

$$T_s = T_\infty - \frac{\Delta h_{lv} M_v}{C_p\, M P\, Le^{\frac{2}{3}}} (P_s - P_\infty) \tag{1.84c}$$

where $M$ is the molecular mass of the mixture of gases, $P$ is the total pressure, and $P_s$ and $P_\infty$ are the pressures of the evaporating species at the interface and in the bulk flow.

In special situation Reynolds or Chilton–Colburn analogies cannot be applied. Thence, correlations for Sherwood number must be established. For example, in the case of mass convection at a vertical wall the buoyancy effect must be considered. In this case, Grashof number for mass transfer (see Table 1.18) can be introduced as follows:

$$Gr = \frac{gL_c^3(\rho_\infty - \rho_s)}{\rho \nu^2}$$
(1.85)

where $\rho_\infty$ is the bulk medium density and $\rho_s$ is the density at the interface.

The problems of transient mass diffusions are very similar to transient heat conduction. Biot and Fourier numbers can be defined for mass diffusion as follows:

$$Bi_m = \frac{h_m L_c}{D}$$
(1.86a)

$$Fo_m = \frac{Dt}{L_c^2}$$
(1.86b)

Therefore, when Biot and Fourier numbers for mass transfer are used, the solutions for transient heat conduction problem can be employed to calculate mass transfer. For example, Eq. (1.74) for the infinite slab, infinite cylinder, sphere, and the solutions for the semi-infinite body given in Table 1.17 can be directly used for mass transfer. The difference consists of replacing temperatures with mass (or molar concentrations). For example, $T(x, t)$ must be replaced with $C_m(x, t)$ or $C(x, t)$, and the surface heat flux must be replaced with $j_m(x=0,t)$ or $j_n(x=0,t)$ depending on the choice for species transport – either mass basis or molar basis. Furthermore, thermal diffusivity $a$ must be replaced with the mass diffusivity coefficient $D$. The analytical solutions are given in Table 1.19.

**Table 1.19**  Analytical solutions for transient mass transfer through the semi-infinite solid

| Boundary and initial conditions | Temperature and surface heat flux solution |
| --- | --- |
| $C_m(x>0,t=0)=C_{m,i}$ <br> $C_m(x=0,t)=C_{m,i}+\xi t^{0.5n}$ <br> $C_m(x=\infty,t)=C_{m,i}$ | $C_m(x,t)=C_{m,i}+4\xi\Gamma(1+0.5n)\left[i^n\mathrm{erfc}\left(\frac{0.5x}{\sqrt{Dt}}\right)\right]$ <br><br> $j_m(x=0,t)=\frac{2^{n-1}}{\sqrt{D}}t^{0.5(n-1)}\rho D^2\Gamma(1+0.5n)\left[i^{n-1}\mathrm{erfc}(0)\right]$ |
| $C_m(x>0,t=0)=C_{m,i}$ <br> $j_m(x=0,t)=j_0$ <br> $C_m(x=\infty,t)=C_{m,i}$ | $C_m(x,t)=C_{m,i}+2j_0\frac{\sqrt{Dt}}{\rho D}\left[i^1\mathrm{erfc}\left(\frac{0.5x}{\sqrt{Dt}}\right)\right]$ |
| $C_m(x>0,t=0)=C_{m,i}$ <br> $-D\frac{\partial C_m(x=0,t)}{\partial x}$ <br> $=h_m[C_{m,\infty}-C_m(x=0,t)]$ <br> $C_m(x=\infty,t)=C_{m,i}$ | $\frac{C_m(x,t)-C_{m,i}}{C_{m,\infty}-C_{m,i}}=\mathrm{erfc}\left(\frac{0.5x}{\sqrt{Dt}}\right)-\exp\left[\frac{h_m x}{\rho D}+\frac{h_m^2 Dt}{(\rho D)^2}\right]\mathrm{erfc}\left[\frac{h_m\sqrt{Dt}}{\rho D}+\frac{0.5x}{\sqrt{Dt}}\right]$ |

## 1.9   Concluding Remarks

In this chapter, some fundamental aspects of thermodynamics, and heat and mass transfer are introduced and discussed through various processes, systems and applications. These are intentionally presented to prepare the readers before proceeding to the second chapter to discuss the moisture transfer aspects with drying processes. These will also be helpful later for system analysis and performance assessment. The fundamental properties and quantities are introduced first, and the main notions from thermodynamics such as pressure, temperature, volume, work, heat, energy enthalpy, and so on are discussed. The state diagrams are presented to give a true idea about the processes. Ideal and real gas equations are given as they are of importance. Therefore, the chapter introduces the ideal gas law, van der Waals law and compressibility factor charts as well as it discusses the properties of gas mixtures. The laws of thermodynamics and related analysis methods with energy and exergy are presented in detail, particularly through the balance equations. Both energy and exergy efficiencies for various are defined for later use. The psychometric processes are presented and discussed for analysis. Basic heat and mass transfer aspects in steady and unsteady forms are introduced, and the analogies between these are presented through various quantities and correlations.

## 1.10   Study Problems

1.1     What represents boundary work and what is its mathematical equation?
1.2     Is it possible that a closed system exchanges mass with the surroundings?
1.3     A closed system is in a state of internal non-equilibrium. What will eventually happen is the system is isolated?
1.4     A closed thermodynamic system containing 1 mol of ideal gas expands two times generating work against standard atmosphere. Determine the useful work amount which can be extracted.
1.5     What is the difference between gauge and absolute pressures?
1.6     A chunk of 1 kg of subcooled ice at $-10\,°C$ is heated under direct sunlight until water reaches $15\,°C$ under 1 atm. Calculate the amount of heat required to melt ice. If the average sunlight intensity is $500\,W/m^2$ and the ice area exposed to light is $100\,cm^2$, how fast the ice can be melted.
1.7     Calculate the quality of water-vapor mixture at $25\,°C$ with density of $500\,kg/m^3$.
1.8     Air is compressed isentropically with a pressure ratio of 2.5. Calculate the discharge temperature and the required work if the intake temperature is $18\,°C$.
1.9     What is the difference between thermal and caloric equation of state?
1.10    How much is the value of compressibility factor for ideal gas?
1.11    A real gas has the compressibility factor of 0.4 and is maintained in a closed box with fixed volume of $1000\,cm^3$. Heat is added to the box such that the gas warms up from 300 to 600 K. Calculate the required heat amount and the final pressure in the box.
1.12    Based on the description given in Table 1.5 derive an expression for the entropy of van der Waals gas.
1.13    Consider a mixture of ideal gases enclosed in a box of $1\,m^3$ in which the mass of oxygen is 100 g. Calculate the molar fraction of oxygen provided that the mixture is at standard pressure and 250 K.
1.14    Explain the difference between internal energy and enthalpy.

1.15  Explain the role of expansion coefficient for determining the work exchange during adiabatic process.

1.16  What is the difference between externally reversible and internally reversible processes?

1.17  What is the relationship between entropy change of a system, the entropy of its surroundings and of generated entropy?

1.18  Using thermodynamic tables or EES software calculate the thermo-mechanical exergy of saturated steam at 10 MPa. Assume standard values for temperature and pressure of the environment.

1.19  A cistern vehicle transports 10 t of butane and runs with 80 km/h, being exactly in the top of a hill of 100 m altitude with respect to seal level. Calculate the total exergy stored in the cistern if the datum for potential energy is at sea level.

1.20  Stack gas of a power plant comprises 10% carbon dioxide by volume and the rest is made of other gaseous combustion products. The pressure of stack gas is 1.25 atm and temperature of 400 K. Using Gibbs free energy calculate the amount of reversible work needed to separate carbon dioxide.

1.21  Calculate chemical exergy of butane based on standard exergy of the elements.

1.22  Write balance equations for a wall assuming that exterior temperature is −5 °C, interior temperature is 20 °C, the wall thermal conductivity is 1.1 W/m K, the heat transfer coefficient is 12 W/m$^2$ K on both sides. Calculate the entropy generation and exergy destruction per unit of wall surface.

1.23  Describe any advantages of exergy analysis over energy analysis.

1.24  Is there such a thing as reversible heat transfer? Explain.

1.25  In a diffuser air of rate 1 kg/s is compressed from standard state to a pressure of 1.5 atm. Write mass, energy, entropy, and exergy balances and determine device energy and exergy efficiency under reasonable assumptions.

1.26  A turbine with expands saturated toluene at 2.8 MP. The isentropic efficiency of the turbine is 0.8. Write balance equations and calculate expansion ratio, energy, and exergy efficiency of the turbine provided that the expanded vapor saturation temperature is 40 °C.

1.27  Consider adiabatic mixing of 10 g/s moist air at 35 °C with relative humidity of 65% with 2 g/s of moist air at 9 °C with relative humidity of 42%. Write balance equation and determine the temperature and relative humidity of air for the mixed stream, entropy generation, exergy destruction, and second law efficiency for the mixer. EES software or a Mollier diagram can be used.

1.28  Refrigerant R134a boils convectively in a horizontal pipe at a local pressure of 75 psi absolute. Determine the local heat transfer coefficient under reasonable assumptions provided that vapor quality is 50%. Use heat transfer correlation described in Eq. (1.65).

1.29  Repeat problem 1.28 for the case of condensation on horizontal tube; use Eq. (1.67).

1.30  Consider a material having specific heat of 2 kJ/kg K, density of 2 kg/l, thermal conductivity of 20 W/m K, temperature of 400 K. The surrounding air is at 300 K. Determine the time required for 1 kg of material to reach 310 K.

1.31  The ground surface is fixed at 260 K while the temperature in the soil is relatively uniform at 385 K. Freezing propagates in the soil. The thermal conductivity and the thermal diffusivity of the soil are given as 0.4 W/m K and 10$^{-7}$ m$^2$/s. Determine the depth where freezing front arrives after 90 days.

1.32  Determine the molar fraction of oxygen dissolved in water at 25 °C for two conditions: (i) under standard atmosphere and (ii) under deep vacuum, of 10 mbar absolute.

1.33   Determine the water evaporation rate in standard atmosphere from a wet cylindrical wick having diameter of 5 mm and height of 10 mm of dry bulb temperature is 20 °C and wet bulb temperature is 15 °C.

# References

Bejan A., Kraus A.D. 2003. Heat Transfer Handbook. Wiley: Hoboken, NJ.

Cengel Y.A., Boles M.A. 2011. Fundamentals of Thermodynamics. McGraw Hill: New York.

Chato J.C. 1962. Laminar condensation inside horizontal and inclined tubes. *ASHRAE Journal* 4:52–60.

Dincer I., Rosen M.A. 2012. Exergy: Energy, Environment and Sustainable Development. Elsevier: Oxford, UK.

EES 2013. Engineering Equation Solver. F-Chart Software (developed by S.A. Klein): Middleton, WI.

Gorenflo D. 1993. Pool Boiling. In: VDI-Heat Atlas. VDI-Verlag: Düsseldorf.

Haynes W.M., Lide D.R. 2012. CRC Handbook of Chemistry and Physics. 92nd ed., Internet version. CRC Press: New York.

ISU 2006. The International System of Units (SI). 8th ed., International Bureau of Weights and Measures: Sèvres, France.

Lin Z., Cleland A.C., Serrallach G.F., Cleland D.J. 1993. Prediction of chilling times for objects of regular multi-dimensional shapes using a general geometric factor. *Proceedings of the International Meeting on Cold Chain Refrigeration Equipment by Design*, 15–18 November, Palmerston, New Zealand, 259–267.

Mujumdar A.S. Ed. 2006. Handbook of Industrial Drying. 3rd ed. Taylor and Francis.

Rivero R., Grafias M. 2006. Standard chemical exergy of elements updated. *Energy* 31:3310–3326.

Steiner D., Taborek J. 1992. Flow boiling heat transfer in vertical tubes correlated by an asymptotic model. *Heat Transfer Engineering* 13:43–69.

Wepfer W.J., Gaggioli R.A., Obert E.F. 1979. Proper evaluation of available energy for HVAC. *ASHRAE Transactions* 85:214–230.

# 2

# Basics of Drying

## 2.1 Introduction

By definition, drying is thermally driven process. Therefore, during drying, at least two main processes occur simultaneously: heat transfer from the surrounding environment toward the moist material to evaporate the moisture at material surface and transfer of internal moisture to the surface of the solid. Other processes may occur in some cases, namely, physical and chemical transformations which may induce changes in the structure of solids and fluids involved.

The fundamental mechanism of moisture transfer within the solid has received much attention in the literature. There appear to be four major modes of transfer: (i) capillary flow of moisture in small interstices, (ii) moisture diffusion due to concentration gradients, (iii) vapor diffusion due to partial pressure gradients, and (iv) diffusion in liquid layers adsorbed at solid interfaces. The mechanisms of capillarity and liquid diffusion have received the most detailed treatment. In general, capillarity is most applicable to coarse granular materials, while liquid diffusion rules single-phase solids with colloidal or gel-like structure. In many cases, the two mechanisms may be applicable to a single drying operation, that is, capillarity dominating moisture movement in the early stages of drying, while taking over at lower moisture contents (Brennan et al., 1976).

Knudsen diffusion is also important in some drying technologies when low temperature and pressure are used to freeze the material and extract humid vapors. Hydrostatic phenomena and surface diffusion may occur sometimes at the surface of solids subjected to drying.

Another important aspect is the mathematical modeling of the drying processes and equipment. Its purpose is to allow design engineers to choose the most suitable operating conditions and then size the drying equipment and drying chamber accordingly to meet desired

*Drying Phenomena: Theory and Applications*, First Edition. İbrahim Dinçer and Calin Zamfirescu.
© 2016 John Wiley & Sons, Ltd. Published 2016 by John Wiley & Sons, Ltd.

operating conditions. The principle of modeling is based on formulating a set of mathematical equations for the process and on the assumption of boundary and initial conditions, which can adequately characterize the system. The solution of these equations must allow for the prediction of the process parameters as a function of time at any point in the drying equipment.

In this chapter, the basics of drying are reviewed. The chapter outlines various aspects related to drying of solids, the structure of porous media (e.g., food products), and their characterization and properties. The shrinkage, the porosity of porous materials, and the heat and moisture transfer analyses of drying in porous materials were also analyzed. The chapter focuses on the identification of fundamental methods of analysis and prediction of the drying process of porous materials.

## 2.2  Drying Phases

In a drying process, the moisture content is generally determined on a dry material basis. More exactly, when the weights of the moist and dry materials are compared, the mass of moisture is determined. The ratio between the mass of moisture ($m_m$) and the mass of dry material ($m_s$) represents the moisture content in percent on a dry basis; this is also denoted with the symbol $W$. Note that in some cases moisture can be heavier than dry bone material. Therefore, the moisture can be superior to unity; for example, in the case of gelatins, $W > 2$. The variation of dry basis moisture content is usually measured versus the drying time.

Figure 2.1a shows the typical variation of moisture content during drying for a general case when a moist solid loses moisture. During the drying of a moist solid in heated air, the air supplies the necessary sensible and latent heat of evaporation to the moisture and also acts as a carrier gas for the removal of the water vapor formed from the vicinity of the evaporating surface. In the diagram from Figure 2.1a, the first part of the process, represented by the curve A–B, occurs by mass transfer from the solid surface. This is a stage of warming up of the solid(s) during which the solid surface conditions come into equilibrium with the drying air. In terms of energy consumption, period A–B often represents a negligible proportion of the overall drying cycle requirements, but in some cases this may be significant. During this period,

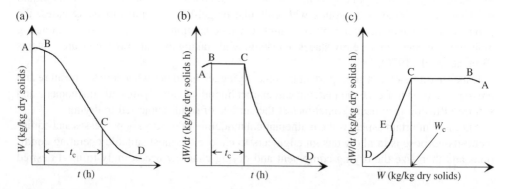

**Figure 2.1**  The drying periods for a solid. (a) Moisture content versus time. (b) Drying rate versus drying time. (c) Drying rate versus moisture content. (The curves are for moist material dried at a constant temperature and relative humidity)

the moist material can even take more humidity instead of being dried, depending on the actual conditions. Eventually, the surface reaches such conditions that the humidity diffuses out of the product. The period A–B is governed mostly by transient heat transfer processes which, in function of Biot number value, can be diffusively or convectively controlled.

The process B–C shows a reduction of the moisture content which is approximately linear in time. During this phase, the area of the saturated surface decreases gradually. This is the period of drying during which the rate of water removal per unit of drying surface is essentially constant. Point C, where the constant rate period ends, is known as the point of critical moisture content.

During B–C period, the movement of moisture within the solid is rapid enough to keep a saturated condition at the surface, and the drying rate is controlled by the rate at which heat is transferred to the evaporating surface. The surface of the solid remains saturated with liquid water (or moisture) by virtue of the fact that movement of water within the solid to the surface takes place at a rate as great as the rate of evaporation from the surface. This stage is controlled by the heat and/or moisture transfer coefficients, the area exposed to the drying medium, and the difference in temperature and relative humidity between the drying air and the wet surface of the solid.

Once the saturated moisture at the surface is completely eliminated, the phase C–D follows during which there is no evaporation at the surface, but rather a diffusion process of the moisture within the solid followed by a convective mass transfer at the solid surface. From point C onward, the surface temperature begins to rise and continues to do so as drying proceeds, approaching the dry-bulb temperature of the air as the material approaches dryness. Therefore, when the initial moisture content is above the critical moisture content, the entire drying process occurs under the constant rate conditions. If it is below the critical moisture content, the entire drying process occurs in the falling rate period.

Figure 2.1b shows the changes in drying rate versus moisture content on dry basis of the solid. This curve is obtained by differentiation of the $W(t)$ curve from Figure 2.1a. Remark that during the period B–C, the rate of drying is constant, which corresponds with the linear reduction of moisture specific to the evaporative drying. Furthermore, the diffusive drying shows a relatively sharp decrease of the moisture elimination rate. Furthermore, Figure 2.1c shows the variation of drying rate versus drying time.

Drying kinetics refers to the changes of average material moisture content and average temperature with time, unlike drying dynamics which describes changes in the temperature and moisture profiles throughout the drying body. Drying kinetics enables calculating the amount of moisture evaporated, drying time, energy consumption, and so on. These depend to a considerable extent on the physicochemical properties of the material. Nevertheless, the changes in material moisture content and temperature are usually controlled by heat and moisture transfer between the body surface, the surroundings, and the internal structure of the drying material. The change in moisture content with time is influenced significantly by the parameters of the drying process, for example, temperature, humidity (pressure), relative velocity of air, or total pressure.

## 2.3  Basic Heat and Moisture Transfer Analysis

In drying analysis, it is important to model the rate moisture content decrease as well as the rate of heat and mass transfer at the surface of the material under consideration. In some cases, the liquid moisture within the moist material is physically and/or chemically bound to solid matrix so as to exert a vapor pressure lower than that of pure liquid at the same temperature. At a given temperature and pressure, the moisture content within the hygroscopic material is in equilibrium with the gas–vapor mixture. This denotes the equilibrium moisture content, $W_e$. For nonhydroscopic materials, $W_e$ is zero. The moisture content which is free to be removed is that which is in excess of the equilibrium moisture content. Also, the unbound moisture exerts a vapor pressure equal to that of pure liquid at the same temperature.

During the constant rate period (B–C), the rate of drying depends on the rate of heat transfer toward the drying surface since it is controlled by the external heat and mass transfer. Mathematically, the rate of moisture transfer at the solid surface under constant drying conditions can be expressed as

$$\frac{dW}{dt} = -j_m A \tag{2.1}$$

where $dW/dt$ is the drying rate, $j_m$ represents the moisture removal flux given in kg/m² s, and $A$ is the drying surface area. Equation (2.1) can be further developed from the more general mass flux equation at the evaporating surface given by Eq. (1.80a) which is based on Chilton–Colburn analogy for mass transfer. Here, the mass concentrations are given by the partial pressures. Equation (2.1) can also be written as

$$\frac{dW}{dt} = \frac{h_m A}{\rho\, C_p\, Le^{\frac{2}{3}} R} \left( \frac{P_s}{T_s} - \frac{P_a}{T_a} \right) \tag{2.2}$$

where $P_s$ is the partial pressure of evaporating moisture (water) at the surface of the material and $P_a$ is the partial pressure of moisture in bulk air (drying agent), $h_m$ is the mass transfer coefficient given in 1/s, $R$ is the universal gas constant, $Le$ is the Lewis number, $\rho C_p$ refers to the drying agent (e.g., bulk air), and $T$ is temperature.

The rate of heat transfer to the drying surface $dQ/dt$ can be expressed in function of the heat transfer coefficient by convection $h$, the surface area $A$, the dry-bulb temperature of air $T_a$, and the surface temperature of the material $T_s$ as follows:

$$\frac{dQ}{dt} = hA(T_a - T_s) \tag{2.3}$$

Note that in some situations the heat transfer $h$ must be taken as a linearized heat transfer coefficient that accounts for combined heat transfer by conduction, convection, and radiation. In addition, if only convective heating is considered, then $T_s$ is generally approximated with the wet-bulb temperature of the air. At steady operation, equilibrium must exist between the rate of heat transfer to the material which is dried and the rate of moisture transfer away from the body. Therefore, the two rates are proportional; thence

$$\frac{dW}{dt} m_s \Delta h_{lv} = -\frac{dQ}{dt} \tag{2.4}$$

where $\Delta h_{lv}$ is the latent heat of evaporation at $T_s$ and $m_s$ is the mass of dry bone solid, and $W = m_m/m_s$. Here, $m_m \Delta h_{lv} dt = dQ$ because moisture evaporates due to heat addition. Combining Eqs. (2.3) and (2.4), one obtains

$$\frac{dW}{dt} = -\frac{hA}{m_s \Delta h_{lv}}(T_a - T_s) \tag{2.5}$$

Assume a tray contains moist material for drying under exposure to air of lower humidity content. If the depth of the moist material in the tray is $d$ and no shrinkage occurs, then the total latent heat produced by evaporation from the upper surface is $m_s = \rho_{dr} A d$, where $\rho_{dr}$ is the density of the dry bone material. Therefore, the rate of drying is in this case as follows:

$$\frac{dW}{dt} = -\frac{h}{\rho_{dr} \Delta h_{lv} d}(T_a - T_s) \tag{2.6}$$

Consequently, the drying time $t_{CR}$ in the constant rate period (B–C) – denoted with the critical drying time – can be obtained by the integration of Eq. (2.7) as follows:

$$t_{CR} = \frac{(W_i - W_c)\rho_{dr} d}{h(T_a - T_s)} \Delta h_{lv} \tag{2.7}$$

where $W_i$ is the initial moisture content of the solid and $W_c$ is the moisture content at the end of constant rate period.

Note that $W_c$ depends on the type of the material and on the drying rate. At the lower bound, the critical moisture content $W_c$ is bounded by the equilibrium moisture content $W_e$. Equilibrium moisture content is the moisture of a given material that is in equilibrium with the vapor contained in the drying agent under specific conditions of air temperature and humidity. This is also called the "minimum hygroscopic moisture content." It changes with the temperature and humidity of the surrounding air. However, at low drying temperatures (e.g., 15–50 °C), the equilibrium moisture content becomes independent of temperature, and it becomes zero at zero relative humidity. The equilibrium moisture content depends strongly on the nature of the material to be dried. For nonporous (i.e., nonhygroscopic materials), the equilibrium moisture content is essentially zero at all temperatures and humidities. For hygroscopic materials (e.g., wood, food, paper, soap, chemicals), it varies regularly over a wide range with temperature and relative humidity. For materials such as sand and rock, the equilibrium moisture content is in general inferior to 0.1 kg water/kg dry solid. For paper and soil, $W_e$ is in average 0.3 kg/kg and for food is in general superior to 0.5 kg/kg. The equilibrium moisture content decreases when temperature increases according to $\ln(W_{e,2}/W_{e,1}) \cong -0.075(T_2 - T_1)$.

In sum, the rate controlling factors during the constant rate period are the drying surface area, the difference in temperature or humidity between the air and the drying surface, and the heat or moisture transfer coefficients. For estimating the drying rates, the use of heat transfer coefficients is considered to be more reliable than the use of moisture transfer coefficients. For many cases, the heat transfer coefficient can be calculated from $Nu$–$Re$ correlations. Thus, the air velocity and system dimensions influence the drying rates during the constant rate period. Alternative expressions for $h$ are used where the airflow is not parallel to the drying surface or for through-flow situations. When heat is supplied to the material by radiation and/or conduction in addition to convection, then an overall heat transfer coefficient that takes this into account must be substituted for $h$ in Eq. (2.7). Under these circumstances, the surface temperature during the constant rate period of drying remains constant, at some value above the wet-bulb temperature of the air and below the boiling point of water.

The "falling rate period" period C–D usually consists of two zones: the zone of unsaturated surface drying and the zone in which the controlling mechanism is the movement of moisture inside the moist material. At point E represented in Figure 2.1c, the entire exposed surface becomes completely unsaturated, and this marks the start of the drying process during which the rate of internal moisture movement controls the drying rate.

In Figure 2.1c, C–E is defined as the first falling rate period, and E–D is the second falling rate period. In the falling rate periods, the rate of drying is influenced mainly by the rate of movement of moisture within the solid, and the effects of external factors, in particular air velocity, are reduced, especially in the latter stage. Usually, the falling rate period represents the major proportion of the overall drying time. For systems where a capillary flow mechanism applies, the rate of drying can often be expressed with reasonable accuracy by an equation of the type as given below:

$$\frac{dW}{dt} = -h_m(W - W_e) \tag{2.8}$$

From here, it results that

$$h_m = -\frac{dW/dt}{W_c - W_e} \tag{2.9}$$

where $-dW/dt$ is the rate of drying at time $t$ from the start of the falling rate period, $W$ is the moisture content of the material at any time $t$, and $W_e$ is the equilibrium moisture content of the material at air temperature and humidity.

Accounting for Eqs. (2.2), (2.6), and (2.9), it can be shown that

$$\frac{dW}{dt} = -\frac{h(T_a - T_s)(W_c - W_e)}{\rho_s d \Delta h_{lv}} \frac{(W_c - W_e)}{(W - W_e)} \tag{2.10}$$

Integrating the above equation from $t = 0$ to $t$ and from $W = W_c$ to $W$, the drying time results as follows:

$$t = \frac{\rho_s d(W_c - W_e)\Delta h_{lv}}{h(T_a - T_s)} \ln \frac{(W_c - W_e)}{(W - W_e)} \tag{2.11}$$

Note that the drying Eqs. (2.6), (2.7), (2.10), and (2.11) are applicable when drying takes place from one side only. In cases where drying occurs from both sides or surfaces, $d$ will be taken as the half thickness.

In order to analyze the transient diffusion-controlled rate drying, let us assume for simplicity that the moist material has the geometrical form of a slab. In this case, the characteristic length

$L_c$ for its Biot and Fourier numbers is the half thickness. The transient mass diffusion through the material becomes important. In this case, the governing equation for moisture transfer is

$$D\frac{\partial^2 \phi}{\partial z^2} = \frac{\partial \phi}{\partial t}$$

(2.12)

where $\phi$ is the dimensionless moisture content and $z$ represents the coordinate across the slab. The dimensionless moisture content is defined by

$$\phi = \frac{W - W_e}{W_i - W_e}$$

(2.13)

In the case when $0.1 < Bi < 100$, the simplified solution to this problem is given elsewhere, for example, in Dincer and Dost (1996), in the following approximated form:

$$\phi = LF \exp\left[-\arctan^2(0.64\,Bi + 0.38)\frac{Dt}{L_c^2}\right]$$

(2.14)

where $D$ is the diffusion coefficient, $t$ is time, and LF is the lag factor given by

$$LF = \exp\left(\frac{0.2533\,Bi}{Bi + 1.3}\right)$$

(2.15)

As detailed in Dincer and Dost (1996), the mass transfer coefficient in diffusion-controlled phase is based on diffusion coefficient of moisture through the solid according to the following equation:

$$h_m = \frac{D}{L_c}\frac{1 - 3.94813\ln LF}{5.13257\ln LF}$$

(2.16)

The drying time during transient period can be determined from integration of the equation $d\phi = -h_m\phi\,dt$ which expresses that the incremental change of dimensionless moisture content is due to a mass transfer at the product surface. This equation is a dimensionless form of Eq. (2.8). It follows that the drying time is

$$t = -\frac{1}{h_m}\ln\left(\frac{\phi_f}{\phi_i}\right)$$

(2.17)

where $\phi_f$ is the final dimensionless moisture content of the dry product and $\phi_i$ is the initial dimensionless moisture content of the product.

## Example 2.1

A potato slice 20 mm thick at an initial temperature of 20 °C is dried in air at 65 °C dry and 20% relative humidity. Calculate the duration of the initial period of drying and the moisture removed. Consider that the heat transfer coefficient is $20\,\mathrm{W/m^2\,K}$ and potato's $a = 1.77 \times 10^{-7}\,\mathrm{m^2/s}$ and $k = 0.5\,\mathrm{W/m\,K}$.

- Assume that for the initial period of drying, the potato behaves as a semi-infinite solid that is warmed up; its surface reaches the wet-bulb temperature.
- The wet-bulb temperature can be easily found from Engineering Equation Solver (EES) or psychometric tables, based on dry-bulb temperature and relative humidity, namely, $T_s = 38.2\,°\mathrm{C}$.
- The solution of heat transfer at semi-infinite solid is given in Table 1.16. For the potato's surface, we take $x = 0$; for bulk air temperature, we have $T_\infty = 65\,°\mathrm{C}$; and for initial temperature, $T_i = 25\,°\mathrm{C}$. Thence, the surface temperature is

$$\frac{T(t) - 25}{65 - 25} = \mathrm{erfc}(0) - \exp\left(\frac{20^2 \times 1.77 \times 10^{-7}}{0.5^2} t\right) \mathrm{erfc}\left(\frac{20\sqrt{1.77 \times 10^{-7}\,t}}{0.5}\right)$$

- In the above equation, one imposes that $T(t) = T_s = 38.2\,°\mathrm{C}$ and solve for $t$. The result is obtained with EES, namely, $t = 574.2\,\mathrm{s}$ or 9.6 min.
- We will now calculate the temperature at the potato slice bottom – which is at 20 mm apart from the surface – at the moment when the surface reached $38.2\,°\mathrm{C}$. This is done according to the solution given in Table 1.16, namely,

$$\frac{T - 25}{65 - 25} = \mathrm{erfc}\left(\frac{0.5 \times 0.02}{\sqrt{574.2a}}\right) - \exp\left[\frac{20 \times 0.02}{0.5} + \frac{20^2 574.2a}{0.5^2}\right] \mathrm{erfc}\left[\frac{20\sqrt{574.2a}}{0.5} + \frac{0.5 \times 0.02}{\sqrt{574.2a}}\right]$$

- The result is $T_\mathrm{bottom} = 26.35\,°\mathrm{C}$ which is close to the initial temperature of $25\,°\mathrm{C}$ and confirm that the semi-infinite model is a good approximation.
- The average heat flux during the initial time interval of 574.2 s results from the integral

$$\bar{q} = \frac{1}{574.2} \int_0^{574.2} 20\,(65 - T(t))\mathrm{d}t = 609\,\mathrm{W/m^2}$$

- The average temperature of product surface results from $\bar{q} = h(T_\infty - \bar{T}) \rightarrow \bar{T}_s = 34.5\,°\mathrm{C}$.
- Assume that during this transient period the moisture is saturated at the average surface temperature. Thence, the latent heat of water becomes $\Delta h_\mathrm{lv} = 2.4 \times 10^6\,\mathrm{J/kg}$. The amount of evaporated water results from the energy balance equation $\bar{q}tA = m\Delta h_\mathrm{lv}$. Thence, the mass of water evaporated per unit of slice volume is

$$\Delta W = \frac{m}{m_\mathrm{dry}} = \frac{m}{\rho_s V} = \frac{m}{A\rho_s d} = \frac{\bar{q}t}{d\rho_s \Delta h_\mathrm{lv}} = \frac{609 \times 574.2}{0.02 \times 820 \times 2.4 \times 10^6} = 0.9\%$$

## Example 2.2

Calculate the drying time during constant rate period at drying of a 20 mm potato slice from Example 2.1 from 50% humidity to 46% humidity under warm dry air atmosphere at 65 °C and relative humidity of 20%. The density of potato is given as 693 kg/m³.

- We will assume that the surface is at wet-bulb temperature under standard pressure.
- The latent heat of water at the wet-bulb temperature is $\Delta h_{lv} = h_v - h_l = 2{,}535 - 79.12 = 2456\,MJ/kg$.
- Equation (2.7) is directly applied with $d = 0.02\,m$, $W_i = 0.5$, $W_c = 0.45$, $h = 20\,W/m^2 K$, and $T_a = 35°C$:

$$t_{CR} = \frac{693 \times 0.02 \times 2456 \times 10^9 \times (0.5 - 0.46)}{20 \times (65 - 38.2)} = 29\,\text{days}$$

## Example 2.3

For the same data as in Example 2.2, consider now that the product enters in the falling rate period C–E and the equilibrium moisture content in the moist material is 45%. Determine the additional time require to dry the material down to 45.5% humidity.

- The humidity at the beginning of falling rate period is assumed from the previous example to be 35%: $W_c = 0.35$.
- Using Eq. (2.13) for $W_e = 0.45$, $W = 0.455$, one obtains

$$t = \frac{693 \times 0.02 \times (0.46 - 0.45) \times 2456 \times 10^9}{20 \times (65 - 38.2)} \ln\left(\frac{0.46 - 0.45}{0.455 - 0.45}\right) = 5\,\text{days}$$

## Example 2.4

A rectangular wood slab with dimensions of $0.2\,m \times 0.1\,m \times 0.04\,m$ is dried in air at 70 °C with 35% relative humidity. The material humidity is 0.173 kg/kg and must be well dried for 8 months; the equilibrium humidity is 0.05 kg/kg. Biot number is 0.71 and the diffusion coefficient is $3.2 \times 10^{-9}\,m^2/s$. Calculate the final humidity.

- Calculate the lag factor with Eq. (2.15): LF = 1.094.
- From the lag factor, the initial humidity result is $LF(W_i - W_e) = W_{(t=0)} - W_e \rightarrow 1.094(W_i - 0.05) = 0.173 - 0.05$. It results that $W_i = 0.1625\,kg/kg$. Thence, during the lag period, the wood absorbs humidity.
- Replace the known values in Eq. (2.14) and obtain $\phi(t) = 1.094\exp(-7.8 \times 10^{-8} t)$.
- The dimensionless moisture content after 8 months is $\phi(8 \times 3600) = 0.2152$.
- From definition of $\phi$, the final humidity result is as follows: $W_f = W_e + \phi_f(W_i - W_e) = 0.05 + 0.2152(0.1625 - 0.05) = 0.0742\,kg/kg$.

## 2.4 Moist Material

So-called moist materials have different physical, chemical, structural, mechanical, biochemical, and other properties, which result from the state of water within it. Although these parameters can influence significantly the drying process and determine the selection of the drying technique and technology, the most important in practice are the structural–mechanical properties, the type of moisture in the solids, and the material–moisture bonding. The moisture content of the material can be defined on a wet basis ($W^*$) as follows:

$$W^* = \frac{m_m}{m} = \frac{m_m}{(m_m + m_s)} \tag{2.18}$$

where $m$ is total mass of moist material (kg). The moisture content on dry and wet basis is connected by the following relationships:

$$W = \frac{W^*}{(1 - W^*)} \tag{2.19}$$

$$W^* = \frac{W}{(1 + W)} \tag{2.20}$$

In addition, the volumetric moisture content ($X_V$) is defined as

$$X_V = \frac{V_{ml}}{(V_s + V_{ml} + V_{mv})} \tag{2.21}$$

where $V_{ml}$ is the volume of liquid moisture (m$^3$), $V_{mv}$ is the volume of vapor moisture (m$^3$), and $V_s$ is the volume of dry material (m$^3$). The percentage saturation of moist material ($\Lambda_V$) becomes

$$\Lambda_V = \frac{V_{ml}}{(V_{ml} + V_{mv}) \times 100} \tag{2.22}$$

Moist materials are classified into three categories based on their behavior with respect to drying (Strumillo and Kudra, 1986): (i) colloidal materials, (ii) capillary-porous materials, (iii) and colloidal-capillary-porous materials. Figure 2.2 shows this classification. The colloidal materials change size but preserve their elastic properties during drying (e.g., gelatin, agar). Capillary-porous materials, which become brittle, shrink slightly and can easily be ground after drying (e.g., sand, charcoal, coffee). In fact, in drying, all moist materials whose pore radius is smaller than $10^{-5}$ m can be treated as capillary porous with various pore diameter distributions. Note that the moisture in such bodies is maintained mainly by surface tension forces. If the pore size is greater than $10^{-5}$ m, then, in addition to capillary forces, gravitational forces also play a role. These bodies are called "porous." Colloidal-capillary-porous materials have the properties of the above two types. The walls of the capillaries are elastic, and they swell during drying (e.g., food, wood, paper, leather). These materials are capillary porous as far as their structure is concerned and colloidal as far as their properties are concerned.

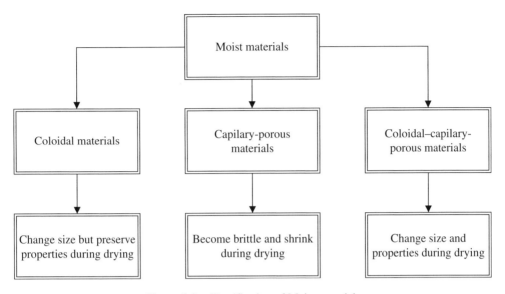

**Figure 2.2**   Classification of Moist materials

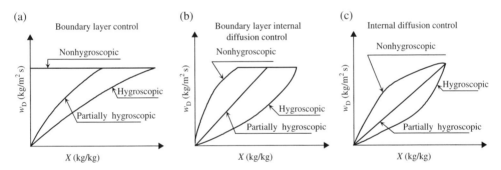

**Figure 2.3**   Classification of moist materials according to the types of their drying rate curves: (a) boundary layer control, (b) boundary layer and internal diffusion, and (c) internal diffusion control

There is an alternative classification of dried materials, which is based on the states of moisture in the solids according to Figure 2.3, namely, (i) nonhydroscopic materials, (ii) partially hygroscopic materials, and (iii) hygroscopic materials. Nonhygroscopic materials contain no bound moisture. These materials include nonporous or porous bodies with pore radii greater than $10^{-7}$ m. Partially hygroscopic materials include macroporous bodies that, although they also have bound moisture, exert a vapor pressure that is only slightly lower than that exerted by a free water surface. Hygroscopic materials may contain bound moisture. These materials cover mainly microporous bodies in which liquid exerts a vapor pressure that is less than that of the pure liquid at the given temperature. If the moisture content in the hygroscopic body exceeds the hygroscopic moisture content, that is, if it contains unbound water, then it behaves like a nonhygroscopic material up to the moment when this unbound water is removed.

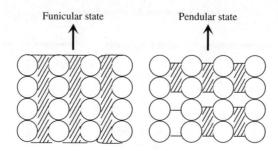

**Figure 2.4** Two forms of unbound moisture. Funicular state is that condition when during a drying process of a capillary porous material air is absorbed into the pores due to capillary suction forces. Pendular state is the state of a liquid in a porous solid when a continuous film of liquid no longer exists around and between discrete particles so that flow by capillary cannot occur. This state succeeds the funicular state

In summary, the drying periods are two for nonhygroscopic materials and three for hygroscopic materials. Karel (1975) defined a nonhygroscopic material as one in which the partial pressure of water in the material is equal to the vapor pressure of water. In hygroscopic materials, however, the partial pressure of water becomes less than the vapor pressure of water at some critical level of moisture. Note that as the moisture content of the material decreases, the transport through capillaries and pores occurs primarily in the vapor phase. Note also that the shape of drying curve in the falling rate period depends strongly on the moisture transfer conditions, where the drying behavior is affected by internal and external conditions. No constant drying rate period is commonly observed during the drying of hygroscopic solids under the condition that the drying is controlled internally.

There are several types of moisture in moist materials. *Surface moisture* is liquid which forms as an external film on the material due to surface tension effects. *Unbound, free or capillary moisture* is unbound moisture in a hygroscopic material in excess of the equilibrium moisture content corresponding to saturation humidity. All moisture in a nonhygroscopic material is unbound water. Note that in a hygroscopic material where actual vapor pressure is a function of the saturated state, there is moisture in excess of the equilibrium moisture content corresponding to saturation relative humidity.

Unbound moisture can be categorized into two forms, as shown in Figure 2.4: funicular form, where a continuous liquid state exists within the porous body, and pendular form, where the liquid around and between discrete particles is discontinuous, and moisture is interspersed by air bubbles. In the funicular state, liquid movement to the external surface of the material takes place by capillary action. As moisture is removed, the continuity of the liquid phase is interrupted due to suction of air into pores, which leaves isolated pockets of moisture (pendular form). Consequently, capillary flow is possible only on a localized scale. When the material is close to the bone dry (solid) state, the moisture is held as a monolayer of molecules on the pore walls, so it is removed by vapor flow mainly.

*Bound, hygroscopic, or dissolved moisture* is the liquid which exerts a vapor pressure less than that of the pure liquid at the given temperature. Liquid may become bound by retention in small capillaries, by solution in cell or fiber walls, by homogenous solution throughout the solid, and by chemical or physical adsorption on solid surfaces.

A useful parameter for the classification of material–moisture bonding is the value of the work needed for the removal of 1 mol of moisture from the given material (Rebinder, 1972): chemical moisture bonding (stoichiometric), in which the ionic and molecular bond energy is of the order of 5000 J/mol; physicochemical moisture bonding (nonstoichiometric), in which the adsorptive and osmotic bond energy is of the order of 3000 J/mol; and physico-mechanical moisture bonding (undetermined proportions), in which the bond energy is about 100 J/mol.

In particular, physicomechanical moisture bonding is of great importance in drying and classified into structurally bound moisture (gels with about 1% of solid) that can be removed by evaporation or mechanical dewatering, capillary bound moisture (capillary-porous bodies) that can be removed by evaporation from capillaries into ambient air, and unbound moisture (nonporous hydrophilic bodies) that can be removed by surface evaporation into ambient air.

A common feature of porous systems is the presence in the material of void spaces called capillaries or pores. These can have complex shapes and different geometric dimensions. Several descriptive parameters of the structure of dried materials are used. *Basis weight* ($G$) is one of the important structural parameter of dried material defined as the ratio of the dry mass ($m$) to its area ($A$):

$$G = \frac{m}{A} \tag{2.23}$$

*Porosity* ($\varphi$) is defined as the ratio of the total void space volume ($V_V$) to the total volume of the material ($V_T$), $\varphi = V_V/V_T$. An alternative to $\varphi$ is the absolute porosity ($\bar{\varphi}$), defined for a moist material as follows:

$$\bar{\varphi} = \frac{V_{vo}}{V_{ap}} = \frac{\left(V_{ap} - V_{ce}\right)}{V_{ap}} = 1 - \frac{V_{ce}}{V_{ap}} \tag{2.24}$$

where $V_{vo}$ is the void volume or pore volume, $V_{ap}$ is the apparent or total volume, and $V_{ce}$ is the volume of cellulose. As an example, the absolute porosity is thus determined from material sheet thickness at each level of deformation and of the ovendry mass of the sheet:

$$\varphi = 1 - \left(\frac{m_o}{Ad\rho_c}\right) \tag{2.25}$$

where $m_o$ is the ovendry mass of the sheet (kg), $A$ is the strained area (m$^2$), $d$ is the thickness of the sample (m), and $\rho_c$ is the density of cellulose (~1550 kg/m$^3$).

*Tortuosity* ($T^*$) is defined as the ratio of body dimension in a given direction ($L$) to the length of the path traversed by the component in the diffusion process ($L_D$), $T^* = L/L_D$. *Pore shape factor* ($\pi$) is the parameter that characterizes the deviation of the diffusion channel shape when compared to a cylinder. In drying practice, porosity is a function of pore diameter (radius). Strumillo and Kudra (1986) indicated that at the maximum value of radius ($r_{max}$), the integral curve of pore diameter intersects the line corresponding to the total pore volume in the total body volume $\phi_{max}$. Therefore, the total pore volume in the total body (e.g., maximum porosity) corresponds to

$$\varphi_{max} = \int_{r_{min}}^{r_{max}} \frac{d\varphi}{dr} dr \tag{2.26}$$

Most drying processes in practice involve fluid flow through porous media. In the case of a completely saturated porous medium, the flow is described well by Darcy's law. When the porous medium is only partially saturated with liquid, the flow through the pores is much more complicated. In such cases, the porous material can be regarded as a bundle of capillaries, and the diffusion coefficient can be predicted as a function of moisture content.

The majority of porous materials have a very intricate solid structure and complex chemical inhomogeneity of the solid, especially at small length scales of the order of the pores. To arrive at a deterministic method of describing moisture transport, the porous material is described with a minimum number of parameters. One parameter is the permeability ($K$) of the saturated porous material which estimates the flow through the porous material when a pressure gradient $dP/dx$ is applied. Darcy's law can be written as follows:

$$\frac{\dot{V}}{A} = \frac{K}{\mu} \frac{\Delta P}{d} \tag{2.27}$$

where $\dot{V}$ is the volumetric flow rate ($m^3/s$), $A$ is the surface area ($m^2$), $K$ is the permeability ($m^2$), $\mu$ is the dynamic viscosity at liquid temperature (Pa s), $\Delta P$ is the pressure drop across the body (Pa), and $d$ is the thickness of the body at a given level of deformation (m). The ratio $\dot{V}/A$ is called the superficial liquid velocity (m/s). The permeability is an intrinsic parameter that can be extracted from Eq. (2.27): $K = (A/\dot{V}\mu)(\Delta P/d)$. Note that the mechanical properties of porous materials may change during the drying process.

**Example 2.5**
The permeability of a porous material is $K = 10^{-13}\,m^2$. The material is subjected to drying under vacuum at 0.5 atm (absolute) using dry air of 10% relative humidity and 60 °C dry-bulb temperature. The thickness of the slab is $d = 0.05\,m$. It is required to calculate (i) the rate of drying per unit of slab surface, (ii) the heat rate required to remove the humidity, and (iii) the superficial heat transfer coefficient.

- From $P = 0.5\,atm$, $T_d = 60\,°C$ (dry bulb), and $\varphi = 0.1$, the wet-bulb temperature $T_s = 11.2°C$ results.
- Saturation pressure of water at wet-bulb temperature is $P''(11.2°C) = 1.33\,kPa$ (absolute).
- From Eq. (2.23), the rate of water migration per unit of surface of porous slab can be calculated under the assumption that water permeated through the material is subjected to a pressure difference equal to 0.5 atm at one side and $P''$ at the other face. The viscosity of liquid water at dry-bulb temperature can be found to $\mu = 7.2 \times 10^{-4}\,Pa\,s$. Thence

$$\frac{\dot{V}}{A} = \frac{K}{\mu d}(P - P'') = \frac{10^{-7}}{7.2 \times 10^{-4} \times 0.05}(50,660 - 1332) = 0.14\,mm/s$$

- From Eq. (2.8) and the definition relationship for humidity, the following identity results for the linear drying rate period:

$$\frac{\dot{V}}{A} = \frac{d}{\rho_s t_{CR}}(W_i - W_c) = \frac{h(T_a - T_s)}{\rho_s^2 \Delta h_{lv}}$$

- In the equation above, $\Delta h_{lv} = 2474\,\text{MJ/kg}$ and $\rho_s = 0.010\,\text{kg/m}^3$ (saturated water vapor density) are taken at the partial pressure corresponding to wet-bulb temperature. Solving the equation for the heat transfer coefficient results to

$$0.014 \times 10^{-3} = \frac{h(60-11.1)}{0.01^2 \times 2474 \times 10^6} \rightarrow h = 144.8\,\text{W/m}^2$$

- The heat transfer flux density required to do the drying is

$$q = h(T_a - T_s) = 144.8 \times (60-11.2) = 3.4\,\text{kW/m}^2$$

## 2.5  Types of Moisture Diffusion

Moisture in a moist material can be transferred in both liquid and gaseous phases. Several modes of moisture transport can be distinguished. Transport by liquid diffusion assumes that liquid diffuses through the moist material. In such a case, the liquid moisture transfer rate by diffusion is proportional to the gradient of moisture concentration inside the material as follows:

$$j = -h_m A \frac{\partial C_m}{\partial x} \tag{2.28}$$

where $j$ is measured in kg/m$^2$ s, $C_m$ is the mass concentration of liquid in kg/m$^3$, and $A$ is the surface area.

*Transport by Vapor Diffusion.* This is the main mechanism of vapor moisture transfer. It takes place in materials where the characteristic diameter of the free air spaces is greater than $10^{-7}$ m. The qualitative effect of this transfer can be described by an equation of the Fick type, using instead of the kinematic diffusion coefficient ($D$) the effective diffusion coefficient ($D_{ef}$) in capillary-porous materials:

$$D_{ef} = D_e D \tag{2.29}$$

Here, $D_e$ is the equivalent coefficient of diffusion in the capillary-porous material. A general relation for $D_e$ is given by Van Brakel and Heertjes (1974) as a function of structural parameters:

$$D_e = \frac{\varphi \ell}{T^*} \tag{2.30}$$

where $T^*$ is the tortuosity introduced earlier and $\ell$ is a constrictivity factor which accounts for the constrained pathways in a porous medium.

*Transport by Effusion (or Knudsen-Type Diffusion).* This takes place when a characteristic dimension of the void space in a capillary-porous material is smaller than $10^{-7}$ m. The mass rate of vapor in this case can be obtained from

$$j_l = D_{Eef} \frac{\partial C_m}{\partial x} \tag{2.31}$$

where $D_{Eef}$ is the effective effusion which can be defined for gel-type materials as follows:

$$D_{Eef} = 3^{-1/2}\varphi^2 D_e \tag{2.32}$$

*Transport by Thermodiffusion.* This can be described by the following equation:

$$j = \rho \frac{D_T}{T}\frac{\partial T}{\partial x} = \rho\beta\frac{D}{T}\frac{\partial T}{\partial x} \tag{2.33}$$

where $D_T$ is the thermodiffusion coefficient and $\beta \approx D_T/D$ is the thermodiffusion constant. Written for gas phase, the above equation becomes

$$j = \beta\frac{\partial T}{\partial x} \tag{2.34}$$

*Transport by Capillary Forces.* If capillaries form interconnected channels, a difference of capillary pressure takes place, and this causes the continuous redistribution of moisture from the large capillaries to small ones.

*Transport by Osmotic Pressure.* Osmotic pressure is a function of moisture content in the material. The osmotic moisture transfer can be then described on the basis of liquid diffusion.

*Transport due to Pressure Gradient.* This results from the internal pressure difference due to the local evaporation of liquid or local condensation of vapor. Moisture movement equalizes the pressure in accordance with Darcy's law. In addition to the types of moisture movement listed above, mass transport can also be caused by internal pressure, shrinkage, or external pressure.

**Example 2.6**

Determine the mass flux due to moisture diffusion at drying of a porous slab of 0.01 m having the tortuosity of 1.4, the constrictivity factor of 1, and porosity of 0.25 when subjected to a difference of concentration between the two faces of 1 kg/m$^3$ producing Knudsen diffusion.

- From Eq. (2.30), the equivalent coefficient of diffusion is calculated as follows:

$$D_e = \frac{\varphi\ell}{T^*} = \frac{0.25}{1.4} = 0.178$$

- From Eq. (2.32), the effective effusion coefficient is $D_{Eef} = 3^{-1/2}\varphi^2 D_e = 0.064$.
- From Eq. (2.31), the mass flux of moisture is

$$j_l = D_{Eef}\frac{\Delta C_m}{d} = 0.064\frac{1}{0.01} = 0.64\,\text{g/sm}^2.$$

## 2.6  Shrinkage

During drying, the nonuniform temperature and moisture content fields induce thermal and shrinkage stresses inside the moist material. Because drying materials are so diverse, we may expect several mechanisms of interaction between water and solids. We may have unsaturated solutions, saturated solutions with crystals, supersaturated solutions, amorphous hydrophilic solids with limited swelling capacity, porous nondeformable solids with various levels of porosity, porous deformable solids with various levels of porosity, cellular solids, and/or any combination of the above items.

Note that food products undergo a certain degree of shrinkage during drying by all the drying methods, with the possible exception of freeze-drying. Colloidal materials also shrink, for example, meat during cooking. In the early stages of drying, at low drying rates, the amount of shrinkage bears a simple relationship to the amount of moisture removed. Toward the end of drying, shrinkage is reduced so that the final size and shape of the material are essentially time independent before drying is completed.

The bulk density and porosity of dried vegetable pieces depend to a large extent on the drying conditions. At high initial drying rates, the outer layers of the pieces become rigid, and their final volume is fixed early in the drying process. As drying proceeds, the tissues split and rupture internally, forming an open structure. The product in this case has a low bulk density and good rehydration characteristics. At low initial drying rates, the pieces shrink inward and give the food product a high bulk density. Shrinkage of foodstuffs during drying may influence their drying rates because of the changes in drying surface area and the setting up of pressure gradients within the material. Some work indicates that shrinkage does not affect drying behavior (Brennan et al., 1976). Van Brakel and Heertjes (1974) suggest that shrinkage cannot always be explained based solely on the amount of moisture evaporated. Shrinkage is specific for each body, depending on the material type and on the characteristic cell and tissue structure.

Bilbao et al. (2000) have studied the porosity profile of a cylindrical apple as a function of the distance to the interface. They found that raw apple exhibits porosity between 18% and 24%, but as drying proceeds, moisture removal is accompanied by the formation of a more porous structure. A gradual increase in porosity is observed from the inner part of the cylinder, where the tissue is less dry, to the outer regions, but near the interface, the porosity decreases as a result of the hardening of the case.

Bilbao et al. (2000) also showed that the volume changes in apple tissue during air-drying appear not to be affected by the air temperature. Nevertheless, volume changes are greatly affected by air velocity, and an equation can be used to predict volume changes in the gas phase as a function of air velocity. The obtained porosity values confirm that food systems could undergo porosity increase as drying proceeds, depending on process variables and system characteristics. Observations of fresh apple tissue usually show the turgid cell walls of parenchyma tissue as bright regions and the small intercellular spaces between cells as dark regions, but drying conditions promote great structural changes. Figure 2.5 shows the appearance of dried apple tissue revealing that cell walls are greatly shrink, leaving wide spaces between neighboring cells. This can be observed throughout the tissue except in the first 0.5 mm under the skin where the tissue is more compact and collapsed.

Structural changes and shrinkage during drying occur, and phenomena such as shrinkage determine the properties and quality of food products. Quality can be characterized by color, texture, taste, porosity, and other physical properties such as density and specific volume. The quality of dried products changes during drying, depending on the type of drying method and drying conditions. For this reason, it is important to be able to anticipate the effect of drying method and drying conditions on the final quality. This can be done by constructing a

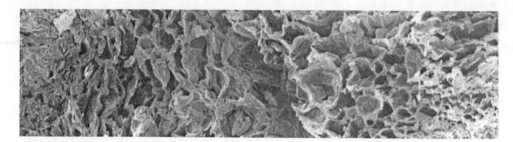

**Figure 2.5**    Dried apple tissue after 2 h at 70 °C (Photograph from Bilbao et al. (2000))

**Table 2.1**    Mathematical models for property equations

| Properties | Equations | Unit |
|---|---|---|
| Particle density | $\rho_p = (1+X)/(1/\rho_s + X/\rho_w)$ | kg/m$^3$ |
| Bulk density | $\rho_b = (1+X)/(1/\rho_{b0} + \beta'X/\rho_w)$ | kg/m$^3$ |
| Porosity | $\phi = 1 - \rho_b/\rho_p$ | N/A |
| Specific volume | $v = 1/\rho_{b0} + \beta'X/\rho_w$ | m$^3$/kg d.b. |

| Parameters | Definitions | Unit |
|---|---|---|
| $X$ | Material moisture content | kg/kg d.b. |
| $\rho_w$ | Enclosed water density | kg/m$^3$ |
| $\rho_s$ | Dry solid density | kg/m$^3$ |
| $\rho_{b0}$ | Dry solid bulk density | kg/m$^3$ |
| $\beta'$ | Volume-shrinkage coefficient | N/A |

*Source*: Krokida and Maroulis (1997).
Factors affecting the parameters: type of material, drying method, and drying conditions.

mathematical model to estimate the properties of the product versus material moisture content and to examine the various features of the model (e.g., see Table 2.1).

Krokida and Maroulis investigated the effect of drying method on bulk density, particle density, specific volume, and porosity for five drying methods: freeze, microwave, conventional, vacuum, and osmotic drying. Each method was used to dry apples, carrots, potatoes, and bananas under typical conditions (Krokida and Maroulis, 1997). The approach sketched in Table 2.1 was used to determine the bulk density, particle density, porosity, and specific volume versus material moisture content. The effect of each of the drying methods on the above properties was examined.

The equation $\rho_p = (1+W)/(1/\rho_s + W/\rho_w)$ is used for determining the particle density ($\rho_p$) as a function of moisture content ($X$). The particle density ranges between the dry solid density ($\rho_s$) and the density of water ($\rho_w$). This equation corresponds to a model with two phases in series. The bulk density ranges between the bulk density of dry solids ($\rho_{b0}$) and the density of the enclosed water ($\rho_w$) and can be calculated with $\rho_b = (1+W)/(1/\rho_{b0} + \beta'W/\rho_w)$ which is based on the two-phase structural model. The total porosity is a function of bulk density, and particle density is given by $\phi = 1 - \rho_b/\rho_p$. The equation for specific volume is $v = 1/\rho_{b0} + \beta'W/\rho_w$ and

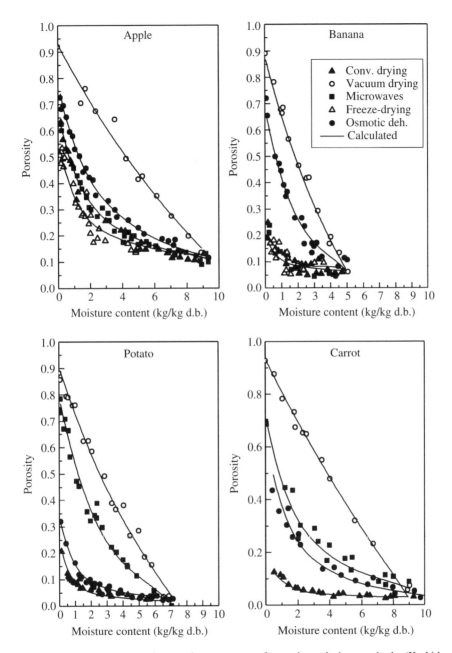

**Figure 2.6** Porosity versus product moisture content for various drying methods (Krokida and Maroulis (1997))

involves three parameters: the bulk density of dry solids ($\rho_{b0}$), the enclosed water density ($\rho_w$), and the shrinkage coefficient $\beta'$. Factors such as material, drying method, and drying conditions affect these parameters and, consequently, the four properties.

The shrinkage coefficient can be defined in terms of temperature by an equation of the Arrhenius type:

$$\beta' = C\exp\left(-\frac{E_a}{RT}\right) \tag{2.35}$$

where $C$ is a constant determined experimentally for each product, $E_a$ is the activation energy of shrinkage, $R$ is the universal gas constant, and $T$ is the absolute drying temperature.

Krokida and Maroulis (1997) determined the values of the required parameters by fitting the proposed model to the experimental data for four types of produce, for example, as shown in Figure 2.6. The porosity of freeze-dried materials is always higher (80–90%) when compared with other dehydration processes. The porosities of microwave-dried potato and carrot are high (75%), while microwave apple and banana do not develop high porosities (60% and 25%). Vacuum-dried banana and apple developed high porosities (70%), while for vacuum-dried carrot and potato, the porosities were lower (50% and 25%). Gabas et al. (1999) investigated the effect of drying temperature on the shrinkage of grapes subjected to pretreatment with various substances, as shown in Figure 2.7. They found that the small temperature effect on shrinkage observed in practice may be attributed to the temperature difference of elastic and mechanical properties.

## 2.7  Modeling of Packed-Bed Drying

The process of drying of grained materials in a packed bed permeated by a gaseous agent is complex because it is time dependent, and the three fundamental transfer phenomena appear simultaneously: flow of the gaseous phase through layer interstices, heat transfer, and transfer of water vapor from grains to gaseous phase. The drying mechanism, the flow of the gaseous phase through the bed, the heat transfer between the thermal agent and the grained material, and the drying of a monogranular layer are topics of current interest (Moise and Tudose, 2000).

The packed-bed drying of a granular material takes place in three stages as shown in Figure 2.8. In the first stage, there are two zones in the layer: zone A, in which the material is kept at the initial humidity, and zone B, where the drying takes place. The height of zone B increases continuously at the expense of zone A. In the drying zone B, the material moisture varies in time and is a function of the axial coordinate.

In the second stage, the new zone C appears. In this zone, the material is dried at the equilibrium humidity. Zone A decreases until it disappears, while the height of zone C increases. In the third stage, only zones B and C remain in the packed bed. The height of zone C increases until the whole bed reaches the equilibrium humidity. In the case of small beds or when drying is made with gaseous thermal agent with low temperature, then the second stage – in which the three zones exist simultaneously – does not appear anymore. Apparently, the packed-bed drying process takes place in three stages: (i) stage 1, where the zones A and B coexist with no drying yet; (ii) stage 2, where three zones exist with dried part; and (iii) stage 3, where two zones exist without A. The "plug flow and external transfer model" for packed-bed drying

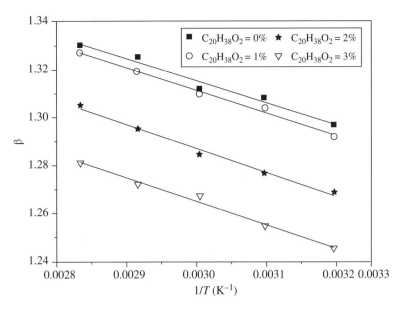

**Figure 2.7** Effects of drying temperature on the shrinkage coefficient of grapes (Krokida and Maroulis (1997))

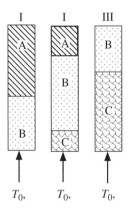

**Figure 2.8** Three stages of packed-bed drying (A material zone with initial moisture, B: drying zone, and C: dried material zone)

of granular materials was initially extended to time-dependent operation, with specific terms accounting for water evaporation and water vapor transfer by Moise and Tudose (2000). So the model consists of the equations

$$\rho_g \varphi \frac{\partial x}{\partial t} = F_v a - F_g \frac{\partial x}{\partial z} \qquad (2.36)$$

$$\rho_s(1-\varphi)\frac{\partial w}{\partial t} = -F_v a \tag{2.37}$$

$$\rho_g C_p \varphi \frac{\partial T_g}{\partial t} = -F_g C_{p_g}\frac{\partial T_g}{\partial z} - h_g a(T_g - T_s) - F_v a C_{p_v}(T_g - T_s) - h_w a_w(T_g - T_m) \tag{2.38}$$

$$\rho_s C_{p_s}(1-\varphi)\frac{\partial T_s}{\partial t} = h_g a(T_g - T_s) - F_v a L \tag{2.39}$$

where

$$F_v = \frac{k_g}{R_g T_g}(P_s - P_v)\frac{1}{1 + Bi \cdot f(w)} \tag{2.40}$$

The function $f(w)$ is measured experimentally, depending on the material characteristics. The initial conditions are as follows: The relative humidity of drying air is constant and equal to the atmospheric value, $x = x_0$. The moisture content of the bed material is uniformly distributed and equal to the initial moisture, $w = w_0$. The air temperature in the bed and in the grains interstices is equal to the environment temperature, $T_g = T_s = T_m$. The coordinate $z$ is measured in the direction of flow, where $z = 0$ is the entrance to the packed bed. The conditions at $z = 0$ are that the relative humidity of drying air is constant, $x = x_0$, and the air temperature is constant, $T_g = T_0$. The value of $h_g$ (between gas and particles) generally varies between 46 and 106 W/m$^2$ K, while $k_g$ varies between 0.6 and 0.14 (cf. Moise and Tudose (2000)). Further details may be found in Achenbach (1995). Equations (2.36)–(2.40) can be solved numerically to obtain the unsteady-state moisture content distributions as a function of material properties and experimental drying conditions.

## 2.8   Diffusion in Porous Media with Low Moisture Content

Although there are many numerical and experimental works focused on drying of porous materials on higher moisture contents where the liquid phase is mobile and liquid-phase transport is dominant, there are a number of applications where drying to low moisture content is important. These include modeling material behavior under fire and combustion conditions, debinding of green ceramic preforms, moisture transport around nuclear waste repositories, and drying of pharmaceuticals (cf. Plumb et al. (1999)).

When the fraction of the pore volume occupied by liquid (the liquid-phase saturation) reaches a certain value, the liquid phase is no longer continuous and becomes immobile. This value of liquid-phase saturation is called irreducible saturation. Thus, moisture cannot be transported to the drying surface or a drying front in the liquid phase as a result of capillary action of pressure gradients in the gas phase. The irreducible saturation for most granular porous materials is 10–20%. Below the irreducible saturation, the liquid phase is not continuous but exists as liquid islands at the contact points between the individual grains composing the bulk material. Under these conditions, diffusion in the vapor phase is the dominant mechanism for drying.

Vapor-phase diffusion in porous media is generally quantified on the basis of Fick's law of diffusion in a continuous medium (cf. Plumb et al. (1999)):

$$j = D_C \rho_g \nabla W + D_T \nabla T \tag{2.41}$$

Note that the diffusion effect due to temperature gradients (or Soret effect) is included in the equation above. For diffusion in a continuum, the Soret effect is generally negligible, because the diffusion coefficient ($D_T$) is small in comparison to the coefficient for diffusion driven by gradients in concentration or partial pressure. For diffusion in a dry porous medium, an effective diffusion coefficient must be defined so that it accounts for the reduced area for diffusion or blockage which results from the presence of the solid phase. This is done by assuming that the effective diffusion coefficient can be obtained by multiplying the diffusion coefficient for free space by the porosity. Furthermore, the increased path length for diffusion is accounted for by including the tortuosity coefficient. The effective diffusion coefficient for a dry porous medium becomes

$$D_{\text{eff}} = T^* \varphi D_C \tag{2.42}$$

The mathematical details for the derivation of the effective diffusion coefficient for a dry porous medium based on volume averaging can be found in Eidsath et al. (1983). The tortuosity coefficient is usually assumed to have a value in the range of 0.6–0.8. When immobile liquid is present in the porous medium, the effective diffusion coefficient is further reduced to account for the blockage due to the presence of the liquid phase. This blockage is accounted for by including the gas-phase saturation $S_g$ (fraction of pore volume occupied by gas):

$$D_{\text{eff}} = T^* \varphi S_g D_C \tag{2.43}$$

In order to be consistent with Eq. (2.42), for a dry porous medium, the tortuosity is also corrected for the presence of the immobile liquid. This line of reasoning leads (cf. Millington and Quirk (1961)) to

$$D_{\text{eff}} = T^{*^{10/3}} D_C \tag{2.44}$$

An additional correction factor is included in the definition of the effective diffusion coefficient because diffusion is enhanced as a result of the presence of the liquid phase. The factor is as follows:

$$D_{\text{eff}} = \beta T^* \varphi S_g D_C \tag{2.45}$$

where $\beta$ is an enhancement factor defined as the ratio of the vapor-phase diffusion coefficient in a porous medium to that in free space.

Although the values of $\beta$ in liquid-phase saturation vary between 0.1 and 0.2, the actual value of the enhancement factor ranges between 1 and 4. For further details, see Plumb et al. (1999). It is important to quantify the effect of liquid-phase saturation on the vapor-phase diffusion coefficient. Many of the experimental studies have been indirect, inferring the vapor-phase diffusion coefficient by measuring the temperature and using the fact that the effect of saturation on the thermal conductivity is known. Plumb et al. (1999) focused on the experimental measurement for quantification of the enhancement factor and its impact on the drying process. The measured tortuosity and the enhancement factor defined in Eq. (2.41) are presented.

## 2.9   Modeling of Heterogeneous Diffusion in Moist Solids

Consider now the modeling of heterogeneous diffusion in capillary-porous materials during drying. The governing heat and mass transfer equations for liquid as well as vapor flow are based on the work done by Dietl et al. (1995). There are two cases: moisture transfer only and heat transfer (i.e., conduction) only. The models and the numerical solutions of the resulting differential equations use as inputs the moisture and the temperature-dependent thermophysical properties of the product. All the equations are given in spherical coordinates, but the numerical calculations are extended to cover other geometries as well.

*Mass Transfer.* With reference to the differential element as shown in Figure 2.9, taking a mass balance between the incoming and outgoing moisture leads to

$$-\frac{\partial(j_r A_r)}{\partial r}dr = \rho_{dr}\frac{\partial W}{\partial t}dV \tag{2.46}$$

Here, $j_r$ is the total mass flux at the radius $r$, $A_r$ is the elemental area at the radius $r$, $dr$ is the infinitesimal thickness, $\rho_{dr}$ is the bone dry density, $W$ is the moisture content (kg/kg) at the time $t$, and $dV$ is the differential volume corresponding to the radius $r$. The moisture transport in the vapor phase is described by the Fick's law according to the following equation:

$$j_v = -\frac{D}{\mathfrak{M}}\frac{1}{R_v T}\frac{\partial P_v}{\partial r} \tag{2.47}$$

where $j_v$ is the mass flux of vapor, kg/m$^2$ s; $D$ is the diffusion coefficient of water vapor in air, m$^2$/s; $R_v$ is the gas constant for water vapor; $P_v$ is the partial vapor pressure at point $r$; and $\mathfrak{M}$ is a dimensionless diffusion resistance factor (cf. Dietl et al. (1995)). The diffusion coefficient is given by

$$D = \frac{22.06 \times 10^{-6}}{P}\left(\frac{T_m}{273\,K}\right)^{1.81} \tag{2.48}$$

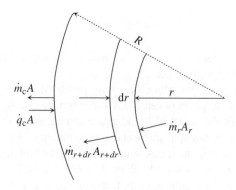

**Figure 2.9**   Mass balance at the product element

Here, $P$ is the total pressure expressed in bar, and $T_{\mathrm{m}}$ is the mean absolute temperature of product surface and drying air, ranging from $-30$ to $120\,^{\circ}\mathrm{C}$. For transport in the sorption (liquid) phase, one has

$$j_{\mathrm{t}} = -\rho_{\mathrm{dr}} D_{\mathrm{m}(W,T)} \frac{\partial W}{\partial r} \tag{2.49}$$

where $j_{\mathrm{t}}$ is the mass flux of liquid that is being transferred; $D_{\mathrm{m}(W,T)}$ is the moisture diffusivity as a function of moisture content and temperature $T$, $\mathrm{m^2/s}$; and $W$ is the moisture content at point $r$, kg/kg.

The transport in the vapor phase and the sorption phase is coupled to each other because the vapor pressure in a pore determines the equilibrium sorption moisture content. The equilibrium moisture content is a function of the water activity $a_{\mathrm{w}}$ and the product temperature $T$. The corresponding saturated vapor pressure $P_{\mathrm{v}}^{*}$ is related to the product temperature $T$ and the total pressure $P$ as follows:

$$\frac{\partial P_{\mathrm{v}}}{\partial r} = \frac{\partial P_{\mathrm{v}}}{\partial W}\frac{\partial P_{\mathrm{v}}}{\partial r} = \frac{\partial a_{\mathrm{w}}}{\partial W} P_{\mathrm{v}}^{*} \frac{\partial W}{\partial r} \tag{2.50}$$

The total mass flux is then given by

$$j_{r} = j_{\mathrm{l}} + j_{v} = -\left(\rho_{\mathrm{dr}} D_{\mathrm{m}(M,T)} + \frac{D}{\mathfrak{M}}\frac{P_{\mathrm{v}}^{*}}{R_{v}T}\frac{\partial a_{\mathrm{w}}}{\partial W}\right)\frac{\partial W}{\partial r} \tag{2.51}$$

By substituting Eq. (2.47) into Eq. (2.42), we obtain the equation for the mass balance at the infinitesimal level:

$$\frac{\partial}{\partial r}\left[A_{r}\left(-\rho_{\mathrm{dr}} D_{\mathrm{m}(W,T)} + \frac{D}{\mathfrak{M}}\frac{P_{\mathrm{v}}^{*}}{R_{v}T}\frac{\partial a_{\mathrm{w}}}{\partial W}\right)\frac{\partial W}{\partial r}\right]\mathrm{d}r = \rho_{\mathrm{dr}}\frac{\partial W}{\partial t}\mathrm{d}V \tag{2.52}$$

*Heat Transfer.* For products with high porosity, the drying rate is essentially determined by high mass transfer rates. The heat balance is now made at the boundary of the product rather in a differential element. One can write

$$A_{r} = R\left[\dot{q} - j_{r=R}\Delta h_{\mathrm{lv}(T_{\mathrm{o}})}\right] = \rho_{\mathrm{dr}}V\left[C_{p\mathrm{dr}} + C_{p\mathrm{w}}\overline{M}_{(t)}\right]\frac{\partial T}{\partial t} \tag{2.53}$$

where

$$\dot{q} = \dot{q}_{\mathrm{c}} = \frac{h\Delta h_{\mathrm{lv}(T_{\mathrm{o}})}}{c_{pv}}\ln\left[1 + \frac{c_{pv}(T_{\mathrm{a}} - T_{\mathrm{o}})}{\Delta h_{\mathrm{lv}(T_{\mathrm{o}})}}\right] \tag{2.54}$$

$\dot{q} = \dot{q}_{\mathrm{c}}$ is the convection heat flux, $h$ is the convective heat transfer coefficient, $\Delta h_{\mathrm{lv}(T_{\mathrm{o}})}$ is the vaporization enthalpy at product surface temperature $T_{\mathrm{o}}$, $C_{pv}$ is the specific heat capacity of water vapor, and $T_{\mathrm{a}}$ is the drying air temperature. Furthermore,

$$j_{r=R} = h_{\mathrm{o}}\frac{P}{R_{v}T_{\mathrm{m}}}\ln\left(\frac{P - P_{\mathrm{va}}}{P - P_{\mathrm{vo}}}\right) \tag{2.55}$$

where $j_{r=R}$ is the convective mass flux, $h_D$ is the convective mass transfer coefficient, $P_{va}$ is the partial vapor pressure in the drying air, $P_{vo}$ is the partial vapor pressure over the product surface, $C_{pdr}$ is the specific heat capacity of bone dry product, and

$$\bar{m}_{(t)} = \frac{1}{V_{r=R}} \int\limits_{V_{r=0}}^{V_{r=R}} m_{(r,T)} \, dV \tag{2.56}$$

where $\bar{m}(t)$ is the average moisture mass at time $t$.

For products with low porosity, the drying rate is essentially determined by high energy transfer rates. In this case, the heat balance must be made in analogy with mass balance, that is, at the product element:

$$\frac{\partial}{\partial r} \left[ A_r \left\{ \dot{q}_r - j_r \Delta h_{lv(T_{avg})} \right\} \right] dr = \left[ C_{pdr} + C_{pw} M_{(t)} \right] \frac{\partial T}{\partial t} dV \tag{2.57}$$

where

$$\dot{q}_r = \frac{q_k}{A} = -k_{(M,T)} \frac{\partial T}{\partial r} \tag{2.58}$$

Here, $\dot{q}_k$ is the heat flux due to conduction, $k_{(M,T)}$ is the thermal conductivity as a function of moisture content $M$ and temperature $T$, and

$$j_r = j_v = \frac{D}{\mathfrak{M} R_v T} \frac{1}{\partial r} \frac{\partial P_v}{\partial r} \tag{2.59}$$

*Boundary Conditions.* The initial conditions for solving the equations are $j_{(r,\, t=0)} = j_{in}$, $T_{(t=0)} = T_{in}$, and $T_{(r,\, t=0)} = T_{in}$ for all $r$. There is no moisture gradient at the center of the product; therefore, the mass flux is $j_{(r=0)} = 0$ for all $t$. At the surface of the product, the vapor pressure must be in equilibrium with the partial vapor pressure of air; therefore, the mass flux at the surface is

$$j_{r=R} = h_m \frac{P}{R_v T_m} \ln \left[ \frac{P - P_{va}}{P - P_{vo}} \right] \quad \text{with } P_{v(M,T)} = a_{w(W,T)} P_v^* \tag{2.60}$$

*Numerical Analysis.* The listed equations were solved numerically using the finite difference procedure. In accordance with Figure 2.10, the product was divided into $n_{max}$ layers of thickness $\Delta r$. One can use this approach for the standard geometries: flat plate, cylinder, and sphere; a detailed discussion on numerical methods for heat and moisture transfer is presented in Chapter 6. For an element of thickness $\Delta r$, one obtains

$$\begin{aligned} V_n &= A \Delta r \quad \text{for plate} \\ &= \pi \left( r_n^2 - r_{n-1}^2 \right) h \quad \text{for cylinder} \\ &= (4\pi/3) \left( r_n^3 - r_{n-1}^3 \right) \quad \text{for sphere} \end{aligned} \tag{2.61}$$

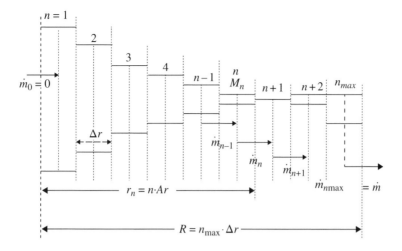

**Figure 2.10** Numerical model for solving the differential equations

The corresponding position of the surface, $r_n$, is

$$
\begin{aligned}
A_n &= A \quad \text{for plate} \\
&= \pi r_n \quad \text{for cylinder} \\
&= 4\pi r_n^2 \quad \text{for sphere}
\end{aligned}
\tag{2.62}
$$

At a particular instant in time $t$, the product has the moisture content $W_n$ in the $n$th element. The flow of moisture from the $n$th element to the $(n+1)$th element is $_nA_n$. The mass flux becomes

$$
j_n(t) = -\left[\left\{\rho_{\mathrm{dr}}D_{\mathrm{m}(M,T)} + \frac{D}{\mu}\frac{P_v^*}{R_vT}\frac{\partial a_{\mathrm{w}}}{\partial W}\right\}\frac{W_{n+1}-W_n}{\Delta r}\right]_t
\tag{2.63}
$$

If the product temperature $T$ and moisture content $W_n$ are known at time $t$, the moisture content at time $t + \Delta t$ can be calculated for all $n$ elements from the finite difference equation for the mass balance at the product element:

$$
W_{n(t+\Delta t)} = W_{n(t)} - \left(\frac{\dot{m}_n A_n - \dot{m}_{n-1} A_{n-1}}{\rho_{\mathrm{dr}}\Delta V_n}\right)_t \Delta t
\tag{2.64}
$$

By using the same procedure for the case where heat conduction is not considered, the following equation for the product temperature at time $t + \Delta t$ is found:

$$
T_{\mathrm{o}(t+\Delta t)} = T_{\mathrm{o}(t)} + A_{\max}\left[\frac{\dot{q} - \dot{m}_{\max}\Delta H_{v(T_o)}}{\rho_{\mathrm{dr}}V\left[C_{p\mathrm{dr}} + C_{p\mathrm{w}}W_{(t)}\right]}\right]_t \Delta t
\tag{2.65}
$$

where

$$\overline{W}_{(t)} = \sum_{n=1}^{n=n_{max}} \frac{W_n V_n}{V_n} \tag{2.66}$$

When heat conduction is taken into account, the corresponding expression is

$$T_{n(t+\Delta t)} = T_{n(t)} + \left[ \frac{\dot{q}_n A_n - \dot{q}_{n-1} A_{n-1} - (\dot{m}_n A_n - \dot{m}_{n-1} A_{n-1}) \Delta H_v (T_{avg})}{\rho_{dr} \Delta V [C_{pdr} + C_{pw} W_{(t)}]} \right]_t \Delta t \tag{2.67}$$

*Heat and Mass Transfer Coefficients.* The heat transfer during drying is characterized by a small surface area of the body (temperature $T_0$) that comes in contact with an air medium, which affects the temperature at the surface layers of the product. The average temperature of the drying air is therefore not much different than its initial temperature. The heat transfer coefficient $h$ can be calculated as follows:

$$h = \frac{kNu}{L_c} \tag{2.68}$$

where $Nu$ is the Nusselt number, $k$ is the thermal conductivity of air–vapor mixture, and $L_c$ is a characteristic length of the product. For individual particles in flow, the Nusselt number $Nu$ can be estimated (cf. Kessler (1981)) with

$$Nu = Nu_{min} + \left( Nu_{lam}^2 + Nu_{turb}^2 \right)^{1/2} \tag{2.69}$$

where

$$Nu_{lam} = 0.664 Re^{0.5} Pr^{0.33} \quad Nu_{turb} = \frac{0.037 Re^{0.81} Pr}{1 + 2.443 Re^{-0.1} (Pr^{0.67} - 1)} \tag{2.70}$$

$$\begin{aligned} Nu_{min} &= 0 \text{ for flat plate in parallel flow} \\ &= 0.3 \text{ for cylinder in crossflow} \\ &= 2 \text{ for sphere immersed in free stream} \end{aligned} \tag{2.71}$$

Here, $Re = vL_c/v$, $v$ is the velocity of drying air; $v$ is the kinematic viscosity; $Pr = v/a$ is the Prandtl number; $a$ is the thermal diffusivity of drying air, $\alpha = k/\rho C_p$; $\rho$ is the density of air–vapor mixture; and $C_p$ is the specific heat capacity of air–vapor mixture.

Using the boundary layer analogies as in Chapter 1.8, the mass transfer coefficient $h_o$ is given by (see also Dincer and Hussain, 2002)

$$h_m = \frac{h}{\rho C_p (a/D)^{1-n}} \tag{2.72}$$

where $n$ is a positive exponent equal to 1/3. Note that the values $\rho$, $C_p$, $a$, and $D$ are evaluated at the interface between the product surface and the drying air.

In drying, heat is transferred from hot air to the produce, whereas the moisture is transferred from the produce to hot air. To account for the coupling between the two fluxes, one defines the corrected transfer coefficients $h^*$ and $h^*_m$, which are related to the original heat and mass transfer coefficients by

$$\frac{h^*}{h} = \frac{\ln(1+B)^\gamma}{(1+B)^\gamma - 1} \quad \text{and} \quad \frac{h^*_m}{h_m} = \frac{\ln(1+B)}{B} \tag{2.73}$$

where $B$ is known as the driving force defined by

$$B = \frac{P_{vo} - P_{va}}{P - P_{vo}} \tag{2.74}$$

and

$$\gamma = \frac{C_{pv}M_v}{C_pM}\left(\frac{a}{D}\right)^{-(n-1)} \quad \text{and} \quad \bar{C}_pM = \left(1 - \frac{P_{vm}}{P}\right)M_aC_{pa} + \left(\frac{P_{vm}}{P}\right)M_vC_{pv} \tag{2.75}$$

where $\bar{C}_pM$ is the mean specific heat capacity of moist air, $P_{vm} = 0.5(P_{vo} + P_{va})$; $M_a$ is the relative molecular mass of air; $C_{pa}$ is the specific heat capacity of drying air; and $M_c$ is the relative molecular mass of water vapor.

Table 2.2 lists the corresponding equations to calculate fluid properties (e.g., air, vapor, and air–vapor mixture). Because of pure convective heat flow, the heat current $q_c$ and mass flow rate $m_c$ are related by

$$q_c = m_c\Delta h_{lv}(T_0) \tag{2.76}$$

where the enthalpy of vaporization $\Delta h_{lv}(T_o)$ is determined at the temperature of the product surface. Using the above relations, we find that the heat flow $q_c$ can be expressed in terms of the mass flow $m_c$ as follows:

$$q_c = \frac{hA\Delta H_{v(T_o)}}{C_{pv}}\ln\left[1 + \frac{C_{pv}(T_a - T_o)}{\Delta H_{v(T_o)}}\right] \tag{2.77}$$

The corresponding product surface temperature $T_0$ is then given by the expression

$$T_o = T_a - \frac{\Delta H_{v(T_o)}}{C_{pv}}\left[(1+B)^\gamma - 1\right] \tag{2.78}$$

## Example 2.7

In a 5 m tray, there is placed a moist material which is dried in air. The temperature of the material is constant at 30 °C. Determine the mass transfer coefficient. Assume turbulent flow with $Re = 5 \times 10^6$.

- Assume that humid air is saturated at the air–material interface at a temperature equal to that of the material, whereas the total pressure is the standard pressure. Therefore, for

**Table 2.2** Fluid properties for numerical calculations

| Specific heat (kJ/kg K) | |
| --- | --- |
| Air | $C_{pa} = 1.006\left(1 + 5 \times 10^{-7} T_m^2\right)$ |
| Vapor | $C_{pv} = 1.88\left(1 + 2.2 \times 10^{-4} T_m\right)$ |
| Air–vapor mixture $(P_{am} = P - P_{vm})$ | $C_p = \dfrac{P_{am} M_a C_{pa} + P_{vm} M_v C_{pv}}{P_{am} M_a + P_{vm} M_v}$ |
| Water | $C_{pw} = 4.178 + 9 \times 10^{-6}(T - 35)^2$ |
| *Thermal conductivity (W/m K)* | |
| Air | $k_a = 0.02454\left(\dfrac{T_m}{273.15\,\mathrm{K}}\right)^{0.83}$ |
| Vapor | $k_v = 0.0182\left(\dfrac{T_m}{273.15\,\mathrm{K}}\right)^{(0.87 + 0.001\theta_m)}$ |
| Air–vapor mixture | $k = k_v \dfrac{P_{vm}}{P} + k_a\left(1 - \dfrac{P_{vm}}{P}\right)$ |
| *Dynamic viscosity (Pa/s)* | |
| Air | $\eta_a = 17.2 \times 10^{-6}\left(\dfrac{T_m}{273.15\,\mathrm{K}}\right)^{0.7}$ |
| Vapor | $\eta_v = 8.1 \times 10^{-6}\left(\dfrac{T_m}{273.15\,\mathrm{K}}\right)^{1.25}$ |
| Air–vapor mixture | $\eta = \dfrac{\eta_a P_{am} M_a^{1/2} + \eta_v P_{vm} M_v^{1/2}}{P_{am} M_a^{1/2} + P_{vm} M_v^{1/2}}$ |
| Density (kg/m$^3$) | $\rho = \dfrac{P_{am} M_a + P_{vm} M_v}{R_m T_m}$ |
| Latent heat of vaporization (kJ/kg) | $\Delta H_v(T) = 2501 - \left(c_{pw} - c_{pv}\right)T$ |
| Saturated vapor pressure (Pa) | $P_v^* = 610.7 \times 10^{\left(\frac{7.5T}{237 + T}\right)}$ |
| Relative molecular mass (kg/kmol) | $M = M_v \dfrac{P_{vm}}{P} + M_a\left(1 - \dfrac{P_{vm}}{P}\right)$ |

*Source*: Dietl et al. (1995). Properties are expressed in the following units: $T$ in K, $P$ in Pa.

$\varphi = 1$, one obtains from thermodynamic tables that $\rho = 1.116\,\mathrm{kg/m^3}$, $C_p = 1057\,\mathrm{J/kg\,K}$, $a = 22 \times 10^{-6}\,\mathrm{m^2/s}$, $\nu = 16.8 \times 10^{-6}\,\mathrm{m^2/s}$, and $Pr = 0.761$.

- Use Eq. (2.44) and determine $D = 22.6 \times 10^{-6}\,\mathrm{m^2/s}$.
- From Table 1.14 for flat plate, $Nu_Y = 0.332\,Re_Y^{0.5} Pr^{0.33} = 0.332 \times 10^{3.5} 0.761^{0.33} = 678$.
- Heat transfer coefficient results from $h = Nu\,k/Y = 3.5\,\mathrm{W/m^2\,K}$.
- From Eq. (2.68), $h_m = 3.4\,\mathrm{mm/s}$.

## Example 2.8

For the same date from Example 2.4, it is given in addition that the relative humidity of the drying air is 5%, while air temperature is 80 °C. Determine the corrected mass transfer coefficient.

- The humidity ratio of water vapor in drying air is given in function of 5% relative humidity and 80 °C air temperature; thence $\omega_a = 0.38\%$.
- The partial pressure of water vapor in drying air results from total pressure $P_0 = 1$ atm and humidity ratio and the molecular masses of water and dry air, $P_{va} = 613.5$ Pa.
- The humidity ratio of air at product surface considered saturated results from $T_s = 30°C$ and $\varphi = 1$ which first give an humidity ratio of $\omega_s = 27.2\%$. Thence, the partial pressure of vapor is $P_{v0} = 4252$ Pa.
- The driving force for diffusion is according to Eq. (2.71)

$$B = \frac{P_{v0} - P_{va}}{P_0 - P_{v0}} = \frac{4252 - 613.15}{100,000 - 4252} = 0.037$$

- Finally, from Eq. (2.69),

$$h_m^* = \frac{\ln(1 + B)}{B} h_m = 3.3 \, \text{mm/s}$$

## 2.10 Conclusions

This chapter has outlined the fundamental aspects of the drying of solids, the structure and drying characteristics of porous materials (particularly food products), and some key drying equipment. Shrinkage, the porosity of porous materials, and the heat and moisture transfer analyses of drying in porous materials are also discussed. The emphasis is put on the fundamental mechanisms and methods for analyzing and predicting the drying of porous materials. Some examples are presented to show how to use basic models and equation for drying processes.

## 2.11 Study Problems

2.1   How many phases a drying process can comprise? Describe each phase.

2.2   Define the humidity parameter of a moist product.

2.3   How much is the drying time of a product provided that the process occurs at constant rate and the thickness of the slab is 10 mm. The humidity must be reduced from 55% to 20%. The drying medium is nitrogen circulated at 40 °C. Consider that the heat transfer coefficient by convection is 25 W/m² K.

2.4   How much must be the entrance temperature of dry air of 5% relative humidity to dry at constant rate some humid products on a tray of 6 m provided that the gas velocity is 10 m/s? The humidity of the product must be reduced with 50%. Make reasonable assumptions.

2.5   The drying time of a product is 1 h during the constant rate period and 10 min during the falling rate period. Provided that the product is dried with 5% during the falling rate period, determine the amount with which the final humidity is superior to the equilibrium moisture content.

2.6   Explain the classification of moist materials according to the types of their drying rate curves.

2.7   Determine the required thickness of a moist porous material to reduce its humidity with 40% in one hour if the permeability is $10^{-12}\,m^2$ and the total pressure is 2 bar and if the drying is from both sides and the product is maintained at 25 °C.

2.8   Explain what Knudsen diffusion represents.

2.9   Explain what the effective diffusion coefficient represents.

2.10  Rework Example 2.4 for moist materials in the form of horizontal cylinder.

2.11  In a 10 m tray, there is placed a moist material with a spherical form of 5 mm radius. The temperature of the material is constant at 25 °C. Assume turbulent flow with $Re = 10^6$ and the drying air relative humidity and temperature of 5% and 80 °C, respectively. Determine the corrected mass transfer coefficient.

# References

Achenbach E. 1995. Heat and flow characteristics of packed beds. *Experimental Thermal and Fluid Science* 10:17–27.

Bilbao C., Albors A., Gras M., Andres A., Fito P. 2000. Shrinkage during apple tissue air-drying: macro and micro-structural changes. *Proceedings of the 12th International Drying Symposium*. K.P.J.A.M. Kerkhof et al. Eds., Paper No. 330, 28–31 August.

Brennan J.G., Butters J.R., Cowell N.D., Lilly A.E.V. 1976. Food Engineering Operations. Applied Science Publishers Limited: London.

Dietl C., George O.P., Bansal N.K. 1995. Modeling of diffusion in capillary porous materials during the drying process. *Drying Technology* 13:267–293.

Dincer I., Dost S. 1996. Determination of moisture diffusivities and moisture transfer coefficients for wooden slabs subject to drying. *Wood Science and Technology* 30:245–251.

Dincer I., Hussain M.M. 2002. Development of a new Bi–Di correlation for solids drying. *International Journal of Heat Mass Transfer* 45:3065–3069.

Eidsath A., Carbonell R.G., Whitaker S., Hennann L.R. 1983. Dispersion in pulsed systems – III. Comparison between theory and experiments for packed beds. *Chemical Engineering Sciences* 38:1803–1816.

Gabas A.L., Menegalli F.C., Telis-Romero J. 1999. Effect of chemical pretreatment on the physical properties of dehydrated grapes. *Drying Technology* 17:1215–1226.

Karel M. 1975. Dehydration of foods. In: Physical Principles of Food Preservation. M. Karel et al. Eds. Marcel Dekker: New York.

Kessler, H.G. 1981. Food Engineering and Dairy Technology. Springer-Verlag: Berlin.

Krokida, M.K., Maroulis, Z.B. 1997. Effect of drying method on shrinkage and porosity. *Drying Technology* 15:2441–2458.

Millington R.J., Quirk J.P. 1961. Permeability of porous solids. *Transaction of Faraday Society* 57:1200–1207.

Moise A., Tudose R.Z. 2000. A study on the drying of the packed beds circulated by the thermal agent. *Proceedings of the 12th International Drying Symposium*. K.P.J.A.M. Kerkhof et al. Eds., Paper No. 34, 28–31 August.

Plumb O.A., Gu L., Webb S.W. 1999. Drying of porous materials at low moisture content. *Drying Technology* 17:1999–2011.

Rebinder P.A. 1972. Physical-Chemical Principles of Food Production. Pishchepromizda: Moscow.

Strumillo C., Kudra T. 1986. Drying: Principles, Applications and Design. Gordon and Breach Science Publishers: New York.

Van Brakel J., Heertjes P.M. 1974. Analysis of diffusion in macro-porous media in terms of a porosity, a tortuosity and a constrictivity factor. *International Journal of Heat and Mass Transfer* 17:1093–1099.

# 3

# Drying Processes and Systems

## 3.1 Introduction

There are many types of drying equipment and systems used for products drying in food processing, pharmaceuticals, construction materials, and other industries. Drying equipment may in fact be simple or more complex engineered systems as designed to conduct mass transfer operations optimally, such that the drying rate is maximized (or drying time is minimized) in a cost-effective manner. In order to achieve the design objective, the dying equipment must be conceived such that the transport processes are optimized. Furthermore, the operating conditions must be well chosen. The drying agent must be rationally selected.

During drying system design, many design details and aspects require careful selection, analysis, and evaluation. For instance, the moist material must be positioned such that the access of the drying agent at its surface is facilitated. The moist material can be continuously fed into the drying chamber using conveyers or rolling bands. As an alternative, batch drying can be applied when the moist material is staked in such a way that the moisture removal is facilitated.

The size of the drying material has to be determined in accordance with its interstices structure. If there is sufficient porosity or capillary ducts, the size of the material can be larger. Otherwise, the moist material should be cut in smaller pieces (when possible) in order to reduce the required concentration gradients for moisture diffusion. In some cases, pulverized materials are subjected to drying. In conjunction with this, the fluidized bed systems are developed. In some instances, humidity can be absorbed in liquid form through a solid absorbent that is put in contact with the moist material, and hence, some hydrostatic phenomena may become important. Furthermore, if vapor pressure is relatively too low, then freeze-drying must be applied. Therefore, in design of such systems, the Knudsen diffusion should be accounted for. In fact, the drying technology should be designed such that both the transport of internal moisture from material core to the surface and the transport of surface moisture into the drying agent are balanced.

*Drying Phenomena: Theory and Applications*, First Edition. İbrahim Dinçer and Calin Zamfirescu.
© 2016 John Wiley & Sons, Ltd. Published 2016 by John Wiley & Sons, Ltd.

The main goal of any drying technology – including its processes and systems – is to enhance product quality with less energy consumption and higher productivity. Of course, the environmental impact should be minimized and the system must be cost-effective. Developing sustainable and clean technologies for drying is the right engineering desiderate. Therefore, combining several types of renewable energies for system supply and integration of drying with other processes for system hybridization is one of the main pathways for advanced drying systems development.

In this chapter, drying systems are classified from multiple points of view. In this regard, the most common drying systems, such as forced drying, spray drying, freeze-drying, and vacuum drying, are discussed and evaluated in detail. A thorough presentation of drying processes includes models and assumptions as well as simulation techniques which are essentially useful, and in some instances required, for design and analysis of drying systems.

## 3.2   Drying Systems Classification

The main criteria based on which the drying equipment can be classified are as follows: drying agent, application type, main drying processes, process continuity (batch or continuous flow), material handling mechanism, and heat transfer method. The general drying agent is air; however, steam is often used. Some technologies use supercritical steam or carbon dioxide as drying agent. In direct dryers, heat transfer between the surface of the moist material and drying agent is accomplished by direct contact. This heat transfer induces superficial evaporation in forced or natural convection. In indirect dryers, there is a wall placed between a heat transfer fluid (or heating medium) and the moist material. Therefore, the moist material is in contact with a hot (warm) surface from where it receives heat.

Figure 3.1 shows the classification criteria of the drying systems. In this figure, the classification with respect to heat transfer mode shows one more category besides direct and indirect-heat transfer through electromagnetic field. This can involve two subcategories: infrared rays and electric field through the moist material. For the infrared radiant-heat dryers, the drying operation depends on the generation, transmission, and absorption of infrared rays. The dielectric-heat dryers operate on the principle of heat generation within the solid by placing the latter in a high-frequency electric field.

Figure 3.2 shows a classification of direct-contact dryers. Due to a direct-contact heat transfer in direct dryers, the obtained temperatures can be very high (up to ~1000 K in general); this can be beneficial in some specific applications. In general, it is preferred that temperature level in direct dryers is above the boiling point at the operating conditions in order to avoid a reduction of the rate of drying due to unfavorable partial pressure gradients. In order to move the products, the continuous-tray dryers use metal belts, or vibrating trays utilizing hot gases, or vertical turbo dryers. In continuous sheeting dryers, a continuous sheet of material passes through the dryer either as festoons or as taut sheet stretched on a pin frame. In pneumatic-conveying dryers, drying is often done in conjunction with grinding. The material is conveyed in high-temperature high-velocity gases to a cyclone collector. In rotary dryers, the material is conveyed and showered inside a rotating cylinder through which hot gases flow. The dryer feed in spray dryers must be capable of atomization by either a centrifugal disk or a nozzle. Through-circulation dryers have the moist material held on a continuous conveying screen, and hot air is blown through it. If the particles are too small, direct dryer systems may require very bulky

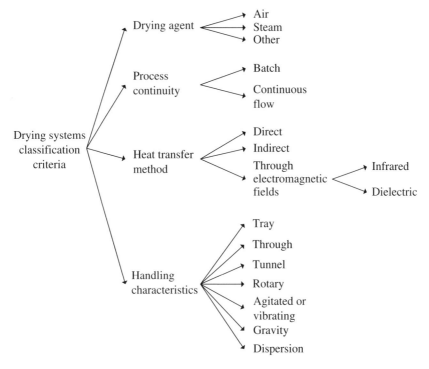

**Figure 3.1** Classification criteria of drying equipment and systems

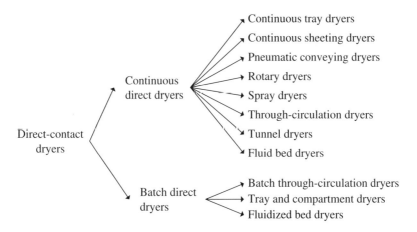

**Figure 3.2** Classification of direct-contact dryers

particle separators which may be inconvenient and can increase the costs. In the continuous tunnel dryers, the material on trucks is moved through a tunnel in contact with hot gases. In fluid bed dryers, solids are fluidized in a stationary tank which may also have indirect-heat coils.

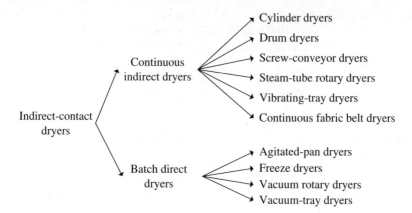

**Figure 3.3**   Classification of indirect-contact dryers

The batch direct dryers are designed to operate on a defined size of batch of wet feed for given time cycles. In these dryers, the conditions of moisture content and temperature continuously change at any point in the dryer. In the batch through-circulation dryers, the wet material is held on screen-bottom trays through which hot air is blown. In tray and compartment dryers, the material is supported on trays that may be or may not be on removable trucks, and air (or drying agent) is blown across the material on trays. For fluidized bed dryers, particulate solids with sizes of up to 5 mm are fluidized with gas in a stationary cart which has a filter mounted above.

Figure 3.3 shows the classification of indirect-contact dryers. As mentioned previously, in indirect type of dryers, heat for drying is transmitted to the wet solid through a retaining wall. Due to this reason, the maximum process temperature in indirect dryers is lower with about 200 K than for the direct ones. However, with indirect dryers, better control of operating conditions can be obtained. Furthermore, indirect drying is probably the best option when sensitive materials must be handled using inert gas solvents or condensing solvents that must be recycled. The vaporized liquid is removed independently of the heating medium. Rate of drying depends on the contact of the moist material with hot surfaces. They are also termed as conduction or contact dryers. In the continuous type, drying is accomplished by the material passing through the dryer continuously and is in contact with a hot surface.

The cylinder dryers are used for drying of continuous sheets such as paper, cellophane, and textile piece goods. These sheets are rolled on cylinders which are generally steam heated and rotate. The drum dryers have rotating drums which are heated with steam or hot water. Screw-conveyor dryers operate continuously under a vacuum to facilitate the drying and solvent recovery. Steam-tube rotary dryers operate also on slight negative pressure to permit solvent recovery with drying if desired; hot water can be used also instead of steam.

In vibrating-tray dryers, heating is accomplished by steam or hot water. Other special types of dryers exist such as continuous fabric belt moving dryers where the material is in close contact with a steam-heated platen. The moist material lies on the belt and receives heat by thermal contact.

Batch indirect dryers operate under vacuum and they can be classified into agitated and non-agitated types. Agitated-pan dryers operate either atmospherically or under vacuum and can

**Table 3.1**  Categorization of numerous moist materials for drying applications

| Material type | Description and examples |
|---|---|
| Liquids | All types of liquids including solutions, colloidal solutions, emulsion such as milk, blood, liquor, latex, aqueous sodium chloride, or other inorganic salts solutions |
| Slurries | These are pumpable suspensions (solid particles in liquids) such as detergents, calcium carbonate slurry, clay slurry, some metal chlorides, some pigments |
| Pastes | Suspensions of granular materials in liquids behaving as non-Newtonian fluids (e.g., Bingham fluid) such as putty, toothpaste, starch, mustard |
| Sludge | A semisolid material with large amounts of interstitial liquid (water) which is normally carried by liquid water such as waste water or sewage or solids separated from suspension in a liquid |
| Free-flowing powders | This is fine dusty powder of less than 100 mesh (particle size smaller than 0.149 mm) such as cement, clay, centrifuged precipitates |
| Granular solids | Conglomeration of discrete solids with size of more than 0.15 mm, for example, rayon staple |
| Crystalline solids | This is a solid material consisting of a well-organized structure of atoms (molecules, ions) such as NaCl |
| Large solids | Any kind of larger material with moisture content such as bricks, lumber, rayon cakes, shotgun shells |

handle small production of nearly any form of moist material, that is, solid, liquids, slurries, pastes, or granular solids. Freeze dryers use frozen moist material. The moist material is frozen prior to drying. Drying in frozen state is then done under very high vacuum. In vacuum-rotary dryers, the moist material is agitated in a horizontal, stationary shell. Vacuum may not always be necessary. In addition to the shell heating, the agitator may be steam heated. Vacuum-tray dryers use a heating system done by contact with steam-heated or hot-water-heated shelves on which the moist material lies. No agitation is involved.

With respect to moist material handling criteria, there are many systems available. The material handling depends on the nature of material subjected to drying. In this respect, the feedstock of moist material can be categorized as given in Table 3.1. Very common are the batch tray dryers and the tunnel dryers. Many other types exist including through, rotary, vibrating, and other. Often, moist materials are preformed before drying. In this operation ("preforming"), the products are processed with the aim of increasing their surface area and the contact area between their surface and the drying agent. The following types of preforming techniques may be applied, depending on the case:

- Granulation – this is a milling process which eventually produces granules from larger size material; this technique is applied to some ceramics and clays (kaolin), cellulose, and starch.
- Flaking – which is a breaking process of the material in smaller pieces; for example, soap can be flaked.
- Squeezing – this is compacting the materials under applied pressure forces, for example, cornstarch.
- Briquetting – in this operation, briquettes are formed which is a small-size compressed material such as sawdust, soda ash, and so on.
- Finned Drums Preforming – in this process, the feed is pushed over hot finned drums and is partially dried and deformed; this applies, for example, to calcium or magnesium carbonates.

• Extrusion – in this process, the material is pushed through a die of certain profile; this applies, for example, to zinc or aluminum stearate.

Table 3.2 gives the dryer types and their descriptions categorized based on the moist material handling method. Figure 3.4 shows the heat demand ranges for the main types of direct-contact

**Table 3.2**  Dryer types categorized with respect to the moist material handling method

| Handling method | Dryer type and description |
| --- | --- |
| Tray handling | *Batch tray dryers*: A tray or compartment dryer is an enclosed, insulated lousing in which solids are placed upon tiers of trays in the case of particulate solids or stacked in piles or upon shelves in the case of large objects |
| Through circulation | *Batch through-circulation dryers*: This type is similar to a standard tray dryer except that hot air passes through the wet solid instead of across it |
| Tunnel/truck handling | *Continuous tunnel dryers*: Continuous tunnels are in many cases batch truck or tray compartments, operated in series |
| Handling on rotating drums | *Rotary dryers*: A rotary dryer consists of a cylinder, rotated upon suitable bearings and usually slightly inclined to the horizontal |
| Handling on agitated shelves | *Agitated dryers*: An agitated dryer is defined as one on which the housing enclosing the process is stationary while solids movement is accomplished by an internal mechanical agitator |
| Vibrating conveyor | *Direct-heat vibrating-conveyor dryer*: The vibrating-conveyor dryer is a modified form of fluid bed equipment, in which fluidization is maintained by a combination of pneumatic and mechanical forces |
| Free-fall particulate matter handling | *Gravity dryers*: A body of solids in which the particles, consisting of granules, pellets, or briquettes, flow downward by gravity through a vessel in a moving bed in contact with hot gases |
| Gas/dilute phase solids handling | *Dispersion dryers*: A gas–solids contacting operation in which the solid phase exists in a dilute condition is termed a dispersion system |

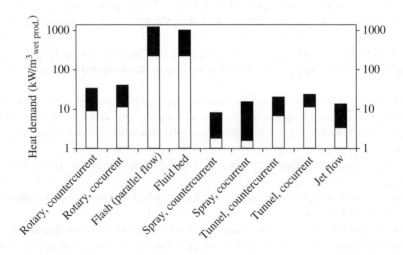

**Figure 3.4**  Heat demand ranges for the main types of direct-contact dryers (data from Mujumdar (2006))

**Table 3.3** Heat transfer parameters and heat demand of two types of indirect dryers

| Dryer type | $U$ (W/m$^2$ K) | $\Delta T$ (°C) | $q''$ (kW/m$^2$) |
|---|---|---|---|
| Drum dryer | 115–230 | 50–80 | 5.8–18 |
| Agitated through dryer | 60–150 | 50–100 | 3.5–15 |

dryers. The heat demand is given with respect to the volume of the wet material. Table 3.3 gives the heat transfer parameters and heat demand for two types of indirect dryers. For these types of dryers, the heat demand is given in rapport to the heat transfer surface area. Some of the most used dryer equipment is described next.

## 3.3   Main Types of Drying Devices and Systems

In this section, the main types of drying devices and systems are described, illustrated, and discussed under various criteria. The key focus is placed on explaining the operation principle and applications. The equipment is introduced here based on the material handling criteria. Some important operational parameters and performance factors are presented to highlight various advantages and disadvantages as well as specific details to operate them in a better way.

### 3.3.1   Batch Tray Dryers

As described in Table 3.1, the heat transfer in batch tray dryers may be direct from gas to solids by circulation of large volumes of hot gas or indirect by use of heated shelves, radiator coils, or refractory walls inside the housing. Because of the high labor requirements usually associated with loading or unloading the compartments, batch compartment equipment is rarely economical. Furthermore, because of the nature of solid–gas contacting, which is usually by parallel flow and rarely by through circulation, the process of transferring heat and mass is comparatively inefficient. For this reason, the use of tray and compartment equipment is restricted primarily to ordinary drying and heat-treating operations.

Figure 3.5 shows a batch tray dryer that used forced air circulation and direct heating. In this configuration, dry air is circulated above the shelves with a wet product. For recirculation purpose, an air handling unit is used which processes the humid air to reduce its humidity; the air handling unit introduces 5–15% fresh air in the loop. Similar indirect drying equipment exists, a case in which the heating medium heats the pans or shelves carrying the products. In some designs, the shelves are heated using electrical wires. However, most indirect tray dryers use high-pressure steam or thermal oil for heating. The size of the dryer is relatively small, with up to about 20 shelves each loaded with few tens of kilograms of moist material per square meter, in general (Berk, 2009). The free space between two loaded shelves is of the order of 3–4 cm, whereas the layer of moist material can be up to 10 cm in height. The drying rate in batch tray dryers is given per unit of tray surface area. As given in Green and Perry (2008) for commercial system, the area-specific drying rate is in the range of 0.1–65 kg/m$^2$ h. A special attention must be given in designing the air distribution system for batch tray dryers. It is technically very challenging to achieve a uniform air distribution with batch tray dryers.

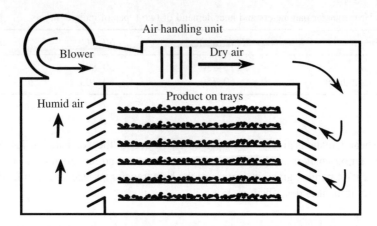

**Figure 3.5**   Batch tray dryer with direct contact (forced air circulation)

One major advantage of batch system is that the drying chamber can be well sealed and vacuumed such that the drying process can be conducted at lower temperature. This technique can be applied to various types of heat-sensitive materials. Provided that the vacuum demand is not very deep, a steam ejector can be used for this purpose. However, the size of the drying chamber cannot be large; thence, batch tray dryers are restricted to small scales of production, and this can be considered a disadvantage. Another disadvantage is related to the high operation costs due to the intensive labor required to set the materials on trays, to load, and unload.

Lower labor costs are possible when the moist material is in form of free-flowing powder or pellets, a case in which the loading and unloading can be automatized in some respect. Some similar equipment exists for drying this type of materials which may be a better choice than batch tray dryers – this is the rotating batch dryer. This is in fact a sealed cylindrical drying chamber in which a batch of power or granulate material is placed. The dryer rotates to facilitate the process, while in the same time vacuum can be applied and a heating agent is circulated in a double shell for indirect-heat transfer. However, the rotating batch dryer cannot be applied to materials that stick to each other, whereas the batch tray dryer can. Provided that the production capacity is not high, many sludge and pastes are suited to batch drying in trays as well as free-flowing powders, granular, or fibrous solids.

## 3.3.2   Batch Through-Circulation Dryers

In this equipment, the wet product is placed on trays and the drying agent (air) is forced through the wet solid instead of across it; of course, the wet solid bed must be permeable. All the material is enclosed in a drying chamber for batch operation. Heated air passes through a stationary permeable bed of the moist material placed on removable screen-bottom trays suitably supported in the dryer. The pressure drop through the bed of material does not usually exceed about ~2 mbar. In some batch through dryers, the products are placed on perforated shelves which allow for airflow and penetration through the bed. The drying rate is in the range of $1\text{–}20\,\text{kg/m}^2\,\text{h}$.

In some models, deep perforated bottom trays are placed on top of plenum chambers in a closed-circuit hot-air-circulation system. In food-drying batch through equipment, the material

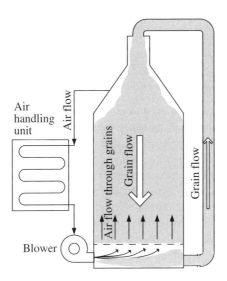

Air handling unit

Air flow

Air flow through grains

Grain flow

Grain flow

Air flow

Blower

**Figure 3.6** Batch through-recirculation dryer

is placed in finishing bins with perforated bottoms; heated air passes up through the material and is removed from the top of the bin, reheated, and recirculated. The convenient drying agent for batch through dryers is steam, which carries the moisture and can be recycled. Therefore, when drying process is at low temperature, all system is maintained in vacuum such that the steam drying agent is slightly superheated. Small-scale equipment of this type with vacuum drying is used in pharmaceutical industry. The drying rate with commercial batch through dryers operating under vacuum is approximately $30 \, \text{kg/m}^2 \, \text{h}$.

Figure 3.6 shows a batch through-recirculation dryer. In the figure, a dryer for grain draying is exemplified. The drying agent is air which is circulated forcedly through the grain bed. Grain is pushed pneumatically up and feeds through the top; any other mechanism of grain feeding at the top can be used. Grains settle in the storage chamber, while warm, dry air is circulated from the bottom to top. Air at the top is humid since it grabs the product moisture. This air is evacuated by an air handling unit which dries the air and recycles it. Other similar configurations of batch through dryers exist.

The bed of particulate solids must be in one of the following flow conditions: dilute, fluidized, moving, or static. In the static conditions, the bed of particles is compact with no relative motion of solids but rather with only gas flowing through the interstices. This is the case depicted in Figure 3.6 for the batch through-recirculation dryer.

For the moving bed, the particles are able to flow due to the presence of gas phase. The direction of movement is influenced in large measure by the gravity forces and mechanical agitation forces when applied. However, the gas-phase velocity is not so high in moving bed such that the true fluidization phenomenon does not occur yet. In a fluidized bed, the particle bed expands even more such that drag forces manifest at the level of each individual particles. In fluidized bed regime, the gas and solid phases manifest in equal measure producing a turbulent flow configuration with gas eddies and up–down particle flow. The particles cannot be conveyed continuously by the gas phase since the upward gas velocity is less than the settling velocity of solid particles. Conveying of solid by the gas phase can be obtained in the dilute flow regime, when the particles are fully dispersed in the gas.

### 3.3.3   Continuous Tunnel Dryers

Various types of trucks or trays are used to transport the products through continuous tunnel dryers. In this equipment, the trucks advance in cocurrent or countercurrent with the drying agent in a continuous or semicontinuous manner, depending on the tunnel design. The operation is often semicontinuous except for some specific designs such as screen-conveyor and belt-conveyor types. In the semicontinuous operation tunnels, there are three process phases: truck loading and entrance, truck progression through the tunnel, and truck exit and downloading. In continuous tunnel design, there are no inlet and outlet phases of the process, but rather the products are continuously conveyed through the tunnel.

Drying tunnels are very versatile with respect to process configuration and the type of products. If permeable materials are dried, then perforated trays may be installed on trucks which allow for through flow configuration. If nonpermeable materials are used, the trucks carry trays on which the products are arranged for drying.

There are several flow configurations with tunnel dryers: parallel flow (or cocurrent), countercurrent, and dual or multizone tunnels which combine or alternate cocurrent and countercurrent flow configurations. The schematic diagram of a countercurrent tunnel dryer is shown in Figure 3.7. As seen in the figure, the airflow direction and the movement of trays are in countercurrent. Another type of tunnel dryer is with parallel flow configuration as shown in Figure 3.8. Near the exit, a portion of the humidified air is combined with fresh dry air and recirculated. This scheme improves the energy efficiency of the equipment. Even better efficiency is obtained with dual-zone tunnel dryers which combine the countercurrent and parallel flow configurations.

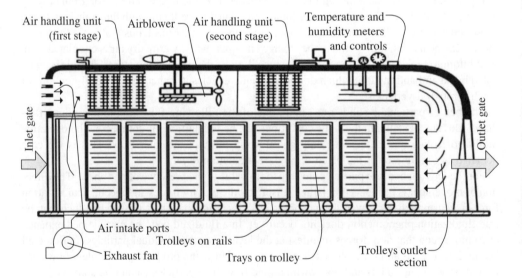

**Figure 3.7**   Countercurrent tunnel dryer

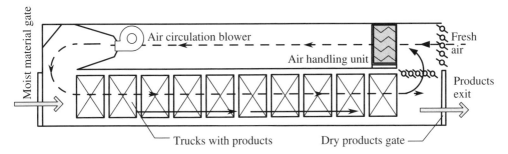

**Figure 3.8**  Parallel flow tunnel dryer

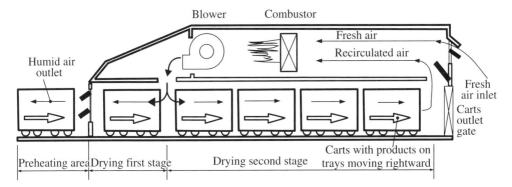

**Figure 3.9**  Dual-zone tunnel dryer with entrance side exhaust

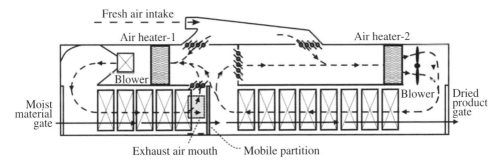

**Figure 3.10**  Dual-zone tunnel dryer with central exhaust

Figure 3.9 shows a dual-zone tunnel dryer. The countercurrent scheme is applied in the entrance section of the trucks, with no recirculation. A parallel flow scheme with recirculation is used for the rest of the drying pathway. Another advanced scheme of tunnel dryer is shown in Figure 3.10 for a dryer with a central exhaust. This scheme uses a countercurrent configuration with recirculation at the entrance zone. After the entrance zone, the air is refreshed and a parallel

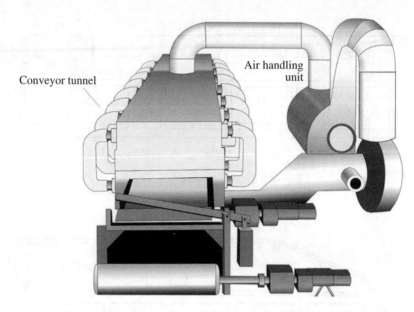

Conveyor tunnel

Air handling
unit

**Figure 3.11**   Sketch of a tunnel dryer of conveyor-screen type

flow configuration follows with recirculation. The energy efficiency is higher than for other systems.

As suggested in Figure 3.8, tunnel dryers can be equipped with air handling units which can be heated by steam or can be part of a heat pump system which regulates the temperature and humidity of the air (or drying agent). Finned coil heat exchangers need to be used for better heat transfer characteristics. Another design option is shown in Figure 3.9 where a gas burner is shown in the system.

The sketch of a tunnel dryer is shown in Figure 3.11 of conveyor-screen type. In this system, the material is placed in a layer on a conveyor tape which is in fact a perforated screen. The drying agent is circulated in perpendicular direction with respect to product movement to apply through circulation when possible. Depending on the type of material, the drying rate in continuous through-circulation tunnel dryers is about 10–50 kg/m$^2$ h with a specific capital investment of approximately \$200/m$^2$ of screen (Green and Perry, 2008).

Applications of tunnel equipment are essentially the same as for batch tray and compartment units previously described, namely, practically all forms of particulate solids and large solid objects. In operation, they are more suitable for large-quantity production, usually representing investment and installation savings over multiple batch compartments.

## 3.3.4   Rotary Dryers

In this type of dryer, a metallic cylinder of a kiln rotating around its symmetry axis is used. The cylinder is sustained on special bearings and is slightly inclined (1–5°) with respect to the horizontal. The length of the cylinder may range from 4 to more than 10 times its diameter. The diameter of industrial rotary dryers (kilns) may vary from less than 0.3 m to more than 3 m.

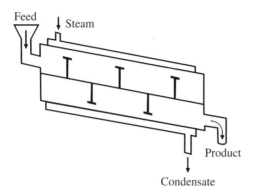

**Figure 3.12**   Simplified sketch illustrating the operating principle of indirect rotary dryer

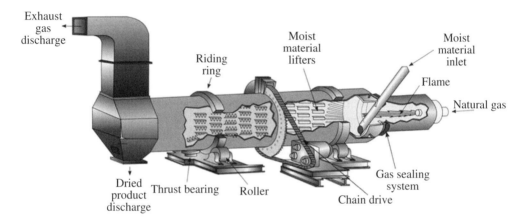

**Figure 3.13**   Sketch showing the mechanical construction of a direct-heat rotary dryer

Feed moist materials in solid form are fed into one end of the cylinder and progress through it by virtue of rotation, head effect, and slope of the cylinder and discharge as finished product at the other end. Most of the rotary dryers operate continuously, but there is also some batch rotary dryer equipment. In any of the possible configurations, the product must be of free-flowing type, or if not, then its flowing characteristics must be improved by scrappers, retarders, or by mixing of the feed with a part of the dry product which is to be recycled.

Figure 3.12 shows the construction of an indirect rotary dryer. In this system, heat is added through the walls as the kiln has double wall and in the annular space steam or hot combustion gases flow through while releasing heat to the process. The heat flux at the cylindrical wall of indirect dryers has values of up to ~6000 W/m². At this heat flux, the heated surface area required per kilogram of product is in the range of 0.1–0.4 m². In indirect rotary dryers, low-pressure steam (3–10 atm) is generally flowed in tubes attached to the exterior side of the dryer's cylinder. In some designs, double shells are used. The inlet and outlet ports of the thermal agent are of special design of revolving type.

In the cases when the contact between combustion gases, hot air or other drying agent, and the moist material is permitted, then direct-contact rotary dryers can be applied. Figure 3.13

**Figure 3.14**   Sketch showing a perspective view of a rotary dryer

shows the mechanical construction of a direct-heat rotary dryer. Gases flowing through the cylinder may retard or increase the rate of solids flow, depending upon whether gas flow is countercurrent or cocurrent with solids flow. Rotary dryer is applicable to batch or continuous processing solids which are relatively free flowing and granular when discharged as product. Materials which are not completely free flowing in their feed condition are handled in a special manner, either by recycling a portion of final product and premixing with the feed in an external mixer to form a uniform granular feed to the process or by maintaining a bed of free-flowing product in the cylinder at the feed end and performing a premixing operation in the cylinder itself.

Figure 3.14 shows a sketch of a rotary dryer. The direct-heat rotary units are the simplest and most economical in construction and are employed when direct contact between the solids and flue gases or air can be tolerated. Because the total heat load must be introduced or removed in the gas stream, large gas volumes and high gas velocities are usually required. These dryers use interior flights of approximately one-tenth of the dryer diameter which guide the materials along the kiln and impede clogging. The flights have lips with variable angle in accordance with their position along the dryer with about 45° in the middle zone and 90° near the exit zone. The angular speed of the cylinder is determined based on the peripheral speed which is of 0.2–0.5 m/s for industrial dryers.

Based on the cylinder surface area of the direct rotary dryer, the evaporation rate of the commercial equipment is in the range of 4.5–5.5 kg/m² h and the specific price is in the range of $2.4–$5.4K/m². The chart from Figure 3.15 shows the correlation of the surface area of the rotary dryer cylinder with the area-specific drying rate and the area-specific price of equipment. Figure 3.16 shows the relative drying obtained for various types of materials in direct rotary dryers. The relative drying is determined based on the initial and final moisture content $X$ of the product, as follows:

$$RD = \frac{X_{initial} - X_{final}}{X_{initial}} \times 100 \qquad (3.1)$$

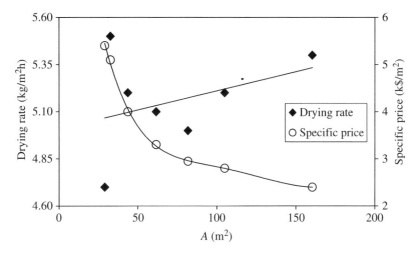

**Figure 3.15** Drying rate and specific price of industrial direct rotary dryers correlated with surface area (data from Green and Perry (2008))

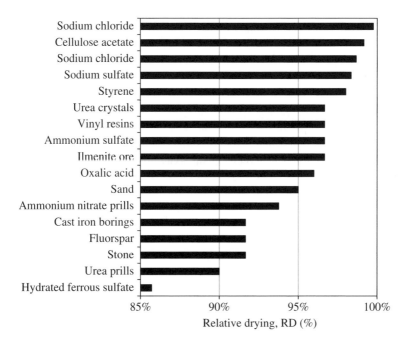

**Figure 3.16** Relative drying of various materials in direct heated rotary dryers (data from Green and Perry (2008))

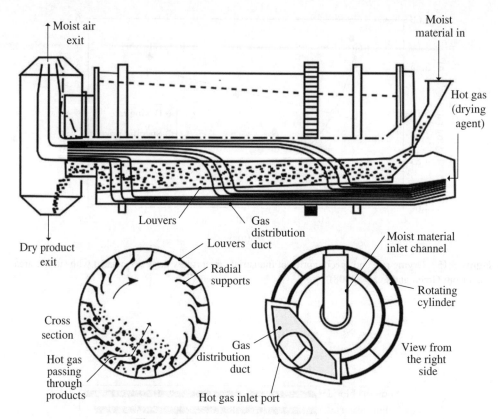

**Figure 3.17**  Schematics of a Roto-Louvre dryer

In a special construction, known as Roto-Louvre dryer, the cylinder has louvers to admit the drying agent (hot air or combustion gases) as shown in Figure 3.17. The drying agent (gaseous) is distributed through a double shell around the cylinder. The gas is pressurized such that it flows through the bed of products and eventually exits through the core of the cylinder from where it is guided through an exhaust stack.

### 3.3.5  Agitated Dryers

In this equipment, an internal mechanical for the movement of wet solids is used. The agitated dryers can be designed for both batch and continuous modes of operation. Table 3.4 gives a classification of agitated dryers. There are seven established designs of agitated dryers as given in the table. Among those, two of the designs most used are, namely, the standard rotating dryer and the double-cone dryer. Agitated dryers require similar types of materials as feedstock such as the rotary dryers. The essential feature of the materials is the free-flowing behavior. With respect to rotary dryers, the agitated ones offer better gas-sealing feature with minimized

**Table 3.4** Types of agitated dryers

| Agitated dryer | Description |
| --- | --- |
| Standard rotating dryer | This type of system operates typically in batch mode and in vacuum. It uses a rotating horizontal cylindrical chamber. Problem: crust formation on the heat exchange surface and stalling of the rotor |
| Double-cone vacuum dryer | This is a batch rotary dryer of a double-cone geometry made with a gas-tight design. Therefore, the system can be used for vacuum drying. Steam is used in a double shell or tubes to transfer heat to the process |
| Turbo tray dryers | This equipment consists of a stack of annular-shape shelves which rotate in the same direction. The moist material (powders or slurries) is fed through the top and falls on the first rotating shelf. After one turn, the material is being spread uniformly to the next lower shelf and so on. The drying agent is hot air which is circulated forcedly between shelves by turbofans placed in the axis. The mode of operation is continuous |
| Plate dryers | This is a continuous dryer consisting of a vertical cylindrical enclosure having inside a number of plates of the shape of a disk. The plates are heated to provide indirect-heat input to the wet products sustained on them. Over each plates, there are arms and plows which are rotated by a central vertical shaft. The product enters from above and while flowing down is moved sequentially from the periphery toward the axis and reversely |
| Conical mixer dryer | This is a batch drying equipment for drying moist free-flowing powders or solvents. The organ that generates product agitation is a screw mounted inside the drying chamber which is of conical shape |
| Reactotherm | This is a heavy-duty dryer and thermal processor used in various industries for drying pasty products. The product transport is done by mixing bars. The system can operate in batch or continuous. The wet fed is generally introduced from the top and discharged from the bottom |
| Hearth furnace | These are furnaces having one or more hearths made of refractory materials where the products are heated with hot combustion gases (mostly by radiation heat transfer at high temperature), and in the same time, they are agitated by mechanical means such as stirrers. The feedstock is charged in the center and discharged at the periphery. These systems are applied to drying processes in chemical industry |

leakage of gaseous drying agent. In addition, the produced shear forces enhance the process. Figure 3.18 shows a sketch of a standard rotating agitated dryer.

The agitated dryer shown in Figure 3.18 consists of a cylindrical vessel which includes agitator blades placed on a central rotating shaft. The vessel can be sealed and operated under vacuum if required. The heat addition is obtained indirectly by flowing steam in a double shell. The condensate is collected at the bottom. As seen in the figure, the feed is supplied at the top and travels through the vessel under the impingement forces generated by the agitation mechanism.

One very common design is the double-cone vacuum dryer which is heated indirectly in a double shell and/or by internal coils. The drying rate specific to agitated dryers is in the range of $1–7 \text{ kg/m}^2 \text{ h}$. Figure 3.19 shows how the specific price varies with the heat transfer area for the

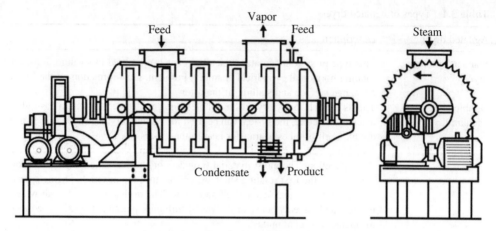

**Figure 3.18** Sketch of an agitated rotating dryer

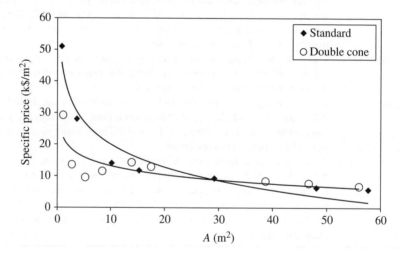

**Figure 3.19** Correlation between heat transfer surface area and specific equipment price for two agitated dryers

double-cone and standard rotating dryer. The price tends to decrease with the system size, with the double-cone system being cheaper at lower scales and slightly more expensive at larger scales.

### 3.3.6 Direct-Heat Vibrating-Conveyor Dryers

In the vibrating-conveyor dryer, the fluidization of the particle bed is maintained by a combination of pneumatic and mechanical vibration forces. The mechanical forces are produced by the vibrating deck which requires few hundred watts per square meter. The pneumatic forces

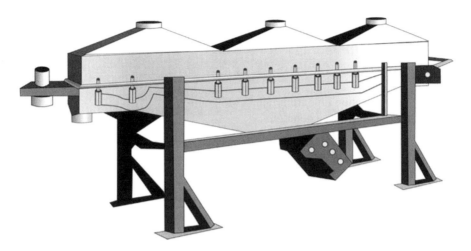

**Figure 3.20** Sketch of a direct-heat vibrating conveyer dryer

are generated by the direct action of the drying agent, which is a gas blown from the bottom through a perforated screen. The flow rate of gas is such that the superficial velocity is of the order of 1 m/s which will insure sufficient fluidization without significant entrainment of dust.

The superficial velocity limits actually the drying capacity of the system. In the same time, all the structure vibrates. After the direct-contact drying process, the gas carrying product moisture is evacuated at the top where exhaust hoods are installed. The type of products suitable for the vibrating-conveyor dryer is granulate materials of 100 mesh or larger which are not sticky.

Figure 3.20 shows a sketch of a direct-heat vibrating-conveyor dryer. The hot gas is blown from the bottom, the moist material is poured from one side of the deck, the product is extracted from the other side of the deck, and the humid gases are removed by the fume hoods placed above. The deck vibration helps handling the nonfluidizable fractions. The nonfluidizable fractions are mechanically channeled toward the exit port of the equipment. In typical beds, particle retention is in the approximate range of 50–350 ms. In some units, in order to enhance the heat input to the process, the external walls of the drying chambers may be heated, or heating coils can be immersed inside the chamber. When more heat is added per unit of volume, the drying rate increases.

### 3.3.7 Gravity Dryers

In this type of equipment, moist materials flow downward gravitationally as a moving bed in contact with the gaseous drying agent. Gas flow is through the solid bed and may be cocurrent or countercurrent and, in some instances, cross-flow. The equipment comprises a vertical vessel having an intake port for wet solids at the top and an extraction port for products at the bottom. The gas flows through the solids that have to be a permeable bed. Because the pebbles (granules) advance in the gravity field, they also convey heat to the process.

The gravity dryers are very versatile with respect to the process temperature; their applications cover drying process from large temperatures to high. Note that shaft furnaces can be used for drying and chemical operations at high temperatures. Shaft furnaces are lined inside with refractory bricks. Some of the advantages of gravity dryers are listed as follows:

- Selection versatility of gas–solids flow rates
- Flexible operation
- Adjustability of retention time from minutes to hours
- Configurable for multistage operation
- Simplicity
- Compactness
- Easy control for start-up and shutdown

The aspect ratio of gravity dryers must be small, in terms of diameter versus height. This shape allows for maintenance of a uniform downward flow of solids across the whole section. If the cross section is too large, then the flow of solids becomes nonuniform which creates problems with solids extraction and produces a variation in the retention time, and moisture content of final product rate of solids movement downward over the entire cross section of the bed is one of the most critical operating problems encountered.

Table 3.5 gives the types to gravity dryers and their descriptions. There are three main classes of gravity dryers: pellet dryers, shaft furnaces, and spouted bed dryers. The flow configuration differs, depending on the dryer type. In pellets dryers, a cross-flow is used with gas flowing from the periphery toward the core. The shaft furnaces are high-temperature units in which the flow is in countercurrent. The spouted system uses a combination of cocurrent, countercurrent flows. In the annulus, the flow is in countercurrent configuration, whereas in the core, the particles are entrained by the gas, and therefore, the gas and particles flow in cocurrent.

Figure 3.21 shows the comprehensive sketch of a gravity dryer illustrating its principle of operation. This is a two-stage drying system. The material enters through the top as shown in the figure and advances downward on a slope through the annular space. Air is blown through the hot-air ducts which flow in countercurrent with the moving bed of particulate matter. While moving down, the material is dried in the first stage and eventually accumulates in the receiving hopper. From there, the partially dried materials conveyed upward through the

**Table 3.5**  Description of gravity-type dryers

| Dryer type | Description |
| --- | --- |
| Pellet dryers | Extruded pellets are dried in downward movement while drying air is circulated in cross-flow. In this respect, air is fed through the outer wall which is louvered, permeates through solid bed, and it is discharged through the central section |
| Shaft furnaces | This is a vertical cylinder lined at the inside with refractory bricks in which the moving bed of solid pellets move downward in countercurrent with hot gases blown forcedly from the bottom. These furnaces are employed for various drying, calcination, and chemical operations in industry |
| Spouted beds | In this equipment, a downward flowing shell of granules/pellets is formed under the action of an upward flowing gas. The flowing gas forms a spout at the center where gas mixes with particulate solids from a loosely packed bed. The spoutable depth is higher for larger bed diameters |

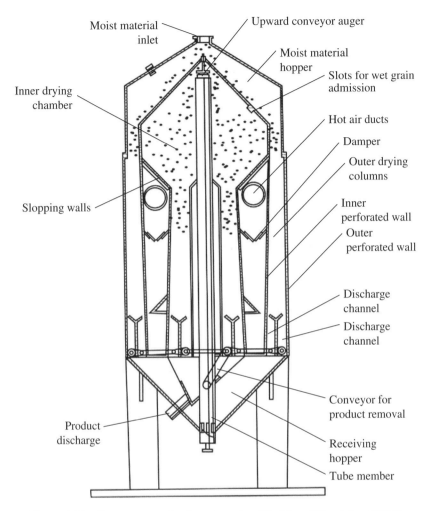

Moist material inlet

Upward conveyor auger

Moist material hopper

Slots for wet grain admission

Inner drying chamber

Hot air ducts

Damper

Outer drying columns

Inner perforated wall

Outer perforated wall

Slopping walls

Discharge channel

Discharge channel

Product discharge

Conveyor for product removal

Receiving hopper

Tube member

**Figure 3.21**   Two-stage gravity flow dryer (modified from Westelaken (1981))

central tube member with the help of a conveyor auger. The material reaches the inner drying chamber at the top where it is mixed with a smaller amount of moist material through the admission slots. Further, while receiving heat input through the walls, the products move downward and eventually reach the bottom where the conveyor removes the dried products.

### 3.3.8   Dispersion Dryers

A gas–solid heat and mass transfer process is conducted in dispersion dryers in which the solid phase exists in a dispersed (dilute) condition. The particle is relatively so small with respect to the surrounding gas phase such that there cannot be either any interaction among particles or essential resistance to flow passage. Therefore, in dispersion dryers, the thermal and mass diffusion processes through the particle core are negligible with respect to the superficial heat and mass transfer. This case refers to Biot number which is much smaller than 1, and hence, lumped capacity model can be applied to model heat and mass transfer processes.

In order to form the dispersion phase, the solids must be finely grinded and the velocity of gas must be set to sufficiently high values such that the solids are conveyed upwardly. The void fraction of solids is obviously very small. The retention time of the particle in the dispersion dryer is of the order of seconds, which is very small as compared to other equipment. The lag factor is insignificant, while the particles are dried completely. Because of these features, dispersion dryers can be applied to explosive, flammable particles or thermally sensitive materials or materials that can be rapidly oxidized.

There are a two established classes of dispersion dryers: flash dryers and spray dryers. The flash dryers are of three main types: pneumatic-conveyor, flash dryer, and agitated flash dryer. The flow configuration of particles and gas in flash dryer is cocurrent. For spray dryers, there is a countercurrent configuration, with sprayed particles moving downward and gas flowing upward. Subsequently, the flush and then the spray dryers are illustratively described.

### 3.3.8.1 Agitated Flash Dryer

Figure 3.22a shows the sketch illustrating the operating principle of an agitated flash dryer. Hot air (drying agent) is distributed from the bottom side. The granular matter is fed at the top and

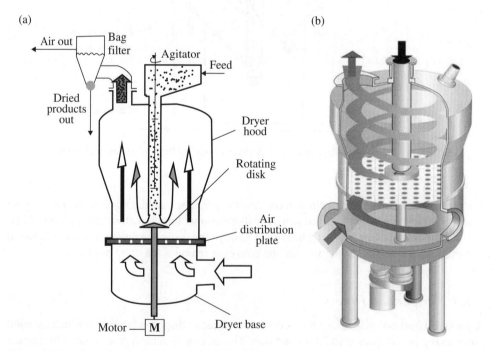

**Figure 3.22** Dispersion dryer: (a) operation principle and (b) sketch showing the spiral path of air and particles

falls down gravitationally through a downspout. An agitator may be used in the feeding tank (at the top) or inside the downspout. The wet granules are eventually dispersed at the bottom of drying chamber due to a swirling generated by a rotating disintegrator and the airflow. The size of the particles is controlled by the speed of the disintegrator. The fine particles flow upward dispersed in the hot gas flow with a swirling motion. Finally, the particles are separated in a cyclone equipped with a bag filter at the top.

Figure 3.22b shows a sketch of the dryer indicating the air and product patterns. The equipment is suitable for materials which are granular and free flowing when dispersed in the gas stream, so they do not stick on the conveyor walls or agglomerate. Due to the continuous agitation and dispersion, viscous liquids, sludge, thick pastes, and filter cakes can be dried in single stage without the necessity of back mixing. The heat and mass transfer processes agitated flash dryers are very effective since the solid phase is well dispersed and gas–solid interface is maximized. The drying chamber is maintained in a slight vacuum with the help of an extraction fan. The air speed inside the chamber is typically of the order of 2–4 m/s.

### 3.3.8.2 Pneumatic-Conveyor Flash Dryer

Another type of flash dryer is the pneumatic-conveyor flash dryer shown in Figure 3.23. This is a very common type of flash dryer comprising an air blower and heater, a feeder for moist material, and a vertical drying chamber. The solid becomes dispersed in the large stream of air. Therefore, they are maintained in suspension and conveyed rapidly upward where drying occurs mostly. Further, the solid materials are separated in a cyclone, where due to centrifugal forces, the solids flow at the periphery, while gas is extracted from the center.

An air lock consisting of a pile of dry solids is formed at the base of the cyclone. A part of the product can be recycled. In this respect, classifiers can be used to extract some of the still-moist product and recirculate it for better drying. Figure 3.24 shows a sketch of the flash duct of a pneumatic-conveyor flash dryer.

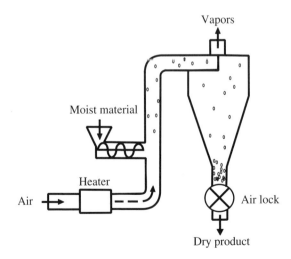

**Figure 3.23**  Pneumatic-conveyor flash dryer schematics

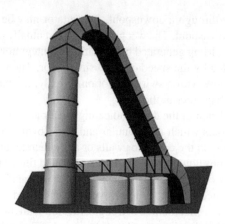

**Figure 3.24**   Sketch showing the 3D view of an industrial flash dryer platform

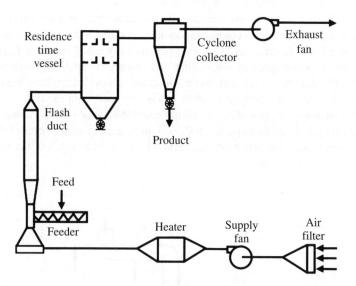

**Figure 3.25**   System schematics of an extended residence time flash dryer (modified from GEA (2014))

The classifiers are usually installed after each drying column for recycling the heavier particles which have more moisture content. In addition, vessels for extended residence time at
drying can be installed downstream the drying column, which will improve the process effectiveness. Figure 3.25 shows an extended residence time flash dryer. The residence time vessel is
interposed between the flash duct and cyclone collector. The vessel has an enlarged cross
section and includes internal baffles, and it is well insulated to maintain the process
temperature.

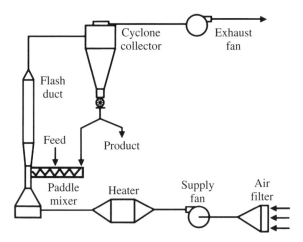

**Figure 3.26**   Flash drying system with partial product recirculation

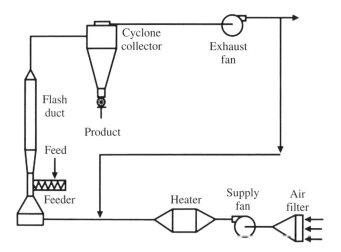

**Figure 3.27**   Flash drying system with partial air recirculation

Another flash dryer design is that with product recycling. In this design, a part of the product is mixed with the feed such that better free-flow characteristics are obtained and the average moisture content of the feed is reduced. Figure 3.26 shows the diagram of a flash drying system with product recirculation. More energy savings are obtained if the drying agent (air) is partially recirculated.

Figure 3.27 shows a flash drying system with partial air recirculation. As seen in the figure, a flow splitter is used to divert a part of the exhaust air back to the process. The recirculated air is injected just after the fresh air heater. The fresh air intake can be reduced with up to 75% if recirculation is applied, which results in very consistent energy savings. The system allows

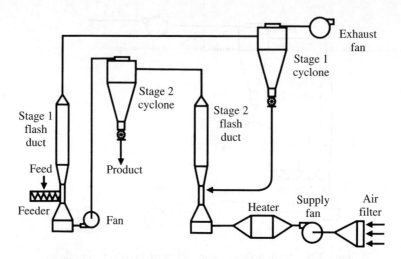

**Figure 3.28**    Countercurrent two-stage flash drying system

for the control of air humidity and oxidizing species, especially when a direct heating with combustion is applied.

The drying process with pneumatic-conveyor dryers can be made in multiple stages. In multistage systems, the dry product generated at the first stage is further dried in a subsequent stage and so on. Two-stage flash dryers are widely used systems which can be applied to many processes, for example, in paper mills, for drying cellulose pulp. Figure 3.28 shows the schematics of a two-stage flash dryer system. In this system, there is opportunity to recycle the drying agent (typically air). Basically, the fresh air is used for drying the products in the second drying stage rather than in the first stage. The input product for the second stage is in fact the output product of the first stage. This product is incompletely dried. After the second-stage flash duct, the final product is separated by the stage two cyclone. The resulting air from this separation is further used as drying agent for the first-stage flash duct.

Another possibility of multistaging is shown in Figure 3.29. Here, a flash dryer is combined with a fluid bed dryer. The flash dryer acts as a first-stage dryer. This removes all superficial moisture. In the second stage, the solids are discharged in a fluid bed system, where the residence time is longer and the bonded moisture can be better removed. Water-bonding polymers can be dried in this type of multistaged system.

Figure 3.30 shows the relative drying obtained with pneumatic-conveyor dryers for a number of products. The relative drying obtained in this equipment is generally higher than 60%. The equipment is relatively bulky with a volumetric heat transfer coefficient of $2 \, \text{kW/m}^3 \, \text{K}$, and due to a large external surface, it requires consistent thermal insulation to reduce heat losses. The commercial units exist for drying rate from few hundred kilograms per hour up to $10 \, \text{t/h}$.

### 3.3.8.3    Ring Dryer

The ring dryer is a special type of flash dryer which includes a centrifugal manifold classifier which has a series of adjustable deflector blades which allow for separation of the larger and

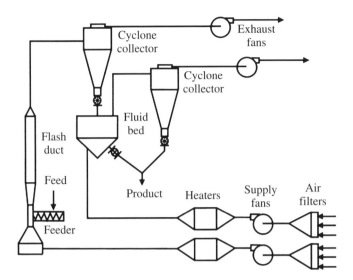

**Figure 3.29**   Flash fluid bed drying system (modified from GEA (2014))

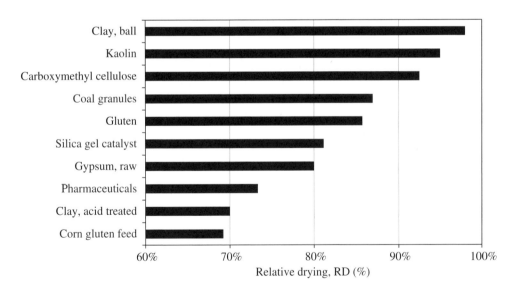

**Figure 3.30**   Relative drying of some materials in pneumatic-conveyor dryer (data from Green and Perry (2008))

heavier (therefore wetter) granules which are returned back in the drying loop for extended residence time. Furthermore, ring dryers may include a mill or disintegrator which allows for simultaneous grinding and drying.

The addition of the manifold results in lower product moisture and better product uniformity than in a simple flash dryer. A large variety of products such as silica, zeolites, and calcium

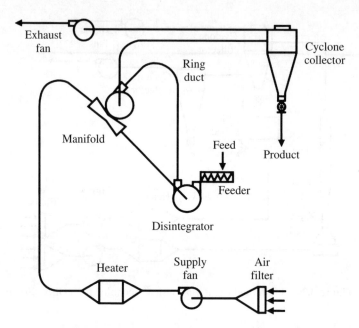

**Figure 3.31**  Schematic diagram of a ring dryer system

carbonate can be dried in ring dryers. Figure 3.31 shows a "full manifold ring dryer" of horizontal configuration. As seen in the figure, the manifold returns the coarser particles to the disintegrator from where the particles are routed toward the ring duct for extended drying time. The full manifold ring dryer is a very versatile configuration allowing for controlling of residence time in correlation with particles size.

If the moist material is heat sensitive, then a "P-type ring flash dryer" can be used, as shown in Figure 3.32. The particles are dispersed with hot air blown though a Venturi flash drying column after which they enter the manifold. Here, the particles which are not completely dried are recycled several times – for an extended residence time – until the desired level of dryness is achieved. Then, the dried particles are directed toward the cyclone collector where they separate at the bottom. Commercial ring dryers exist with drying rate from 500 kg/h to 20 t/h and prices up to $1M for the largest units; the specific evaporation rate (per unit of volume of dryer) is of the order of 10 kg/m$^3$ h.

### 3.3.8.4  Spray Dryer

One of the widely used types of dispersion dryers is the spray dryer. Figure 3.33 shows the operating principle of the countercurrent spray dryer. The moist material (in form of slurry or aqueous solutions or pastes that can be dispersed) is atomized and sprayed in countercurrent with hot drying gas. The process occurs in a tall cylindrical chamber or long horizontal chamber (a case in which screw conveyors are used to remove the dried products). The particles segregate and dry during their free fall. Evaporation occurs at a faster rate at particle surface such that the dried particles are collected at the bottom. The gas exits eventually from above and entrains

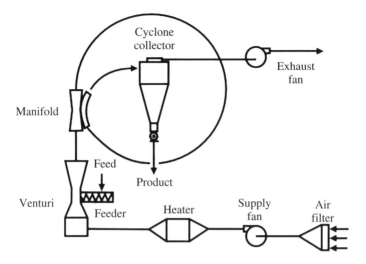

**Figure 3.32** Schematics of a P-type ring dryer system

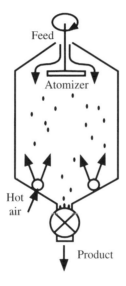

**Figure 3.33** Sketch illustrating the operation principle of spray dryer

the finest particles. These particles can be separated from the gas in cyclones, bag collectors, or wet scrubbers. In some special situation, where there is no particle adherence to the walls, the spray drying can be run in batch mode. However, in most of the situation, spray dryers have continuous operation.

Other configurations of spray dryers do exist, which use a cocurrent flow or mixed flow rather than countercurrent flow. Regardless the flow configuration, the process in spray dryers evolves in three phases, namely, atomization, gas–liquid mixing, and liquid evaporation (drying). The droplet size is in the range of few micrometers to few hundreds of micrometer, depending on the application. The shape of product particles is essentially spherical which is a peculiarity of spray drying technology. However, the particle size is not uniform as it depends on many parameters (liquid viscosity and density, solid fraction of droplet, spraying rate) and on the atomizer operating conditions.

The cost of spray dryers is relatively high, and they are not economically competitive expect for those materials which are heat sensitive, can agglomerate and stick to each other, or cannot be dewatered mechanically. Materials such as clays or other non-Newtonian fluids remain fluid at small moisture content, and therefore, they cannot be dried in other equipment except spray dryers. One of the most costly components of the spray dryer is the atomizer. Currently, two types of atomizers are used: centrifugal disks and nozzles.

Centrifugal disks operate as shown in Figure 3.22 for the dispersion dryer. They rotate at thousand revolutions per minute, while wet slurry is poured in thin sheets on their surface. The material is accelerated centrifugally and discharged at high speeds according to the disk radius which typically is in the range of 3–15 cm. The speed of the particles at the disk periphery influences the atomization: the particle size is smaller when the speed is higher. The rotating disks are active means of atomization, and therefore, they do not clog as they are continuously turned. The erosion is relatively low in centrifugal disk atomizer as opposing to the erosion rate in nozzle atomizers which is high. This is why the disk atomizers have an extended lifetime of few thousands of hours of operation. These atomizers offer a uniform distribution of droplet sizes and are able to produce coarse particles.

Nozzle atomizers can be divided in two subtypes: (i) two-fluid nozzles and (ii) high-pressure nozzles. The two-fluid nozzles use air or steam as dispersing agent. The slurry or liquid subjected to drying is entrained by a higher-pressure gas (the dispersion agent) when it passes through the nozzle orifice. Various hard alloys are used for nozzle construction to prevent the erosion, for example, carbides of tungsten or silicon. The high-pressure nozzles use a high-pressure pump to force the slurry/liquid through a small orifice with submillimeter diameter. Due to small diameter, relatively large pressure drop is observed in flash dryers.

## 3.3.9 Fluidized Bed Dryers

Fluidized bed processors are widely used in many applications involving gas–solid heat and mass transfer and chemical reaction. The types of solids handled in fluidized bed systems are that of granular materials, slurries, pastes, and suspensions. Fluidized bed dryers are characterized by a longer residence time and good solid mixing as compared with typical dispersion dryers. This is why in fluidized bed dryers the moisture from particle core can be extracted. In fact, fluidized bed dryers can be combined or integrated with dispersion dryers (e.g., spray dryers) for a better drying process. In this integration, the first stage is dispersion drying in which the superficial moisture is removed, and the second stage is fluid bed drying in which the interior moisture is removed.

The particle sizes applicable to fluidized bed dryer are in a large range, namely, 0.05–2 mm. The particles are fed in a fluidized bed dryer consisting of a large vessel (usually cylindrical) in which a gas is passed from the bottom through the bed of particles. The gas is distributed by a special "distributor plate" which impedes particles to fall down into the air distribution ducts.

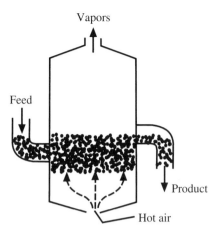

**Figure 3.34** Sketch illustrating the operation principle of well-mixed continuous-flow fluidized bed dryer

The gas must have a higher speed than that required for the minimum fluidization when the weight of the bed is totally supported by the pressure forces exerted by the gas.

Figure 3.34 shows the sketch of a continuous-flow, well-mixed fluidized bed dryer. The particles are fed continuously by a feeding mechanism and once they enter in the drying chamber are fluidized. They reside in the bed for long periods – as long as 60 min – depending on the operating parameters. The dried products are extracted from the other side of the equipment, whereas the gas carrying water vapors exits from above, as shown in the figure. The system operation can be continuous, semicontinuous, or in batch mode. Figure 3.35 shows a tridimensional view of a fluidized bed dryer.

A gas cleaning system is included in fluidized bed drying systems. This can be either a cyclone, or a bag filter, or a scrubber which separate particles from the gas. Partially, recycling of the cleaned gas can be applied, which will save energy. The overall system requires also an air handling unit or a blower–heater system which prepares the hot/dry normally used as drying agent. The fluidized bed itself must be in a fluidization regime of either "smooth" or "bubbling fluidization." The height of the reactor is correlated with the height of the bed such that it facilitates the separation of coarse particles which cannot be pneumatically entrained. The room is high enough so that most of the particles carried by the gas can fall back in the fluidized bed and in this way the particle entrainment is minimized.

Various types and configurations of fluidized bed dryers exist beside the most used type – the well-mixed fluidized dryer – shown in Figure 3.34. There are plug-flow fluidized dryers which compound multiple fluid beds separated by vertical baffles. The particles eventually advance horizontally from one bed to the other as drying progresses. Spray fluidized dryer is an example of a hybrid system which integrates a spray dryer and a fluidized bed dryer. The spraying system generates a dispersed particle flow which in fact falls on the top of a fluid bed where drying continues for an extended time. Vibrating and agitated fluidized bed dryers were mentioned previously: in these systems, the fluidization is assisted by mechanical vibration and/or agitation. Recirculated fluidized bed dryer is also a common type among the fluidized bed systems.

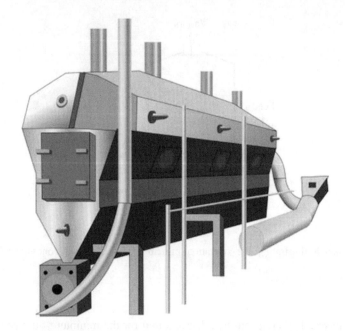

**Figure 3.35**  Sketch showing the system construction of fluid bed dryer

In this system, a draft tube is used which realizes a flow pattern of particles which flow up and then come downward in a downcomer zone.

## 3.3.10   Drum Dryers

There is a very wide application of drum dryers in food industry and pharmaceuticals to dry viscous slurry and pastes of various materials. This is an indirect drying system where heat transfer mechanism is essentially through thermal conduction. The drum dryers are very versatile being able to operate under atmospheric, pressurized, and vacuum conditions.

Figure 3.36 shows the sketch illustrating the operating principle of a drum dryer having a single drum. The moist material is spread (using various mechanisms) over the surface of a (steam) heated drum. Once the slurry material is applied, it soon solidifies and forms a quasiuniform sheet adherent to the drum surface. The drum rotates and the sheet advances while drying. When drying is complete, the sheet is scraped off.

There are two types of drum system configurations used for drying of liquids/slurry materials: single drum and double drum. The single-drum system is that shown in Figure 3.36a. For the double drum, there are two identical drums that rotate in opposing directions, and a gap exists at the nip between them. In double-drum dryer, the liquid material is poured (spread) at the nip where eventually boils and solidifies and forms two sheets, each of the sheet being adherent to a drum. The sheets are relatively thin and the heat transfer by conduction intense; therefore, the drying process evolves at high rate. In pharmaceuticals and food processing, the drying is often accompanied by other structural transformations of the materials, including precooking or cooking, gelatinization, and so on.

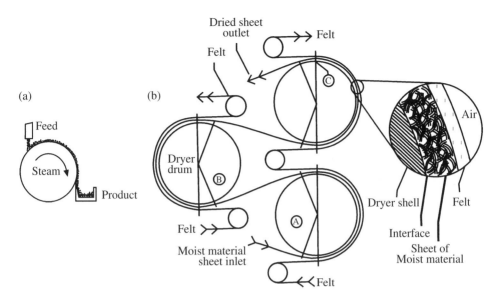

**Figure 3.36** Sketch showing the operation principle of drum dryers: (a) single-drum system and (b) sheets drying system

Drum dryers can be applied to dry sheet material such as paper. Figure 3.36b shows a drum system for sheet material. In this system, which is typically used for paper drying, felt sheet is used to enhance the mass transfer and moisture absorption. The moist material is turned over the drum A, where it is covered by a dry felt sheet. After turning more than 180°, the material passes to a second drum where the next drying stage takes place. Here, the material is covered by a fresh dried felt sheet which absorbs moisture. Similarly, the material passes over the third drying drum C and is eventually discharged as a dried sheet. In the zoom from Figure 3.36b, the drying configuration can be observed. Here, at the interface between the drum and the sheet material, a heat conduction process occurs which provides heat to the material to enhance the pressure of the water vapors. The felt sheet absorbs the vapors which pass through its porous tissue. Further, the moisture is absorbed by the surrounding air.

The drum diameter in practical drum drying systems goes from 0.4 m to approximately 1.5 m with lengths of up to 3 m. Regarding the slurry/liquid products, there are several mechanisms that spread them on the surface of the drums, as follows:

- *Pendulum Nozzle* – in this system, a traveling nozzle system is used which spreads the slurry at the nip between two drum or atop of a single-drum system.
- *Multiple Nozzle Header* – which contains several nozzles along a header.
- Applicator *Rollers* – which are smaller diameter rollers on which the feed is poured and adheres to their surface, and it is carried out to the nip between the roller and drum.
- *Dip* Applicator *Roller* – these are rollers immersed in a tray with feed material in slurry phase (or paste); as the roller rotates, it spreads a layer of moist material at the nip between the roller and the drum.

- *Feed Tray* – in this system, the drum itself is slightly immersed in a tray from where wet product in the form of paste or slurry adheres and forms a thin layer sheet on the drum surface.
- *Splasher* – this is a wheel with cups or fins that turns as it is immersed partially in a tray containing wet feed; as it turns, the splasher throws the material on the surface of the drum from where a layer is formed which adheres to the surface.
- *Sprayer* – this system is similar with the splasher with the only difference that a shower is throwing the slurry over the drum surface. A tray with slurry is placed underneath the drum, which supplies the spray system with material; the tray is placed such that any material which does not adhere sufficiently to the drum falls back and can be recirculated.

The drum(s) turns with angular speed inferior to approximately 30 rpm. The thickness of the formed sheets is in general below 1 mm and so is the space between drums or between drums and rollers. According to Daud (2006), the area-specific evaporation rate generated by industrial drum dryers is in the range of 10–50 kg/m² h. The scraping of the dried sheet of the product is made with a help of a blade placed downstream from the feed point. Once scrapped, the product forms flakes or powders which fall in a trough from where it is transported with adequate means for further processing.

## 3.3.11   Solar Drying Systems

There are many types of solar drying systems which in general use greenhouse effect in enclosures. Simple cabinets are constructed with a tilted glass windows exposed to solar radiation. Figure 3.37 shows a sectional view of a cabinet dryer. The products are placed on a perforated tray.

The cabinet has a number of holes to allow for natural ventilation and refreshing the air (which carries the product moisture). Air is ventilated naturally through the products and heat transfer is by natural convection. An alternative design to the solar cabinet dryer is the staircase solar dryer. Figure 3.38 shows the design of a staircase solar dryer. In this system, air flows through successive trays with wet product where drying occurs. A glass of a transparent polycarbon material is used to entrap as much as possible the solar radiation. The back side is thermally insulated to impede heat losses.

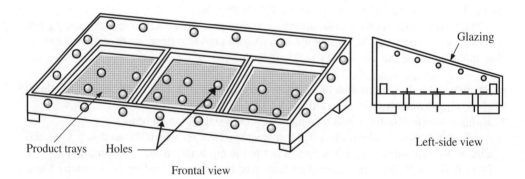

**Figure 3.37**   Solar cabinet dryer

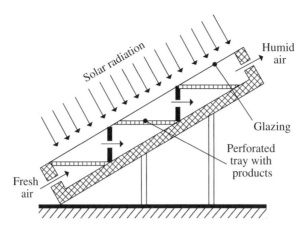

**Figure 3.38**   Staircase-type solar dryer

More advanced solar dryer concepts involve the use of a solar air heater in conjunction with a drying chamber. The solar air heater prepares the warm/hot air which is then fed into the drying chamber. The drying chamber itself can be exposed or not to solar radiation, depending on the type of design. Goyal and Tiwari (1997) proposed the concept of reverse flat collector solar dryer as shown in Figure 3.39 which integrates a solar air heater with a drying chamber. This system uses nonconcentrated optics to collect and divert the sunrays in a vertically upward direction below the dryer where an absorber plate is placed. The absorbed plate becomes hot under the exposure to solar radiation. Air moves by natural convection in a horizontal direction along the plate and warms up. Further, the warmed air flows upward through the permeable bed of moist material where drying occurs; eventually, the humidified air leaves the drying chamber from above.

Figure 3.40 shows the schematic diagram of a solar tunnel dryer which comprises an air collector and a drying chamber connected in series. In the simplest arrangements, these tunnels can operate in natural convection. In more complicated configurations, air can be flown using active elements such as fans and blowers. In some arrangements, cross-flows with or without recirculation are used. The benefit of forced circulation of drying agent is due to the ability of drying rate enhancement and therefore reducing the drying time. Figure 3.41 shows a solar chimney dryer which is a design version of the solar tunnel dryer. The chimney has the role to enhance the natural convection process. The solar collector of large scale placed horizontally heats air which then flows over the products placed on trays and further flows upward through the chimney.

A smaller-scale version of the solar chimney dryer is the solar through dryer with separate air heater which is a typical design as shown in Figure 3.42. This system uses natural convection. The material is arranged on perforated trays which allow for air passage or permeation. Air is heated in a tilted solar collector facing the equator. After passing through the permeable bed or product trays (whichever is used), the moist air leaves the drying chamber as it is naturally drafted into the chimney. Since the flow is driven by natural convection, the pressure drop across the system must be relatively small. Therefore, the bed must be sufficiently thin and its permeability must be high or the product load must be low. For thicker product beds, fans or blowers must be added to the system to create a forced convection.

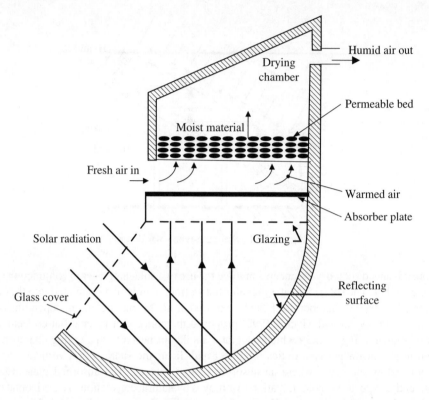

**Figure 3.39** Schematic diagram of the reverse absorber cabinet dryer (modified from Goyal and Tiwari (1997))

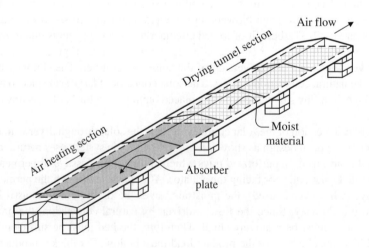

**Figure 3.40**   Solar tunnel dryer

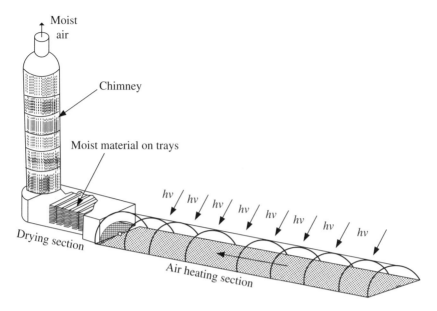

**Figure 3.41**  Solar chimney dryer

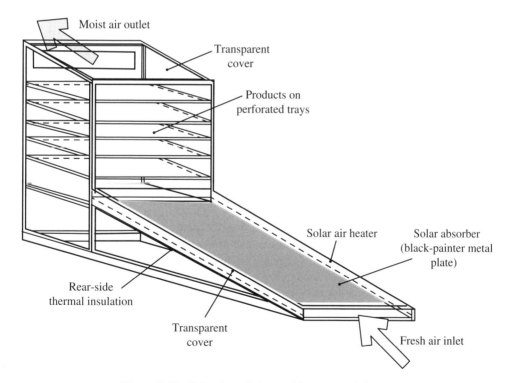

**Figure 3.42**  Solar through dryer with separate air heater

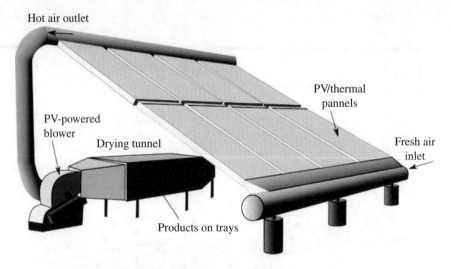

**Figure 3.43**   PV/T-assisted solar drying tunnel

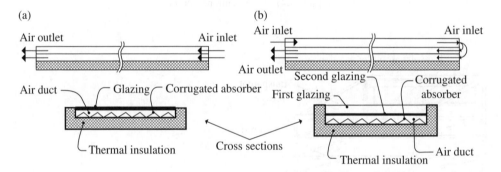

**Figure 3.44**   Solar air heaters with corrugated absorber: (a) single glazing and (b) double glazing

A solar dryer with forced convection is shown in Figure 3.43. In this system, the air heater is a PV/thermal system (PV/T). Here, the PV arrays are used to generate electricity which is used to power the blowers and fans which are used both for air supply into the drying tunnel and for extraction of humid air out of it. The air heater is in form of a channel duct just below the PV arrays. Inside the tunnel, the product is disposed on multiple trays and air flows above the products before being eventually extracted by the exhaust fan. Note that when a solar drying system comprises devices such as blowers, fans, conveyors, agitators, and so on, then the common term used for this system is of active solar dryer. When drying is done by natural convection, only then the solar dryer is of passive type.

Figure 3.44 shows two types of solar air heaters with corrugated absorber that can be integrated with solar dryers. The air heater from Figure 3.44a is a single-glazing collector. Air circulates beneath and above the corrugated absorber plate as shown in the figure. In order to increase the heat gain by the air, the solar collector can be elongated such that sufficient area of heat transfer and light absorption exists. Figure 3.44b shows a double-glazing solar air heater. In this system, the intake air flows first in the space between the two glasses in order to preheat. Thereafter, the preheated air flows though the space below the second glazing where the corrugated absorber is placed. The soar collector for air heating must be well insulated at the sides and underneath.

## 3.4 Processes in Drying Systems

In this section, the main processes in drying systems are reviewed according to the type of the equipment. In this respect, the drying systems are grouped in two categories, namely, natural drying, forced drying. Each of these types of dryer category has its specific processes, and therefore, for each of the case, some specific parameters are relevant for the design and analysis.

### 3.4.1 Natural Drying

Natural drying is one of the most basic processes which is generally conducted by exposing materials to atmospheric air, to wind, or to solar radiation. Some well-known examples of natural drying are laundry drying, foodstuff drying on shelves or cribs, grass, crop and grain drying, or drying of paint, bricks, concrete, and other construction materials. Natural drying does not involve high costs (it may not need any equipment); therefore, it is a very convenient method to dry biomasses that eventually are used as fuels. Many types of biomasses and farming products are dried using natural drying. For example, peat moisture can be reduced at half with natural drying (Wimmerstedt, 2006). Natural drying of tree logs stored in piles can reduce the humidity content below 35% which is the threshold required for combustion (Filbakk et al., 2011).

Natural drying essentially depends on climacteric conditions, season, position, staking and storage methods of the moist material, and the type and properties of the moist material (its humidity, its surface area and surface properties, its porosity). In natural drying, the ventilation is in general by natural draft or by wind pressure. However, many crop dryers that use natural air are equipped with fans to enhance the ventilation and drying rate. Exposure of the materials to solar radiation and to winds is desirable in natural drying processes because it may reduce the drying time.

In this natural drying, the "natural" air is not heated prior to its contact with the wet products, but the products can be exposed to solar radiation which facilitates drying due to the established thermal gradients. No mechanical ventilation is applied, but sometimes, the moist material can be stirred or turned to facilitate the process. Natural drying is slow (sometimes of the order of few weeks for crops and tree logs) but in many cases is economically advantageous because it does not involve costs related to energy consumption, and furthermore, in fact, in many situations (e.g., crop drying), it stores the products in the drying facility for sufficient period to concord better with the market demand. One important advantage of natural drying is due to the fact that temperature and concentration gradients are small; this allows for a smooth process which does not deteriorate the products and not induce mechanical stresses; therefore, despite a longer drying time, in many applications, better products are obtained using natural drying. However, natural drying is highly dependent on the method of stacking which frequently leads to a not uniform drying of the batches. It also depends highly on the climacteric conditions and humidity of natural air.

The natural drying can be conducted generally in three types of configurations: (i) thin horizontal layers with natural convection from above, (ii) vertical sheets exposed to natural convection at both sides, and (iii) packed porous beds with through flow of natural air ventilated mechanically or by natural draft. Regardless the actual configuration, the main modeling assumptions are given as follows:

- Airflow can be assumed one-dimensional.
- The change in thermophysical properties of air is negligible.
- The heat transfer occurs only between the wet product and the air stream.
- The material is uniformly distributed and a bulk density can be determined.

According to the general drying process with natural air shown schematically in Figure 3.45, the mass balance equation stating that the rate of moisture lost by the wet product is equal to the humidity gain of the natural air stream is written as follows:

$$\rho_p \frac{dW}{dt} = -G \frac{d\omega}{dl} \tag{3.2}$$

where $W$ is the moisture content on dry basis of the product and $\rho_p$ is the bulk density of the product.

Also in Eq. (3.2), the mass velocity of air (in kilogram per square meter and second) is denoted with $\dot{m}_a''$ and the humidity ratio of air is $\omega$; as shown in the figure, $l$ is the longitudinal coordinate along the air stream. The energy balance equation states that the heat lost by the product due to moisture elimination and temperature change must equal the heat gained by air due to air temperature change and humidification by water evaporation. Mathematically, this is written as follows:

$$\rho_p \left( C_p \frac{dT_p}{dt} + C_w \left( W \frac{dT_p}{dt} + T_p \frac{dW}{dt} \right) \right) = -\dot{m}_a'' \left( C_a \frac{dT_a}{dl} + C_w \left( \omega \frac{dT_a}{dt} + T_a \frac{d\omega}{dt} \right) + \Delta h_{lv} \frac{d\omega}{dt} \right)$$

$$\tag{3.3}$$

where $T_p$ is the product temperature, $T_a$ is the air temperature, $C_p$ is the specific heat of the product, $C_a$ is the specific heat of dry air, $C_w$ is the specific heat of water, and $\Delta h_{lv}$ is the latent heat of water evaporation.

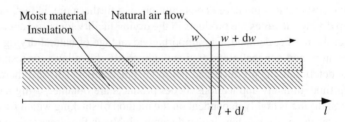

**Figure 3.45**   Schematic illustration for modeling natural drying process

In Eqs. (3.2) and (3.3), the rate of drying is related to the equilibrium moisture content according to the drying rate constant $k$, namely:

$$\frac{dW}{dt} = -k(W - W_e) \tag{3.4}$$

where $X_e$ is the equilibrium moisture content.

In order to solve Eqs. (3.2)–(3.4), correlations are necessary for the drying constant and the equilibrium moisture content. These types of correlations are empirical; they are based on experimental data and are valid for defined conditions. Hossain et al. (2003) developed such correlations for natural drying of maize placed on horizontal cribs. When the products are subjected to natural drying, then the moisture content decreased from 38% (db) to 16% (db). The correlations recommended for the equilibrium moisture and drying rate constant are from Ameobi (1992):

$$\begin{cases} |W_e = 7.7819 - 8.4938 \ln(1 - \varphi) \\ |k = (1.5954E10)\, m_{10}^{-1.304} \exp\left(\dfrac{-7334.643}{T_a}\right) \end{cases} \tag{3.5}$$

where $\varphi$ is the relative humidity of air and $m_{10}$ is the average mass of 10 maize cobs.

In many situations, natural drying occurs under the outside wind conditions; the solar irradiance, air humidity, and rain influence the process when the products are exposed to the atmosphere. This is the case of grass drying or of willow stems dried by natural drying. For example, in the case of grass drying, this is first cut and collected in homogeneous rows and spread uniformly over a perimeter. The heat input necessary to evaporate water is due to solar radiation mainly and partially is due to the natural convection heat transfer caused by the winds. As discussed in Atzema (1992) after a period of intense drying, a layer of approximately 5 cm is formed with dried grass. Then, the grass is mixed thoroughly and a second phase of drying follows. As it is the case with many agricultural products, both the free water and absorbed water must be removed by drying.

The drying rate of products exposed to natural drying due to combined action of wind, solar radiation, and evaporation can be determined with the help of evaporation rate. The evaporation rate per unit of surface (given in kilogram of water per second and square meter) depends on the aerodynamic drag due to wind speed and height of the moist material layer. This is determined based on the wind speed (taken in general at 10 m height from the field), the von Kármán constant (0.41), and other geometrical parameters. Based on the considerations from Atzema (1992), the following semiempirical equation can be derived for the area-specific evaporation rate:

$$\dot{m}_{ev}'' = \frac{\left(I_g - \dot{Q}_{loss}''\right)\dfrac{\Delta h_{lv}}{T(v'' - v')} + 0.41^2 V_w (\rho C)_a \dfrac{P_s'' - P_a''}{\left[\ln\left(\dfrac{10 - 0.63\,z}{0.13\,z}\right)\right]^2}}{\dfrac{\Delta h_{lv}}{T(v'' - v')} + \dfrac{C_a P}{0.622\,\Delta h_{lv}}\left\{1 + \dfrac{0.41^2 V_w\, r_{wm}}{\left[\ln\left(\dfrac{10 - 0.63\,z}{0.13\,z}\right)\right]^2}\right\}} \tag{3.6}$$

where $I_g$ is the global radiation incident on earth, $\dot{Q}''_{loss}$ is the rate of heat losses from the products, $P$ is the atmospheric pressure, $P''_s$ is the partial pressure of water vapor at the material surface, $P''_a$ is the partial pressure of water vapor in bulk air, $v''$ and $v'$ are the specific volume of saturated water vapor and saturated liquid water at air temperature $T$ at product surface, $z$ is the thickness of product layer subjected to natural drying, $\rho$ is the air density, and $V_w$ is the wind velocity at 10 m height above the ground.

In Eq. (3.6), the parameter $r_{wm}$ represents an aerodynamic resistance of the moist material expressed in (s/m). The value of $r_{wm}$ is correlated in Atzema (1992) for the case of natural drying of grass with the product moisture content and with the product load per unit of surface (expressed in kg of product per square meter). For small product load of the order of 1 kg/m², the average value of $r_{wm}$ is 250 s/m. For larger loads of 2.5 kg/m², this becomes 50 s/m. The moister is the product, the lower is the value of the $r_{wm}$ parameter.

Another superficial process through which water is lost is transpiration; this applies only to plants and is relevant for natural drying. During transpiration, moisture from the plant tissue migrates to the surface. The green leaves have stomata spread underneath through which water reaches the surface. Other vegetal products as well as meats have transpiration through their skin. During long-term crop drying in natural air, water condensation can occur. This is a mechanism through which the products gain temporarily humidity. In addition, precipitation may occur when crops are dried uncovered; this will temporarily lead to an increase of moisture content.

If the mass transfer at product surface is significant (e.g., due to evaporation), then the boundary condition for mass transfer at the product–air interface can be written as follows:

$$-D_e \rho_m \frac{\partial W}{\partial \delta} = k_m (C'' - C_\infty) \tag{3.7}$$

where $\rho_m$ is the density of the material, $\delta$ is the coordinate along the boundary air thickness, $D$ is the effective diffusivity, $k_m$ is the mass transfer coefficient, $C''$ is the water vapor concentration assumed saturated at the product surface, $C_\infty$ is the concentration of the water vapor in air, and $D_e$ is the effective diffusivity which is defined according to the Arrhenius equation:

$$D_e = D_0 \exp\left(-\frac{T_{act}}{T}\right) \tag{3.8}$$

The mass transfer coefficient results from Sherwood ($Sh$) number as follows:

$$k_m = Sh \frac{D_{AB}}{d} \tag{3.9}$$

where $d$ is the characteristic length (or diameter) of the product and $D_{AB}$ is the water diffusivity in air which is determined from Eq. (3.8) where $D_0 = 1.47E{-}4\,\text{m}^2/\text{s}$ and $T_{act} = 523.78\,\text{K}$. The Sherwood number is correlated with the Grashof ($Gr$) and Schmidt ($Sc$) numbers as follows:

$$Sh = a(Gr\,Sc)^n \tag{3.10}$$

where

$$Gr = \frac{g\rho_\infty (|\rho_{int} - \rho_\infty|)\Delta h_{lv}}{\mu^2}; \; Sc = \frac{\mu}{\rho_\infty D_{AB}}$$

where $\rho_\infty$ is the density of the drying air, $\rho_{int}$ is the air density at the interface, $\mu$ is the dynamic viscosity of drying air, and $\Delta h_{lv}$ is the latent heat of water evaporation.

In the Sherwood number correlation (3.10), the constants $a, n$ can be determined based on empirical methods. These constants take in account the actual geometrical configuration and relevant operating parameters. Clemente et al. (2007) determined $a = 1.36$, $n = 0.001$ for natural convection drying at low temperatures of salted meat. If the product is exposed to wind, then the Sherwood number correlation depends on Reynolds number instead of Grashof number. This is the case of natural drying of crops in piles or of wood logs or willow steams, and so on. As given in Gigler et al. (2000), the Sherwood number correlation at natural drying of willows in piles, exposed to winds, is as follows:

$$Sh = a(\mathrm{AR})^{-0.33}(Re\,Sc)^n \tag{3.11}$$

where $a = 1.86$ and $n = 0.33$ and $\mathrm{AR} = L/d$ is the aspect ratio given in terms of length divided by diameter. Here, the Reynolds number is given with respect to the stem diameter $d$ and air velocity within the pile. According to Gigler et al. (2000), the air velocity within the pile of stem can be assumed of the order of 0.1 m/s.

Natural drying of grain, rice, or other agricultural products is often made on concrete floors under direct exposure to atmospheric air, precipitation, wind, and solar radiation. Figure 3.46 shows a sketch illustrating how the granular products are laid on concrete slabs. Laying the grains on concrete slabs represents a good solution to avoid contamination by germs and other living systems which may deteriorate the product.

Also, the concrete acts as a solar heat absorber which helps the drying process. Another option to expose the farming products to natural drying is shown in Figure 3.47. Here, the

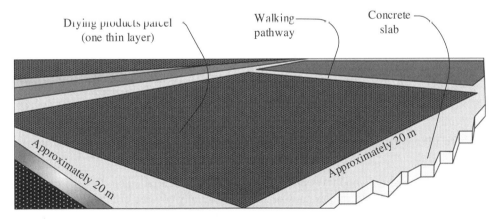

**Figure 3.46**   Natural drying of agricultural products (grains, rice) in open air on concrete slabs

**Figure 3.47** Natural drying of fruits in trays directly exposed to solar radiation (Good Humus Farm (2014))

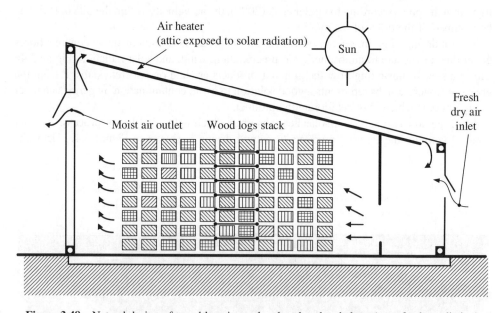

**Figure 3.48** Natural drying of wood logs in stacks placed under shelters (use of solar radiation)

products are placed on wood trays on a field with intense solar radiation. Depending on the climate, some products are traditionally dried under partially covered or totally covered shelters, particularly in villages. Here, in Figure 3.48, the typical arrangement for natural drying of logs in shelter is shown.

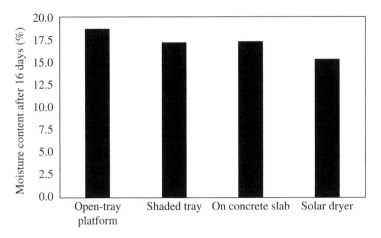

**Figure 3.49** Comparison of four natural drying technologies for cassava (data from Olufayo and Ogunkunle (1996))

The shelter has a tilted roof heated by solar radiation. A double was installed as attic which has the role of heating the air while forcing the flow downward. Recirculated warmed air and dry fresh air are mixed at one side and enter the drying zone through a lower opening. Due to buoyancy, the air flows upward.

A part of the moist, air flows toward outside and another part is recirculated. Natural drying of logs covered by shelters decreases the final moisture content with approximately 10% after about 1 year of drying. Some species such as alder or birch dry better with approximately 30% moisture reduction, whereas birch reduces its moisture with up to 20% from its initial value.

A comparison of natural drying on concrete slab (Figure 3.46), on open trays (Figure 3.47), on shaded trays/piles (Figure 3.48) and using solar dryer with natural convection (Figure 3.45), is presented by Olufayo and Ogunkunle (1996) as shown in Figure 3.49. It appears that drying on concrete slabs is more effective than drying on open trays, although more initial investment is required for construction of the concrete slab. Shaded tray technique is also reasonably effective and offers better protection to precipitation and animals/birds. The best technique looks to be the solar drying which maximizes the drying rate, although the initial investment is the highest.

In Figure 3.50, natural drying of several types of products in solar dryer is compared. The geometrical shape, the size, and the thermophysical properties of the products affect the process. As seen, products such as sultana grapes – which have a good protective skin – dry slowly than other type of vegetables such as peppers and beans.

As mentioned previously, many types of construction materials require natural drying (pains, concrete, etc.). Here, an example is given for drying of concrete, as shown in Figure 3.51. The drying time varies between 40 and 150 days depending on the concrete type and depending on the thickness. The diffusion coefficient of moisture through cement pores is highly nonlinear with respect to the relative humidity; however, in a low relative humidity, the diffusion coefficient remains constant.

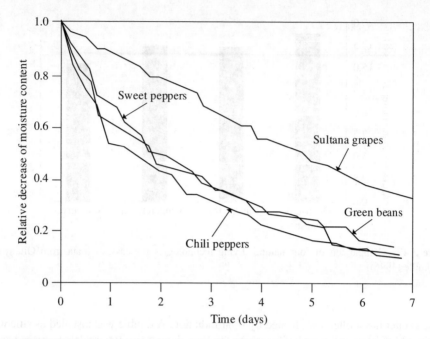

**Figure 3.50**  Relative decrease of moisture content (with respect to initial value) at natural drying of some fruits and vegetables (data from Tiris et al. (1994))

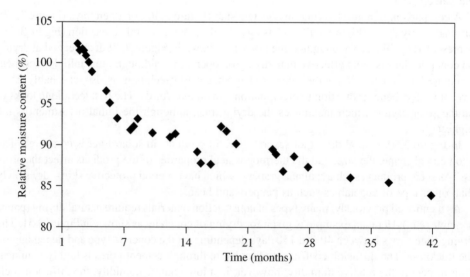

**Figure 3.51**  Drying of concrete (data from Kofoed (2006))

## 3.4.2   Forced Drying

Forced convection is extensively used in drying system; this requires the use of active flow system such as blowers and fans. Many of the systems discussed in Section 3.3, including tunnel dryers, batch tray dryers, or some of the solar dryers, use forced convection which increase the drying rate on the expense of some low power input required by the blowers or fans; however, the initial investment is much higher with respect to natural drying systems.

The forced convection drying can be modeled based on heat and mass transfer equations written at the interface between the product surface and the drying gas. The heat flux per unit of heat transfer surface ($\dot{Q}_h''$) and the mass flux ($J_m$) at the interface influence the process. The following equation describes the heat transfer process:

$$\left\{ (C_p\rho)_a - \Delta h_{lv} l \left( \rho_s + W \frac{\partial \rho_s}{\partial W} \right) \xi_T \right\} \frac{\partial T}{\partial t} - \Delta h_{lv} l \left( \rho_s + W \frac{\partial \rho_s}{\partial w} \right) \xi_M \frac{\partial \mu_m}{\partial t} = -\operatorname{div}\left( \dot{Q}_h'' \right) \quad (3.12)$$

whereas the mass transfer process is described by the following equation:

$$\rho_s W \frac{\partial \rho_s}{\partial W} \left( \xi_M \frac{\partial \mu_m}{\partial t} + \xi_T \frac{\partial T}{\partial t} \right) = \operatorname{div}(J_m) \quad (3.13)$$

Here, $l$ is the characteristic length of the product, $\xi_T = (\partial W / \partial T)_{\mu_m}$ is a specific temperature coefficient of the product measured in kilogram of moisture per kilogram of dry solids and Kelvin, $\mu_m$ is the chemical potential for mass transfer, and $\xi_M = (\partial W / \partial \mu_m)_T$ is defined as in Akiyama et al. (1997) as the specific mass capacity. The chemical potential is defined with respect to a reference chemical potential ($\mu^0$) as follows:

$$\mu = \begin{cases} \left| \mu^0 + RT \ln a_w, & \text{for } W < W_e \\ \left| \mu^0 \dfrac{W - W_e}{1 + \left( W_e - W_e^0 \right)}, & \text{for } W \geq W_e \end{cases} \quad (3.14)$$

where $a_w$ is the water activity, $W_e$ is the equilibrium moisture content for 100% relative humidity, and $W_e^0$ is the equilibrium moisture content for 0% relative humidity.

The results of the heat and mass transfer analysis can be represented by the temperature variation in time as well as by the moisture content variation in time. An exemplary result of forced convection drying modeling through heat and mass transfer is shown in Figure 3.52.

The forced convective heat and moisture coefficients are calculated from the average surface temperature and the drying rate during the constant rate period of convection drying. Experimental results from loss of moisture and surface temperature indicate that calculated heat transfer coefficients are twice of systems with only heat transfer. The coupled heat and mass transfer results are correlated in terms of the dimensionless Nusselt and Sherwood numbers. For porous cylinders subjected to forced convection drying, the $Nu$ and $Sh$ numbers can be calculated as follows:

$$\begin{cases} |Nu = 0.056 \, Re^{0.65} Gu^{-0.43} \\ |Sh = 0.047 \, Re^{0.67} Gu^{-0.42} \end{cases} \quad (3.15)$$

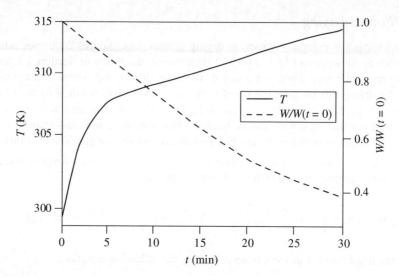

**Figure 3.52** Temporal variation of central temperature and moisture content hydrated high amylose starch powders (data from Akiyama et al. (1996))

where $Gu$ is the Gukhman number, which represents a dimensionless temperature defined as the ratio between $\Delta T = T_m < T_g$ and the absolute temperature of the drying gas $(T_g)$ in contact with the moist surface at temperature $T_m < T_g$.

In many forced convection drying system, heat pumps are used. Heat pump dryers are considered state-of-the-art drying equipment in terms of energy efficiency and the high quality of dried food products produced. Unlike conventional dryers, they allow close control of air velocity, temperature, and humidity in the dryer. In their study, they also presented drying data from two different scales of heat pump dryer.

An air-loop heat-pump dryer is shown in Figure 3.53. This system combines a refrigeration unit with electrical heaters to achieve a temperature and humidity control of the hot air directed toward the drying chamber. The system operates in batch mode and has a low capacity of production. Figure 3.54 shows a forced air drying system with heat pump applied to a batch tray dryer of larger capacity. The humidity and temperature control method using heat pump-based air handling unit is illustrated with the help of the diagram shown in Figure 3.55. In these systems, a cooling section is used first. As shown for the example in the figure, the air temperature is first reduced to 15 °C along a constant relative humidity process. However, during this process, the humidity ratio of the air decreases. Further, the air is heated to 45 °C and thence its relative humidity decreases substantially to 18%.

Similar systems can be applied for drying air or other agents at higher temperatures. Figure 3.56 shows a system implementation where an advanced heat recovery option is implemented. The fresh air is preheated with the rejected air. A controlled flow of humid air is subjected to dehumidification. The condenser unit is supplemented by an additional heat exchanger for air heating.

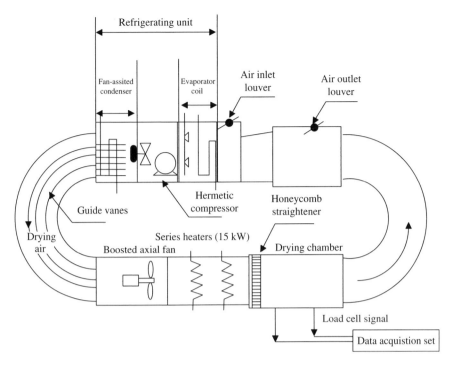

**Figure 3.53** Forced drying with an air-loop heat-pump dryer

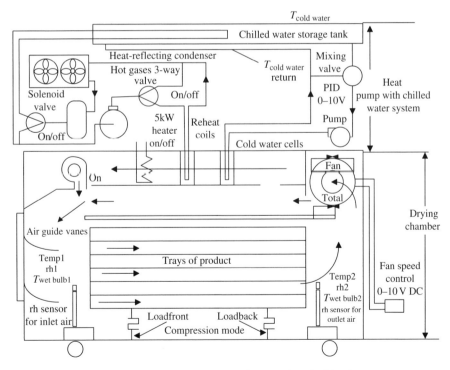

**Figure 3.54** Forced drying with a batch tray dryer equipped with a heat pump system

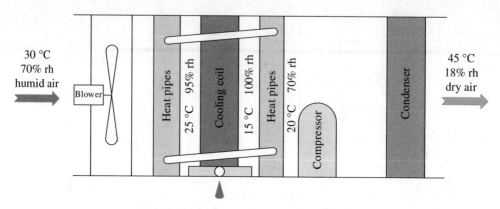

**Figure 3.55**  Temperature and humidity control with heat pump-based air handling units

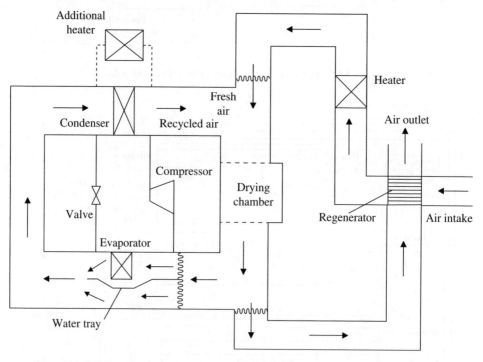

**Figure 3.56**  Forced drying system with heat pump and advanced heat recovery option

## Example 3.1

In a parallel current tunnel dryer (Figure 3.8), a granular material is dried on trucks. The mass specific surface area of the material powder in contact with the air is $A = 0.06 \, \text{m}^2/\text{kg}$, and the dying rate is $k = 140 \, \text{kg/m}^2 \, \text{h}$. The rate of moist material that enters into the tunnel is

$\dot{m}_1 = 3000\,\text{kg/h}$. The initial moisture content of the material is $W_1 = 45\%$, the critical moisture content is $W_c = 20\%$, and the equilibrium moisture content is $W_e = 1\%$. The material must be dried to a moisture content of $W_2 = 4\%$, at tunnel exit, when the material temperature is given, $T_m = 40\,°C$. The drying agent is hot air with the inlet temperature and humidity ratio of $T_1 = 130\,°C$ and $x_1 = 7\,\text{g/kg}$, respectively. The air temperature at exit is $T_2 = 40\,°C$. Determine the required flow rate of dry air and the drying time knowing that the constant rate drying period is twice larger than the variable rate drying time.

- Since the moisture content is given in kilogram water per kilogram of dry material, the rate of dry material passing through the tunnel can be calculated based on the given inlet conditions:

$$\dot{m}_{\text{dry}} = \frac{\dot{m}_1}{1 + W_1} = \frac{3000}{1 + 0.41} = 2069\,\frac{\text{kg}_{\text{dry}}}{\text{h}}$$

- The flow rate of water which is carried out by the dried product at exit can be further determined based on the given outlet moisture content:

$$\dot{m}_{w,2} = \dot{m}_{\text{dry}}\,W_2 = 2069 \times 0.04 = 82.76\,\frac{\text{kg}_w}{\text{h}}$$

- The mass flow rate of water absorbed by the drying agent must thence be

$$\dot{m}_w = \dot{m}_{\text{dry}}\,W_1 - \dot{m}_{w,2} = 2069 \times 0.45 - 82.76 = 848.29\,\frac{\text{kg}_w}{\text{h}}$$

- The rate of moisture carried by the powder material at the moment when the critical conditions are reached

$$\dot{m}_{w,c} = \dot{m}_{\text{dry}}\,W_c = 2069 \times 0.2 = 413.8\,\frac{\text{kg}_w}{\text{h}}$$

- The rate of water removed in the first drying period

$$\dot{m}_{w,\text{I}} = \dot{m}_{\text{dry}}\,W_1 - \dot{m}_{w,c} = 517.25\,\frac{\text{kg}_w}{\text{h}}$$

- The rate of water removed in the second drying period

$$\dot{m}_{w,\text{II}} = \dot{m}_w - \dot{m}_{w,\text{I}} = 848.29 - 517.25 = 331.04\,\frac{\text{kg}}{\text{h}}$$

- Assume that the psychometric process during the air drying is isenthalpic; the enthalpy change of air is nil, namely, $0 = C_{p,a}(T_1 - T_2) + \Delta h_{lv}(x_1 - x_2) + C_{p,v}(T_1 x_1 - T_2 x_2)$; taking the dry air specific heat $C_{p,a} = 1\,\text{kJ/kg\,K}$, $\Delta h_{lv} = 2500\,\text{kJ/kg}$, and specific heat of vapor $C_{p,v} = 2\,\text{kJ/kg\,K}$, the previous equation can be solved for $x_2$ for the given values of $T_1, x_1, T_2$. The result is $x_2 = 29.7\,\text{g/kg}$.

- The required airflow rate is therefore

$$\dot{m}_{air,1} = \frac{\dot{m}_w}{x_2 - x_1} = 37,369.6 \frac{kg}{h}$$

- According to the mass balance equation, at the end of first drying period (constant rate drying), the humidity ratio in the drying air becomes

$$x_I = x_1 + \frac{\dot{m}_{w,I}}{\dot{m}_{air,1}} = 20.8 \frac{g}{kg}$$

- The effective mass of material dried in the first phase depends among other factors on the saturation humidity of air; this is determined for a relative humidity of 100% and a temperature equal to that of the material $T_m$, and it is $x_s = 60\,g/kg$. The mass of material is

$$m_I = \frac{\dot{m}_{air,1}}{kA} \ln\frac{x_s - x_1}{x_s - x_I} = \frac{80.51}{0.06} = 1341.8\,kg$$

- The drying time for the first period is therefore

$$t_I = \frac{m_I}{\dot{m}_{dry}} = 0.65\,h$$

- The drying time for the variable rate period is given as half of $t_I$. Therefore, the total drying time is

$$t = 1.5\,t_I = 0.975\,h$$

**Example 3.2**

A porous material is dried for $t = 6h$ while its moisture content decreases from $W_1 = 30\%$ to $W_2 = 6\%$; the critical humidity is $W_c = 16\%$ and the equilibrium humidity is $W_e = 2\%$. Determine the drying time provided that the initial moisture was 35% and the final 4%.

- The drying time for the constant drying period is related to the drying constant and the change in moisture content as follows:

$$t_I = \frac{W_1 - W_c}{k} = \frac{0.14}{k}$$

- For the variable rate period, the drying time is

$$t_{II} = \frac{W_c - W_e}{k} \ln\frac{W_c - W_e}{W_2 - W_e} = \frac{0.175}{k}$$

- The total drying time is

$$t = t_I + t_{II} = \frac{0.14 + 0.175}{k} = 6$$

- Then, the drying constant becomes

$$k = 0.052 \frac{kg_{water}}{kg_{dry} \, h}$$

- The drying time in the new conditions becomes

$$t = \frac{0.35 - 0.16}{0.052} + \frac{0.16 - 0.02}{0.052} \ln \frac{0.16 - 0.02}{0.04 - 0.02} = 8.89 \, h$$

## 3.5   Conclusions

This chapter describes and discusses the main types of drying equipment and systems which are classified under various criteria related to drying materials, operational and technical aspects, efficiency and effectiveness, and so on. The details on their types, operations, and applications for various drying materials are explained and illustrated. The specific equations for system modeling in natural and forced convection drying processes are briefly introduced and discussed. Some illustrative examples are presented to show how to make respective drying calculations.

## 3.6   Study Problems

3.1   In a continuous type dryer, 2000 kg/h of material with 50% moisture content is dried such that its moisture content is reduced with 10%. The drying agent is hot air at 100 °C and a humidity ratio corresponding to a partial pressure of 9 mmHg. The exit parameters of the air are 40 °C and 50% relative humidity. Determine the required flow rate of air.

3.2   In a continuous dryer, a material with 5% critical moisture content is dried at a rate of 1000 kg/h while its moisture content decreases from 60% to 40%. Air at 110 °C with 10 g/kg humidity content exits the dryer at 50 °C with 60% relative humidity. Make reasonable assumptions to determine the material temperature at exit.

3.3   For the previous problem, determine the heat losses from the dryer in the ambient air assuming that the moist material temperature at entrance is 10 °C and its specific heat is 1 kJ/kg K.

3.4   A material with 18% moisture content enters a dryer at a rate of 2000 kg/h and is dried until a moisture content of 10%. The temperature of the material is 10 °C at entrance and 100 °C at exit, and its specific heat is 1000 J/kg K. Air from outside conditions at 20 °C and 60% relative humidity is heated to 150 °C and used as drying agent; the air takes moisture until its relative humidity reaches 80%. Determine the required heat input for the dryer.

3.5 A heat pump dryer produces 300 kg dried material/h (dry base). The drying agent is air recirculated at 10 t/h (no fresh air input) which is heated from 20 to 35 °C where the humidity ratio is 34 g/kg. The final moisture content of the material is 3%. Determine the air temperature at heater exit, the wet bulb temperature of air at dryer exit, and the moisture content of the moist material.

3.6 Redo problem from Example 3.1 for other data set of your choice.

3.7 Rework Example 3.2 of other set of data of your choice.

# References

Akiyama T., Liu H., Hayakawa K.I. 1997. Hygrostress-multicrack formation and propagation in cylindrical viscoelastic food undergoing heat and moisture transfer processes. *International Journal of Heat and Mass Transfer* 40:1601–1609.

Ameobi J.B. 1992. Maize drying with natural air. PhD Thesis, University of Newcastle upon Tyne, UK.

Atzema A.J. 1992. A model for the drying of grass with real time weather data. *Journal of Agricultural Engineering Research* 53:231–247.

Berk Z. 2009. Food Processing Equipment and Technology. Academic Press: London, UK.

Clemente G., Bon J., García-Pérez J.V., Mulet A. 2007. Natural convection drying at low temperatures of previously frozen salted meat. *Drying Technology* 25:1885–1891.

Daud W.R.W. 2006. Drum dryers. In: Handbook of Industrial Dryers. A.S. Mujumdar Ed., 3rd ed. Taylor and Francis Group, LLC.

Filbakk T., Høibø O., Nurmi J. 2011. Modelling natural drying efficiency in covered and uncovered piles of whole broadleaf trees for energy use. *Biomass and Bioenergy* 35:454–463.

GEA 2014. GEA Process Engineering Inc. Internet source accessed on June 9, 2015 http://www.niroinc.com/food_-chemical/ flash_drying_concepts.asp

Gigler J.K., van Loon W.K.P., van den Berg J.V., Sonneveld C., Meedrink G. 2000. Natural wind drying of willow stems. *Biomass and Bioenergy* 19:153–163.

Good Humus Farm 2014. Internet source (public domain) accessed on September 11th: http://goodhumus.com/website %20photos/drying%20trays%201med.jpg

Goyal R.K., Tiwari G.N. 1997. Parametric study of a reverse flat plate absorber cabinet dryer: a new concept. *Solar Energy* 60:41–48.

Green D.W., Perry R.H. 2008. Solids-drying fundamentals. In: Perry's Chemical Engineers' Handbook, 8th ed. McGraw-Hill Professional: AccessEngineering.

Hossain M.A., Bala B.K., Satter M.A. 2003. Simulation of natural air drying of maize in cribs. *Simulation Modelling Practice and Theory* 11:571–583.

Kofoed B.P. 2006. Esbenderup Hospital successfully treated. *Concrete* 40:32–33.

Mujumdar A.S. 2006. Principles, classification, and selection of dryers. In: Handbook of Industrial Dryers. A.S. Mujumdar Ed., 3rd ed. Taylor and Francis Group, LLC.

Olufayo A.A., Ogunkunle O.J. 1996. Natural drying of cassava chips in the humid zone of Nigeria. *Bioresource Technology* 58:89–91.

Tiris C., Ozbalta N., Tiris M., Dincer I. 1994. Experimental testing of a new solar dryer. *International Journal of Energy Research* 18:483–490.

Westelaken C.M.T. 1981. Gravity flow drying for particulate material having channelled discharge. US Patent 4,423,557.

Wimmerstedt R. 2006. Drying of peat and biofuels. In: Handbook of Industrial Dryers. A.S. Mujumdar Ed. 3rd ed. Taylor and Francis Group, LLC.

# 4

# Energy and Exergy Analyses of Drying Processes and Systems

## 4.1 Introduction

Despite the fact that a large number of experimental and theoretical studies were published on drying, only few works treat the energy and exergy analyses of drying processes and systems. Thermodynamic analysis through energy and exergy is described in general terms in Chapter 1. In this type of analysis, the balance equations for mass, energy, entropy, and energy are written for each of the system subunits and for the overall system. Then energy- and exergy-based efficiency formulations are used to assess the system performance. Other assessment parameters can be also formulated and concurrently used such as relative irreversibility (or relative exergy destructions of system subunits or subprocesses), sustainability index (SI), improvement potential, greenization factor, emission factors, specific moisture extraction ratio, coefficient of performance (COP) (when applicable), and so on.

Drying processes are complex because they involve beside heat and mass transfer which are highly nonlinear, phase change processes, mechanical interactions, cracking, fracture, product shrinking, material sticking, and so on. Thence, appropriate methods are required to provide best designs and optimal solutions to drying problems.

Energy and exergy analyses when combined into a unitary method can help at identification and reduction of irreversibilities at drying. Furthermore, this can lead to an enhanced system efficiency and therefore to a reduced energy consumption. Exergy analysis provides a better accounting of the loss of availability of heat in drying, it provides more meaningful and useful information than energy analysis on drying efficiency, and it provides a good understanding whether or not and by how much it is possible to design more efficient drying systems by reducing the inefficiencies in the existing units. Exergy analysis is thus useful for determining optimum drying conditions.

*Drying Phenomena: Theory and Applications*, First Edition. İbrahim Dinçer and Calin Zamfirescu.
© 2016 John Wiley & Sons, Ltd. Published 2016 by John Wiley & Sons, Ltd.

In addition, there is a demonstrated connection between the exergy utilization and the environmental impact (direct or indirect) of the studied system. More specifically, in drying, which in most of the cases uses fossil fuels for energy supply, there is associated an important environmental impact. Provided that the drying system is optimized based on exergy methods, then less waste emissions are produced leading to lower environmental impact.

In this chapter, the energy and exergy analysis of drying processes and systems is introduced. The concept of moist solid is introduced for the purpose of energy and exergy modeling of drying. The focus of the chapter is on drying operations with water liquid removed by evaporation and air employed as purge gas, because this case reflects the vast majority of practical drying processes. The goal in drying is to use a minimum amount of exergy for maximum moisture removal such as to achieve the desired final conditions of the product.

## 4.2   Balance Equations for a Drying Process

In this section, the balance equations and other equations relevant to modeling and analysis of drying processes through energy and exergy methods are introduced. Here, a link is made between the analysis method through energy and exergy developed in Chapter 1 and the specific equations that model heat and mass transfer during drying processes introduced in Chapter 2. In essence, through these equations, the heat and moisture transfer rates are related to the velocity and temperature of the drying gas and to the temperature and physical characteristics of the material subjected to drying.

Assume a stream of moist solid subjected to drying in a flow of dry air (other gaseous drying agents may be considered). In this respect, let us assume a drying device designed in a form of a drying tunnel. The drying tunnel can be modeled as a control volume as shown in Figure 4.1 in which both the stream of moist solids and dry air are introduced. The materials that exit the control volume are a stream of dried material and a flow of humidified air. In addition, the control volume receives work for driving auxiliary components and loses heat through this boundary.

We model the humid air as a mixture of dry air and water vapor; denote $\dot{m}_a$ the flow rate of dry air supplied at port 1 of the system from the figure. The humidity ratio in state 1 is given as $\omega_1$. Denote the mass flow rate of product with moisture completely removed with $\dot{m}_p$. In this

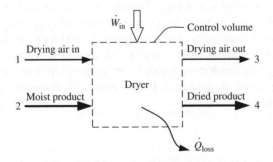

**Figure 4.1**   Thermodynamic model schematics for a drying process

case, the mass balance equations (MBEs) can be written for each of the substances – dry air, dry material, and water. The MBE for dry air is

$$\dot{m}_{a,1} = \dot{m}_{a,3} = \dot{m}_a \tag{4.1}$$

The MBE for dry product (i.e., the product without any water/moisture content) is defined as

$$\dot{m}_{p,2} = \dot{m}_{p,4} = \dot{m}_p \tag{4.2}$$

The MBE written for the water content is written as follows:

$$\omega_1 \dot{m}_a + \dot{m}_{w,2} = \omega_3 \dot{m}_a + \dot{m}_{w,4} \tag{4.3}$$

where $\dot{m}_w$ is the moisture content in the moist product at the entrance ($\dot{m}_{w,2}$) and exit ($\dot{m}_{w,4}$), respectively.

The overall MBE can be written by combining the above given three balance equations as follows:

$$(\omega_1 + 1)\dot{m}_a + \dot{m}_{w,2} + \dot{m}_{p,2} = (\omega_3 + 1)\dot{m}_a + \dot{m}_{w,4} + \dot{m}_{p,4} \tag{4.4}$$

One denotes the mass flow rate of humid air at ports 1 and 3 with $\dot{m}_1 = (\omega_1 + 1)\dot{m}_a$ and $\dot{m}_3 = (\omega_3 + 1)\dot{m}_a$. Furthermore, one denotes the moisture content of material with $W_2$ for entrance and $W_4$ for exit. Therefore, the mass flow rate at port 2 becomes $\dot{m}_2 = \dot{m}_{p,2} + \dot{m}_{w,2} = \dot{m}_{p,2}(1 + W_2)$, and at port 4, it is $\dot{m}_4 = \dot{m}_{p,4}(1 + W_4)$. With these notations, the overall MBE becomes

$$\dot{m}_1 + \dot{m}_2 = \dot{m}_3 + \dot{m}_4 \tag{4.5}$$

The energy balance equation (EBE) is written as follows: $\dot{E}_1 + \dot{E}_2 = \dot{E}_3 + \dot{E}_4 + \dot{Q}_{\text{loss}}$. The energy rate in state 1 depends on the specific enthalpy of the humid air, namely, $\dot{E}_1 = \dot{m}_{a,1}h_1$. Since at the inlet port 1 there is a mixture of dry air and water vapor, the specific enthalpy can be taken from the psychometric chart or it can be calculated with $h_1 = h_{a,1} + \omega_1 h_{v,1}$, where $h_{a,1}$ is the enthalpy of dry air in state 1 and $h_{v,1}$ the water vapor enthalpy. Similarly, for the exit port 3, one has $h_3 = h_{a,3} + \omega_3 h_{v,3}$. In summary, the EBE becomes

$$\dot{m}_{a,1} h_1 + \dot{m}_{p,2}h_{p,2} + \dot{m}_{w,2}h_{bw,2} + \dot{W}_{in} = \dot{m}_{a,3} h_3 + \dot{m}_{p,4} h_{p,4} + \dot{m}_{w,4}h_{bw,4} + \dot{Q}_{\text{loss}} \tag{4.6}$$

where $(h_{bw})_{1,2}$ is the specific enthalpy of moisture bounded within the solid at the entrance and exit, respectively.

The entropy balance equation can be written based on the specific entropies of the dry air, water, and dry product as follows (see Table 1.12):

$$\dot{m}_{a,1}s_1 + \dot{m}_{p,2}s_{p,2} + \dot{m}_{w,2}s_{bw,2} + \dot{S}_g = \dot{m}_{a,3}s_3 + \dot{m}_{p,3}s_{p,3} + \dot{m}_{w,4}s_{bw,4} + \frac{\dot{Q}_{\text{loss}}}{\bar{T}_{\text{loss}}} \tag{4.7}$$

where $\bar{T}_{loss}$ is the average temperature of the control volume boundary (see Figure 4.1).

The entropy balance equation (Eq. (4.7)) can be solved to determine the entropy generation rate $\dot{S}_g$. Once this is determined, the exergy destruction results from $\dot{Ex}_d = T_0\dot{S}_g$, where $T_0$ is the reference temperature in Kelvin. Alternatively, the exergy destruction can be determined from the exergy balance equation (ExBE). Writing the ExBE requires the derivation of an expression for the specific exergy of humid air which depends on the specific enthalpies and entropies of the dry air and water vapor. The specific entropy of humid air is given by the following equation:

$$s = s_0 + \left(C_{p,a} + \omega C_{p,v}\right)\ln\frac{T}{T_0} - \left(\frac{1}{M_a} + \frac{\omega}{M_w}\right)R\left(\ln\frac{P}{P_0} + \ln\frac{M_w + M_a\,\omega_0}{M_w + M_a\,\omega}\right) - \frac{\omega}{M_w}R\ln\frac{\omega}{\omega_0} \quad (4.8)$$

where $C_{p,a}$ is the specific heat of air at temperature (assumed constant), $C_{p,v}$ is the specific heat of water vapor, $s_0$ is the specific entropy of air at the reference state, $R$ is the universal gas constant, $M_a = 28.97\,\text{kg/kmol}$ is the molar mass of air, and $M_w = 18\,\text{kg/kmol}$ is the molar mass of water. The specific heat of the humid air – assuming constant specific heats – becomes

$$h = h_0 + \left(C_{p,a} + \omega C_{p,w}\right)\left(T - T_0\right) \quad (4.9)$$

where $h_0$ is the specific heat of air at the reference state, and the specific enthalpies are given in kilojoule per kilogram of dry air.

Now, the specific exergy of humid air results from Eqs. (4.8) and (4.9) according to the known formula for thermomechanical exergy of a fluid, $ex = h - h_0 - T_0(s - s_0)$, namely,

$$ex = \left(C_{p,a} + \omega C_{p,v}\right)\left(T - T_0\right) - T_0\left[\left(C_{p,a} + \omega C_{p,v}\right)\ln\frac{T}{T_0} - \left(\frac{1}{M_a} + \frac{\omega}{M_w}\right)R\ln\frac{P}{P_0}\right] \quad (4.10)$$

Assuming that the reference state for water corresponds to standard temperature and pressure, $T_0$ and $P_0$, the specific exergy of water bound into the moist solid is given according to Dincer and Rosen (2013) as follows:

$$ex_{bw} = h_{bw} - h_w''(T_0) - T_0\left[s_{bw} - s_w''(T_0) + \frac{R}{M_w}\ln\phi_0\right] \quad (4.11)$$

where $h_w''$, $s_w''$ are the specific enthalpy and entropy of saturated vapor, $\phi$ is the relative humidity of air and the subscript 0 refers to the reference state, and $h_{bw}$, $s_{bw}$ represent the specific enthalpy and entropy of the moisture within the solid.

The equation for $h_{bw}$ must account for the specific enthalpy of liquid at the temperature of the solid and partial pressure of the vapor and the displacement work of the extracted moisture $v[P - P''(T)]$. Therefore, the specific enthalpy of the bounded moisture is

$$h_{bw} = h_w(T, \phi P''(T)) + v(T, \phi P''(T))[P - P''(T)] \quad (4.12)$$

where $P''$ is the saturated vapor pressure. The specific entropy of the bounded moisture within the solid is given by the following equation:

$$s_{bw} = s_w(T, \phi P''(T)) + \frac{R}{M_w} \ln \phi_0 \tag{4.13}$$

Further, assuming the reference state at $T_0$ and $P_0$, the specific exergy of the solid material is given by the following equation:

$$ex_p = h_p(T,P) - h_p(T_0,P_0) - T_0 \left[ s_p(T,P) - s_p(T_0,P_0) \right] \tag{4.14}$$

Here, the specific enthalpy of the product can be derived from its specific heat; in addition, the specific entropy is related to the ratio of specific heat and the absolute temperature, and it can be determined by integration.

The ExBE can be written as follows:

$$\dot{E}x_1 + \dot{E}x_2 + \dot{W} = \dot{E}x_3 + \dot{E}x_4 + \dot{Q}_{loss}\left(1 - \frac{T_0}{\overline{T}_{loss}}\right) + \dot{E}x_d \tag{4.15}$$

where $\dot{E}x_d$ is the destroyed exergy.

Equations (4.10)–(4.14) can be now used to write expressions for the terms $\dot{E}x_i$, $i = 1, \ldots, 4$ required by the balance equation (Eq. (4.15)). One has $\dot{E}x_1 = \dot{m}_{a,1}ex_1$ with $ex_1$ from Eq. (4.10), $\dot{E}x_2 = \dot{m}_{p,2}ex_{p,2} + \dot{m}_{w,2}ex_{w,2}$ with Eq. (4.14) for $ex_{p,2}$ and Eq. (4.11) for $ex_{bw,2}$, $\dot{E}x_3 = \dot{m}_{a,3}ex_3$ with $ex_3$ from Eq. (4.10), and $\dot{E}x_4 = \dot{m}_{p,4}ex_{p,4} + \dot{m}_{w,4}ex_{bw,4}$ with Eq. (4.14) for $ex_{p,4}$ and Eq. (4.11) for $ex_{bw,4}$. Accounting also on MBEs, Eq. (4.15) can be solved for $\dot{E}x_d$, and after some algebraic manipulations, it gives

$$\dot{E}x_d = \dot{m}_p\left(ex_{p,2} + W_2 ex_{bw,2} - ex_{p,4} - W_4 ex_{bw,4}\right) + \dot{m}_a\left[(\omega_1 + 1)ex_1 - (\omega_3 + 1)ex_3\right] - \dot{Q}_{loss}\left(1 - \frac{T_0}{\overline{T}_{loss}}\right) \tag{4.16}$$

The balance equations, namely, Eqs. (4.4), (4.7), and (4.16), can be completed with other heat and mass transfer equations to form a closed system which can be solved to determine the irreversibility. The following types of auxiliary equations can be considered, depending on the case:

- Specific heat correlations with temperature, for drying agent, moisture, vapor, and moist material
- Heat transfer coefficient correlations specific to the analyzed system configuration
- Mass transfer correlations
- Correlations for the diffusion coefficients
- Relevant heat and mass transfer equations

The system of equation can be simplified based on various assumptions. Of major importance in exergy analysis are assumptions that affect the magnitude of irreversibility.

For example, for dryers that use air as drying agent – which is very common in industry – one can assume that the exit air reaches the wet bulb temperature. However, one must account for the fact that at higher wet bulb temperatures, significant exergy is destroyed (lost) with the exiting air. Other relevant assumptions for processes of drying with air are given as follows:

- The exergy carried by the dried product is relatively small, sometimes negligible, with respect to the exergy input; this is due to the fact that the input exergy is transferred mostly to the moisture.
- In tunnel dryers or any other form of enclosed dryers with continuous or batch operation, the exergy loss from the dryer wall to the surrounding atmosphere is significant; hence, this irreversibility cannot be neglected, although if the dryer is thermally insulated.
- The type of the dryer is important when the heat loss through the wall is considered. For spray dryers, the typical exergy destruction due to heat losses through the walls accounts for 25% of energy input. Smaller dryer types such as jet ring dryer shows much more exergy destruction due to heat loss because the dimensions of its enclosure is much smaller with respect to a spray dryer of equivalent capacity.

The drying system with direct contact such as drum dryers has a different feature than indirect dryers such as air dryers. In drum dryers, the heat loss through convection at the product surface does not produce significant irreversibility. Rather, this irreversibility is comparable to the exergy lost with the solid products and often can be neglected. Some other assumptions relevant to exergy analysis of indirect dryers are as follows:

- In steam-heated processes, the exergy destroyed due to exhaust of steam is large and cannot be neglected.
- Steam condensation in steam-heated processes is a source of high irreversibility.
- There is a high potential to recover heat from the condensate and reuse it within the process.

Another drying process with important exergy destruction is the freeze-drying. When this system is analyzed through exergy method, the following assumptions can be considered:

- Exergy destruction due to radiative heat transfer is negligible.
- The exergy destroyed by the condenser of the refrigeration subunit is significant, and it cannot be neglected. The exergy destruction in condenser is typically equivalent to the exergy required to remove 1 kg of water. However, the heat rejected by the condenser can be reclaimed and used within the process to reduce the irreversibility.
- The heat dissipated by the vacuum pump is significant and leads to important exergy destructions.

The energy and exergy analysis of drying processes usually requires the following parameters at input (see Figure 4.1): $\dot{m}_a$, $\omega_1$, $T_1$, $T_2$, $\dot{m}_p$, $\dot{m}_{w,2}$, $\dot{m}_{w,4}$, $T_0$, $P_0$, $\omega_0$. These parameters can be determined based on assumptions and analysis, depending on the case. Once these parameters are known, the humidity ratio $\omega_3$ at air exit can be calculated from EBE for air. Further, the temperatures at the exit for air and products as well as the pressures are determined. From those, the heat loss rate can be calculated. Next, using correlations or tabulated data for specific heats and an assumed value for the average temperature of the system boundary $T_{loss}$, the ExBE can be solved to determine the exergy destruction rate.

## 4.3  Performance Assessment of Drying Systems

The main system assessment parameters in energy analyses are the energy and exergy efficiencies, respectively. General formulations for energy and exergy efficiency were introduced previously in Chapter 1, Section 1.5.2. Here, efficiency definitions which are specific to drying systems are given. Furthermore, other assessment parameters for drying systems are introduced here. Such parameters are useful not only for assessment but also as selection criteria for specific applications of drying.

### 4.3.1  Energy and Exergy Efficiencies

With reference to the drying system model shown in Figure 4.1, the energy efficiency can be defined as the energy used to extract the moisture from the product divided to the energy supplied in the form of hot drying air and in the form of work (mechanical or electrical work consumed by blowers, conveyors, or other auxiliary equipment depending on the actual case). Note that the moisture evaporation rate results as a difference between moisture content of the moist material at the inlet and moisture content of the product at the exit.

The moisture content of the moist material at the inlet, measured in kilogram of moisture per kilogram of dry material, is expressed as $W_2 = \dot{m}_{w,2}/\dot{m}_p$; at the product exit port, the moisture content is $W_4 = \dot{m}_{w,4}/\dot{m}_p$; thence, the moisture removed is $\Delta W = W_2 - W_4$. The rate of moisture removal is therefore written as

$$\dot{m}_{mr} = \dot{m}_p \,\Delta W = \dot{m}_{w,2} - \dot{m}_{w,4} \tag{4.17}$$

The energy required to remove the moisture at the rate $\dot{m}_{mr}$ depends on the specific enthalpy difference of the moisture between states 3 and 2. For state 2, the specific enthalpy is given by Eq. (4.12) applicable to the bounded moisture $h_{bw,2}$. For state 3, the specific enthalpy is given by Eq. (4.9) applicable to the humid air $h_3$. Therefore, the energy rate required to remove the moisture is expressed as follows:

$$\dot{E}_{mr} = \dot{m}_{mr}(h_{v,3} - h_{bw,2}) \tag{4.18}$$

The energy supplied to the system can be expressed as $\dot{E}_{in} = \dot{m}_a(h_1 - h_3 + w)$, where $w$ is the work corresponding to 1 kg of dry air supplied to the system for all auxiliary devices; $\dot{m}_a w = \dot{W}$. According to the definition provided earlier, the energy efficiency of the drying system becomes

$$\eta = \frac{\dot{E}_{mr}}{\dot{E}_{in}} = \left(\frac{\dot{m}_{mr}}{\dot{m}_a}\right)\left(\frac{h_{v,3} - h_{bw,2}}{h_1 - h_3 + w}\right) \tag{4.19}$$

The exergy efficiency definition is analogous to the energy efficiency definition, namely, it represents the ratio between the exergy used to extract moisture from the product (i.e., the reversible work required to conduct the process) and the exergy of the supplied drying air plus any work supplied to the system for the auxiliary processes (i.e., the actual exergy supplied to the process). The exergy used to extract the moisture is expressed as follows:

$$\dot{E}x_{mr} = \dot{m}_{mr}(ex_{v,3} - ex_{bw,2}) \tag{4.20}$$

where $ex_{v,3}$ is the specific exergy of water vapor in state 3 and $ex_{bw,2}$ is the specific exergy of the bounded moisture in state 2 calculated with Eq. (4.11).

The exergy of the supplied drying air is given by the product $\dot{m}_a ex_1$; therefore, the exergy efficiency is expressed by the following equation:

$$\psi = \frac{\dot{E}x_{mr}}{\dot{E}x_{in}} = \left(\frac{\dot{m}_{mr}}{\dot{m}_a}\right)\left(\frac{ex_{v,3} - ex_{bw,2}}{ex_1 + w}\right) \tag{4.21}$$

Here, the exergy of the air exiting the system $\dot{m}_a ex_3$ is considered an exergy destroyed (lost) due to the interaction of the system with the environment. Provided that the temperature of the exiting air is not very high, this destroyed exergy is negligible. However, in some specific applications at high temperature, this destroyed exergy can be high, and therefore, it can be reclaimed using a recovery system which recycles the exhaust air. This is also the case when heat pump dryers are considered. Therefore, the following exergy efficiency formulation may be useful in such cases:

$$\psi = \left(\frac{\dot{m}_{mr}}{\dot{m}_a}\right)\left(\frac{ex_{v,3} - ex_{bw,2}}{ex_1 - ex_3 + w}\right) \tag{4.22}$$

Another type of energy formulations can be preferred in some cases as discussed next. Assume that only a part of energy lost by the drying system is recovered, and this is $\dot{E}_{rec} = \dot{m}_{mr}e_{rec}$, where $e_{rec}$ stands for the energy recovered corresponding to 1 kg of moisture extracted. The associated exergy recovered is denoted with $\dot{E}x_{rec} = \dot{m}_{mr}ex_{rec}$. In this case, the energy efficiency can be defined as follows:

$$\eta = \left(\frac{\dot{m}_{mr}}{\dot{m}_a}\right)\left(\frac{h_{v,3} - h_{bw,2} + e_{rec}}{h_1 - h_3 + w}\right) \tag{4.23}$$

and the exergy efficiency becomes

$$\psi = \left(\frac{\dot{m}_{mr}}{\dot{m}_a}\right)\left(\frac{ex_{v,3} - ex_{bw,2} + ex_{rec}}{ex_1 + w}\right) \tag{4.24}$$

A particular situation for efficiency definition is for heat pump-driven dryers. In this case, the energy input is in the form of work required to drive the refrigeration compressor and auxiliaries. Therefore, in energy efficiency definitions from Eqs. (4.19) and (4.24), the denominator $h_1 - h_3 + w$ is replaced by $w$ (where $w$ is the specific work required by the system in order for any 1 kg of drying agent). As such, the exergy efficiency expressions from Eqs. (4.19) and (4.24) will have at the denominator a $w$ instead of the $ex_1 + w$.

### 4.3.2 Other Assessment Parameters

The generic drying system shown in Figure 4.1 is a black-box representation. Obviously, the system comprises many subunits, for example, fans, blowers, combustors, heaters, heat pumps, pumps, conveyors, and passive components such as ducts, pipes, tubing, and so on. Each of the system subunits destroys a fraction of the total exergy destruction, $\dot{Ex}_d$. The sum of exergy destroyed by the subunits must equal the total exergy destruction, $\sum \dot{Ex}_{d,i} = \dot{Ex}_d$. The relative irreversibility can be defined for each of the system subunits as the fraction of exergy destroyed from the total exergy destroyed by the overall system. Mathematically, the relative irreversibility is expressed as follows:

$$RI_i = \frac{\dot{Ex}_{d,i}}{\dot{Ex}_d} \qquad (4.25)$$

Regarding Eq. (4.25), it is obvious that $\sum RI_i = 1$. The relative irreversibility is useful for identification of major exergy destruction within the system. Another assessment parameter is the exergetic SI previously introduced in Rosen et al. (2008). The SI represents the ratio between the exergy input to the system and the exergy destroyed by the system. The systems or processes with high SI are more sustainable than for low SI. Mathematically, the SI is expressed as follows:

$$SI \equiv \frac{\dot{Ex}_{in}}{\dot{Ex}_d} = \frac{1}{1-\psi} \qquad (4.26)$$

Note that in Eq. (4.26), the exergy destruction includes both the exergy destruction within the system and the exergy destruction at the interaction between the system and the surroundings. Consequently, the ExBE for the overall system is written concisely as $\dot{Ex}_{in} = x_{mr} + \dot{Ex}_d$. Now, if this expression is divided by $\dot{Ex}_{in}$, then one gets, according to Eq. (4.21), that $1 = \psi + \dot{Ex}_d/\dot{Ex}_{in}$, which proves that the SI is the reciprocal of $1-\psi$.

We now introduce a new parameter here, namely, the drying effectiveness (DE). This parameter expresses the relative increase of humidity ratio in the exit air with respect to the inlet air. In fact, the drying process is treated as a kind of humidification process of air. It is obvious that the higher the humidification for drying air, the better the DE. In this regard, the DE is defined as follows:

$$DE = \frac{\omega_{out}}{\omega_{in}} = \frac{\omega_3}{\omega_1} \qquad (4.27)$$

where the state point notations from Figure 4.1 are used, that is, $\omega_{in} = \omega_1$ and $\omega_{out} = \omega_3$. The humidity ratios can be easily correlated with the air parameters at the inlet and outlet ports, respectively. This can be done using common psychometric relationships, for example,

$$\omega = \frac{C_{p,a}(T_{wb} - T) + \omega'' \Delta h_{fg}}{h_v - h_1''(T_{wb})}$$

where $T_{wb}$ is the wet bulb temperature and $\omega''$ is the humidity ratio at saturation

$$\omega'' = \frac{0.622P''}{P-P''}$$

Another new parameter is introduced as the drying quality (DQ) which is defined as the ratio of mass of moisture removed (e.g., evaporated water) to the mass of the moist product. In this regard, the DQ is expressed as follows:

$$DQ = \frac{\dot{m}_{mr}}{\dot{m}_{p,2} + \dot{m}_{w,2}} = \frac{\dot{m}_{mr}}{\dot{m}_p(1 + W_2)} \tag{4.28}$$

where the mass rate of the moist product at the inlet of the dryer is represented by the sum $\dot{m}_{p,2} + \dot{m}_{w,2}$.

## 4.4 Case Study 1: Analysis of Continuous-Flow Direct Combustion Dryers

Here, the analysis based on energy and exergy methods of continuous-flow direct combustion dryers is presented. This process is specific to continuous-flow direct-type dryers where combustion gases are used as a drying agent. Many types of industrial dryers such as tunnel, rotary, tray, and sheeting through circulation dryers use this process. It is reasonable to assume that combustion gases can be modeled as humid air as a drying agent; however, the analysis presented herein can be made more accurate when the actual composition of exhaust gases are considered. The method of analysis shown here is representative for any continuous drying process with air or other gases and drying agent.

Figure 4.2 shows the schematics of a general purpose dryer featuring a continuous-flow process. Because the combustion gases are used as drying agent and heat transfer agent altogether, the dryer is categorized as a direct type. As shown in the figure, the system includes a gas turbine power generator to supply its own power. This is generally justified due to the typically large scale of the dryer systems in industry and large heating demand.

The generated power is required mainly to supply the moist material feeder/conveyor (state 19) and the fan (state 20); additional power in (state 21) is available for local demand. The "generic" drying system comprises 10 subunits, namely, the gas turbine/generator unit, the supplementary combustor, the mixer for recirculated air, the material feeding/conveying unit, the dryer unit, the filter, the fan, the cyclone separator, and the air recirculation flap. The state points shown on the diagram from Figure 4.2 are given in Table 4.1.

The system analysis with the combined energy and exergy methods requires that the balance equations for mass, energy, entropy, and exergy are written for each of the system components (subunits). For the 10-component generic drying system analyzed here, the balance equations are given in Table 4.2. The following assumptions are relevant for the considered system:

- The air enters the gas turbine at standard state $T_0, P_0$.
- A reasonable value for the relative humidity $\phi_0$ of standard atmosphere must be assumed.
- Combustion process is adiabatic for both the gas turbine and supplementary combustor.

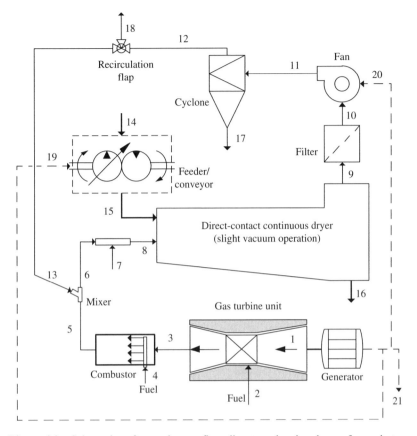

**Figure 4.2**    Schematics of a continuous-flow direct combustion dryer of generic type

**Table 4.1**    State point descriptions for the generic drying system described in Figure 4.2

| State | Flow | State | Flow |
|---|---|---|---|
| 1 | Fresh air | 11 | Pressurized air |
| 2 | Fuel | 12 | Particle-free air |
| 3 | Combustion gases | 13 | Recycled air |
| 4 | Fuel | 14 | Moist material |
| 5 | Combustion gases | 15 | Conveyed and preprocessed moist material |
| 6 | Flue gas | 16 | Dried product |
| 7 | Fresh air | 17 | Recovered particulate matter |
| 8 | Hot drying agent | 18 | Exhausted air |
| 9 | Humid air and conveyed particles | 19 | Electric power for feeder/conveyor |
| 10 | Humid air and fine particles | 20 | Electric power for fan(s) |

**Table 4.2** State point descriptions for the generic drying system described in Figure 4.2

| Component and process | Balance equations |
|---|---|
| Gas turbine/generator | MBE: $\dot{m}_1 + \dot{m}_2 = \dot{m}_3$ |
| | EBE: $\dot{m}_1 h_1 + \dot{m}_2 h_2 = \dot{m}_3 h_3 + \dot{W}_{19} + \dot{W}_{20} + \dot{W}_{21}$ |
| | EnBE: $\dot{m}_1 s_1 + \dot{m}_2 s_2 + \dot{S}_{g,gt} = \dot{m}_3 s_3$ |
| | ExBE: $\dot{m}_1 ex_1 + \dot{m}_2 ex_2 = \dot{m}_3 ex_3 + \dot{W}_{19} + \dot{W}_{20} + \dot{W}_{21} + \dot{Ex}_{d,gt}$ |
| Supplementary combustor | MBE: $\dot{m}_3 + \dot{m}_4 = \dot{m}_5$ |
| | EBE: $\dot{m}_3 h_3 + \dot{m}_4 h_4 = \dot{m}_5 h_5$ |
| | EnBE: $\dot{m}_3 s_3 + \dot{m}_4 s_4 + \dot{S}_{g,sc} = \dot{m}_5 s_5$ |
| | ExBE: $\dot{m}_3 ex_3 + \dot{m}_4 ex_4 = \dot{m}_5 ex_5 + \dot{Ex}_{d,sc}$ |
| Air mixer | MBE: $\dot{m}_5 + \dot{m}_{13} = \dot{m}_6$; $\dot{m}_{w,5} + \dot{m}_{w,13} = \dot{m}_{w,6}$ |
| | EBE: $\dot{m}_5 h_5 + \dot{m}_{13} h_{13} = \dot{m}_6 h_6$ |
| | EnBE: $\dot{m}_5 s_5 + \dot{m}_{13} s_{13} + \dot{S}_{g,am} = \dot{m}_6 s_6$ |
| | ExBE: $\dot{m}_5 ex_5 + \dot{m}_{13} ex_{13} = \dot{m}_6 ex_6 + \dot{Ex}_{d,am}$ |
| Air penetration process 6–7–8 | MBE: $\dot{m}_6 + \dot{m}_7 = \dot{m}_8$; $\dot{m}_{w,6} + \dot{m}_{w,7} = \dot{m}_{w,8}$ |
| | EBE: $\dot{m}_6 h_6 + \dot{m}_7 h_7 = \dot{m}_8 h_8$ |
| | EnBE: $\dot{m}_6 s_6 + \dot{m}_7 s_7 + \dot{S}_{g,ap} = \dot{m}_8 s_8$ |
| | ExBE: $\dot{m}_6 ex_6 + \dot{m}_7 ex_7 = \dot{m}_8 ex_8 + \dot{Ex}_{d,ap}$ |
| Dryer | MBE: $\dot{m}_8 + \dot{m}_{15} = \dot{m}_9 + \dot{m}_{16}$; $\dot{m}_{w,8} + \dot{m}_{bw,15} = \dot{m}_{w,9} + \dot{m}_{bw,16}$ |
| | EBE: $\dot{m}_8 h_8 + \dot{m}_{15} h_{15} = \dot{m}_9 h_9 + \dot{m}_{16} h_{16} + \dot{Q}_{loss}$ |
| | EnBE: $\dot{m}_8 s_8 + \dot{m}_{15} s_{15} + \dot{S}_{g,dryer} = \dot{m}_9 s_9 + \dot{m}_{16} s_{16} + \dfrac{\dot{Q}_{loss}}{\overline{T}_{loss}}$ |
| | ExBE: $\dot{m}_8 ex_8 + \dot{m}_{15} ex_{15} = \dot{m}_9 ex_9 + \dot{m}_{16} ex_{16} + \dot{Q}_{loss}\left(1 - \dfrac{T_0}{\overline{T}_{loss}}\right) + \dot{Ex}_{d,dryer,int}$ |
| Filter | MBE: $\dot{m}_9 = \dfrac{dm_{fi}}{dt} + \dot{m}_{10}$; $\dot{m}_{w,9} = \dot{m}_{w,10}$ |
| | EBE: $\dot{m}_9 h_9 = \dfrac{d(me)_{fi}}{dt} + \dot{m}_{10} h_{10}$ |
| | EnBE: $\dot{m}_9 s_9 + \dot{S}_{g,fi} = \dfrac{d(ms)_{fi}}{dt} + \dot{m}_{10} s_{10}$ |
| | ExBE: $\dot{m}_9 ex_9 = \dfrac{d(m\,ex)_{fi}}{dt} + \dot{m}_{10} ex_{10} + \dot{Ex}_{d,fi}$ |
| Fan | MBE: $\dot{m}_{10} = \dot{m}_{11}$ |
| | EBE: $\dot{m}_{10} h_{10} + \dot{m}_{20} = \dot{m}_{11} h_{11}$ |
| | EnBE: $\dot{m}_{10} s_{10} + \dot{S}_{g,fa} = \dot{m}_{11} h_{11}$ |
| | ExBE: $\dot{m}_{10} ex_{10} + \dot{W}_{20} = \dot{m}_{11} ex_{11} + \dot{Ex}_{d,fa}$ |
| Cyclone | MBE: $\dot{m}_{11} = \dot{m}_{12} + \dot{m}_{17}$ |
| | EBE: $\dot{m}_{11} h_{11} = \dot{m}_{12} h_{12} + \dot{m}_{17} h_{17}$ |
| | EnBE: $\dot{m}_{11} s_{11} + \dot{S}_{g,c} = \dot{m}_{12} s_{12} + \dot{m}_{17} s_{17}$ |
| | ExBE: $\dot{m}_{11} ex_{11} = \dot{m}_{12} ex_{12} + \dot{m}_{17} ex_{17} + \dot{Ex}_{d,c}$ |
| Recirculation flap | MBE: $\dot{m}_{12} = \dot{m}_{13} + \dot{m}_{18}$ |
| | EBE: $\dot{m}_{12} h_{12} = \dot{m}_{13} h_{13} + \dot{m}_{18} h_{18}$ |
| | EnBE: $\dot{m}_{12} s_{12} + \dot{S}_{g,rf} = \dot{m}_{13} s_{13} + \dot{m}_{18} s_{18}$ |
| | ExBE: $\dot{m}_{12} ex_{12} = \dot{m}_{13} ex_{13} + \dot{m}_{18} ex_{18} + \dot{Ex}_{d,rf}$ |
| Feeder/conveyor | MBE: $\dot{m}_{14} = \dot{m}_{15}$ |
| | EBE: $\dot{m}_{14} h_{14} + \dot{W}_{19} = \dot{m}_{15} h_{15} + \dot{Q}_{dissip}$ |
| | EnBE: $\dot{m}_{14} s_{14} + \dot{S}_{g,fc} = \dot{m}_{15} s_{15} + \dfrac{\dot{Q}_{dissip}}{T_0}$ |
| | ExBE: $\dot{m}_{14} ex_{14} + \dot{W}_{19} = \dot{m}_{15} ex_{15} + \dot{Ex}_{d,fc}$ |

- The fuel temperature at inlet port is assumed to be $T_0$.
- The gas mixing (5–13–6 and 6–7–8) and splitting (12–13–18) processes are adiabatic.
- The power provided to the feeder dissipates integrally as heat into the environment ($\dot{Q}_{\text{dissip}} = \dot{W}_{19}$).
- The dryer can be modeled at steady state based on the method described in Section 4.1.
- The dryer process is nonadiabatic.
- The pressure drop across the dryer can be neglected.
- The air leaks into the dryer are assumed proportional to the magnitude of vacuum.
- If the moist material is in powder, pellet, or grain form, then dry material escape in state 9 must be considered.
- The filter is modeled in nonsteady-state regime, assuming a reasonable accumulation rate.
- The pressure drop across the filter cannot be neglected; a reasonable value for it must be assumed.
- The filter process is adiabatic.
- A reasonable isentropic efficiency must be assumed for the blower.
- The cyclone process is assumed adiabatic.
- The pressure drop in all ducts is negligible.
- Due to local pressure drop, the recirculation flap reduces the pressure to atmospheric value in state 18.
- Due to local pressure drop, the recirculation flap reduces pressure such that $P_{13} = P_5 = $ subatmospheric.
- The humidity ratio of drying gas in state 6 results from mass balances for gas turbine, combustor, and mixer.
- Complete combustion can be assumed for the purpose of calculating the humidity ratio in state 6.
- The drying agent provided in state 8 is modeled as humid air.
- The humidity ratio in state 8 results from the humidity ratios in states 6 and 7 according to MBE.

The analysis based on balance equations and assumptions will lead to the formulation of a complex, nonlinear algebraic system of equations which eventually can be solved to determine the thermodynamic parameters and flow rates at each state point. In addition, the exergy destroyed by each of the system components is determined. Further, the system efficiencies and other assessment parameters can be computed. In particular, for the dryer subunit, the energy and exergy efficiencies can be calculated using Eqs. (4.15) and (4.17), respectively.

In addition, the overall system efficiencies can be defined for energy and exergy. Here, the input energy and exergy is that of fuels in states 2 and 4. One has $\dot{E}_{\text{in}} = (\dot{m}_2 + \dot{m}_4)\text{LHV}_{\text{f}}$, where $\text{LHV}_{\text{f}}$ is the specific enthalpy of fuel (including the formation enthalpy) and $\dot{\text{Ex}}_{\text{in}} = (\dot{m}_2 + \dot{m}_4)\text{ex}_{\text{f}}^{\text{ch}}$.

Based on the aforementioned considerations, the energy efficiency of the overall system becomes

$$\eta = \frac{(\dot{m}_{\text{w},15} - \dot{m}_{\text{w},16})(h_9 - h_{\text{bw},15}) + \dot{W}_{21}}{(\dot{m}_2 + \dot{m}_4)\text{LHV}_{\text{f}}} \tag{4.29}$$

The state point numbers from Eq. (4.29) correspond to Figure 4.2. Also, the exergy efficiency is mathematically expressed as follows:

$$\psi = \frac{(\dot{m}_{w,15} - \dot{m}_{w,16})(ex_9 - ex_{bw,15}) + \dot{W}_{21}}{(\dot{m}_2 + \dot{m}_4)ex_f^{ch}} \tag{4.30}$$

**Example 4.1**

Here, a numerical example is given to illustrate the analysis procedure through energy and exergy. The system considered is that represented in Figure 4.2. This example is a modification of the case study from Coskun et al. (2009) which analyzed a woodchip dryer system. Here, we use the same numerical values for the assumed parameters as in Coskun et al. (2009), but the system is modified by the following: recirculation is considered for exhaust air, the effect of local pressure drop in filter and recirculation flap is included in the analysis, and the gas turbine is included in the analysis. The specific heat of the dry wood can be calculated with the equation: $C_{p,p} = 0.1031 + 0.003867\,T\,(K)$, where the result is in kJ/kg K. Table 4.3 shows the assumed data input for this example. The exergy destructions and relative irreversibility are to be determined for each of the system components. The energy and exergy efficiencies are calculated.

- Dryer:
  - Based on the table data and Eq. (4.9), the energy rate at state 8 is calculated as $\dot{E}_8 = 90{,}385\,kW$; here, the specific heats of air and water vapor can be obtained from the Engineering Equation Solver (EES); for hand calculations, these specific heats can be found in the tables: $C''_{p,w} \cong 2.1\,kJ/kgK$ and $C_{p,a} \cong 1.08\,kJ/kgK$.
  - Based on moist material temperature in state 15 and specific heat of wood calculated with the provided relation, one has $C_{p,p,15} = 1.217\,kJ/kgK$. Accounting also for the moisture content for which one calculates with Eq. (4.12) that $h_{bw,15} = 62.85\,kJ/kgK$, the energy rate at state 15 becomes $\dot{E}_{15} = 1322\,kW$.

**Table 4.3** Assumptions for Example 4.1

| Parameter | Value | Parameter | Value |
|---|---|---|---|
| $T_0$ (K) | 288 | $\omega_9$ | 0.2169 |
| $P_0$ (kPa) | 101.3 | $\dot{m}_{w,16}$ (kg/s) | 0.079 |
| $\omega_0$ | 0.007 | $T_{16}$ (K) | 363 |
| $\dot{m}_{a,8}$ (kg/s) | 81.53 | $\dot{W}_{19}$ | $0.06\% \times (\dot{E}_8 + \dot{E}_{15})$ |
| $\omega_8$ | 0.072 | $\dot{W}_{20}$ | $7.7\% \times (\dot{E}_8 + \dot{E}_{15})$ |
| $T_8$ (K) | 739 | $\dot{Q}_{lost}$ | $0.41\% \times (\dot{E}_8 + \dot{E}_{15})$ |
| $\dot{m}_{p,15}$ (kg/s) | 14.09 | $\bar{T}_{loss}$ (K) | 296 |
| $\dot{m}_{w,15}$ (kg/s) | 11.98 | $\dot{m}_7$ (kg/s) | 2.27 |
| $T_{15}$ (K) | 288 | $T_3$ (K) | 533 |
| $T_9$ (K) | 403 | $P_8 = P_9$ | $0.8\,P_0$ |

- The energy rate at state 9 (humid air exit from the dryer) is determined with Eq. (4.9) for the given temperature $T_9$ and $\omega_9$; the specific enthalpy of humid air in state 9 becomes $h_9 = 2720 \text{kJ/kg}$; thence, the energy rate associated with stream state 9 is $\dot{E}_9 = 88,610 \text{kW}$.
- The energy rate in state 16 results similarly as for state 15, namely, $\dot{E}_{16} = 3479 \text{kW}$.
- The energy rate input in the dryer subunit is $\dot{E}_{\text{in,dryer}} = \dot{E}_8 + \dot{E}_{15} = 91,707 \text{kW}$.
- The work rate consumed by the conveyor/feeder is $\dot{W}_{19} = 0.06\% \left( \dot{E}_{\text{in,dryer}} \right) = 55 \text{kW}$.
- The work rate consumed by the fan(s) is $\dot{W}_{20} = 7.7\% \left( \dot{E}_{\text{in,dryer}} \right) = 710 \text{kW}$.
- The rate of moisture removal is $\dot{m}_{\text{mr}} = \dot{m}_{w,15} - \dot{m}_{w,16} = 11.98 - 0.079 = 11.9 \text{kg/s}$.
- The energy efficiency of the dryer becomes

$$\eta_{\text{dryer}} = \frac{\dot{m}_{\text{mr}}(h_9 - h_{\text{bw},15})}{\dot{E}_{\text{in,dryer}}} = \frac{11.9\,(2720 - 62.85)}{91,707} = 34.5\%$$

- The specific exergy in state 8 as calculated with Eq. (4.10) is $\text{ex}_8 = 252.2 \text{kJ/kg}$; therefore, the energy rate of stream state 8 becomes $\dot{\text{Ex}}_8 = 20,526 \text{kW}$.
- Since the moist material in state 15 is at reference temperature, the specific exergy content $\text{ex}_{p,8} = 0$ (chemical exergy is neglected in this problem); furthermore, the bounded moisture may be assumed close to the reference conditions, therefore in equilibrium with the environment; therefore, $\text{ex}_{\text{bw},8} \cong 0$. Consequently, the exergy rate in state 15 is nil, $\dot{\text{Ex}}_{15} = 0$.
- The exergy rate associated with the heat lost by the dryer is $\dot{\text{Ex}}^{Q_{\text{loss}}} = \dot{Q}_{\text{loss}}(1 - T_0/\bar{T}_{\text{loss}}) = 10 \text{kW}$.
- The specific exergy in state 9 calculated with Eq. (4.10) is $\text{ex}_9 = 146.74 \text{kJ/kg}$.
- The exergy rate in state 9 becomes $\dot{\text{Ex}}_9 = 11,964 \text{kW}$.
- The exergy rate in state 16 calculated with Eqs. (4.9) and (4.14) is $\dot{\text{Ex}}_{16} = 323 \text{kW}$, and $\text{ex}_{\text{bw},16} = 75.94 \text{kJ/kg}$.
- The exergy used by the process, internally, is $\dot{\text{Ex}}_{\text{used}} = \dot{m}_{\text{mr}}(\text{ex}_9 - \text{ex}_{\text{bw},15}) = 11.9 \times 14,674 = 1746 \text{kW}$.
- The exergy input into the dryer subunit is $\dot{\text{Ex}}_{\text{in,dryer}} = \dot{\text{Ex}}_8 = 20,256 \text{kW}$.
- The exergy balance on dryer subunit is written according to Table 4.2 as follows.

$$\dot{\text{Ex}}_8 + \dot{\text{Ex}}_{15} = \dot{\text{Ex}}_9 + \dot{\text{Ex}}_{16} + \dot{\text{Ex}}^{Q_{\text{loss}}} + \dot{\text{Ex}}_{\text{used}} + \dot{\text{Ex}}_{\text{d,dr,int}}$$

or

$$20,526 + 0 = 11,964 + 323 + 10 + 1746 + \dot{\text{Ex}}_{\text{d,dr,int}}$$

- The above equation solves for $\dot{\text{Ex}}_{\text{d,dr,int}}$ which is the exergy destroyed by the internal processes of the dryer subunit; one has $\dot{\text{Ex}}_{\text{d,dr,int}} = 6483 \text{kW}$.
- The total exergy destroyed by the dryer unit is $\dot{\text{Ex}}_{\text{d,dr}} = \dot{\text{Ex}}^{Q_{\text{loss}}} + \dot{\text{Ex}}_9 + \dot{\text{Ex}}_{16} + \dot{\text{Ex}}_{\text{d,dr,int}} = 18,780 \text{kW}$.
- The exergy efficiency of the dryer subunit becomes

$$\Psi_{\text{dryer}} = \frac{\dot{Ex}_{\text{used}}}{\dot{Ex}_{\text{in,dryer}}} = \frac{1746}{20,526} = 8.5\%$$

- Feeder/conveyor:
  - Since all power input is dissipated as heat in the feeder, one has $\dot{Q}_{\text{dissip}} = \dot{W}_{19} = 55\text{kW}$; therefore, the balance equations (see Table 4.2) require that $h_{14} = h_{15}$.
  - All work is therefore destroyed: $\dot{Ex}_{\text{d,fc}} = \dot{W}_{19} = 55\text{kW}$.
- Heat penetration process 6–7–8:
  - From MBEs and input data, one obtains the mass flow rates in state 6; then one calculates the humidity ratios $\omega_6 = 0.0739, \omega_7 = 0.007175, \omega_8 = 0.07207$; then the enthalpies of humid air are calculated with EES or humid air chart; this allows for writing the EBE which determines $h_6 = 748\text{ kJ/kg}$.
  - The ExBE determines $\dot{E}_6 = 20,763\text{kW}$ and $\dot{Ex}_{\text{d.ap}} = 237\text{kW}$.
- Mixer process 5–13–6:
  - Since the process 9–10–11–12–13 is adiabatic according to the assumptions, $T_{13} = T_9 = 403\text{K}$ and $\omega_{13} = \omega_9 = 0.2169 \rightarrow h_{13} = 726.2\text{kJ/kg}$.
  - Assume the air recirculation ratio $f = 0.1$; then $\dot{m}_{13} = f\dot{m}_{12} \cong f\,\dot{m}_9 = 8.153\text{kg/s} \rightarrow \dot{m}_{a,5} = 71.15\text{kg/s}$.
  - The EBE determines the parameters of state 5 $\rightarrow T_5 = 798.1\text{K}, h_5 = 750.3\text{kJ/kg}$.
  - The exergy destruction becomes $\dot{Ex}_{\text{d,m}} = 623\text{kW}$ and $\dot{Ex}_5 = 20,995\text{kW}$.
- Combustor:
  - Excess air is required both in the combustor and gas turbine; the relative humidity of air at state 3 is well below 1%. Simple combustion calculation leads to the conclusion that the relative humidity is somewhere between 0.6% and 0.7%. Assume that $\phi_3 = 0.65\%$. In this case, the thermodynamic parameters in state 3 can be computed as follows: $h_3 = 429\text{kJ/kg}, s_3 = 6.886\text{kJ/kg K}$.
  - The required heat input for the combustor is $\dot{Q}_c = \dot{m}_a(h_5 - h_3) = 22,890\text{kW}$; assume that the fuel is natural gas with LHV = 50,000\text{kJ/kg}; then the required mass flow rate is $\dot{m}_4 = 0.46\text{kg/s}$.
  - The ExBEs gives $\dot{Ex}_{\text{d,c}} = 7191\text{kW}$ and $\dot{Ex}_3 = 4379\text{kW}$.
- Gas turbine:
  - It is reasonable to assume that the gas turbine efficiency is 50% of the ideal efficiency, whereas the ideal efficiency is given by $\eta_{\text{gt,id}} = r^{-k} - 1$; practical gas turbines have a compression ratio of around 10, $r = 10$. Assuming that the working fluid is air, then $k = 0.28$ and $\eta_{\text{gt}} = 26.2\%$.
  - The turbine inlet temperature is calculated from the isentropic relationship $\text{TIT} = r^k T_3 = 1016\text{K}$.
  - The compressor exit temperature results from a similar equation $\text{TEC} = r^k T_0 = 548.8\text{K}$.
  - Based on the given $\omega_0$ of the surrounding air, on the combustion pressure $P_c = r P_0 = 1013\text{kPa}$, and on the temperatures TIT and TEC, one can determine the specific enthalpies at the inlet and outlet of the combustion chamber as follows: $h_{\text{EC}} = 301.3\text{kJ/kg}$ (compressor exit) and $h_{\text{IT}} = 818.7\text{ kJ/kg}$ (turbine inlet).
  - The required heat input for the gas turbine results from $\dot{Q}_{\text{gt}} = \dot{m}_1(h_{\text{IT}} - h_{\text{EC}}) = 22,890\text{kW}$. Here, the mass flow rate can be approximated as $\dot{m}_1 = \dot{m}_3$.
  - The mass flow rate of fuel results from the equation $\dot{Q}_{\text{gt}} = \dot{m}_2 \text{LHV} \rightarrow \dot{m}_2 = 0.72\text{kg/s}$.

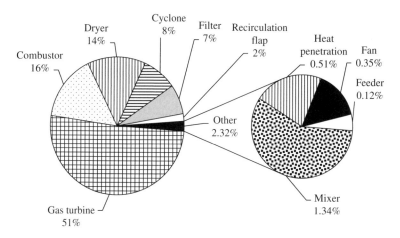

**Figure 4.3**  Relative irreversibility for the drying system studied in Example 4.1

- ○ Also, the work generated by the gas turbine is $\dot{W}_{gt} = \eta_{gt}\dot{Q}_{gt} = 9498\,\text{kW}$; therefore, the additional power supplied by the gas turbine is $\dot{W}_{21} = 9498 - 765 = 8733\,\text{kW}$.
- ○ From the ExBE, one determines the exergy destruction by the gas turbine $\dot{Ex}_{d,gt} = 23,767\,\text{kW}$ and the exergy efficiency of the gas turbine $\psi_{gt} = 25.2\%$.
- Remaining components (filter, blower, cyclone, and recirculation flap):
  - ○ The procedure of analysis goes further in a similar manner and determines the exergy destructions for the remaining components: filter, blower, cyclone, and recirculation flap.
  - ○ This has been pursued with an EES code, and the results are $\dot{Ex}_{fi} = 3227\,\text{kW}$, $\dot{Ex}_{fa} = 161\,\text{kW}$, $\dot{Ex}_{c} = 3872\,\text{kW}$, $\dot{Ex}_{rf} = 807\,\text{kW}$.
- Assessment parameters:
  - ○ The total energy input is $\dot{E}_{in} = 59,087\,\text{kW}$.
  - ○ Because energy is recycled within the system, an improved energy efficiency is obtained $\eta = 58.6\%$.
  - ○ Total exergy destruction is $\dot{Ex}_{d} = 46,423\,\text{kW}$.
  - ○ Total exergy input is $\dot{Ex}_{in} = \dot{Ex}_{2} + \dot{Ex}_{4} = 61,450\,\text{MW}$.
  - ○ Exergy efficiency of the overall system $\psi = 1 - \dot{Ex}_{d}/\dot{Ex}_{in} = 24.4\%$.
  - ○ Sustainability index $SI = 1.54$.
  - ○ The relative irreversibility is represented as a pie chart as shown in Figure 4.3.
  - ○ For 1 kJ heat input, 0.127 g of moisture is removed.

## 4.5  Analysis of Heat Pump Dryers

Heat pump dryers are very attractive from thermodynamic point of view because of their ability to recover heat and to remove moisture from the drying agent (air). In fact, with a heat pump dryer, an excellent match of temperature levels between supply side and demand side is obtained. Minimization of the temperature differences leads to reduced exergy destructions. The humidity control is achieved by the heat pump systems mainly with the help of evaporators which are able to extract the moisture before the drying air is recirculated. Controlling the

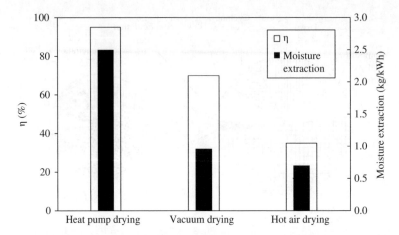

**Figure 4.4** Comparison of heat pump drying with other conventional drying methods in terms of energy efficiency (η) and moisture extraction for kilojoule of heat input

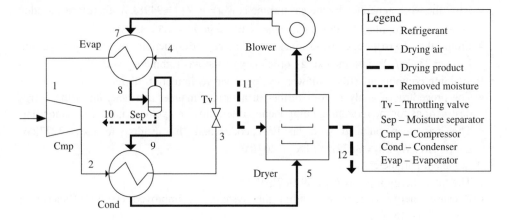

**Figure 4.5** Heat pump drying system of simple configuration

drying conditions is critical especially when the product quality is important, and precise requirements must meet for a wide range of products.

The heat pump dryers are introduced in Chapter 3, Section 3.4.2. Heat pump dryers can be applied to any conventional dryer which operate with forced air and moderate drying temperatures. Figure 4.4 shows a bar chart comparison of heat pump dryers with systems that use hot air drying and vacuum drying, in terms of drying efficiency. This comparison clearly demonstrates the superiority of heat pump drying.

A heat pump drying system of simple configuration is shown in Figure 4.5. For this system, the energy and exergy analyses are demonstrated subsequently. This system consists of three circuits: the refrigerant circuit (closed loop), the drying air circuit (semiclosed loop; it allows for moisture extraction in state 10), and drying product circuit (open circuit). The moisture separator is normally in form of a tray placed below the evaporator. Moisture condenses on the cold surface of the evaporators and is collected in the tray. At steady-state operation, the moist materials enter in state 11 and interact with the dry air supplied in state 5, which extracts the moisture; the dried product leaves the dryer in state 12. The humid air is "pumped" over the

**Table 4.4** Balance equations for the heat pump dryer shown in Figure 4.5

| Component | Balance equations |
|---|---|
| Compressor | MBE: $\dot{m}_1 = \dot{m}_2 = \dot{m}_r$ |
| | EBE: $\dot{m}_1 h_1 + \dot{W}_{cmp} = \dot{m}_2 h_2$ |
| | EnBE: $\dot{m}_1 s_1 + \dot{S}_{g,cmp} = \dot{m}_2 s_2$ |
| | ExBE: $\dot{m}_1 ex_1 + \dot{W}_{cmp} = \dot{m}_2 ex_2 + \dot{Ex}_{d,cmp}$ |
| Condenser | MBE: $\dot{m}_2 = \dot{m}_3 = \dot{m}_r; \dot{m}_9 = \dot{m}_5$ |
| | EBE: $\dot{m}_2 h_2 + \dot{m}_9 h_9 = \dot{m}_3 h_3 + \dot{m}_5 h_5$ |
| | EnBE: $\dot{m}_2 s_2 + \dot{m}_9 s_9 + \dot{S}_{g,cond} = \dot{m}_3 s_3 + \dot{m}_5 s_5$ |
| | ExBE: $\dot{m}_2 ex_2 + \dot{m}_9 ex_9 = \dot{m}_3 ex_3 + \dot{m}_5 ex_5 + \dot{Ex}_{d,cond}$ |
| Throttling valve | MBE: $\dot{m}_3 = \dot{m}_4 = \dot{m}_r$ |
| | EBE: $\dot{m}_3 h_3 = \dot{m}_4 h_4$ |
| | EnBE: $\dot{m}_3 s_3 + \dot{S}_{g,tv} = \dot{m}_4 s_4$ |
| | ExBE: $\dot{m}_3 ex_3 = \dot{m}_4 ex_4 + \dot{Ex}_{d,tv}$ |
| Evaporator | MBE: $\dot{m}_4 = \dot{m}_1 = \dot{m}_r; \dot{m}_7 = \dot{m}_8$ |
| | EBE: $\dot{m}_4 h_4 + \dot{m}_7 h_7 = \dot{m}_1 h_1 + \dot{m}_8 h_8$ |
| | EnBE: $\dot{m}_4 s_4 + \dot{m}_7 s_7 + \dot{S}_{g,evap} = \dot{m}_1 s_1 + \dot{m}_8 s_8$ |
| | ExBE: $\dot{m}_4 ex_4 + \dot{m}_7 ex_7 = \dot{m}_1 ex_1 + \dot{m}_8 ex_8 + \dot{Ex}_{d,evap}$ |
| Dryer | MBE: $\dot{m}_5 + \dot{m}_{11} = \dot{m}_6 + \dot{m}_{12}; \dot{m}_{w,5} + \dot{m}_{w,11} = \dot{m}_{w,6} + \dot{m}_{w,12}$ |
| | EBE: $\dot{m}_5 h_5 + \dot{m}_{11} h_{11} = \dot{m}_6 h_6 + \dot{m}_{12} h_{12} + \dot{Q}_{loss}$ |
| | EnBE: $\dot{m}_5 s_5 + \dot{m}_{11} s_{11} + \dot{S}_{g,dryer} = \dot{m}_6 s_6 + \dot{m}_{12} s_{12} + \dfrac{\dot{Q}_{loss}}{\overline{T}_{loss}}$ |
| | ExBE: $\dot{m}_5 ex_5 + \dot{m}_{11} ex_{11} = \dot{m}_6 ex_6 + \dot{m}_{12} ex_{12} + \dot{Q}_{loss}\left(1 - \dfrac{T_0}{\overline{T}_{loss}}\right) + \dot{Ex}_{d,dryer}$ |
| Blower | MBE: $\dot{m}_6 = \dot{m}_7; \dot{m}_{w,6} = \dot{m}_{w,7}$ |
| | EBE: $\dot{m}_6 h_6 + \dot{W}_{blower} = \dot{m}_7 h_7$ |
| | EnBE: $\dot{m}_6 s_6 + \dot{S}_{g,blower} = \dot{m}_7 s_7$ |
| | ExBE: $\dot{m}_6 ex_6 + \dot{W}_{blower} = \dot{m}_7 ex_7 + \dot{Ex}_{d,blower}$ |
| Moisture separator | MBE: $\dot{m}_8 = \dot{m}_9 + \dot{m}_{10}; \dot{m}_{w,9} = \dot{m}_9 - (\dot{m}_9 - \dot{m}_{w,8})$ |
| | EBE: $\dot{m}_8 h_8 = \dot{m}_9 h_9 + \dot{m}_{10} h_{10}$ |
| | EnBE: $\dot{m}_8 s_8 + \dot{S}_{g,sep} = \dot{m}_9 s_9 + \dot{m}_{10} s_{10}$ |
| | ExBE: $\dot{m}_8 ex_8 = \dot{m}_9 ex_9 + \dot{m}_{10} ex_{10} + \dot{Ex}_{d,sep}$ |
| Overall system | MBE: $\dot{m}_{11} = \dot{m}_{10} + \dot{m}_{12}$ |
| | EBE: $\dot{m}_{11} h_{11} + \dot{W}_{cmp} + \dot{W}_{blower} = \dot{m}_{10} h_{10} + \dot{m}_{12} h_{12} + \dot{Q}_{loss}$ |
| | EnBE: $\dot{m}_{11} s_{11} + \dot{S}_{g,sys} = \dot{m}_{10} s_{10} + \dot{m}_{12} s_{12} + \dfrac{\dot{Q}_{loss}}{\overline{T}_{loss}}$ |
| | ExBE: $\dot{m}_{11} ex_{11} + \dot{W}_{cmp} + \dot{W}_{blower} = \dot{m}_{10} ex_{10} + \dot{m}_{12} ex_{12} + \dot{Q}_{loss}\left(1 - \dfrac{T_0}{\overline{T}_{loss}}\right) + \dot{Ex}_{d,sys}$ |

evaporator where the moisture is extracted and heat is recovered. Further, the dry air in state 9 is heated by the condenser and delivered to the dryer at the desired temperature and humidity. The heat pump can be of a simple, single-stage vapor compression type, or more complex and high performance heat pump systems can be applied.

From the ExBE for the overall system (see Table 4.4), the total exergy destruction $\dot{Ex}_{d,tot}$ can be determined. This is the sum of exergy destruction due to the irreversibility $\dot{Ex}_{d,sys}$ and the exergy destruction (lost) at the interaction of the system with the reference environment.

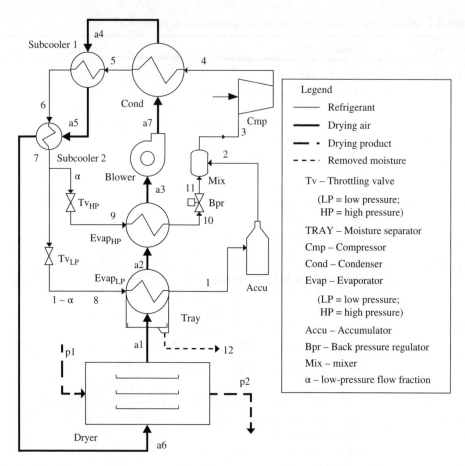

**Figure 4.6** Heat pump drying system with two-stage evaporators (modified from Chua and Chou (2005))

A more complex heat pump dryer is studied next. It includes a heat pump, having two-stage evaporators. The system was initially proposed in Mujumdar (2006) and later developed, built, and tested by Chua and Chou (2005) for agricultural product drying. In this system, the high-pressure evaporator cools the air, and the low-pressure evaporator is used to remove latent heat and moisture. This system was further analyzed by energy and exergy methods by Acar and Dincer (2014). Figure 4.6 shows the system schematics. The system has one high- and one low-pressure evaporators, two subcoolers, one condenser, one compressor, two expansion valves, one accumulator, and one drier. The high-pressure evaporator has a tray underneath where the condensate is collected. A back pressure regulator (BPR) is required in the system to control the pressure at the high-pressure evaporator. The regulator adjusts the flow such that the pressure at state 10 remains constant.

The balance equations for this system are given in Table 4.5. Note that the MBEs must reflect the conservation of the general mass but also the conservation of each species; since here, there are no chemical reaction. Therefore, the mass of moisture, the mass of dry product, the mass of dry air, and the mass of refrigerant are all conserved.

**Table 4.5**  Balance equations for the heat pump dryer shown in Figure 4.5

| Component | Balance equations |
|---|---|
| Compressor | MBE: $\dot{m}_3 = \dot{m}_4 = \dot{m}_r$ |
| | EBE: $\dot{m}_3 h_3 + \dot{W}_{cmp} = \dot{m}_4 h_4$ |
| | EnBE: $\dot{m}_3 s_3 + \dot{S}_{g,cmp} = \dot{m}_4 s_4$ |
| | ExBE: $\dot{m}_3 ex_3 + \dot{W}_{cmp} = \dot{m}_4 ex_4 + \dot{Ex}_{d,cmp}$ |
| Condenser | MBE: $\dot{m}_4 = \dot{m}_5 = \dot{m}_r$; $\dot{m}_{a7} = \dot{m}_{a4}$; $\omega_{a7} = \omega_{a4}$ |
| | EBE: $\dot{m}_4 h_4 + \dot{m}_{a7} h_{a7} = \dot{m}_5 h_5 + \dot{m}_{a4} h_{a4}$ |
| | EnBE: $\dot{m}_4 s_4 + \dot{m}_{a7} s_{a7} + \dot{S}_{g,cond} = \dot{m}_5 s_5 + \dot{m}_{a4} s_{a4}$ |
| | ExBE: $\dot{m}_4 ex_4 + \dot{m}_{a7} ex_{a7} = \dot{m}_5 ex_5 + \dot{m}_{a4} ex_{a4} + \dot{Ex}_{d,cond}$ |
| Subcooler 1 | MBE: $\dot{m}_5 = \dot{m}_6 = \dot{m}_r$; $\dot{m}_{a4} = \dot{m}_{a5}$; $\omega_{a4} = \omega_{a5}$ |
| | EBE: $\dot{m}_5 h_5 + \dot{m}_{a4} h_{a4} = \dot{m}_6 h_6 + \dot{m}_{a5} h_{a5}$ |
| | EnBE: $\dot{m}_5 s_5 + \dot{m}_{a4} s_{a4} + \dot{S}_{g,sc1} = \dot{m}_6 s_6 + \dot{m}_{a5} s_{a5}$ |
| | ExBE: $\dot{m}_5 ex_5 + \dot{m}_{a4} ex_{a4} = \dot{m}_6 ex_6 + \dot{m}_{a5} ex_{a5} + \dot{Ex}_{d,sc1}$ |
| Subcooler 2 | MBE: $\dot{m}_6 = \dot{m}_7 = \dot{m}_r$; $\dot{m}_{a5} = \dot{m}_{a6}$; $\omega_{a5} = \omega_{a6}$ |
| | EBE: $\dot{m}_6 h_6 + \dot{m}_{a5} h_{a5} = \dot{m}_7 h_7 + \dot{m}_{a6} h_{a6}$ |
| | EnBE: $\dot{m}_6 s_6 + \dot{m}_{a5} s_{a5} + \dot{S}_{g,sc2} = \dot{m}_7 s_7 + \dot{m}_{a6} s_{a6}$ |
| | ExBE: $\dot{m}_6 ex_6 + \dot{m}_{a5} ex_{a5} = \dot{m}_7 ex_7 + \dot{m}_{a6} ex_{a6} + \dot{Ex}_{d,sc2}$ |
| Throttling valve, LP | MBE: $\dot{m}_7 = \dot{m}_8 + \dot{m}_9 = \dot{m}_r$; $\dot{m}_9 = \alpha\,\dot{m}_7$; $\alpha \in [0,1]$ |
| | EBE: $(1-\alpha)\dot{m}_7 h_7 = \dot{m}_8 h_8$ |
| | EnBE: $(1-\alpha)\dot{m}_7 s_7 + \dot{m}_{g,tv,LP} = \dot{m}_8 s_8$ |
| | ExBE: $(1-\alpha)\dot{m}_7 ex_7 = \dot{m}_8 ex_8 + \dot{Ex}_{d,tv,LP}$ |
| Evaporator, LP | MBE: $\dot{m}_8 = \dot{m}_1$; $\dot{m}_{a1} = \dot{m}_{a2} + \dot{m}_{12}$; $\dot{m}_{a1}(1-\omega_{a1}) = \dot{m}_{a2}(1-\omega_{a2})$ |
| | EBE: $\dot{m}_8 h_8 + \dot{m}_{a1} h_{a1} = \dot{m}_1 h_1 + \dot{m}_{a2} h_{a2} + \dot{m}_{12} h_{12}$ |
| | EnBE: $\dot{m}_8 s_8 + \dot{m}_{a1} s_{a1} + \dot{S}_{g,evap,LP} = \dot{m}_1 s_1 + \dot{m}_{a2} s_{a2} + \dot{m}_{12} s_{12}$ |
| | ExBE: $\dot{m}_8 ex_8 + \dot{m}_{a1} ex_{a1} = \dot{m}_1 ex_1 + \dot{m}_{a2} ex_{a2} + \dot{m}_{12} ex_{12} + \dot{Ex}_{d,evap,LP}$ |
| Throttling valve, HP | MBE: $\alpha\dot{m}_7 = \dot{m}_9$ |
| | EBE: $\alpha\dot{m}_7 h_7 = \dot{m}_9 h_9$ |
| | EnBE: $\alpha\dot{m}_7 s_7 + \dot{S}_{g,tv,HP} = \dot{m}_9 s_9$ |
| | ExBE: $\alpha\,\dot{m}_7 ex_7 = \dot{m}_9 ex_9 + \dot{Ex}_{d,tv,HP}$ |
| Evaporator, HP | MBE: $\dot{m}_9 = \dot{m}_{10}$; $\dot{m}_{a2} = \dot{m}_{a3}$; $\omega_{a2} = \omega_{a3}$ |
| | EBE: $\dot{m}_9 h_9 + \dot{m}_{a2} h_{a2} = \dot{m}_{10} h_{10} + \dot{m}_{a3} h_{a3}$ |
| | EnBE: $\dot{m}_9 s_9 + \dot{m}_{a2} s_{a2} + \dot{S}_{g,evap,HP} = \dot{m}_{10} s_{10} + \dot{m}_{a3} s_{a3}$ |
| | ExBE: $\dot{m}_9 ex_9 + \dot{m}_{a2} ex_{a2} = \dot{m}_{10} ex_{10} + \dot{m}_{a3} ex_{a3} + \dot{Ex}_{d,evap,HP}$ |
| Accumulator | MBE: $\dot{m}_1 = \dot{m}_2$ |
| | EBE: $\dot{m}_1 h_1 = \dot{m}_2 h_2$ |
| | EnBE: $\dot{m}_1 s_1 + \dot{S}_{g,accu} = \dot{m}_2 s_2$; $\dot{S}_{g,accu} \cong 0$ |
| | ExBE: $\dot{m}_1 ex_1 = \dot{m}_2 ex_2 + \dot{Ex}_{d,accu}$; $\dot{Ex}_{d,accu} \cong 0$ |
| Back pressure regulator | MBE: $\dot{m}_{10} = \dot{m}_{11}$ |
| | EBE: $\dot{m}_{10} h_{10} = \dot{m}_{11} h_{11}$ |
| | EnBE: $\dot{m}_{10} s_{10} + \dot{S}_{g,bpr} = \dot{m}_{11} s_{11}$ |
| | ExBE: $\dot{m}_{10} ex_{10} = \dot{m}_{11} ex_{11} + \dot{Ex}_{d,bpr}$ |
| Mixer | MBE: $\dot{m}_2 + \dot{m}_{11} = \dot{m}_3 = \dot{m}_r$ |
| | EBE: $\dot{m}_2 h_2 + \dot{m}_{11} h_{11} = \dot{m}_3 h_3$ |

*(continued overleaf)*

**Table 4.5**  *(continued)*

| Component | Balance equations |
|---|---|
| Blower | EnBE: $\dot{m}_2 s_2 + \dot{m}_{11} s_{11} + \dot{S}_{g,mix} = \dot{m}_3 s_3$ <br> ExBE: $\dot{m}_2 ex_2 + \dot{m}_{11} ex_{11} = \dot{m}_3 ex_3 + \dot{Ex}_{d,mix}$ <br> MBE: $\dot{m}_{a3} = \dot{m}_{a7}$; $\omega_{a3} = \omega_{a7}$ <br> EBE: $\dot{m}_{a3} h_{a3} + \dot{W}_{blower} = \dot{m}_{a7} h_{a7}$ <br> EnBE: $\dot{m}_{a3} s_{a3} + \dot{S}_{g,blower} = \dot{m}_{a7} s_{a7}$ <br> ExBE: $\dot{m}_{a3} ex_{a3} + \dot{W}_{blower} = \dot{m}_{a7} ex_{a7} + \dot{Ex}_{d,blower}$ |
| Dryer | MBE: $\dot{m}_{a6} + \dot{m}_{p1} = \dot{m}_{a1} + \dot{m}_{p2}$; $\dot{m}_{w,a6} + \dot{m}_{w,p1} = \dot{m}_{w,a1} + \dot{m}_{w,p2}$ <br> EBE: $\dot{m}_{a6} h_{a6} + \dot{m}_{p1} h_{p1} = \dot{m}_{a1} h_{a1} + \dot{m}_{p2} h_{p2} + \dot{Q}_{loss}$ <br> EnBE: $\dot{m}_{a6} s_{a6} + \dot{m}_{p1} s_{p1} + \dot{S}_{g,dryer} = \dot{m}_{a1} s_{a1} + \dot{m}_{p2} s_{p2} + \dfrac{\dot{Q}_{loss}}{\overline{T}_{loss}}$ <br> ExBE: $\dot{m}_{a6} ex_{a6} + \dot{m}_{p1} ex_{p1} = \dot{m}_{a1} ex_{a1} + \dot{m}_{p2} ex_{p2} + \dot{Q}_{loss}\left(1 - \dfrac{T_0}{\overline{T}_{loss}}\right) + \dot{Ex}_{d,dryer}$ |
| Overall system | MBE: $\dot{m}_{p1} = \dot{m}_{p2} + \dot{m}_{12}$; $\dot{m}_{w,p1} = \dot{m}_{w,p2} + \dot{m}_{12}$ <br> EBE: $\dot{m}_{p1} h_{p1} + \dot{W}_{cmp} + \dot{W}_{blower} = \dot{m}_{p2} h_{p2} + \dot{m}_{12} h_{12} + \dot{Q}_{loss}$ <br> EnBE: $\dot{m}_{p1} s_{p1} + \dot{S}_{g,sys} = \dot{m}_{p2} s_{p2} + \dot{m}_{12} s_{12} + \dfrac{\dot{Q}_{loss}}{\overline{T}_{loss}}$ <br> ExBE: $\dot{m}_{p1} ex_{p1} + \dot{W}_{cmp} + \dot{W}_{blower} = \dot{m}_{p2} ex_{p2} + \dot{m}_{12} ex_{12} + \dot{Q}_{loss}\left(1 - \dfrac{T_0}{\overline{T}_{loss}}\right) + \dot{Ex}_{d,sys}$ |

In the analyses, the potential and kinetic energy changes during the course of operation are neglected within the system. All processes are assumed to be steady state steady flow. The pressure drop in tubing connecting the components, valves, and passive component can be neglected. However, for modeling the valve that regulates pressure such as throttling valves and BPR, the pressure drop must be specified. Furthermore, all the tubing, piping elements, valves, and accumulator are assumed adiabatic. All heat exchangers (condenser and two evaporators) are assumed well insulated such that there is no thermal interaction with the surroundings. However, heat losses must be considered for modeling the dryer.

The COP of the heat pump is a parameter of interest which can be useful for system assessment. The general definition for COP is introduced in Chapter 1. Here, with the notations from Figure 4.5, the mathematical expression for the energetic COP becomes

$$COP_{en} = \frac{\dot{m}_9 (h_5 - h_9)}{\dot{W}_{cmp}} \tag{4.31}$$

whereas for the exergetic COP, they are defined as follows:

$$COP_{ex} = \frac{\dot{m}_9 (ex_5 - ex_9)}{\dot{W}_{cmp}} \tag{4.32}$$

If the more complex system from Figure 4.6 is considered, then the COPs are defined as

$$\text{COP}_{\text{en}} = \frac{\dot{m}_{a7}(h_{a4} - h_{a7})}{\dot{W}_{\text{cmp}}} \tag{4.33}$$

$$\text{COP}_{\text{ex}} = \frac{\dot{m}_{a7}(\text{ex}_{a4} - \text{ex}_{a7})}{\dot{W}_{\text{cmp}}} \tag{4.34}$$

Note that in Eqs. (4.26) and (4.27), the $\text{COP}_{\text{ex}}$ represents the exergetic COP of the heat pumps. The energy and exergy efficiencies of the system subunits can be defined. For the heat pump drying system with two evaporators, these definitions are given in Table 4.6.

Since compressor power depends strongly on the inlet and outlet pressures, any heat exchanger improvements that reduce the temperature difference will reduce compressor power by bringing the condensing and evaporating temperatures closer together. From a design standpoint, compressor irreversibility can be reduced independently. The compressor irreversibility is partly due to the large degree of superheat achieved at the end of the compression process, leading to large temperature differences associated with the initial phase of heat transfer. The mechanical–electrical losses are due to imperfect electrical, mechanical, and isentropic

**Table 4.6** Energy and exergy efficiencies of system units

| Units | Energy efficiency | Exergy efficiency |
| --- | --- | --- |
| Compressor | $\eta_1 = \dfrac{\dot{E}_4 - \dot{E}_3}{\dot{W}_1}$ | $\psi_1 = \dfrac{\dot{E}x_4 - \dot{E}x_3}{\dot{W}_1}$ |
| Condenser | $\eta_2 = \dfrac{\dot{E}_{a4} - \dot{E}_{a3}}{\dot{E}_4 - \dot{E}_5}$ | $\psi_2 = \dfrac{\dot{E}x_{a4} - \dot{E}x_{a3}}{\dot{E}x_4 - \dot{E}x_5}$ |
| Subcooler 1 | $\eta_3 = \dfrac{\dot{E}_{a5} - \dot{E}_{a4}}{\dot{E}_5 - \dot{E}_6}$ | $\psi_3 = \dfrac{\dot{E}x_{a5} - \dot{E}x_{a6}}{\dot{E}x_5 - \dot{E}x_6}$ |
| Subcooler 2 | $\eta_4 = \dfrac{\dot{E}_{a6} - \dot{E}_{a5}}{\dot{E}_6 - \dot{E}_7}$ | $\psi_3 = \dfrac{\dot{E}x_{a6} - \dot{E}x_{a5}}{\dot{E}x_6 - \dot{E}x_7}$ |
| Throttling valve, LP | $\eta_5 = \dfrac{\dot{E}_9}{\dot{E}_{7}\alpha}$ | $\psi_5 = \dfrac{\dot{E}x_9}{\dot{E}x_{7}\alpha}$ |
| Throttling valve, HP | $\eta_6 = \dfrac{\dot{E}_8}{\dot{E}_7(1-\alpha)}$ | $\psi_6 = \dfrac{\dot{E}x_8}{\dot{E}x_7(1-\alpha)}$ |
| Evaporator, LP | $\eta_7 = \dfrac{\dot{E}_1 - \dot{E}_8}{\dot{E}_{a1} - \dot{E}_{a2}}$ | $\psi_7 = \dfrac{\dot{E}x_1 - \dot{E}x_8}{\dot{E}x_{a1} - \dot{E}x_{a2}}$ |
| Evaporator, HP | $\eta_8 = \dfrac{\dot{E}_{10} - \dot{E}_9}{\dot{E}_{a2} - \dot{E}_{a3}}$ | $\psi_8 = \dfrac{\dot{E}x_{10} - \dot{E}x_9}{\dot{E}x_{a2} - \dot{E}x_{a3}}$ |
| Back pressure regulator | $\eta_9 = \dfrac{\dot{E}_{11}}{\dot{E}_{10}}$ | $\psi_9 = \dfrac{\dot{E}x_{11}}{\dot{E}x_{10}}$ |
| Accumulator | $\eta_{10} = \dfrac{\dot{E}_2}{\dot{E}_1}$ | $\psi_{10} = \dfrac{\dot{E}x_2}{\dot{E}x_1}$ |
| Mixing chamber | $\eta_{11} = \dfrac{\dot{E}_3}{\dot{E}_2 + \dot{E}_{11}}$ | $\psi_{11} = \dfrac{\dot{E}x_3}{\dot{E}x_2 + \dot{E}x_{11}}$ |

efficiencies and emphasize the need for careful selection of this equipment, since components of inferior performance can considerably reduce overall system performance.

### Example 4.2

Consider a heat pump dryer system according to Figure 4.6. For this numerical example, the assumed working fluid of the heat pump is R134a. Here, we give in Table 4.7 the numerical values for the main parameters characterizing the system operation. The summary of each stream, along with their corresponding stream number, component, mass flow rate (kg/s), temperature (°C), pressure (kPa), and state, is listed in Table 4.8. These data allow for conducting the system analysis based on energy and exergy methods:

**Table 4.7** Input data for the system

| Parameter and measurement unit | Symbol | Value |
|---|---|---|
| Refrigerant mass flow rate (kg/s) | $m_r$ | 0.066 |
| Air mass flow rate (kg/s) | $m_a$ | 0.5 |
| The mass ratio of the refrigerant sent to HP evaporator | $\alpha$ | 0.35 |
| Compressor inlet temperature (K) | $T_3$ | 280.6 |
| Compressor inlet pressure (kPa) | $P_3$ | 500 |
| Degree of subcooling in Subcooler 1 (K) | $\Delta T_{SC1}$ | 5 |
| Degree of subcooling in Subcooler 2 (K) | $\Delta T_{SC2}$ | 5 |
| Ambient temperature (K) | $T_0$ | 278.15 |
| Ambient pressure (kPa) | $P_0$ | 101.33 |

*Source*: Adapted from Chua and Chou (2005).

**Table 4.8** Summary of each stream along with their component and properties, for Example 4.2

| Stream | Component | State | $\dot{m}\,(\mathrm{kg/s})$ | $T\,(°C)$ | $P\,(\mathrm{kPa})$ |
|---|---|---|---|---|---|
| 1 | R-134a | Gas | 0.0429 | 7.45 | 500 |
| 2 | R-134a | Gas | 0.0429 | 7.45 | 500 |
| 3 | R-134a | Gas | 0.066 | 7.45 | 500 |
| 4 | R-134a | Liquid | 0.066 | 83.48 | 2500 |
| 5 | R-134a | Liquid | 0.066 | 61.35 | 2500 |
| 6 | R-134a | Liquid | 0.066 | 56.35 | 2500 |
| 7 | R-134a | Liquid | 0.066 | 51.35 | 2500 |
| 8 | R-134a | VLE | 0.0429 | 0.1148 | 500 |
| 9 | R-134a | VLE | 0.0231 | 23.4 | 1000 |
| 10 | R-134a | Gas | 0.0231 | 23.4 | 1000 |
| 11 | R-134a | Gas | 0.0231 | 7.45 | 500 |
| a1 | Air | Gas | 0.5 | 28.05 | 101.3 |
| a2 | Air | Gas | 0.5 | 25.27 | 101.3 |
| a3 | Air | Gas | 0.5 | 23.6 | 101.3 |
| a4 | Air | Gas | 0.5 | 44.21 | 101.3 |
| a5 | Air | Gas | 0.5 | 45.17 | 101.3 |
| a6 | Air | Gas | 0.5 | 46.08 | 101.3 |

- The analysis procedure is similar to Example 4.1; it requires the writing of all balance equations. These are given already in Table 4.5.
- The EES software can be used to solve the problem as it includes the equations of state for R134a, humid air, water, dry air, and so on.
- Starting from the data given in Tables 4.7 and 4.8, the balance equations can be solved and exergy destructions calculated for all components; as well, the energy and exergy efficiencies can be calculated and SI.
- Here, the numerical results are presented in Table 4.9; in addition, a chart illustrating the relative irreversibilities is shown in Figure 4.7.
- It results that the low-pressure evaporator produces the highest irreversibility as it destroys the most of the exergy (24%); also, the condenser destroys 22% of the exergy and is followed by the high-pressure evaporator with 19% irreversibility. In conclusion, over 60% of the

**Table 4.9** Exergy efficiency, rate of exergy destruction, RI, and SI of the major units of the heat pump dryer system

| Unit | $\psi$ (%) | $\dot{Ex}_{dest}$ (kW) | RI | SI |
|------|-----------|------------------------|-----|-----|
| Compressor | 84 | 0.37 | 0.10 | 6.2 |
| Condenser | 59 | 0.83 | 0.22 | 2.42 |
| Subcooler 1 | 79 | 0.02 | 0.5 | 4.78 |
| Subcooler 2 | 87 | 0.01 | 0.5 | 7.49 |
| Throttling valve, LP | 94 | 0.06 | 0.02 | 17.51 |
| Throttling valve, HP | 85 | 0.07 | 0.02 | 6.61 |
| Evaporator, LP | 41 | 0.9 | 0.24 | 1.68 |
| Evaporator, HP | 37 | 0.7 | 0.19 | 1.58 |
| Back pressure regulator | 72 | 0.17 | 0.05 | 3.51 |
| Dryer | 66 | 0.42 | 0.11 | 2.9 |

*Source*: Acar and Dincer (2014).

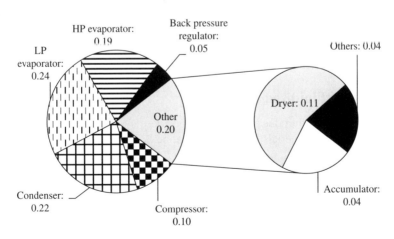

**Figure 4.7** Heat pump drying system with two-stage evaporators (data from Acar and Dincer (2014))

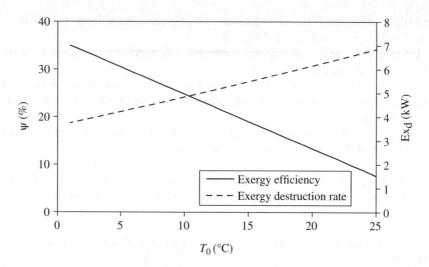

**Figure 4.8** Effect of reference temperature on exergetic performance (data from Acar and Dincer (2014))

exergy destruction of the overall system is caused by the operation of the evaporators and the condenser.

- The energy and the exergy efficiencies of the system are 62% and 35%, respectively.
- The overall SI of the system is 1.54.
- The overall system's performance change with respect to the change in ambient temperature is similar to the performance change of the condenser and the dryer. As shown in Figure 4.8, the exergy efficiency decreases when the reference temperature increases; on the same plot, the exergy destruction rate is shown.
- Based on this analysis, one can conclude that the evaporators destroy the most of the exergy. Therefore, the suggested improvement is a reduction of temperature difference at the evaporators. Even more improvement is obtained if the temperature difference can be decreased at the condenser. Obviously, these require more capital investment in the heat exchangers. Reducing the temperature difference brings evaporator and condensing temperatures together increases the amount of heat recovered per amount of work provided to the system.

## 4.6   Analysis of Fluidized Bed Dryers

Fluidized bed dryers are introduced in Chapter 3. These systems are very important in the industry when granulate materials or powders must be dried. The analyses for fluidized beds present a series of peculiarities, mostly related to the fluidization process. Therefore, some specific correlation must be used, and sometimes, transient modeling equations may be formulated. In this section, the energy and exergy analyses for fluidized bed dryers are presented.

The process of gas–solid fluidization is the most significant in these drying systems. In this process, the solid phase, under fluidization conditions created by a rapidly flowing gas, assumes a "fluidlike" behavior. The drying gas, in most cases air, sustains the fluidization process and, in the same time, extracts moisture from the particles. The most influential parameters on the process are the gas velocity, gas temperature, and temperature gradients inside the bed. Transient modeling may be necessary mostly for the start-up and shutdown procedures.

In fluidized bed dryers, heat, necessary for moisture evaporation, is supplied to the moist particles in the bed and to moisture vapor by heat convection. Further, heat is transferred into the particle. The moisture is transported in the opposite direction as a liquid or vapor; on the surface, it evaporates and is convected to the drying gas which carries it outside the device.

## 4.6.1 Hydrodynamics of Fluidized Beds

The process of fluidization must be studied and understood hydrodynamically and through heat and mass transfer methods; the material properties of the particles or powders must be adequately known. The particles are characterized by their density (e.g., wheat kernels have 1215 kg/m$^3$) and their size (e.g., wheat kernels have an average diameter of 3–4 mm, depending on the type). The density and more importantly the specific heat of the particles vary with their moisture content.

As known, water has a very high specific heat, whereas typical solids have much lower specific heat. Therefore, one of the first aspects in modeling is to correlate the specific heat with the moisture content. Correlations of the type

$$C_p = C_{p,s} + C_{p,w} f(W)$$

are often used, where $C_{p,s}$ is the specific heat of the solid, $C_{p,w}$ is the specific heat of the moisture, and $f(W)$ is a function of moisture content which tends to vanish when the material approaches a dry phase.

The hydrodynamics of the fluidized bed is described by two essential parameters which are the minimum fluidization velocity, usually denoted with $u_{mf}$, and the bed pressure, $\Delta P_0 = N_p m_p g / A_b$, representing the weight of the particles divided by the cross-sectional area of the bed (here, $N_p$ and $m_p$ are the statistically averaged number of particles and the particle mass, $g$ is the gravity constant, and $A_b$ is the area). The fluidization regimes are described qualitatively as shown in Figure 4.9 in correlation with $u_{mf}$ and $\Delta P_0$ parameters.

As seen in the figure, when fluid velocity is small, there is no fluidization; rather, the gas penetrates through the fixed/permeable bed, regime (a). At the onset of fluidization, not all particles are fluidized because of adhesive forces in the bed, and the top layers of the bed usually start fluidizing while the bottom layers are still stationary. Thus, the bed pressure drop is slightly less than the pressure drop equivalent to the weight of bed material. When flow velocity reaches the so-called minimum fluidization velocity $u_{mf}$, the regime of fluidization becomes incipient, regime (b). If the gas velocity continues to increase, the bubbling fluidization regime occurs, regime (c). Increasing the gas velocity further increases the drag force exerted on the particles, which can separate more contact points between particles, thus bringing them to the fluidized state. The pressure drop increases with increasing gas velocity, as more particles need to be suspended. At a certain velocity, all particles are suspended and full fluidization occurs. Once the flow reaches an even higher velocity, equal or superior to the terminal particle velocity, then the drag forces become very important and the pneumatic transport regime occurs, regime (d).

The correlations for the minimum fluidization velocity are of utmost importance for modeling. Some aspects of the process are discussed in Dincer and Rosen (2013). The Archimedes

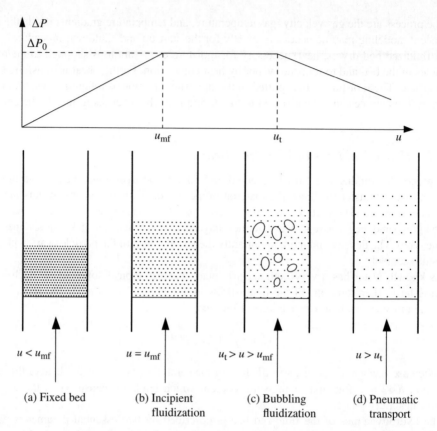

**Figure 4.9**  The correlation between the fluidization regime, flow velocity, and bed pressure drop

number is convenient for describing the fluidization regimes. The higher is $Ar$, the higher the $u_{mf}$. Also, the Reynolds number influences the regime. For higher Reynolds number, the regime displaces toward the pneumatic transport. The Archimedes number and the Reynolds number for minimum fluidization are defined by

$$\begin{cases} |Ar = \dfrac{\rho_g \left(\rho_p - \rho_g\right) d_p^3}{\mu_g} \\[3mm] |Re_{mf} = \dfrac{u_{mf}\, d_p\, \rho_g}{\mu_g} \end{cases} \tag{4.35}$$

A general correlation for minimum fluidization velocity, $\varepsilon_{mf}$, is given by Kunii and Levenspiel (1991). This correlation is given as follows $\varepsilon_{mf}$:

$$\varepsilon_{mf}^3 \Phi_s^2 Ar = 150\left(1 - \varepsilon_{mf}\right) Re_{mf} + 1.75\, Re_{mf}^2 \tag{4.36}$$

where $\varepsilon_{mf}$ represents the bed void fraction at the fluidization velocity and $\Phi_s$ is the dimensionless sphericity factor or the particles.

Once all the required parameters are defined, the equation can be solved for $R_{mf}$, and further, the minimum fluidization velocity can be determined. The minimum fluidization velocity depends on particle moisture content; increasing moisture content increases the minimum fluidization velocity. For wet particle fluidization, the bed pressure drop for velocities above the minimum fluidization point gradually increases with increasing gas velocity.

In the fluidized bed literature, many correlations can be found for the minimum fluidization velocity for specific and well-defined situations. Relating to drying operations, some equations correlate the $u_{mf}$ with the particle moisture content. As one example, we give here the correlation by Hajidavalloo (1998) which gives $u_{mf}$ as follows:

$$u_{mf} = \begin{cases} |0.9813\,W_p + 0.9054, & |\text{for } 0.1 \leq W_p \leq 0.4 \\ |0.1317\ln W_p + 1.42, & |\text{for } 0.4 < W_p \leq 0.6 \end{cases} \tag{4.37}$$

### 4.6.2  Balance Equations

Assume a batch fluidized bed dryer with perfect mixing of particles and uniform pressure gradient across the bed. Further assume that the particles are uniformly distributed within the bed and they have a uniform moisture content, $W_p$ at any moment. Denote the specific enthalpy and entropy of the moist particles with $h_m$ and $s_m$, respectively. Figure 4.10 shows the schematics of the system. The control volume for which the balance equations needs to be written surrounds the fluidized bed.

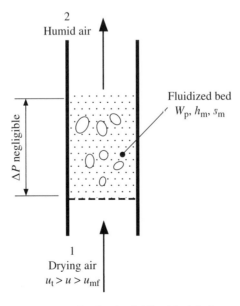

**Figure 4.10**   Schematics for the fluidized bed drying modeling

In the figure, state 1 is the air entrance, while state 2 is the exit; since the model is for batch process, one assumes that no particle enters or exits the control volume. Denote with $\dot{m}$ the mass flow rate of humid air. The general, transient MBE requires that

$$\dot{m}_1 = \frac{dm}{dt} + \dot{m}_2 \tag{4.38}$$

The mass of water must also conserve during the process; therefore, a second MBE can be written. Water enters and exits together with humid air; inside the reactor, water is removed from particles; thence,

$$\dot{m}_a \omega_1 = m_p \frac{dW_p}{dt} + \dot{m}_a \omega_2 \tag{4.39}$$

where $\dot{m}_a$ is the mass flow rate of dry air and $m_p$ is the total mass of dry particles in the bed; this mass does not change because no particles enter or exit. Since it is reasonable to assume that the operation is isobaric, there is no accumulation of air inside the reactor. Equation (4.39) reflects this aspect because it assumes that $\dot{m}_{a1} = \dot{m}_{a2} = \dot{m}_a$.

The EBE must account on heat transfer processes. The main heat transfer is due to the heat of evaporation between the solid and the drying air, and there is also heat transfer with the surroundings. It is reasonable to assume that kinetic and potential energy changes are negligible. Since the mass flow rate of the dry air and the mass of dry material within the control volume remain constant with time, the energy rate balance can be expressed as follows:

$$\dot{m}_a h_1 = m_p C_m \frac{dT_m}{dt} + m_p \Delta h_{fg} \frac{dW_p}{dt} + \dot{m}_a h_2 + \dot{Q}_{loss} \tag{4.40}$$

where $C_m$ is the specific heat of the moist material, assumed constant. In Eq. (4.40), one recognizes that the first term of the RHS represents the sensible heat rate transferred to the articles, whereas the second term of the RHS represents the energy rate for moisture removal that is the rate of evaporation as follows:

$$\dot{Q}_{evap} = m_p \Delta h_{fg} \frac{dW_p}{dt} \tag{4.41}$$

The entropy balance equation becomes

$$\dot{m}_a s_1 + \dot{S}_g = m_p \left( \frac{C_m}{T_m} \right) \left( \frac{dT_m}{dt} \right) + m_p \left( \frac{\Delta h_{lv}}{T_m} \right) \left( \frac{dW_p}{dt} \right) + \dot{m}_a h_2 + \frac{\dot{Q}_{loss}}{\bar{T}_{loss}} \tag{4.42}$$

where $s_1$ and $s_2$ are the specific entropies of moist air as given by Eq. (4.8).

Furthermore, the ExBE is derived as follows:

$$\dot{m}_a ex_1 = m_p C_m \left[ \frac{dT_m}{dt} - \left( \frac{T_0}{T_m} \right) \frac{dT_m}{dt} + \frac{\Delta h_{fg}}{C_m} \left( 1 - \frac{T_0}{T_m} \right) \frac{dW_p}{dt} \right] + \dot{m}_a s_2 + \dot{Q}_{loss} \left( 1 - \frac{T_0}{\bar{T}_{loss}} \right) + \dot{Ex}_{d,int} \tag{4.43}$$

Here, the first term of the RHS represents the exergy rate used for sensible heating of the particles, whereas the second term is the exergy rate used to evaporate the moisture, namely,

$$\dot{Ex}_{evap} = m_p \Delta h_{fg} \left(1 - \frac{T_0}{T_m}\right)\left(\frac{dW_p}{dt}\right) \tag{4.44}$$

The total exergy destruction is represented by the sum of the exergy destroyed due to the internal irreversibilities $\dot{Ex}_{d,int}$ and the exergy destruction due to system interaction with the reference environment. Therefore, the total exergy destruction can be expressed as follows:

$$\dot{Ex}_{d,tot} = \dot{Q}_{loss}\left(1 - \frac{T_0}{T_{loss}}\right) + \dot{Ex}_{d,int} \tag{4.45}$$

## 4.6.3   Efficiency Formulations

The considered case herein represents a nonsteady-state drying operation. Denote the drying time with $t_d$. The final moisture content $W_{p,f}$ and the final moisture temperature $T_{m,f}$ of the product can be determined by integration of the balance equation from $t = 0$ to $t = t_d$. The initial conditions are represented by the specified moisture content $W_{p,i}$ and moist material temperature $T_{m,i}$ at $t = 0$.

The amount of energy used by the process represents the total thermal energy transferred to the moist material in order to perform the drying operation which can be determined with the following equation:

$$E_{used} = m_p\left[\Delta h_{fg}\left(W_{p,i} - W_{p,f}\right) + C_m\left(T_{m,f} - T_{m,i}\right)\right] \tag{4.46}$$

The energy consumed by the dryer (the input energy) during the whole drying period is

$$E_{in} = \dot{m}_a(h_1 - h_0)t_d \tag{4.47}$$

Equations (4.39) and (4.40) permit for the formulation of an energy efficiency of the drying process, namely,

$$\eta = \frac{E_{used}}{E_{in}} = \frac{m_p\left[\Delta h_{fg}\left(W_{p,i} - W_{p,f}\right) + C_m\left(T_{m,f} - T_{m,i}\right)\right]}{\dot{m}_a(h_1 - h_0)t_d} \tag{4.48}$$

An exergy efficiency for the dryer column, which provides a true measure of its performance, can be derived using an exergy balance. In defining the exergy efficiency, it is necessary to identify both a "product" and a "fuel." Here, the product is the exergy evaporation rate, and the fuel is the rate of exergy drying air entering the dryer. In a rate format, the exergy efficiency can be defined as follows:

$$\psi = \frac{\dot{Ex}_{evap}}{\dot{Ex}_{in}} = \frac{m_p \Delta h_{fg}\left(1 - \frac{T_0}{T_m}\right)\left(\frac{dW_p}{dt}\right)}{\dot{m}_a ex_1} \tag{4.49}$$

**Example 4.3**

This example is an illustrative case study of fluidized bed dryer analysis through energy and exergy methods. The example is adapted from the works by Dincer and Rosen (2013) and Syahrul et al. (2002a,b) and is based on the experimental data obtained by Hajidavalloo (1998). The particulate material subjected to fluidized bed drying is of two kinds: wheat and corn. Note that wheat has grains of substantially smaller size than corn; therefore, the analysis is expected to emphasize the differences between the two materials:

- The input data for this case consists of initial temperature of the product, air parameters at the entrance, reference state temperature, mass of the moist material (dry basis), initial moisture content, and air velocity. Two sets of data are considered as given in Table 4.10. The reference temperature is set to $T_0 = 22°C$.
- The results generated by the model (which works based on the equations presented previously) are shown in Figure 4.11 for the variation of moisture content during the drying period.

**Table 4.10**  Input data for Example 4.3

| Air parameters at inlet | | | Moist material parameters | | |
| --- | --- | --- | --- | --- | --- |
| $T_1$ (°C) | $\phi_1$ | $U$ (m/s) | $W_{p,i}$ (d. b.) | $T_{p,i}$ (°C) | $M_p$ (kg) |
| 40.2 | 21.1 | 1.95 | 0.326 | 7.0 | 2.5 |
| 65.0 | 18.5 | 1.95 | 0.317 | 6.0 | 2.5 |

*Source*: Syahrul et al. (2002a).

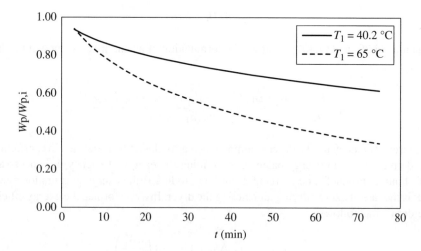

**Figure 4.11**  Relative reduction of the moisture content of wheat grains during the drying process (given with respect to the initial moisture content)

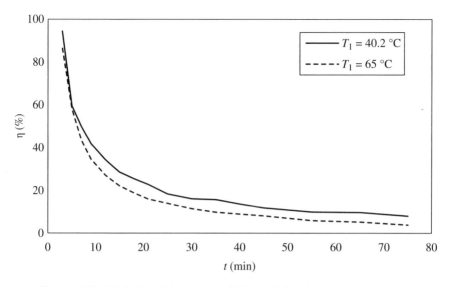

**Figure 4.12**  Variation of the energy efficiency during drying of wheat grains

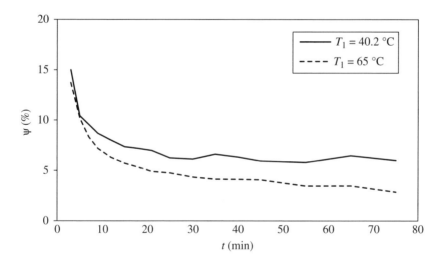

**Figure 4.13**  Variation of the exergy efficiency during drying of wheat grains

- The variation of energy efficiency during wheat drying is shown in Figure 4.12. As observed, the efficiency is better when the drying temperature is lower. Recall that when the drying temperature is lower, the reduction of moisture content is slower.
- The average energy efficiencies are of 20.7% when drying air is at 40.2 °C and 15.7% when the drying air is at 65 °C. This study illustrates the importance of process parameter selection.
- The variation of exergy efficiency during wheat drying is shown in Figure 4.13. As observed, the trend is similar as with energy efficiency: the efficiency is better when the drying

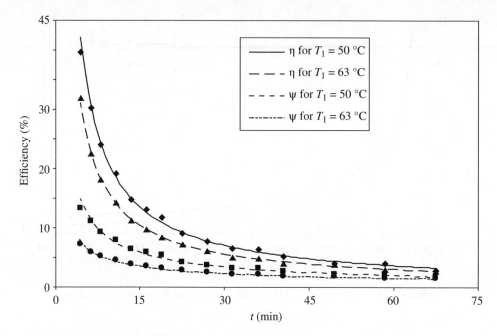

**Figure 4.14**   Variation of the exergy efficiency during drying of wheat grains

temperature is lower. However, the magnitude of exergy efficiency is significantly smaller than for energy efficiency.

- The average exergy efficiencies are of 6.7% when drying air is at 40.2 °C and 4.8% when the drying air is at 65 °C.
- The results for corn in similar conditions are shown in Figure 4.14 in terms of energy and exergy efficiencies. For corn, the temperatures of air at inlet were maintained to 50 °C and 63 °C.
- The efficiencies at drying wheat and corn in fluidized bed dryer are compared in Figure 4.15 for the following fixed conditions: the air temperature at inlet is fixed at 65 °C and the moisture content reduction relative to initial value is 45%. As observed, the efficiencies of drying wheat in similar conditions are much better than for corn. This study illustrates the impact of material properties on drying performance.

Several generalizations can be drawn from thermodynamic analyses through energy and exergy of various aspects of fluidized bed drying systems as described here:

- The effect of inlet air temperature on the efficiencies of fluidized bed dryer systems is important.
- The physical properties of the moist material affect in large measure the energy and exergy efficiencies.
- For wheat grains, the increase of the drying air temperature increases the efficiency nonlinearly because the diffusion coefficient is a function of temperature.
- For corn grains, the increase of drying air temperature does not necessarily increase efficiency because the diffusion coefficient depends on temperature and moisture content.

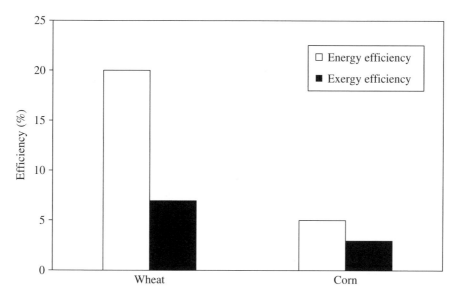

**Figure 4.15**   Comparison of energy and exergy efficiency for wheat and corn drying when input air temperature is fixed at 65 °C and the moisture content reduction relative to initial value is 45%

- Energy and exergy efficiencies are higher at the beginning of the drying process than at the end because the moisture removal rate from wet particles is higher in the beginning.
- The effect of gas velocity on energy and exergy efficiencies depends on the materials.
- For corn, energy and exergy efficiencies do not exhibit any difference at the end of drying.
- For wheat, energy and exergy efficiencies increase with reduced air velocity.
- If the initial moisture contents are high, then the efficiencies are higher.
- If the drying air temperature and velocity through the dryer are properly controlled and varied during the process, then the efficiencies can be improved.

## 4.7   Conclusions

In this chapter, the analysis of drying processes and systems through energy and exergy methods is presented. Exergy efficiencies depend highly on heat and mass transfer parameters. Several examples illustrate the applicability of the method to the drying of moist solids with air and highlight the sensitivities of the results to such parameters as drying air temperature, moisture content, humidity ratio and specific exergy, the exergy difference between inlet and outlet products, and product mass. Exergy analysis is demonstrated to provide a significant tool for design and optimization of drying processes.

## 4.8   Study Problems

4.1   What are the most important irreversibilities at air drying and what measures can be taken to minimize them?

4.2   What are the most important irreversibilities of heat pump dryers and what measures can be taken to minimize them?

4.3   What are the most important irreversibilities at freeze-drying and what measures can be taken to minimize them?

4.4   What are the most important irreversibilities of fluidized bed dryers and what measures can be taken to minimize them?

4.5   Explain how exergy analysis can pinpoint the irreversibility at drying and help to better match the energy source to the end use.

4.6   Briefly explain the effects on the exergy efficiency of the dryer considered in Example 4.1 on mass flow rate of the drying air, temperature and humidity ratio of the drying air, the amount of products entering, and the moisture content of the material at the entrance.

4.7   Rework Example 4.1 using the given input data and try to duplicate the results. If your results differ from those given in the example, discuss why. Propose methods for improving the performance of the system based on reducing or minimizing exergy destruction.

4.8   Obtain a published article on exergy analysis of drying systems. Using the operating data provided in the article, perform a detailed exergy analysis of the system and compare your results to those in the original article. Also, investigate the effect of varying important operating parameters on the system exergetic performance.

4.9   Rework Example 4.2 for another type of refrigerant and different process parameters.

# References

Acar C., Dincer I. 2014. Performance assessment of a two-stage heat pump-drying-system. In: Progress in Sustainable Energy Technology, Vol. 1, Chapter 7.

Chua K.J., Chou S.K. 2005. A modular approach to study the performance of a two-stage heat pump system for drying. *Applied Thermal Engineering* 25:1363–1379.

Coskun C., Bayraktar M., Oktay Z., Dincer I. 2009. Energy and exergy analyses of an industrial wood chips drying process. *International Journal of Low Carbon Technology* 4:224–229.

Dincer I., Rosen M.A. 2013. Exergy: Energy, Environment and Sustainable Development. Elsevier: Oxford, UK.

Hajidavalloo E. 1998. Hydrodynamic and thermal analysis of a fluidized bed drying system. PhD Thesis. Dalhousie University, Daltech: Halifax, Canada.

Kunii D., Levenspiel O. 1991. Fluidization Engineering. Butterworth-Heinemann: Boston.

Mujumdar A.S. 2006. Handbook of Industrial Drying. 3rd ed. CRC Press, Taylor and Francis LLC.

Rosen M.A., Dincer I., Kanoglu M. 2008. Role of exergy in increasing efficiency and sustainability and reducing environmental impact. *Energy Policy* 36:128–137.

Syahrul S., Hamdullahpur F., Dincer I. 2002a. Energy analysis in fluidized-bed drying of large particles. *International Journal of Energy Research* 26:507–525.

Syahrul S., Hamdullahpur F., Dincer I. 2002b. Exergy analysis of fluidized bed drying of moist particles. *Exergy, An International Journal* 2:87–98.

# 5

# Heat and Moisture Transfer

## 5.1   Introduction

Drying is a process of moisture transfer which is essentially driven by heat. That is, in most of practical situations of drying, the temperature field influences the moisture concentration gradient and therefore the moisture transfer process. Moreover, the moisture content influences thermophysical parameters such as thermal conductivity, specific heat, density, and thermal diffusivity, and in some instances, it affects the material shrinkage. As a consequence, the moisture content influences the heat transfer process and therefore the temperature field within the moist material. Although in some practical cases of drying, the heat and mass transfer processes can be modeled with sufficient accuracy in a decoupled manner, in many cases, the mathematical equations for the two processes must be solved simultaneously.

The mass transfer by diffusion and its governing Fick's law as well as the convective mass transfer which manifests at the surface of the material subjected to drying were introduced in Chapter 1. The Biot number for mass transfer indicates if the process is diffusively or convectively controlled or it is controlled by both diffusion within the material and moisture convection at material surface. For diffusively controlled processes, the Fourier number for mass transfer is important in solving the partial differential equations for modeling. Actually, all drying processes (either batch or continuous) are transient, since the moisture content of the material varies continuously during the process.

Regarding the moisture convection at the surface of the material, the analogy between heat and mass transfer plays a crucial role. Various models are available for this analogy as briefly introduced in Chapter 1. Here, the theories for convective moisture transfer are expanded even more by introducing various thin-film models based on theoretical, semiempirical, and empirical approaches.

*Drying Phenomena: Theory and Applications*, First Edition. İbrahim Dinçer and Calin Zamfirescu.
© 2016 John Wiley & Sons, Ltd. Published 2016 by John Wiley & Sons, Ltd.

The geometry of the material subjected to drying is very important and the time-dependent governing equations for moisture and heat transfer must account for it. The analytical solutions for the semi-infinite solid with time-dependent mass transfer by moisture diffusion are given in Chapter 1. In Chapter 2, the analytical solution for time-dependent diffusion mass transfer within infinite slab subjected to drying is presented. The theory of time-dependent moisture diffusion is expanded even more in this chapter to cover two other fundamental geometries with one-dimensional mass transfer beside the infinite slab: the infinite cylinder (IC) and the sphere. Also the cases of convectively controlled process ($Bi_m < 0.1$), diffusively and convectively driven process ($0.1 \leq Bi_m \leq 100$), and diffusively controlled process $Bi_m > 100$ are all covered herein for these fundamental one-dimensional geometries.

Of course, in practice, other regular or irregular geometries of finite size are important. In many geometries, the diffusion process is two or three dimensional. For example, through the rectangular slab with mass convection at all edges, a three-dimensional moisture transfer process occurs during drying. If one of the edges is much longer than the other, the rectangular slab can be modeled as an infinite slab with moisture convection at the two side edges; therefore, the process is two dimensional. In a finite cylinder placed in a uniform medium, the moisture diffusion can be modeled as two dimensional with one direction being the radial and the other the axial.

Various regular shapes are treated herein beside the rectangular slab such as the elliptic, cylinder, and various ellipsoids. However, in the overwhelming majority of cases, materials with irregular geometry are subjected to drying. Methods to approximate irregular shapes with regular shapes exist, as discussed in the chapter, where some shape factors are introduced for various relevant geometries.

A significant part of the chapter is dedicated to the simultaneous heat and moisture transfer analysis and modeling. The governing equations and correlation for moisture and thermal diffusivities in function of temperature and moisture content are given. Some analytical solutions are commented. Most importantly, drying models for heat and moisture transfer during the first and second phase of drying process are introduced. It is of paramount importance to develop appropriate models as tools for the prediction of drying kinetics and to study the behavior of drying processes and their transport phenomena, including moisture diffusion and evaporation used for designing new drying systems as well as selection of optimum drying conditions and for accurate prediction of simultaneous heat and moisture transfer phenomena during drying process.

## 5.2    Transient Moisture Transfer During Drying of Regularly Shaped Materials

Since any drying process involves diffusion in transient regime, it appears logical to treat this topic firstly. Here, we consider a set of simple geometries of the materials subjected to drying and write the transient diffusion equation for each treated case. Further, analytical solutions are given under reasonable assumptions for each case. However, there are two general assumptions in this section as follows:

- The heat and mass transfer processes are decoupled; that is, the temperature field does not essentially influences the moisture concentration gradient. This situation is relevant for some drying situations.
- The thermophysical properties of the moist material and drying agent are constant.

### 5.2.1 Transient Diffusion in Infinite Slab

To start with, consider an infinite slab object subjected to drying in an infinite gaseous medium at constant parameters. This problem is illustrated graphically in Figure 5.1. For the infinite slab, it is reasonable to assume that the diffusion process occurs in one dimension, namely, in the direction across the slab; this is denoted as $z$ in the figure. The infinite slab problem approximates well the mass transfer through a finite slab having all margins perfectly impermeable.

The general one-dimensional mass diffusion equation is similar to Fourier equation for heat conduction which has been given in Chapter 1 as Eq. (1.71b). In this equation, the temperature can be replaced with mass concentration ($C_m$) and the thermal diffusivity $a$ with mass diffusivity $D$. Thence, the transient mass diffusion equation for the infinite slab becomes

$$D\left(\frac{\partial^2 C_m}{\partial z^2}\right) = \frac{\partial C_m}{\partial t}$$

Here, the relationship between moisture concentration $C_m$ (kg moisture/m$^3$) and moisture content $W$ (kg moisture/kg dry material) is given by $W = C_m/\rho_s$, where $\rho_s$ (kg dry material/m$^3$) is the density of the dry material (assumed constant). Therefore, since $C_m$ and $W$ differ by a constant, the moisture diffusion process during drying can be (better) represented by the following equation:

$$D\left(\frac{\partial^2 W}{\partial z^2}\right) = \frac{\partial W}{\partial t} \tag{5.1}$$

In Eq. (5.1), the moisture content is a function of thickness coordinate, $z$, and time, $t$, namely, $W(z, t)$. As shown in the figure, there must be a continuity of moisture flux at the surface; that is, the mass flux that exits the moist material must be taken by the drying agent entirely. This must be reflected as the boundary condition at the surface. Two curves are qualitatively represented in the figure: (i) the variation of moisture content within the moist material and (ii) the variation

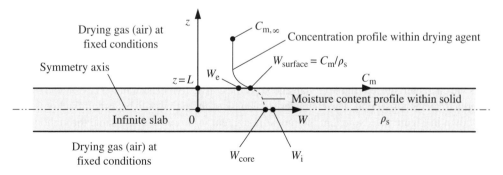

**Figure 5.1** General model for moisture diffusion through the infinite slab

of moisture concentration in the space surrounding the slab, where the drying agent is circulated.

Equation (5.1) can be solved when appropriate initial and boundary conditions are specified. Assume that the moist material and drying conditions are specified. In such situation, the initial moisture content $W_i$ is given and the equilibrium moisture content $W_e$ can be determined. Then the moisture content can be replaced with a dimensionless moisture content – also introduced in Chapter 1 – and given by

$$\phi = \frac{W - W_e}{W_i - W_e}$$

Since in a drying process it is expected that $W \le W_i$ and $W > W_e$, the dimensionless moisture content must be in the domain $\phi \in (0,1)$. With $\phi$ as a subunitary parameter that never cannot be zero, the numerical and analytical solutions for Eq. (5.1) simplify; in fact, since $W_i$ and $W_e$ are fixed, the moisture diffusion equation becomes

$$D\left(\frac{\partial^2 \phi}{\partial z^2}\right) = \frac{\partial \phi}{\partial t} \tag{5.2}$$

where $\phi = \phi(z,t)$ and $t$ is the time variable.

The following initial and boundary conditions are relevant for drying of a slab-shaped material:

• At the initial moment, the moisture content is $W_i$ everywhere throughout the slab, that is,

$$\phi(z,0) = 1 \tag{5.3}$$

• The slab is dried from the two sides as shown in Figure 5.1; therefore, at the median plane, a symmetry contrition must be imposed. The symmetry conditions means that the gradient of moisture content is nil at the symmetry plane. This is the same as writing

$$\frac{\partial \phi(0,t)}{\partial t} = 0 \tag{5.4}$$

• Surface condition, which can be of three kinds, namely:

Case I: *Convection-controlled process*. In this case, the moisture diffusion process throughout the slab is very fast with respect to the moisture convection at the surface. This case is specific for $Bi_m < 0.1$, that is, the mass transfer coefficient multiplied to the characteristic length is significantly smaller than the diffusion coefficient, namely, $h_m L \ll D$, where $L$ is the characteristic length of the slab and $h_m$ is the convective mass transfer coefficient measured in meter per second. The characteristic length of the slab would be its half thickness $L$ as shown in Figure 5.1. Note that the moisture content in this case is the same throughout the solid and lumped capacity model applies $W(z,t) = W(L,t) = W_s$. The boundary condition is

$$-h_m(C_{m,s} - C_{m,\infty}) = L\rho_s \frac{\partial W_s}{\partial t} \tag{5.5}$$

which, based on the remark that the solid cannot be dried below the equilibrium moisture content, namely, $W_s \geq W_e$ and $C_{m,s} = \rho_s W_s$, becomes

$$-h_m \phi(L,t) = L \frac{\partial \phi(L,t)}{\partial t} \qquad (5.6)$$

Case II: *Diffusion-controlled process*. In this case, the moisture convection at the surface of the slab is much faster than the diffusion throughout the slab. This case is specific for $Bi_m > 100$, that is, the mass transfer coefficient is significantly higher than the diffusion coefficient, namely, $h_m L \ll D$. Consequently, the moisture at the surface reaches rapidly the equilibrium content, whereas in the solid core, the moisture reduces much more slowly. This boundary condition writes as follows:

$$\phi(L,t) = 0 \Leftrightarrow W(L,t) = W_e \qquad (5.7)$$

Case III: *Process controlled by both diffusion and convection*. In this case, both processes – the moisture diffusion throughout the solid and the moisture convection at solid surface – are equally important because $0.1 \leq Bi_m \leq 100$. The boundary condition becomes

$$-D \frac{\partial \phi(L,t)}{\partial z} = h_m \, \phi(L,t) \qquad (5.8)$$

The solutions of moisture diffusion problem are given based on similarity with heat conduction problems. In this respect, the analogy between heat transfer parameters and mass transfer parameters must be determined. Once the analogue parameters are known, then the solutions of transient conduction can be easily changed to give the analogous solutions for transient mass diffusion. Table 5.1 gives the analogue parameters of heat transfer and mass transfer applicable for modeling the transient diffusion.

**Table 5.1** Analogue parameters for transient heat transfer and transient mass transfer modeling

| Heat transfer | | Mass transfer | |
|---|---|---|---|
| Temperature ($T$) | $T$ (K) | Concentration ($C_m$)or moisture content ($W$) | $C_m$ (kg/m$^3$), $W$ (kg/kg dry) |
| Dimensionless temperature | $\theta = \dfrac{T - T_\infty}{T_i - T_\infty}$ | Dimensionless moisture content | $\phi = \dfrac{W - W_e}{W_i - W_e}$ |
| Thermal diffusivity | $\alpha$ (m$^2$/s) | Mass diffusivity or effective diffusivity | $D$ (m$^2$/s) |
| Dimensionless length | $\zeta = \dfrac{z}{2\sqrt{\alpha t}}$ | Dimensionless length | $\zeta = \dfrac{z}{2\sqrt{Dt}}$ |
| Biot number for heat transfer | $Bi = \dfrac{hL}{k}$ | Biot number for mass transfer | $Bi_m = \dfrac{h_m L}{D}$ |
| Fourier number (dimensionless time) | $Fo = \dfrac{\alpha t}{L^2}$ | Fourier number for mass transfer | $Fo_m = \dfrac{Dt}{L^2}$ |

The solution of this problem and various other transient mass transfer cases are widely documented in the literature. Here, we give the formulations from Dincer (1996a, 1998a,b) and Dincer and Dost (1996). These solutions give the moisture content at the median plane of the slab rather than the distribution of the moisture content along the thickness, $z$. For simplicity, the dimensionless moisture content at the median plane is denoted with $\Phi$, that is, $\Phi = \phi(0,t)$. Note that $\Phi(t)$ is a function of time only, whereas $\phi = \phi(z,t)$ gives the momentarily distribution of the moisture content across the slab thickness. The moisture content at the median plane of the slab is an important parameter that is useful for determination of the drying time of the product when the final value for $\Phi$ is specified. The simplest analytical formula for $\Phi$ is found for Case I when the lumped capacity modeling can be applied as given below:

$$\Phi(t) = \exp(-Bi_m Fo_m) \tag{5.9}$$

where $Fo_m$ represents the dimensionless time as given in Table 5.1.

For Cases II and III, the general solution is given in form of an infinite series according to

$$\Phi(t) = \sum_{n=1}^{\infty} A_n(\mu_n, Bi_m) B_n(\mu_n^2 Fo_m) \tag{5.10}$$

where $\mu_n$ is the solution of the transcendental characteristic equation for the partial differential equation (5.2) with the initial condition given by Eq. (5.3), the symmetry condition given by Eq. (5.4), and the boundary condition at the surface given by Eq. (5.7) for Case II and by Eq. (5.8) for Case III.

The characteristic equations are given for any $n \geq 1$. Here, we give the characteristic equations for $n = 1$, because these are of the most interest, as commented in Dincer (1998a). For Case II, the characteristic equation is written as

$$2\mu_1 = \pi \quad \text{for } Bi_m > 100 \tag{5.11}$$

whereas for Case III, the characteristic equation at $n = 1$ becomes

$$Bi_m \cot(\mu_1) = \mu_1 \quad \text{for } 0.1 \leq Bi_m \leq 100 \tag{5.12}$$

The factor $A_n(\mu_n)$ from the general solution expressed by Eq. (5.10) is given, as shown in Dincer (1998a), by the following equation:

$$A_n(\mu_n) = \begin{cases} \left| (-1)^{n+1} \dfrac{2}{\mu_n} \right| & \text{for Case II, } Bi_m > 100 \\[4mm] \left| \dfrac{2\sin(\mu_n)}{\mu_n + \sin(\mu_n)\cos(\mu_n)} \right| & \text{for Case III, } 0.1 \leq Bi_m \leq 100 \end{cases} \tag{5.13}$$

and the factor $B_n(\mu_n^2 Fo_m)$ is given by the following equation:

$$B_n(\mu_n^2 Fo_m) = \exp(-\mu_n^2 Fo_m) \tag{5.14}$$

which is valid for $Bi_m \geq 0.1$.

In many practical situations, the drying process takes place for long duration; therefore, it is reasonable to assume that $\mu_1^2 Fo_m > 1.2$, case in which the solution from Eq. (5.10) can be fairly approximated only by the first term of the series expansion, that is,

$$\Phi(t) \cong A_1(\mu_1, Bi_m) B_1\left(\mu_1^2 Fo_m\right) \tag{5.15}$$

where $A_1(\mu_1)$ is denoted as the lag factor (see Chapter 2) and $\mu_1 = 0.5\pi$ for $Bi_m > 100$ and for $0.1 \leq Bi_m \leq 100$ the characteristic root can be obtained by solving Eq. (5.12) for $\mu_1$. An approximate solution is given in Dincer and Dost (1995) for this equation as follows:

$$\mu_1 = \arctan(0.640443 Bi_m + 0.380397) \quad \text{for } 0.1 < Bi_m < 100 \tag{5.16}$$

Based on Eqs. (5.13) and (5.16), one can determine the following expressions for the factor $A_1(\mu_1)$:

$$A_1(\mu_1) = \begin{cases} \left|\dfrac{2}{\mu_1} = \dfrac{4}{\pi}\right| & \text{for Case II, } Bi_m > 100 \\[4mm] \left|\dfrac{2\sin(\mu_1)}{\mu_1 + \sin(\mu_1)\cos(\mu_1)} = \exp\left(\dfrac{0.2533\,Bi_m}{1.3 + Bi_m}\right)\right| & \text{for Case III, } 0.1 \leq Bi_m \leq 100 \end{cases} \tag{5.17}$$

The chart representing $\Phi(t)$ versus dimensionless time $Fo_m$ can be easily obtained for different Biot numbers for mass transfer. This chart can be easily calculated using computer software (EES, 2014). Here, we show in Figure 5.2 the total dimensionless moisture content $\Phi$

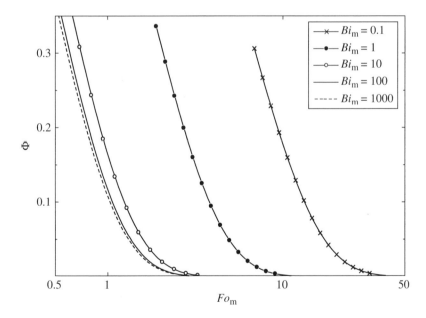

**Figure 5.2** Dimensionless moisture content variation at the infinite slab median plane with Fourier number for mass transfer at drying of an infinite slab for various $Bi_m$

variation with Fourier number for mass transfer at drying of an infinite slab for various Biot number for mass transfer.

In some analyses, it may be important to predict the distribution of moisture content across the slab thickness, that is, to determine the function $\phi(z, t)$. Analytical solution exist for $\phi(z, t)$ of the infinite slab (or finite slab with impermeable sides) which is obtained by analogy with the conduction heat transfer problem. Accordingly, based on Adebiyi (1995), the following solution can be found:

$$\phi(z,t) = \sum_{n=0}^{\infty} \frac{2Bi_m (z/L)^{0.5} \mu_n^{0.5} \cos(\mu_n z/L)}{(\mu_n^2 + Bi_m^2 - (n-1)Bi_m)\mu_n^{-0.5}\cos(\mu_n)} \exp\left(-\mu_n^2 Fo_m\right) \tag{5.18}$$

with the following characteristic equation:

$$\mu_n \tan(\mu_n) = Bi_m \tag{5.19}$$

The average dimensionless moisture content in the infinite can be also determined based on the analytical solution given in Treybal (1981) (Eq. (4.8)) which is valid for diffusion-controlled case, $Bi_m > 100$, namely,

$$\bar{\phi}(t) = \frac{1}{L} \int_0^z \phi(z,t) \mathrm{d}t = \frac{8}{\pi^2} \sum_{n=1}^{\infty} \frac{1}{(2n-1)^2} \exp\left[(2n-1)^2 \frac{\pi^2}{4} Fo_m\right] \tag{5.20}$$

Beside the moisture content at the material center ($\Phi$) and the average moisture content ($\bar{\phi}$), the moisture content at the product surface is also important, especially for some detailed analysis of mass transfer at the interface. An approximate solution for dimensionless moisture content during drying of the infinite slab surface during time-dependent moisture diffusion in diffusion controlled and diffusion- and convection-controlled problems is studied in Dincer (1998b). For the infinite slab, the approximate solution for surface dimensionless moisture transfer $\Phi_s = \phi(L,t)$ is given by the following equation:

$$\Phi_s = \sum_{n=0}^{\infty} A_n B_n \cos \mu_n \cong A_1 \exp\left(\mu_1^2 Fo_m\right) \cos \mu_1 \tag{5.21}$$

with $A_1$ given by Eq. (5.17) and $\mu_1$ given by Eq. (5.11) or Eq. (5.16) depending on the case.

From Eq. (5.21) results that the ratio between $\Phi_s$ and $\Phi$ is in fact equal to $\cos \mu_1$. A ratio of dimensionless moisture content at slab middle plane $\Phi_s$ and at the surface $\Phi$ is shown in Figure 5.3 for a range of Biot number. On the same plot, the dimensionless moisture content at the material surface is represented for three values of Fourier number. In many cases of agricultural product drying, the Biot number is of the order of 10 case in which the surface moisture is ~20% of the middle plane moisture.

Dincer (1996b) developed a dimensionless number as a tool for modeling transient heating and cooling of solids in forced convection which are mathematically similar with transient moisture diffusion problems. The transient cooling or heating of solids can be modeled based on analytical solutions of heat conduction problem which can be generally written in the form

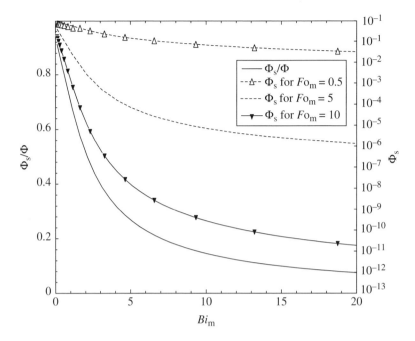

**Figure 5.3**  Dimensionless moisture content at infinite slab center and surface for $Bi_m = 50$

$\theta = LF \exp(-ct)$, where $c$ (s$^{-1}$) is the cooling or heating coefficient. The Dincer number is useful to correlate the flow velocity with the transient heating or cooling coefficient which represent the reciprocal of the time constant of transient.

The Dincer number method has been extended further to model transient mass transfer operations (e.g., moisture transfer in drying) which are described by equations such as Eq. (5.15). From these types of mathematical solution, the following general model for transient moisture transfer results:

$$\Phi = LF \exp(-St) \tag{5.22}$$

where $t$ (s) is time, $S$ (s$^{-1}$) is the moisture transfer coefficient which is also commonly denoted as drying coefficient when a drying operation is conducted, and LF is the dimensionless lag factor (see also Chapter 2).

In a forced convection drying process, the flow velocity must influence the drying coefficient. The Dincer number is a convenient parameter to correlate the flow velocity and solid dimension with the drying coefficient. If one denotes the characteristic length of the solid with $Y$ (which would be the half-width $L$ for the slab and the radius $R$ for the cylinder and sphere) and the flow velocity is denoted with $U$, then the Dincer number is defined as follows:

$$Di = \frac{U}{SY} \tag{5.23}$$

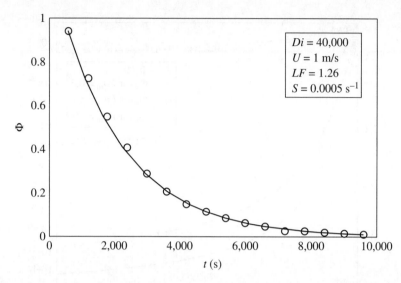

**Figure 5.4** Comparison between experimental data (circles) and the drying model predictions based on Dincer number (Eq. (5.22), continuous line) for a case study of sultana grapes drying assumed with cylindrical shape (data from Sahin and Dincer (2004)). The dimensionless moisture content is plotted against the time

Using the Dincer number, a new form of the dimensionless moisture content at product center is derived in Sahin and Dincer (2004), namely,

$$\Phi = LF \exp\left(-\frac{\tau}{Di}\right) \tag{5.24}$$

where $\tau$ is a newly introduced dimensionless time which is based on the flow velocity and solid's characteristic length. This dimensionless time is given by

$$\tau = \frac{U}{Y}t \tag{5.25}$$

The predictions based on the moisture transfer model expressed by Eq. (5.24) are compared with the experimental determinations for a case study as shown in Figure 5.4. The case study is taken from Sahin and Dincer (2004). The drying air has been provided forcedly with a velocity of 1 m/s having the temperature of 60 °C and a relative moisture of 60%. In these conditions, the Dincer number has been calculated; it is $Di = 40,000$. The method showed good agreement with the experiments.

If in Eq. (5.24) the dimensionless moisture content is divided with the lag factor ($\Phi/LF$), then a quantity denoted as "normalized dimensionless moisture content" is obtained. The plot shown in Figure 5.5 is useful for quick determination of $\Phi/LF$ when the Dincer number and the dimensionless time are specified. This helps the subsequent determination of the actual moisture content variation with drying time.

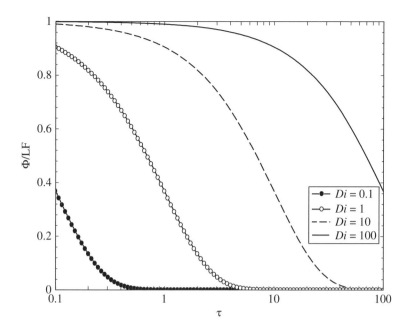

**Figure 5.5**  Variation of the normalized dimensionless moisture content with $\tau$ and Dincer number

## Example 5.1

A sheet of moist material with diffusivity of $5E{-}7\,\mathrm{m^2/s}$ can be approximated as in infinite slab of 10 mm thickness. During the material drying process, the mass transfer coefficient is determined to $h_m = 0.005\,\mathrm{m/s}$. The initial moisture content is $W_i = 1\,\mathrm{kg/kg}$ and the equilibrium moisture content is $W_e = 0.05\,\mathrm{kg/kg}$. Determine the moisture content in the middle plane and at the surface after 36 min:

- The characteristic length of the slab is $L = 0.005\,\mathrm{m}$.
- One calculates the Biot number $Bi_m = h_m L/D = 50$.
- This is therefore Case III, then using Eq. (5.16) $\rightarrow \mu_1 = 1.54$, $A_1 = 1.28$.
- The solution for time-dependent diffusion is valid for $\mu_1^2 Fo_m > 1.2 \rightarrow t > 25.3\,\mathrm{s}$.
- At time $t = 3600\,\mathrm{s}$, the Fourier number is $Fo_m = 0.72$.
- Therefore, the dimensionless moisture contents are $\Phi = 0.232$ and $\Phi_s = 0.00716$.
- The final moisture content in the middle plane is

$$W = W_e + \Phi(W_i - W_e) = 0.05 + 0.232(1 - 0.05) = 0.2705\,\mathrm{kg/kg}.$$

- The moisture content at the surface is

$$W = W_e + \Phi_s(W_i - W_e) = 0.0568\,\mathrm{kg/kg}.$$

## 5.2.2   Drying Time of an Infinite Slab Material

The analytical solutions for transient moisture diffusion bring an opportunity for derivation of relatively simple algebraic expressions for drying time determination. This parameter – the drying time – is one of the most important in designing/conducting drying processes in the industry. Assume that the diffusivity and moisture transfer coefficients can be determined for a particular drying situation. Experimental data or theoretical models can be used in this respect. Once these coefficients are known, the Biot number for mass transfer can be calculated. Furthermore, for regular objects such as the slab, the transient mass diffusion problem can be analytically solved and the moisture content at the solid center predicted. From here, the drying time to achieve the desired moisture content at the center can be determined. This method for drying time determination can be generalized for objects with regular and irregular geometry.

Consider now the infinite slab shown in Figure 5.1 subjected to drying under convection-controlled conditions, Case I. Thence, the dimensionless moisture content at the centerline will vary in time according to Eq. (5.9) deduced above. This equation can be written equivalently as follows:

$$W = W_e + (W_i - W_e)\exp\left(-\frac{h_m}{L}t\right)$$

Once the desired final moisture content at the product middle plane is specified, the above equation can be solved for the required drying time, and for Case I, $Bi_m < 0.1$, it gives

$$t_{slab} = -\frac{L}{h_m}\ln\Phi_f \tag{5.26}$$

where $\Phi_f$ represents the final dimensionless moisture at the slab median plane.

If the boundary condition at the slab material surface is for Case II or Case III, that is, $Bi_m \geq 0.1$, then the moisture content will vary according to Eq. (5.15) which can be put in the following form:

$$W = W_e + (W_i - W_e)A_1\exp\left(-\frac{\mu_1^2 D}{L^2}t\right)$$

which can be solved for $t$ to obtain the drying time for Cases II and III, $Bi_m \geq 0.1$, when $\Phi_f$ is given, namely,

$$t_{slab} = -\frac{L^2}{D\mu_1^2}\ln\left(\frac{\Phi_f}{A_1}\right) \tag{5.27}$$

With the use of Eqs. (5.11) and (5.17) for $Bi_m > 100$, the drying time from Eq. (5.27) becomes

$$t_{slab} = -\frac{\pi L^2}{4D}\ln\left(\frac{4}{\pi}\Phi_f\right) \tag{5.28}$$

With the use of Eqs. (5.16) and (5.17) for $0.1 \le Bi_m \le 100$, the drying time from Eq. (5.27) becomes

$$t_{\text{slab}} = -L^2 \frac{\ln(\Phi_f) - \dfrac{0.2533\,Bi_m}{1.3 + Bi_m}}{D\arctan^2(0.640443\,Bi_m + 0.380397)} \tag{5.29}$$

One can remark that an equivalent expression for drying time of an infinite slab can be obtained if Eqs. (5.22) and (5.27) are combined. This equation is for Cases II and III, $Bi_m \ge 0.1$, as follows:

$$t_{\text{slab}} = -\frac{1}{S}\ln\left(\frac{\Phi_f}{\text{LF}}\right) \tag{5.30}$$

An important measure in the drying process can be given as the so-called half drying time. This represents the time required for a product subjected to drying to reduce its moisture content at the center to a half of the initial value. The final dimensionless moisture content is for the half drying time is therefore set to $\Phi_f = 0.5$. Based on Eq. (5.30), the following equation for the half drying time results:

$$t_{0.5} = -\frac{1}{S}\ln\left(\frac{1}{2\,\text{LF}}\right) \tag{5.31}$$

where the shape factor $E = 1$ for the infinite slab, $S$ ($s^{-1}$) is the drying (or humidification) coefficient, and LF is the lag factor.

## Example 5.2
Sliced onions of 3 mm thick are subjected to drying in air; the mass transfer coefficient is given $h_m = 1.5\,\text{m/s}$ and the moisture diffusivity coefficient is $6\text{E}{-}10\,\text{m}^2/\text{s}$. The initial moisture content is $W_i = 0.6\,\text{kg/kg}$ and the equilibrium moisture content is $W_e = 0.045\,\text{kg/kg}$. Calculate the drying time provided that the onion slice is considered dried when the moisture content at the center plane reaches 0.05 kg/kg:

- Calculate Biot number for mass transfer $Bi_m = h_m L/D = 1.5 \times 0.0015/(6 \times 10^{-10}) = 3.75 \times 10^6 \gg 100$.
- This is therefore Case II, boundary condition by Eq. (5.7) $\rightarrow \mu_1 = \pi/2$, $A_1 = \pi/4$.
- The solution for $\Phi$ is valid for $\mu_1^2 Fo_m > 1.2$ that is for $Fo_m > 0.4863$ or $t > 30.4\,\text{min}$.
- Using Eqs. (5.14) and (5.15), the following is obtained:

$$W = W_e + \frac{\pi}{4}\exp\left(-\frac{\pi^2 D}{4L^2}t\right)(W_i - W_e) = 0.045 + 0.593\exp(-0.000658\,t).$$

- The lag time (Chapter 2) results from $W_i = 0.045 + 0.593\exp(-0.000658\,t_{\text{lag}}) \rightarrow t_{\text{lag}} = 1.68\,\text{min}$.

- The drying time, that is, the time when the moisture at the middle plane reaches 5%, results from

$$0.05 = 0.45 + 0.593 \exp\left(-0.000658\, t_{\text{dry}}\right) \rightarrow t_{\text{dry}} = 121 \text{ min.}$$

### 5.2.3 Transient Diffusion in an Infinite Cylinder

Consider now a permeable infinite cylinder (IC) of radius $R$ in which a moisture diffusion process occurs during a drying operation which drives the moisture from the core to the periphery and from there into the drying gas. The model for this process is shown in Figure 5.6. The diffusion process occurs in radial direction only with axial symmetry. The partial differential equation for the moisture content $W(r, t)$ at drying of an IC. However, the dimensionless moisture content $\phi(r, t)$ where $r \in [0, R]$ and $t \geq 0$ can be used instead of $W$. The equation is given as follows:

$$D\left(\frac{\partial^2 \phi}{\partial r^2} + \frac{1}{r}\frac{\partial \phi}{\partial r}\right) = \frac{\partial \phi}{\partial t} \tag{5.32}$$

The initial and boundary conditions are analogue as for the infinite slab treated previously. Here, we give the solutions for the following two cases which are the most representative for drying, namely, $0.1 \leq Bi_m \leq 100$ and $Bi_m > 100$. In both cases, the same initial and axial symmetry conditions apply, namely:

- Initial condition – the moisture content is the same throughout the cylinder at the initial moment, that is,

$$\phi(r,0) = \phi_i \tag{5.33}$$

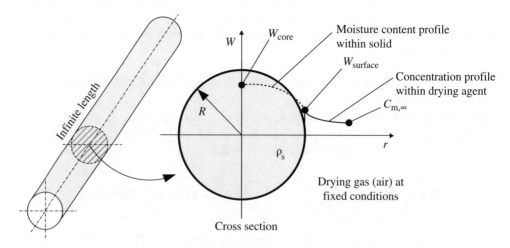

**Figure 5.6** General model for moisture diffusion through the infinite cylinder

- Axial symmetry – there is no preferential direction for diffusion; the process occurs radially from the central axis of symmetry to the periphery of the cylinder at any time. Thence,

$$\frac{\partial \phi(0,t)}{\partial r} = 0 \tag{5.34}$$

- Surface boundary condition – here, we consider the two cases:

  Case I: *Diffusion-controlled moisture transfer process* in which $Bi_m > 100$; in this case, the surface is at the equilibrium moisture content ($\phi = 0$) at all times, therefore

$$-D\frac{\partial \phi(R,t)}{\partial r} = 0 \tag{5.35}$$

  Case II: *Process controlled by both diffusion and convection* in which $0.1 \le Bi_m \le 100$; therefore, it must be a moisture flux continuity condition at the surface satisfied at any time (mass conservation; the surface does not accumulate mass) which is written mathematically as follows:

$$-D\frac{\partial \phi(R,t)}{\partial r} = h_m \phi(R,t) \tag{5.36}$$

The Biot number for mass transfer is based on the cylinder radius $R$, namely,

$$Bi_m = \frac{h_m R}{D}$$

whereas the Fourier number for the IC is

$$Fo_m = \frac{Dt}{R^2}$$

The dimensionless moisture content at the symmetry axis of the IC $\Phi(t)$ can be approximated when $\mu_1^2 Fo_m > 1.2$ with Eq. (5.15) for which the characteristic equation is given as follows:

$$\text{For Case I, } 0.1 \le Bi_m \le 100 : \frac{J_0(\mu_1)}{J_1(\mu_1)} = \frac{\mu_1}{Bi_m} \tag{5.37a}$$

$$\text{For Case II, } Bi_m > 100 : J_0(\mu_1) = 1 - \left(\frac{\mu_1}{2}\right)^2 + \left(\frac{\mu_1^2}{8}\right)^2 - \left(\frac{\mu_1^3}{48}\right)^2 \rightarrow \mu_1 = 0.733 \tag{5.37b}$$

The determination characteristic root $\mu_1$ requires solving Eq. (5.50). Dincer and Dost (1995) determined simpler regression expressions to approximate the $\mu_1$ roots as follows:

$$\mu_1 = \begin{cases} \left| [0.72 \ln(6.796\, Bi_m + 1)]^{1/1.4} \right| & \text{for } 0.1 \le Bi_m \le 10 \\ \left| [\ln(1.737792\, Bi_m + 147.32)]^{1/1.2} \right| & \text{for } 10 < Bi_m \le 100 \end{cases} \tag{5.38}$$

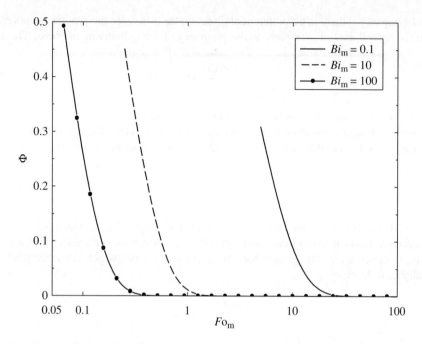

**Figure 5.7** Dimensionless moisture content variation at the infinite cylinder axis with Fourier number for mass transfer at drying of an infinite slab for various $Bi_m$

The factor $A_1$ in Eq. (5.15) for the IC is calculated based on the following regression equation taken also from Dincer and Dost (1995):

$$\text{For Case I}, 0.1 \leq Bi_m \leq 100 : A_1 = \frac{2 Bi_m}{\mu_1^2 + Bi_m^2 + J_0(\mu_1)} \cong \exp\left(0.5066 \frac{Bi_m}{1.7 + Bi_m}\right) \qquad (5.39a)$$

$$\text{For Case II}, Bi_m > 100 : A_1 = \frac{2}{\mu_1 J_1(\mu_1)} \qquad (5.39b)$$

The factor $B_1$ from Eq. (5.15) is calculated according to Eq. (5.14) in which $\mu_1$ is given by Eq. (5.35). Figure 5.7 shows the solution for the dimensionless moisture content at the IC axis when the process is controlled both by diffusion and convection for three Biot numbers. The dimensionless moisture content is plotted against the dimensionless time that is the Fourier number for mass transfer.

The moisture content at the surface of cylindrical materials subjected to drying is a parameter of interest for detailed time-dependent analysis. More importantly, the distribution of moisture within the material can be appreciated based on the moisture content at the axis and the moisture content at the surface. The solution for dimensionless moisture content at the cylinder surface can be derived from Adebiyi (1995). For drying process of an IC, the solution for surface dimensionless moisture content is given in Dincer (1998b) as follows:

$$\Phi_s(t) = \sum_{n=1}^{\infty} A_n(\mu_n, Bi_m) \exp\left(\mu_n^2 Fo_m\right) J_0(\mu_n) \qquad (5.40)$$

or, using the first term approximation, the solution becomes

$$\Phi_s(t) = A_1(\mu_1, Bi_m) \exp(\mu_1^2 Fo_m) J_0(\mu_1) \tag{5.41}$$

The time-dependent solution for moisture diffusion within the IC can be applied to finite length cylinder with the sides impermeable or even to sufficiently long cylinders such that the end effects can be neglected. Furthermore, the transient analysis is useful to determine the drying time. The same general equations for drying time apply as for the infinite slab, namely, Eqs. (5.27) and (5.30). For $0.1 \le Bi_m < 10$, the following equation results for the drying time of a cylinder when Eqs. (5.27), (5.38), and (5.39a) are combined:

$$t_{cyl} = -R^2 \frac{\ln \Phi_f - 0.5066 \dfrac{Bi_m}{1.7 + Bi_m}}{D[0.72 \ln(6.796 Bi_m + 1)]^{1/1.4}} \tag{5.42}$$

If $10 < Bi_m \le 100$, then Eqs. (5.27), (5.38), and (5.39a) give the following expression for the drying time of the IC:

$$t_{cyl} = -R^2 \frac{\ln \Phi_f - 0.5066 \dfrac{Bi_m}{1.7 + Bi_m}}{D[\ln(1.737792 Bi_m + 147.32)]^{1/1.2}} \tag{5.43}$$

If $Bi_m > 100$, then Eqs. (5.27), (5.37b), and (5.39b) can be combined to obtain the following equation:

$$t_{cyl} = -1.364 \frac{R^2}{D} \ln(0.1255 \Phi_f) \tag{5.44}$$

**Example 5.3**

Carrots with diameter of 10 mm are dried in air. The carrot diffusivity is 5.2E–9 m²/s, and the moisture transfer coefficient is 6.6E–7 m/s. Determine the drying time to reduce the initial moisture to 20%:

- Carrots can be assimilated with ICs of characteristic length (the radius) $R = 0.005$ m.
- Biot number is immediately calculated as $Bi_m = h_m R/D = 0.6346$.
- Based on Biot value, Eq. (5.42) is selected to determine the drying time.
- The finial dimensionless drying content is given as $\Phi_f = 0.2$.
- Thence, using Eq. (5.42), the following drying time results: $t_{cyl} = 7363$ s.

## 5.2.4  Transient Diffusion in Spherical-Shape Material

Consider now a third basic case, namely, a spherical-shape material (SSM) submerged in an infinite medium where a gaseous drying agent is maintained at constant conditions. The sphere has radius $R$ and it is assumed symmetry for the mass transfer. The partial differential equation for diffusion determines the variation of the moisture content throughout the sphere at any time, namely, $W(r, t)$ with $r \in [0, R]$. The dimensionless moisture content defined previously is used again to formulate the equation as follows:

$$D\left(\frac{\partial^2 \phi}{\partial r^2} + \frac{2}{r}\frac{\partial \phi}{\partial r}\right) = \frac{\partial \phi}{\partial t} \tag{5.45}$$

The initial and boundary conditions for Eq. (5.45) are written in the same way as for the IC, namely, Eq. (5.35) for diffusion-controlled process (Case I) and Eq. (5.36) when both diffusion and convection control the process (Case II). The dimensionless moisture content at the center of the sphere is given by Eq. (5.15) for which the characteristic equations are as follows:

$$\text{For Case I}, 0.1 \leq Bi_m \leq 100 : \mu_1 \cot(\mu_1) + Bi_m = 1 \tag{5.46a}$$

$$\text{For Case II}, Bi_m > 100 : \mu_1 = \pi \tag{5.46b}$$

Numerical solution is required for Eq. (5.46a); however, approximate solutions exist, as given in Dincer and Dost (1995); namely, the characteristic root for Case I can be approximated with

$$\mu_1 = \begin{cases} \left|[1.1223\ln(4.9\,Bi_m + 1)]^{1/1.4}\right| & \text{for Case I}, \ 0.1 \leq Bi_m \leq 10 \\ \left|[1.667\ln(2.1995\,Bi_m + 152.4386)]^{1/1.2}\right| & \text{for Case II}, \ 10 < Bi_m \leq 100 \end{cases} \tag{5.47}$$

The factor $A_1$ in Eq. (5.15) for the SSM is calculated based on the following equation:

$$\text{For Case I}, 0.1 \leq Bi_m \leq 100 : A_1 = \frac{2Bi_m \sin(\mu_1)}{\mu_1 - \sin(\mu_1)\cos(\mu_1)} \cong \exp\left[0.7599\frac{Bi_m}{2.1 + Bi_m}\right] \tag{5.48a}$$

$$\text{For Case II}, Bi_m > 100 : A_1 = 2 \tag{5.48b}$$

The factor $B_1$ from Eq. (5.15) is calculated according to Eq. (5.14) in which $\mu_1$ is given by Eq. (5.30). Here, in Figure 5.8, the solutions for dimensionless moisture at the center of the sphere are compared with the solutions for dimensionless moisture content at the axis of IC and at the midplane of the infinite slab for $Bi_m = 10$. One can learn from this comparative analysis that the mass transfer process in sphere objects is faster than in cylinder geometry; furthermore, the cylinder geometry allows for relatively faster mass transfer with respect to slab geometry, of course, for the same operating conditions.

The solution of the time-dependent moisture content at the sphere surface for $0.1 \leq Bi_m \leq 100$ during a drying process can be expressed based on the one-term series approximation as given in Sahin and Dincer (2004):

$$\Phi_s(t) \cong \frac{2Bi_m \exp\left(\mu_n^2 Fo_m\right)}{\mu_1^2 + Bi_m^2 - Bi_m} \tag{5.49}$$

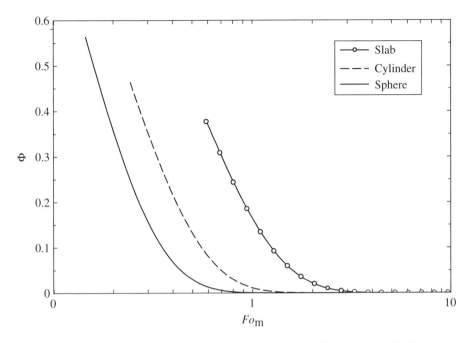

**Figure 5.8** Comparison of dimensionless moisture content for moist materials of spherical shape, infinite cylinder shape, and infinite slab shape for $Bi_m = 10$

The time-dependent solution for sphere is used now to determine the drying time based on Eq. (5.27). For $0.1 \leq Bi_m < 10$, the following equation results for the drying time when Eqs. (5.27), (5.47), and (5.48a) are combined:

$$t_{sph} = -R^2 \frac{\ln \Phi_f - 0.7599 \dfrac{Bi_m}{2.1 + Bi_m}}{D[1.223 \ln(4.9 Bi_m + 1)]^{1/1.4}} \tag{5.50}$$

If $10 < Bi_m \leq 100$, then Eqs. (5.27), (5.47), and (5.48a) give the following expression for the drying time of the spherical-shape material:

$$t_{sph} = -R^2 \frac{\ln \Phi_f - 0.7599 \dfrac{Bi_m}{2.1 + Bi_m}}{D[1.667 \ln(2.1985 Bi_m + 152.4381)]^{1/1.2}} \tag{5.51}$$

If $Bi_m > 100$, then Eqs. (5.27), (5.46b), and (5.48a) can be combined to obtain the following equation:

$$t_{sph} = -\frac{R^2}{\pi^2 D} \ln\left(\frac{1}{2} \Phi_f\right) \tag{5.52}$$

**Example 5.4**

Potatoes assumed spheres with diameter of 18 mm are dried in air. The potato moisture diffusivity is 9.5E–9 m²/s, and the moisture transfer coefficient is 8.3E–6 m/s. Determine the drying time to reduce the initial moisture to 33%:

- The characteristic length (the radius) $R = 0.009$ m.
- Biot number is immediately calculated as $Bi_m = h_m R / D = 7.863$.
- Based on Biot value, Eq. (5.50) is selected to determine the drying time.
- Ggg.
- The finial dimensionless drying content is given as $\Phi_f = 0.33$.
- Thence, using Eq. (5.50), the following drying time results: $t_{sph} = 4977$ s.

## 5.2.5 Compact Analytical Solution or Time-Dependent Diffusion in Basic Shapes

The solution for transient moisture transfer equation for infinite slab, IC, and sphere presented previously can be written in a more compact and complete manner. The general solution for moisture diffusion, applicable to solids drying, equation in these geometries is given in Sahin and Dincer (2004) in the following approximate form:

$$\phi(\zeta, Fo_m) = \frac{2Bi_m}{\varphi(\mu_1)\left[\mu_1^2 + Bi_m^2 - (m-1)Bi_m\right]} \varphi(\mu_1 \zeta)\exp\left(-\mu_1^2 Fo_m\right) \tag{5.53}$$

where $m = 0$ for the infinite slab, $m = 1$ for the IC, $m = 2$ for sphere, $\zeta$ is the dimensionless spatial coordinate defined by $\zeta = z/L$ for the slab and $\zeta = r/R$ for the cylinder and sphere, $L$ is the half width of the slab, $R$ is the radius of the cylinder and sphere, and $\varphi(x)$ is the eigenfunction defined as follows:

$$\varphi(x) = \begin{cases} |\cos(x) & \text{for the infinite slab} \\ |J_0(x) & \text{for the infinite cylinder} \\ \left|\dfrac{\sin(x)}{x}\right| & \text{for the sphere} \end{cases} \tag{5.54}$$

The transcendental characteristic equation which is used to determine the eigenvalues $\mu_1$ is as follows:

$$\frac{\partial \varphi(\mu_1 \zeta)}{\partial \zeta} = -Bi_m \varphi(\mu_1 \zeta) \tag{5.55}$$

Note that the dimensionless moisture content at the solid center $\Phi$ can be determined from Eq. (5.54) by setting $\zeta = 0$. One has therefore $\Phi = \phi(0, Fo_m)$. Also, if $\zeta \rightarrow 0$, then $\varphi(\mu_1 \zeta) \rightarrow 1$ for all three geometries. The dimensionless moisture content at the surface and at the center correlate as follows:

$$\Phi_s \cong \cos(\mu_1)\Phi \quad \text{for the infinite slab} \tag{5.56a}$$

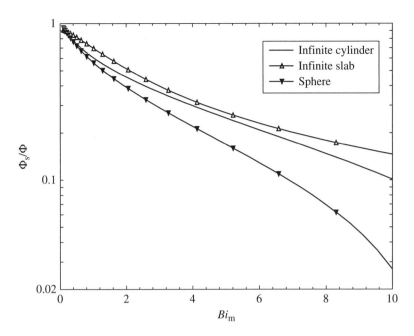

**Figure 5.9** Ratio of dimensionless moisture contents at material surface and the center during a drying process for a range of Biot numbers

$$\Phi_s \cong J_0(\mu_1)\Phi \quad \text{for the infinite cylinder} \tag{5.56b}$$

$$\Phi_s \cong \frac{\sin(\mu_1)}{\mu_1}\Phi \quad \text{for the sphere} \tag{5.56c}$$

The graph from Figure 5.9 shows the variation of the ratio between the dimensionless moisture at the material surface and the dimensionless moisture in the center, $\Phi_s/\Phi$. The three fundamental geometries are studied in the graphs: infinite slab, IC, and sphere. It is apparent that for Biot numbers in the range 0.16–10, the moisture content gradient in the sphere is the highest, followed by the IC and then by the infinite slab. This means that sphere dries the fastest.

## 5.3 Shape Factors for Drying Time

Analytical solution becomes difficult for moisture transfer in solids with complicated geometry. This would make the determination of drying time laborious. In order to give a practical solution for drying time determination of materials with complex geometry, the shape factor is introduced. Here, the shape factor represents the ratio between drying time of a slab product $t_{slab}$ and drying time of a product of other shape, $t_{shape}$. The shape factor is denoted $E$ and is thence defined with

$$E = \frac{t_{slab}}{t_{shape}} \tag{5.57}$$

The shape factors for drying time of various solids with regular geometries were developed in Sahin et al. (2002) based on the analogy of moisture diffusion with heat transfer at cooling (see Hossain et al., 1992). Any other irregular multidimensional solids can be approximated with regular solids. The shape of the regular solid is defined based on three parameters, namely:

- The characteristic length $L$ defined as the smallest distance between the geometrical center and the solid surface. This length stands for "thickness" and is used for determination of Biot number for mass transfer.
- The first semimajor axis $\beta_1 L$ representing the half width.
- The second semimajor axis $\beta_2 L$ representing the half length.

## 5.3.1   Infinite Rectangular Rod of Size $2L \times 2\beta_1 L$

In this geometry, the solid has four permeable faces, whereas along the infinite no mass transfer occurs; therefore, the moisture transfer is two dimensional. This geometry is a good approximation for rods having the length 4–5 times larger than the dimension $2\beta_1 L$. The infinite rectangular rod is shown in Figure 5.10a. The shape factor is given by the following expression:

$$E = \frac{1 + 2Bi_{\mathrm{m}}^{-1}}{1 + 2Bi_{\mathrm{m}}^{-1} - 4\sum_{n=1}^{\infty} \sin\mu_n \left\{ \mu_n^3 \left(1 + Bi_{\mathrm{m}}^{-1}\sin^2\mu_n\right) \left[ Bi_{\mathrm{m}}^{-1}\mu_n \sinh(\beta_1\mu_n) + \cosh(\beta_1\mu_n) \right] \right\}^{-1}}$$

where the eigenvalues are the solutions of the following characteristic equation:

$$Bi_{\mathrm{m}} = \mu_n \tan\mu_n \tag{5.58}$$

Figure 5.11 shows the shape factor for the infinite slab and the infinite square rod. The infinite rectangular rod is an infinite rectangular rod with $\beta = 1$, as defined above. The infinite rod has a smaller drying time than the slab because its permeable surface is larger: mass diffuses at four sides as opposing to the slab which has mass diffusion at two faces only.

## 5.3.2   Rectangular Rod of Size $2L \times 2\beta_1 L \times 2\beta_2 L$

For a rectangular parallelepiped, six faces are exposed to moisture transfer in a surrounding gas and therefore the process is three dimensional. Figure 5.10b shows a rectangular rod of finite size. The analytical solution represents a superposition of solutions for simpler two-dimensional case. There will be two characteristic equations of which one is $Bi_{\mathrm{m}} = \mu_n \tan\mu_n$ and the other is

$$\beta_1 Bi_{\mathrm{m}} = \mu_m \tan\mu_m$$

The eigenvalue is a weighted average of $\mu_n$ and $\mu_m$ given by $\mu_{nm}$ as follows:

$$\mu_{nm} = \beta_2 \sqrt{\mu_n^2 + \left(\frac{\mu_m}{\beta_1}\right)^2}$$

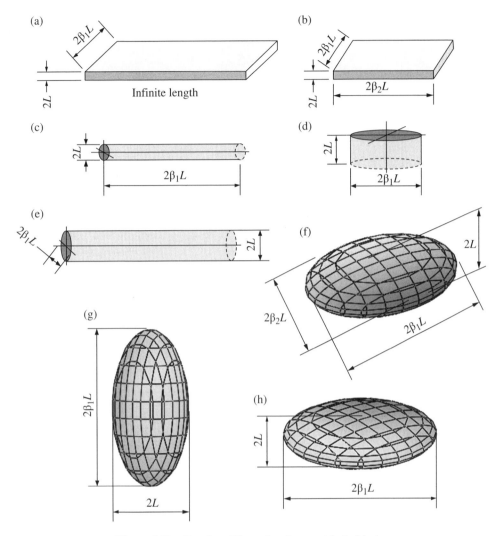

**Figure 5.10**   Regular tridimensional geometrical objects

The expression for the shape factor is given as shown in Sahin et al. (2002), namely,

$$E = \frac{1 + 2Bi_{\mathrm{m}}^{-1}}{1 + 2Bi_{\mathrm{m}}^{-1} - 4F_1 - 8\beta_2^2 F_2}$$

where $F_1$ and $F_2$ are series expansions as given below:

$$F_1 = \sum_{n=1}^{\infty} \sin \mu_n \left\{ \mu_n^3 \left(1 + Bi_{\mathrm{m}}^{-1} \sin^2 \mu_n \right) \left[ Bi_{\mathrm{m}}^{-1} \mu_n \sinh(\beta_1 \mu_n) + \cosh(\beta_1 \mu_n) \right] \right\}^{-1}$$

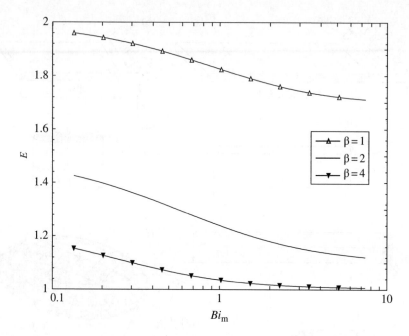

**Figure 5.11**   Variation of shape factor with Biot number for the infinite square rod ($\beta = 1$)

$$F_2 = \sum_{n=1}^{\infty}\sum_{m=1}^{\infty} \frac{\mu_n^{-1}\mu_m^{-1}\mu_{nm}^{-2}\sin(\mu_n)\sin(\mu_m)}{\left(\cosh\beta_1\mu_{nm} + \beta_2^{-1}Bi_{\mathrm{m}}^{-1}\sinh\mu_{nm}\right)\left(1 + Bi_{\mathrm{m}}^{-1}\sin^2\mu_n\right)\left(1 + Bi_{\mathrm{m}}^{-1}\beta_1^{-1}\sin^2\mu_m\right)}$$

### 5.3.3   Long Cylinder of Diameter 2L and Length $2\beta_1$L

For a long cylinder which is shown in Figure 5.10c, the length is much greater than the diameter, that is, $2\beta_1 L > 2L$ or $\beta_1 > 1$. Also, due to the axial symmetry for the cylinder, one has $\beta_2 = \beta_1$ and there is no preferential radial direction for mass transfer. Therefore, the diffusion mass transfer process can be modeled as two dimensional with one dimension being along the cylinder axis and the other being perpendicular on the axis.

The problem is also symmetric with respect to the midplane perpendicular on cylinder axis. This implies that the mass transfer can be modeled as one dimensional at the midplane where the diffusion occurs in radial direction only. The characteristic equation is in this case as follows:

$$\mu_n J_1(\mu_n) = Bi_{\mathrm{m}} J_0(\mu_n)$$

and the shape factor then becomes

$$E = \frac{2 + 4Bi_{\mathrm{m}}^{-1}}{1 + \dfrac{2}{Bi_{\mathrm{m}}} - 8\sum_{n=1}^{\infty}\left[\mu_n^3 J_1(\mu_n)\left(1 + \dfrac{\mu_n^2}{Bi_{\mathrm{m}}^2}\right)\left(\cosh\beta_1\mu_n + \dfrac{\mu_n}{Bi_{\mathrm{m}}}\sinh\beta_1\mu_n\right)\right]^{-1}}$$

Very many moist solids of irregular geometry can be approximated for drying time determination with a regular slender cylinder, for example, wood logs, carrot, cucumber, eggplant, zucchini, concrete poles, and so on.

### 5.3.4 Short Cylinder of Diameter $2\beta_1 L$ and Length $2L$

In this case, the characteristic length of the cylinder is the half-height $L$, because this is the shortest dimension, as shown in Figure 5.10d. The diffusion process is in this case more pronounced in the axial direction than in the radial direction. The analytical solution appears to be less complicated than for the slender cylinder. The characteristic equation is given by Eq. (5.43), whereas the shape factor equation is as follows:

$$E = \frac{1 + 2Bi_m^{-1}}{1 + 2Bi_m^{-1} - 4\sum_{n=1}^{\infty} \sin\mu_n \left\{ \mu_n^2 (\mu_n + \cos\mu_n \sin\mu_n) \left[ I_0(\beta_1\mu_n) + \mu_n Bi_m^{-1} I_1(\beta_1\mu_n) \right] \right\}^{-1}}$$

where $I_0$ and $I_1$ are the Bessel function of second kind of orders 0 and 1, respectively.

### 5.3.5 Infinite Elliptical Cylinder of Minor Axis $2L$ and Major Axis $2\beta_1 L$

The cross section of this cylinder is an ellipse having the minor axis $2L$ and the major axis $2\beta_1 L$ with $\beta_1 > 1$. An infinite elliptical cylinder is shown in Figure 5.10e. The characteristic length is in the case the half of the minor axis, that is, $L$. The shape factor is obtained as follows:

$$E = 1 + \frac{1 + 2Bi_m}{\beta_1^2 + 2\beta_1 Bi_m}$$

### 5.3.6 Ellipsoid Having the Axes $2L$, $2\beta_1 L$, and $2\beta_2 L$

For this solid body which is shown in Figure 5.10f, the diffusion process must be modeled as three dimensional. In general, all the irregular-shape products can be approximated to be ellipsoid:

$$E = 1 + \frac{1 + 2Bi_m^{-1}}{\beta_1^2 + 2\beta_1 Bi_m^{-1}} + \frac{1 + 2Bi_m^{-1}}{\beta_2^2 + \beta_2 Bi_m^{-1}}$$

Two special cases can be of common interest. Many food products appear to have these two shapes. They are as follows:

- Prolate spheroid as shown in Figure 5.10g (egg shape) for which $\beta_1 = 1$ and the shape factor is

$$E = 2 + \frac{1 + 2Bi_m^{-1}}{\beta^2 + 2\beta Bi_m^{-1}}$$

- Oblate spheroid as shown in Figure 5.10h ("hamburger" shape) for which $\beta_1 = \beta_2 = \beta$ and the shape factor is

$$E = 1 + 2\frac{1 + 2Bi_m^{-1}}{\beta^2 + 2\beta Bi_m^{-1}}$$

The shape factor for spheroids is represented as shown in Figure 5.12 in the form of a 3D surface representing the function $E = f(\beta, \ln Bi_m)$. The shape factor of the prolate spheroid is higher than the shape factor of the oblate spheroid. This implies that for similar condition, the solids with geometry of prolate spheroid (egg shape) dry faster.

The ellipsoid geometry is very useful for approximating the shape of many irregular solids; in addition, the elliptical cylinder can be useful for approximating objects with 2D symmetry. For example, a "hamburger" can be approximated with an oblate spheroid and an egg is approximately as a prolate spheroid. A fish can be viewed as an ellipsoid and a cucumber as an elliptical cylinder.

The recommended procedure for drying time determination of irregular products is as follows: (i) determine the geometric center of the product; if the irregular product has a flat surface, this flat surface will be taken as the symmetry surface, so that the characteristic dimension will be measured from this surface to the other outer surface in perpendicular direction to the symmetry surface; (ii) find the characteristic length being the smallest distance between the center and product surface, (iii) determine the volume of the irregular object, (iv) identify the type of regular geometry body which resemble the best with the irregular object; (v) determine the

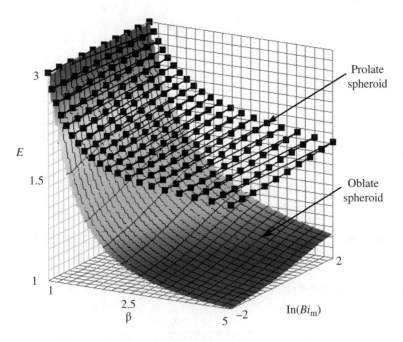

**Figure 5.12**    Shape factors for spheroids (oblate and prolate)

parameters $\beta_1$ and $\beta_2$ of the regular body such that its volume is equal to that of the irregular object; (vi) estimate the Biot number for mass transfer based on the characteristic length; (vii) determine the eigenvalue for the infinite slab problem and then the drying time for the slab according to Eqs. (5.26)–(5.30) whichever applies; (viii) and calculate the shape factor and then the drying time for the irregular object.

**Example 5.5**

A cylindrically shaped moist solid of 8 mm average diameter and 24 mm length are dried in air. The moisture transfer coefficient is $6.6 \times 10^{-9}$ m/s and the diffusivity is $5 \times 10^{-11}$ m/s. The dimensionless moisture at centerline for the dried product is 0.02. Determine the drying time assuming that the carrots are slender cylinders:

- The characteristic length of the carrot is $L = 4$ mm.
- The Biot number for mass transfer is calculated as follows:

$$Bi_{\mathrm{m}} = \frac{h_{\mathrm{m}}L}{D} = 0.528 > 0.1$$

- In order to determine the eigenvalue, Eq. (5.12) must be solved for $\mu_1$, that is,

$$1.056 \cot \mu_1 = \mu_1 \rightarrow \mu_1 = 5.492.$$

- The factor $A_1$ is calculated based on Eq. (5.13)

$$A_1 = \frac{2 \sin(\mu_1)}{\mu_1 + \sin(\mu_1)\cos(\mu_1)} = 0.03426.$$

- The drying time for the slab in similar conditions results from Eq. (5.27), namely,

$$t_{\mathrm{slab}} = -\frac{L^2}{D\mu_1^2} \ln\left(\frac{\Phi_{\mathrm{f}}}{A_1}\right) = 5711 \text{ s.}$$

- The factor $\beta_1$ is $24/8 = 3$.
- The characteristic equation for the shape factor is $\mu_1' J_1(\mu_1') = Bi_{\mathrm{m}} J_0(\mu_1') \rightarrow \mu_1' = 0.9635$.
- The shape factor is calculated using only the first term of the series expansion

$$E = \frac{2 + 4Bi_{\mathrm{m}}^{-1}}{1 + \dfrac{2}{Bi_{\mathrm{m}}} - 8\left[\mu_1^3 J_1(\mu_1)\left(1 + \dfrac{\mu_1^2}{Bi_{\mathrm{m}}^2}\right)\left(\cosh \beta_1 \mu_1 + \dfrac{\mu_1}{Bi_{\mathrm{m}}} \sinh \beta_1 \mu_1\right)\right]^{-1}} = 2.08.$$

- The drying time of the cylindrical product results from Eq. (5.57)

$$t_{\mathrm{dry}} = \frac{t_{\mathrm{slab}}}{E} = 2742 \text{ s.}$$

## 5.4   Moisture Transfer Coefficient and Diffusivity Estimation from Drying Curve

In drying analysis and design, the moisture transfer coefficient and the diffusivity are two of the most important parameters. In drying processes, the diffusivity and moisture transfer coefficient can be determined from drying curve and transient analysis of moisture transfer. The drying curve for a particular process can be determined experimentally based on a record of product weight (or moisture content) during the test time.

Therefore, test data can be obtained in form of moisture content variation in time $W = f(t)$, which expresses the drying curve. This curve can be put in a dimensionless format such as $\bar{\phi} = f(t)$ or $\Phi = f(t)$, where $\phi$ is the average dimensionless moisture content and $\Phi$ is the dimensionless moisture content at the center. Figure 5.13 shows qualitatively an example of drying data for dimensionless moisture content. Regression method can be applied to fit the data to a drying curve equation such as Eq. (5.22), $\Phi = LF\ln(-St)$, as shown in the figure. This process leads to lag factor (LF) and drying coefficient determination ($S$).

Once LF and $S$ are known, transient analysis can be applied to the specific process. The general solution can be approximated with the following expression giving the dimensionless content at the center:

$$\Phi = A_1(\mu_1, Bi_m)\exp\left(-\mu_1^2 Fo_m\right)$$

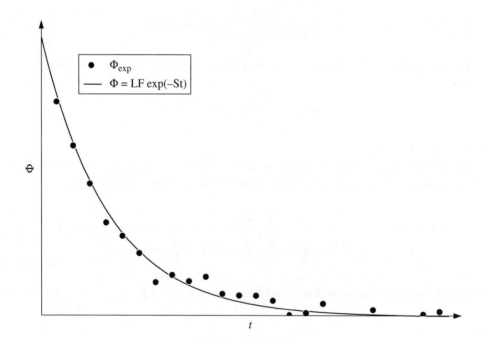

**Figure 5.13**   Data regression in a drying experiment to determine the exponential drying curve

where $\mu_1^2 = Bi_m$ and $A_1 = 1$ for the infinite slab when the lumped capacity model ($Bi_m < 0.1$) and for cases when $Bi_m \geq 0.1$ the specific characteristic equations are given above and can be solved to determine the eigenvalue and subsequently the factor $A_1$. Furthermore, it results in

$$LF = A_1(\mu_1, Bi_m) \tag{5.59}$$

and

$$St = \mu_1^2 \frac{Dt}{L^2} \tag{5.60}$$

The determination of Biot number implies numerical solution of Eq. (5.59) together with the transcendental characteristic equation for the eigenvalue $\mu_1$. The determination of diffusion coefficient requires solving Eq. (5.60) for $D$. This is solved as follows:

$$D = \frac{SL^2}{\mu_1^2} \tag{5.61}$$

Once the diffusivity is known, the moisture transfer coefficient can be determined according to the Biot number definition as follows:

$$h_m = D \frac{Bi_m}{L} \tag{5.62}$$

For $Bi_m < 1$, one sets $\mu_1^2 = Bi_m$ in Eq. (5.61) which gives that $D = SL^2/Bi_m$. Now, when this expression for $D$ is introduced in Eq. (5.62), the following equation for mass transfer coefficient results:

$$h_m = SL \tag{5.63}$$

Note that if $0.1 \leq Bi_m < 100$, then Eq. (5.17) where $A_1 = LF$ can be solved analytical for $Bi_m$, and thus, the following equation is obtained for the moisture transfer coefficient within the infinite slab:

$$h_m = \frac{1.3D\ln(LF)}{L(0.2533 - \ln(LF))} \tag{5.64}$$

Similarly for the IC Eq. (5.69a) and for the sphere Eq. (5.48a) can be solved analytically for $Bi_m$. This allow for determination of the following equations for the moisture transfer coefficient for the IC and SSM as shown in Dincer and Dost (1996), namely,

$$\text{For infinite cylinder}: h_m = \frac{D(1 - 1.974\ln(LF))}{3.3559\,L\ln(LF)} \tag{5.65a}$$

$$\text{For spherical material}: h_m = \frac{D(1 - 1.316\ln(LF))}{2.76396\,L\ln(LF)} \tag{5.65b}$$

Here, $L$ is the radius.

**Table 5.2** Experimental determinations of diffusivity and moisture transfer coefficient for some moist materials

| Shape | Moist material | $L$ (mm) | $Bi_m$ | LF | $S \times 10^4$ $(s^{-1})$ | $D \times 10^{10}$ $(m^2/s)$ | $h_m \times 10^7$ $(m/s)$ |
|---|---|---|---|---|---|---|---|
| Slab | Tomato slice | 2.5 | 0.1–100 | 1.4105 | 2.6 | 6.58 | 230 |
| Slab | Tomato slice | 2.5 | >100 | 1.218 | 5.5 | 23.61 | 2.54 |
| Slab | Wood (Douglas fir) | 20 | 0.7516 | 1.0941 | 394.66 | 32.4 | 2.263 |
| Slab | Prune | 2.5 | 0.4054 | 1.0037 | 3 | 669 | 108.5 |
| Slab | Prune | 7.5 | 0.0745 | 1.0016 | 0.7 | 1989 | 19.76 |
| Slab | Onion slice | 2.5 | 2.421 | 1.1503 | 2 | 12.6 | 12.23 |
| Cylinder | Okra | 3 | 0.0851 | 1.1981 | 1 | 5.675 | 0.161 |
| Cylinder | Starch[*] | 5 | 0.093 | 1.0181 | 6 | 1299 | 2.414 |
| Cylinder | Potato | 13.5 | 0.314 | 1.032 | 0.7 | 0.6717 | 15.62 |
| Sphere | Potato | 9 | 0.312 | 1.0074 | 9 | 9425.9 | 326.65 |
| Sphere | Yam | 30 | 47.9 | 1.2864 | 46 | 15,110 | 24,100 |

*Sources*: Dincer (1998a), Dincer and Dost (1996), Dincer and Hussain (2002, 2004), Dincer et al. (2002).
[*] High amylose starch powder.

**Table 5.3** Dimensionless moisture content data for Example 5.6

| $t$ (min) | 0.01667 | 27.79 | 55.57 | 83.34 | 111.1 | 138.9 | 166.7 | 194.4 | 222.2 | 250 |
|---|---|---|---|---|---|---|---|---|---|---|
| $\Phi_{exp}$ | 1.182 | 0.6079 | 0.298 | 0.1452 | 0.07601 | 0.05781 | 0.01286 | 0.01232 | 0.03069 | 0.02125 |

Table 5.2 give a set of data for experimental determinations of diffusivity and moisture transfer coefficient for 12 moist materials compiled from literature sources. The drying curve has been regressed for each case to determine the lag factor and the drying coefficient. The Biot number has been calculated based on the equation introduced above, namely, $LF = A_1$. Then, based on $Bi_m$, the eigenvalue has been calculated for each case. Further, the diffusivity and moisture transfer coefficient have been determined.

**Example 5.6**
Consider the drying process of a moist solid that can be assimilated with an infinite cylinder. The drying data in terms of dimensionless moisture content is given in Table 5.3. Determine the lag factor, the drying coefficient, the diffusivity, and moisture transfer coefficient provided that the characteristic length (cylinder radius) is 5 mm:

- Data can be regressed using a specialized software to determine the drying curve. Here, we used the EES software (EES, 2014) and obtained $\Phi = 1.129 \exp(-0.0241t)$ with a correlation coefficient as high as 85% and root mean square error of 0.157. Thence, the lag factor is $LF = 1.129$ and the drying coefficient $S = 0.0241 \, s^{-1}$.
- Equation (5.59) must be now solved numerically for $Bi_m$. Assume that $0.1 \leq Bi_m \leq 100$. In this case, Eq. (5.39a) is applicable; therefore, one has

$$LF = 1.129 = \exp\left(0.5066 \frac{Bi_m}{1.7 + Bi_m}\right) \rightarrow Bi_m = 0.536 > 0.1$$

- It is now confirmed that Biot is in the assumed range.
- Determine the eigenvalue according to the approximate Eq. (5.38) $\rightarrow \mu_1 = [0.72\ln(6.796\,Bi_\mathrm{m}+1)]^{1/1.4}$, which gives $\mu_1 = 1.074$.
- The diffusivity results now for $L = 0.005\,\mathrm{m}$ from Eq. (5.61)

$$D = \frac{SL^2}{\mu_1^2} = 5.22 \times 10^{-7}\,\mathrm{m^2/s}$$

- The moisture transfer coefficient results now from Eq. (5.62), namely,

$$h_\mathrm{m} = D\frac{Bi_\mathrm{m}}{L} = 5.6 \times 10^{-5}\,\mathrm{m/s}$$

## 5.5 Simultaneous Heat and Moisture Transfer

In many drying problems, the balance equations for heat and moisture transfer are coupled and must be solved simultaneously. For example, the surface moisture on solids subjected to drying must be evaporated and swept away in the gaseous drying agent (e.g., air). This process requires heat input, and the rate of heat input influences directly the evaporation rate. In many cases, the mass transfer induces deformation such as shrinkage and change of the characteristic length, material volume, and density. This effects impact both the heat and mass transfer processes in a dynamic manner.

The interdependence of mass transfer and heat transfer can manifest in many ways. For example, in many situations, the moisture diffusivity depends on the temperature according to an Arrhenius-type equation as follows:

$$D = D_0 \exp\left(-\frac{E_\mathrm{a}}{\mathcal{R}T}\right) \tag{5.66}$$

where $D_0$ (m²/s) is the preexponential factor, $E_\mathrm{a}$ is the activation energy, $\mathcal{R}$ is the universal gas constant, and $T(K)$ is the temperature. Note that the activation energy for diffusion is of the order of 10 kJ/mol for many types of agricultural and food products and construction materials.

On the other hand, the thermal conductivity of the moist material varies with the moisture content. In general, the presence of the diffused moisture increases the thermal conductivity. If one denotes with $k_0$ the thermal conductivity of the dry material, then the moist material thermal conductivity can be estimated with an equation of the following form:

$$k = k_0(1 + k_\mathrm{w}W) \tag{5.67}$$

where $k_\mathrm{w}$ is a dimensionless correlation coefficient and $W$ is the local moisture content.

In addition, the presence of moisture affects the material density, specific heat, and thermal diffusivity. In general, the thermal diffusivity of the moisture is higher than that of dry material. Therefore, the presence of moisture leads to a thermal diffusivity increase. If one denotes with $a_0$ the thermal diffusivity of the dry material and with $a_\mathrm{w}$ the thermal diffusivity of the moisture, then the moist material thermal diffusivity can be correlated with the moisture content based on an equation such as the following:

$$a(T,W) = a_0(T) + [a_w(T) - a_0(T)]W \qquad (5.68)$$

where both $a_0$ and $a_w$ are functions of temperature.

The values for $k_0$ and the function $a_0(T)$ can be established based on experiments for each specific material. In the study by Yildiz and Dincer (1995), a value of $k_0 = 0.08\,\text{W/mK}$ is given for sausages, whereas $k_w$ is taken 6.5. For the same material, $a_0(T)$ is taken as invariant with temperature, namely, $a_0(T) = 8.8 \times 10^{-8}\,\text{m}^2/\text{s}$.

The time-dependent mass balance equation (MBE) and energy balance equation (EBE) must be written in order to model the simultaneous process of heat and mass transfer. These equations are accompanied with initial and boundary conditions and all necessary constitutive equations to make the problem mathematically solvable. One-, two-, and three-dimensional governing equations can be written for simultaneous heat and mass transfer. These equations are solved for two types of parameters, namely, temperature $T(x, y, z, t)$ and moisture concentration $C_m(x, y, z, t)$. Note that in drying problems, it is often useful to use moisture content $W(x, y, z, t)$ instead of moisture concentration. Depending on the moist solid geometry, the most suitable system of coordinate must be selected to formulate the problem: Cartesian $(x, y, z)$, cylindrical $(r, \theta, z)$, and spherical $(r, \theta, \varphi)$.

Let us consider first the one-dimensional drying process with simultaneous heat and mass transfer. In this case, we denote generically the spatial coordinate with $y$, where $y = z$ (transversal coordinate) for the infinite slab and $y = r$ (radius) for the IC and sphere. Therefore, the temperature and moisture content are functions of $y$ and $t$ (time): $T = T(y,t)$ and $W = W(y,t)$.

The time-dependent heat and moisture transfer equations in one-dimensional rectangular, cylindrical, and spherical coordinates for an infinite slab, IC, and a sphere, respectively, all with assumed central symmetry of heat and mass transfer processes, can be written in the following compact form:

$$\text{For heat transfer}: \frac{a}{y^m} \frac{\partial}{\partial y}\left[y^m\left(\frac{\partial T}{\partial y}\right)\right] = \frac{\partial T}{\partial t} \qquad (5.69a)$$

$$\text{For moisture transfer}: \frac{D}{y^m} \frac{\partial}{\partial y}\left[y^m\left(\frac{\partial W}{\partial y}\right)\right] = \frac{\partial W}{\partial t} \qquad (5.69b)$$

where $m = 0$ for the infinite slab, $m = 1$ for the IC, and $m = 2$ for the sphere.

If one denotes $T_i$ as the initial temperature which is assumed uniform throughout the moist material and $T_\infty$ as the bulk temperature in the drying agent (air), then the dimensionless temperature is introduced as follows:

$$\theta(y,t) = \frac{T(y,t) - T_i}{T_\infty - T_i} \qquad (5.70)$$

Furthermore, the dimensionless moisture content – defined above as $\phi$ – is also used as a function of location and time, $\phi = \phi(y,t)$. Based on $\theta(y,t)$ and $\phi(y,t)$, the one-dimensional time-dependent equations for simultaneous heat and moisture transfer become

$$\text{For heat transfer}: \frac{a}{y^m} \frac{\partial}{\partial y}\left[ y^m \left( \frac{\partial \theta}{\partial y} \right) \right] = \frac{\partial \theta}{\partial t} \tag{5.71a}$$

$$\text{For moisture transfer}: \frac{D}{y^m} \frac{\partial}{\partial y}\left[ y^m \left( \frac{\partial \phi}{\partial y} \right) \right] = \frac{\partial \phi}{\partial t} \tag{5.71b}$$

Let us make a change of variables in Eqs. (5.71) denoting $y = \zeta L$, where $\zeta$ is the dimensionless coordinate and $L$ is the characteristic length. For the infinite slab, $L$ represents the half thickness of the slab, whereas for the IC and the sphere, $L = R$ is the radius of the cylinder/sphere.

Further, other changes of variables are made in Eq. (5.69) for the time. Namely, the time variable is replaced with $t = Fo L/a_0$, where $Fo$ is the Fourier number for heat conduction and $a_0$ is the thermal diffusivity of the dry material. Also in Eq. (5.69b), the time variable is replaced with $t = Fo_m L/D_0$, where $Fo$ is the Fourier number for moisture diffusion and $D_0$ is the moisture diffusivity of the dry material. Note also that $Fo_m = Fo D_0/a_0$. The thermal and moisture diffusivities are modeled based on Eqs. (5.68) and (5.66), respectively. With these changes of variables and under the assumption that $a_0$ and $D_0$ are invariant, Eq. (5.69) rewrites as follows:

$$\frac{(a_1 + a_2\phi)}{\zeta^m} \frac{\partial}{\partial \zeta}\left[ \zeta^m \left( \frac{\partial \theta}{\partial \zeta} \right) \right] = \frac{\partial \theta}{\partial Fo} \quad \text{for heat transfer} \tag{5.72a}$$

$$\frac{1}{\zeta^m} \frac{\partial}{\partial \zeta}\left[ \zeta^m \left( \frac{\partial \phi}{\partial \zeta} \right) \right] = \left( \frac{a_0}{D_0} \right) \exp\left( \frac{E_a}{d_1 + d_2\theta} \right) \frac{\partial \phi}{\partial Fo} \quad \text{for moisture transfer} \tag{5.72b}$$

where $\zeta \in [0,1]$ and $Fo \geq 0$.

The constants $a_1$ and $a_2$ from thermal diffusion equation are derived from Eq. (5.68) as follows:

$$a_1 = 1 + \left( \frac{a_w}{a_0} - 1 \right) W_i$$

and

$$a_2 = \left( \frac{a_w}{a_0} - 1 \right)(W_i - W_e)$$

The constants $d_1$ and $d_2$ from moisture diffusion equation are derived from Eq. (5.66) as follows:

$$d_1 = \mathcal{R} T_i$$

and

$$d_2 = \mathcal{R}(T_\infty - T_i)$$

Equations (5.72) are dimensionless and can be solved simultaneously for $\theta(\zeta, Fo)$ and $\phi(\zeta, Fo)$ using numerical methods. Of course, these equations must be supplemented with initial and boundary conditions for both heat and mass transfer. The following initial conditions can be applied, assuming that the moisture content and temperature is initially uniform throughout the product. Therefore, for $\zeta \in [0,1]$, one has

$$\theta(\zeta,0) = \theta_i \tag{5.73a}$$

$$\phi(\zeta,0) = \phi_i \tag{5.73b}$$

The boundary condition at the center reflects the symmetry of the one-dimensional heat and mass transfer process. Therefore, for all $Fo \geq 0$, namely,

$$\frac{\partial\theta(0,Fo)}{\partial\zeta} = 0 \tag{5.74a}$$

$$\frac{\partial\phi(0,Fo)}{\partial\zeta} = 0 \tag{5.74b}$$

The boundary condition at the surface reflects the continuity or nonaccumulation of neither mass nor of energy at the interface. This conditions can be written for all $Fo \geq 0$ as follows:

$$-\frac{\partial\theta(1,Fo)}{\partial\zeta} = Bi(1-\theta) \tag{5.75a}$$

$$-\frac{\partial\phi(1,Fo)}{\partial\zeta} = Bi_m(\phi-\phi_\infty) \tag{5.75b}$$

where $Bi = h_{tr}L/k$ is the Biot number for heat transfer, $h_{tr}$ is the heat transfer coefficient, $k$ is the thermal conductivity of the moist solid, $Bi_m = h_m L/D$ is the Biot number for mass transfer, $h_m$ is the moisture transfer coefficient, $D$ is the moisture diffusivity, and $\phi_\infty$ represents a dimensionless moisture content in the surrounding gas defined as follows:

$$\phi_\infty = \frac{W_\infty - W_e}{W_i - W_e} = \frac{C_{m,\infty}/\rho_s - W_e}{W_i - W_e}$$

where $W_\infty$ is the bulk moisture content in the surrounding gas expressed in kilogram of moisture per kilogram of dry solid, $\rho_s$ is the density of dry solid, and $C_{m,\infty}$ is the moisture concentration in bulk surrounding gas, expressed in kilogram of moisture per unit of volume.

Recall that Eq. (5.72) represents the mathematical model for the simultaneous heat and mass transfer through thermal conduction and moisture diffusion in solids for the case when the process is one dimensional. The one-dimensional case can be often used to approximate practical drying problems. It applies when the external conditions around the moist solid are kept uniform and the solid has a regular shape with geometrical symmetry. For example, in the infinite

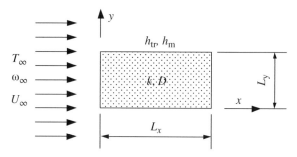

**Figure 5.14**  Sketch for 2D heat and mass transfer time-dependent modeling in Cartesian coordinates

slab, the transfer process occurs across the thickness, and in the IC and the sphere, it occurs radially with no preferential direction.

In many other practical cases, the surface boundary conditions are not uniform. Then, in such cases, multidimensional equations for heat and moisture transfer can be written. In general, for rectangular materials, the Cartesian system of reference is useful for expressing the heat and mass transfer equations. Cylindrical coordinate is useful for objects having an axis of revolution, and spherical coordinate is useful for ellipsoids.

Here, the heat and mass transfer equations are illustrated for the two-dimensional case in Cartesian coordinates. Refer for simplicity to a rectangular, permeable solid immersed in a gas which is forced over the solid with a constant and uniform bulk velocity and temperature. The sketch of this system is shown in Figure 5.14.

The following governing equations for heat and mass transfer can be written for $T(x, y, t)$ and $W(x, y, t)$ as follows:

$$a\left(\frac{\partial^2 T}{\partial x^2} + \frac{\partial^2 T}{\partial y^2}\right) = \frac{\partial T}{\partial t} \tag{5.76a}$$

$$D\left(\frac{\partial^2 W}{\partial x^2} + \frac{\partial^2 W}{\partial y^2}\right) = \frac{\partial W}{\partial t} \tag{5.76b}$$

with the following initial conditions are commonly set (uniform temperature and moisture content throughout the moist material) for all $x \in [0, L_x]$ and $y \in [0, L_y]$:

$$T(x, y, 0) = T_i \tag{5.77a}$$

$$W(x, y, 0) = W_i \tag{5.77b}$$

and with the continuity boundary conditions (that is at product surface) as given in Table 5.4.

A general 3D formulation for governing equations for heat and moisture transfer can be given using the Laplace operator. The two equations, thermal conduction and mass diffusion, are similar. Also in most drying processes, the source terms can be neglected because no phase change occurs within the moist material (note that the moisture evaporation occurs normally at product surface). The partial differential equations states that the rate of temperature change

**Table 5.4** Boundary conditions for simultaneous heat and mass transfer in 2D domain from Figure 5.14

| Edge | Thermal (energy) boundary condition | Mass (moisture) boundary condition |
|---|---|---|
| Left: $x=0, y \in [0, L_y]$ | $-k\dfrac{\partial T(0,y,t)}{\partial x} = h_{\text{tr}}[T(L_x,y,t)-T_\infty]$ | $-D\dfrac{\partial W(0,y,t)}{\partial x} = h_{\text{m}}[W(0,y,t)-W_\infty]$ |
| Right: $x=L_x, y \in [0, L_y]$ | $-k\dfrac{\partial T(L_x,y,t)}{\partial x} = h_{\text{tr}}[T(L_x,y,t)-T_\infty]$ | $-D\dfrac{\partial W(L_x,y,t)}{\partial x} = h_{\text{m}}[W(L_x,y,t)-W_\infty]$ |
| Bottom: $y=0, x \in [0, L_x]$ | $-k\dfrac{\partial T(x,0,t)}{\partial y} = h_{\text{tr}}[T(x,0,t)-T_\infty]$ | $-D\dfrac{\partial W(x,0,t)}{\partial y} = h_{\text{m}}[W(x,0,t)-W_\infty]$ |
| Top: $y=L_y, x \in [0, L_x]$ | $-k\dfrac{\partial T(x,L_y,t)}{\partial y} = h_{\text{tr}}[T(x,L_y,t)-T_\infty]$ | $-D\dfrac{\partial W(x,L_y,t)}{\partial y} = h_{\text{m}}[W(x,L_y,t)-W_\infty]$ |

**Table 5.5** Time-dependent moisture diffusion equation without source terms in 3D coordinate systems

| Coordinate system | Coordinates | Moisture diffusion equation |
|---|---|---|
| Cartesian | $x$ – abscissa<br>$y$ – ordinate<br>$z$ – elevation | $\dfrac{1}{D}\left(\dfrac{\partial W}{\partial t}\right) = \dfrac{\partial^2 W}{\partial x^2} + \dfrac{\partial^2 W}{\partial y^2} + \dfrac{\partial^2 W}{\partial z^2}$ |
| Cylindrical | $r$ – radial coordinate<br>$\theta$ – azimuth<br>$z$ – elevation | $\dfrac{1}{D}\left(\dfrac{\partial W}{\partial t}\right) = \dfrac{1}{r}\left[\dfrac{\partial}{\partial r}\left(r\dfrac{\partial W}{\partial r}\right)\right] + \dfrac{1}{r^2}\left(\dfrac{\partial^2 W}{\partial \theta^2}\right) + \dfrac{\partial^2 W}{\partial z^2}$ |
| Spherical | $r$ – radial coordinate<br>$\theta$ – azimuth<br>$\varphi$ – zenith | $\dfrac{1}{D}\left(\dfrac{\partial W}{\partial t}\right) = \dfrac{1}{r^2}\left[\dfrac{\partial}{\partial r}\left(r^2\dfrac{\partial W}{\partial r}\right)\right] + \dfrac{1}{r^2 \sin\theta}\left[\dfrac{\partial}{\partial \theta}\left(\sin\theta\dfrac{\partial W}{\partial \theta}\right)\right] + \dfrac{1}{r^2 \sin\theta}\dfrac{\partial^2 W}{\partial \varphi^2}$ |

equals the Laplacian of temperature field multiplied to thermal diffusivity, whereas the rate of change of moisture content equals the Laplacian of moisture content field into to moisture diffusivity, namely,

$$\frac{\partial T}{\partial t} = a\nabla^2 T \tag{5.78a}$$

$$\frac{\partial W}{\partial t} = D\nabla^2 W \tag{5.78b}$$

Table 5.5 gives the time-dependent moisture diffusion equation in 3D coordinate system for Cartesian, cylindrical, and spherical coordinates. The heat conduction equation is similar: in Table 5.5, the $W$ must be replaced with $T$ and $D$ is replaced with $a$. In many practical problems, the number of dimensions can be reduced to two. For example, in a finite size cylinder placed in a uniform medium, there is no preferential radial direction for heat and moisture transfer; therefore, there is neither gradient of moisture content nor gradient of temperature in the θ direction. In this situation, the problem is two dimensional in coordinates $(r, z)$. The two-dimensional time-dependent governing equations for heat and moisture transfer become

$$\frac{1}{a}\left(\frac{\partial T}{\partial t}\right) = \frac{1}{r}\left[\frac{\partial}{\partial r}\left(r\frac{\partial T}{\partial r}\right)\right] + \frac{\partial^2 T}{\partial z^2} \qquad (5.79a)$$

$$\frac{1}{D}\left(\frac{\partial W}{\partial t}\right) = \frac{1}{r}\left[\frac{\partial}{\partial r}\left(r\frac{\partial W}{\partial r}\right)\right] + \frac{\partial^2 W}{\partial z^2} \qquad (5.79b)$$

Consider now a spherical material placed in a stream of gas which flows from a fixed direction. In this case, the transfer processes can be assumed occurring with no preferential direction with respect to the zenith angle. This means that $\partial T/\partial \varphi = 0$ and $\partial W/\partial \varphi = 0$. The problems become two dimensional with the following governing equations in spherical coordinates:

$$\frac{1}{a}\left(\frac{\partial T}{\partial t}\right) = \frac{1}{r^2}\left[\frac{\partial}{\partial r}\left(r^2\frac{\partial T}{\partial r}\right)\right] + \frac{1}{r^2\sin\theta}\left[\frac{\partial}{\partial \theta}\left(\sin\theta\frac{\partial T}{\partial \theta}\right)\right] \qquad (5.80a)$$

$$\frac{1}{D}\left(\frac{\partial W}{\partial t}\right) = \frac{1}{r^2}\left[\frac{\partial}{\partial r}\left(r^2\frac{\partial W}{\partial r}\right)\right] + \frac{1}{r^2\sin\theta}\left[\frac{\partial}{\partial \theta}\left(\sin\theta\frac{\partial W}{\partial \theta}\right)\right] \qquad (5.80b)$$

Analytical solutions for simultaneous heat and moisture transfer do not exist, even for simplest geometries and boundary conditions. The sole approach to solve the coupled equations is by using numerical methods. In this way one can account for the shrinkage and for material properties variation with the temperature and moisture content (recall that: moisture diffusivity, thermal diffusivity, thermal conductivity, density, and specific heat depend on temperature and moisture content). This is treated in Chapter 6. However, there are in the open literature some analytical approaches to heat and mass transfer problems applicable to porous moist materials. These aspects are discussed in the next section.

## 5.6   Models for Heat and Moisture Transfer in Drying

Since analytical solutions for the coupled partial differential equations does not exist, some special approaches are established in the literature to model heat and moisture transfer for specific cases through approximate solutions and empirical methods. Although there is no general unified model, there are many published models which give sufficiently good results for specific situations. These models can be classified as theoretical, semiempirical, and empirical. In this section, literature models for heat and moisture transfer applicable to drying processes are reviewed based on the following main sources: Luikov (1975), Barbosa-Cánovas and Vega-Mercado (1996), Kucuk et al. (2014).

## 5.6.1 Theoretical Models

For the first drying period, the heat and mass transfer processes can be modeled as decoupled. In this case, the heat transfer influences the mass transfer and not vice versa. This fact simplifies the modeling. As also treated in Chapter 2 (Eq. (2.5)), the solution for heat and moisture transfer in the first drying period when heat transfer induces moisture evaporation is written as follows:

$$\frac{dW}{dt} = -\frac{h_{tr}A}{m_s \Delta h_{lv}}(T_a - T) = -\frac{h_m \rho_a A}{L}(\omega_{sat}(T) - \omega(T_a)) \tag{5.81}$$

where $A$ is the material surface area, $h_{tr}$ is the heat transfer coefficient, $m_s$ is the mass of dry material, $\Delta h_{lv}$ is the latent heat of moisture evaporation, $\omega$ is the humidity ratio of air (gas) at bulk temperature, $\omega_{sat}$ is the humidity ratio of saturated moist air at surface temperature, $L$ is the characteristic length of the slab, and $\rho_a$ is the density of dry air.

Equation (5.81) expresses in a compact form the mass and energy balance at the interface, where neither mass nor energy can accumulate because the interface is conceived as zero-thickness material. More exactly, this equation expresses the fact that the energy received by heat transfer at the interface is entirely used to evaporate the moisture. Furthermore, the evaporated moisture is entirely absorbed by the drying gas (air) which is humidified in this manner. In order for the equation to be applied for drying rate and the thence the drying curve determination, the mass transfer coefficient must be estimated. There are many correlations available for $h_m$. Here, we give the correlation of Okos et al. (1992) which is valid for laminar flow across slab-shaped material, namely,

$$h_m = 0.664 \frac{D}{L_x} \left(\frac{UL\rho}{\mu}\right)^{0.5} \left(\frac{\mu}{\rho D}\right)^{0.333} \tag{5.82}$$

with $\rho$ and $\mu$ being the density and kinematic viscosity of humid air, $D$ the moisture diffusivity in air, $L$ the characteristic length (slab thickness), and $L_x$ the length of the slab in the direction of air movement.

Note that in this case, the temperature of the moist material remains constant due to the fact that the moisture evaporates at the surface. Therefore, the heat transfer equation is time independent. In addition, only the moisture that is at the material surface evaporates, which means that the moisture content is a function of time only.

For the falling rate period, the moisture content and the temperature are interdependent and the heat diffusion and moisture diffusion equations must be solved in a coupled manner. The theoretical models aim to develop alternative approaches to approximate the solution for simultaneous heat and moisture transfer. The most widely used theoretical models take in account for the external condition (Biot number) and the mechanism of internal movement of moisture and their consequent effects. The derivation of the most of theoretical models is based on the solution of the one-dimensional time-dependent diffusion equation (also known as Fick's second law of diffusion), namely,

$$\frac{\partial C_m}{\partial t} = D \frac{\partial^2 C_m}{\partial z}$$

Luikov (1975) reviewed the theories for simultaneous heat and mass transfer among which the so-called Luikov theory is remarked by employment of irreversible thermodynamics in the derivations of its equations. Luikov calls the temperature and moisture concentration gradients as the driving forces for the simultaneous process of heat and moisture transfer. In Luikov theory, some phenomenological arguments are presented so that to modify the Fick's second law for capillary materials. The one-dimensional equations for simultaneous heat and moisture transfer become

$$\text{For heat transfer:} \frac{\partial T}{\partial t} = a\nabla^2 T + \frac{\varepsilon \Delta h_{\text{lv}}}{C_p}\left(\frac{\partial C_{\text{m}}}{\partial t}\right) \tag{5.83a}$$

$$\text{For moisture transfer:} \frac{\partial C_{\text{m}}}{\partial t} = D\nabla^2 C_{\text{m}} + D\delta\nabla^2 T \tag{5.83b}$$

where $\varepsilon$ is a so-called phase conversion factor of liquid into vapor, $\delta$ is a thermal gradient coefficient based on moisture content difference, $C_p$ is the specific heat of moist body, and $\Delta h_{\text{lv}}$ is the latent heat of moisture evaporation.

As seen from the above equations, the heat and mass transfer processes are highly coupled and require simultaneous solution. If the capillary flow in the moist material is important, then the pressure field cannot be neglected especially because of the fact that moisture evaporation in constrained capillary spaces will increase the pressure. The rate of drying results in this case from the capillary flow theory which states that the rate of moisture transfer is proportional to the gradient of pressure difference between the liquid moisture and gas interface. This model is described as follows:

$$\frac{\partial W}{\partial t} = -\sigma\cos\vartheta_{\text{contact}} f(r)\nabla\left(P_{\text{l}} - P_{\text{g}}\right) \tag{5.84}$$

where $\sigma$ is the surface tension, $\vartheta_{\text{contact}}$ is the contact angle, $f(r)$ is a function representing the distribution of pores radius, $P_{\text{l}}$ is the moisture pressure, and $P_{\text{g}}$ is the gas (air) pressure.

If capillary effects are included in the Luikov model from Eq. (5.83), then a set of three equations for heat, moisture, and momentum transfer will describe the process. Here, the Luikov model is expressed in a compact matrix form as follows:

$$\frac{\partial}{\partial t}\begin{pmatrix} W \\ T \\ P \end{pmatrix} = \nabla^2 \begin{pmatrix} k_{11} & k_{12} & k_{13} \\ k_{21} & k_{22} & k_{23} \\ k_{31} & k_{32} & k_{33} \end{pmatrix} \begin{pmatrix} W \\ T \\ P \end{pmatrix} \tag{5.85}$$

where $k_{ij}$ are coefficients that are estimated based on thermophysical parameters, temperature, moisture content, and pressure. The equations for these coefficients are given in Luikov (1975). Further discussion on Luikov equations, including a numerical example, is given in Chapter 6.

The condensation–evaporation theory as described in Barbosa-Cánovas and Vega-Mercado (1996) assumes that as the moisture diffusing through the moist material reaches the surface, it immediately condenses. However, under the influence of temperature gradient that drives the process, the rate of moisture condensation equals the rate of moisture evaporation so that

**Table 5.6** Sorption isotherms models

| Model | Description | Reference |
|---|---|---|
| BET model | $\dfrac{a_w}{W_e(1-a_w)}=\dfrac{C-1}{W_m C}a_w+\dfrac{1}{W_m C}$ | Barbosa-Cánovas and Vega-Mercado (1996) |
| GAB model | $W_e=\dfrac{W_m C K a_w}{(1-Ka_w)(1-Ka_w+CKa_w)}$ | Kiranoudis et al. (2003), Van den Berg (1985) |
| Chung–Pfost model | $W_e=\dfrac{1}{B}\left(\ln\dfrac{A}{RT}-\ln(-\ln a_w)\right)$ | Chung and Pfost (1967) |
| Halsey model | $W_e=\left(-\dfrac{A}{T\ln a_w}\right)^{\frac{1}{B}}$ | Halsey (1948) |
| Henderson model | $W_e=A\left(\dfrac{a_w}{1-a_w}\right)^{B}$ | Henderson (1952) |
| Oswin model | $W_e=0.01\left(-\dfrac{\ln(1-a_w)}{AT}\right)^{\frac{1}{B}}$ | Oswin (1946) |
| Smith model | $W_e=A-B\ln(1-a_w)$ | Smith (1947) |

no moisture is accumulated at the interface. This model is based on the MBE and EBE as follows:

$$\text{MBE}:\ \frac{\partial W}{\partial t}=\frac{\alpha t D}{\rho_s(1-\alpha)}\nabla^2 C_m-\frac{\alpha}{\rho_s(1-\alpha)}\left(\frac{\partial C_m}{\partial t}\right) \tag{5.86a}$$

$$\text{EBE}:\ \alpha\rho_s C_{p,s}\frac{\partial T}{\partial t}+\Delta h_{lv}\left(\frac{\partial W}{\partial t}\right)=a\nabla^2 T \tag{5.86b}$$

where $\alpha$ represents the volume fraction of air in the material pores, $t$ is a factor that include the effect of the porous material tortuosity, $\rho_s$, $C_{p,s}$ are the dry material density and specific heat, $C_m$ is the moisture concentration by weight, and $\Delta h_{lv}$ is the latent heat of moisture evaporation.

The models given in Table 5.6 are based on sorption isotherms. These isotherms represent the relationship between the equilibrium moisture content and water activity within a material at a fixed temperature and they may refer to both adsorption and desorption. For the special case of drying, the desorption isotherm is the most relevant. Figure 5.15 shows a sketch illustrating the effect of hysteresis in which the sorption and desorption isotherm are different. During the desorption, the moisture content is necessarily higher than during adsorption.

The basic form to express a sorption isotherm is through Langmuir equation given as follows:

$$\frac{a_w}{W_e}=\frac{a_w}{W_m}+\frac{1}{bP_{sat}(T)W_m} \tag{5.87}$$

where $W_e$ is the equilibrium moisture content due to absorbed water, $W_m$ is the moisture content due to one single layer of water molecules (monolayer) adsorbed at material surface, $b$ is a

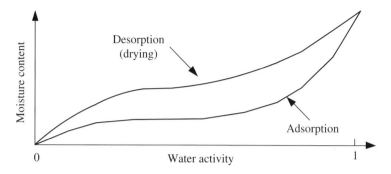

**Figure 5.15** Sketch illustrating the hysteresis effect of sorption isotherms

constant, $a_w$ is the water activity, and $P_{sat}$ is the saturation pressure of water at the temperature considered.

As seen from Eq. (5.87), there is linear dependence of $(a_w/W_e)$ with $a_w$. Experimental data are used to plot the Langmuir isotherms as straight lines in $a_w/W_e = f(a_w)$ format. Then the slope of the curve is used to determine the moisture content in the monolayer since slope $= a_w/W_m$; furthermore, the intercept is used to determine the parameter $b$ since intercept $= [bP_{sat}(T)W_m]^{-1}$.

Other sorption isotherms are given in Table 5.6. They can be used to predict the moisture content in correlation with water activity. One well-known isotherm model is due to Brunauer et al. (1938) – also denoted as BET isotherm. This isotherm provides an effective way to predict the local adsorption rate with respect to other water-binding mechanisms. The BET isotherm represents a straight line correlating $a_w [W_e(1 - a_w)]^{-1}$ with the water activity. An extension of the BET theory is given by the GAB model created by Guggenheim, Anderson, and de Boer (see Table 5.6) which gives better predictions as compared with BET model for water activity less than 0.9. The equilibrium moisture content is very useful for expressing the moisture content in dimensionless format.

### Example 5.7

Assume that a single thin layer of grain material of surface $1 \times 1\,\mathrm{m}^2$ is dried in forced convection air provided at the top side. The air velocity is 5 m/s, the layer thickness is 2.5 mm, the moisture diffusivity in air is 1E–7 m²/s, the temperature of air is 70 °C, and the humidity ratio is 0.019 kg/kg. Calculate the rate of drying in the first drying period:

- We assume that the bottom is well insulated, thence the characteristic length is $L = 0.0025\,\mathrm{mm}$.
- Assume that air pressure is 1.01325 bar.
- Ggg.
- The subsequent calculations are made with EES software which include humid air calculation subroutines; however, moist air tables can be used instead.
- The density of humid air is $\rho(T = 70°C, \omega = 0.019) = 0.9982\,\mathrm{kg/m}^3$.
- The kinematic viscosity of air is $\mu(T = 70°C, \omega = 0.019) = 2.049\mathrm{E}{-}5\,\mathrm{Pa\,s}$.

- Reynolds number is calculated for the characteristic length $L_x = 1\,\text{m} \rightarrow Re = 609 \rightarrow$ laminar flow.
- Eq. (5.82) apply to calculate the mass transfer coefficient $\rightarrow h_m = 9.6489\text{E}{-}6\,\text{m/s}$.
- The wet-bulb temperature of air is determined $T_{wb}(T = 70°\text{C}, \omega = 0.019) = 34$ °C.
- The humidity ratio of saturated air at wet-bulb temperature becomes $\omega_{sat}(T = 34°\text{C}, \text{RH} = 100\%) = 0.03446\,\text{kg/kg}$.
- Eq. (5.81) is applied to obtain the drying rate: $dW/dt = 0.229\,\text{kg/h}$.

**Example 5.8**

During an experiment for BET isotherm determination for a foodstuff at standard temperature, it is found that for water activity of 0.15, the equilibrium moisture content is 1.176 kg/kg. Also, for water activity of 0.05, the equilibrium moisture content is determined to 0.7018 kg/kg. Determine the following: the monolayer moisture content, the BET constant $C$, and the equilibrium moisture content when water activity within the material is 0.1.

- The BET model is a straight line correlating $f(a_w) = a_w[W_e(1-a_w)]^{-1}$ with the water activity.
- For the first experiment, it is determined that $f(a_{w,1}) = 0.15[1.176(1-0.15)]^{-1} = 0.15$.
- For the second experiment, it is determined that $f(a_{w,2}) = 0.05[0.7018(1-0.05)]^{-1} = 0.075$.
- The slope of BET line is given by

$$\text{Slope} = \frac{C-1}{W_m C} = \frac{f(a_{w,1})-f(a_{w,2})}{a_{w,1}-a_{w,2}} = \frac{0.15-0.075}{0.15-0.05} = 0.75.$$

- The intercept of the BET line is given by

$$\text{Intercept} = \frac{1}{W_m C} = f(0) = \frac{a_{w,1}f(a_{w,2})-a_{w,2}f(a_{w,1})}{a_{w,1}-a_{w,2}} = 0.0375.$$

- From the intercept equation, the following results: $W_m C = 26.67$
- Using this result in the slope equation, one has $C - 1 = 26.67 \times 0.75 = 20 \rightarrow C = 21$.
- Therefore, $W_m = 1.27\,\text{kg/kg}$.
- The BET equation becomes

$$\frac{a_w}{W_e(1-a_w)} = 0.75 a_w + 0.0375.$$

- For water activity of 0.1, the BET equation gives

$$\frac{0.1}{W_e(1-0.1)} = 0.75 \times 0.1 + 0.0375 \rightarrow W_e = 0.9877\,\text{kg/kg}.$$

## 5.6.2   Semitheoretical and Empirical Models for Drying

These types of models relate to experimental data taken from thin-layer drying to develop equations which predict the heat and moisture transfer process, the drying rate, and the drying curve. The main parameters to be considered for model fitting to experimental data are process conditions, such as the drying chamber temperature, pressure, air velocity (if the carrier gas is air), relative humidity, and the product retention time.

The semitheoretical models are generally derived from Fick's second law by applying various simplifications and introducing certain constants which can be fitted to experimental data. As a consequence of this approach, the semitheoretical models are only valid under the drying and product conditions for which these models were developed.

Semitheoretical and empirical models consider only the external resistance to moisture transfer between the product and air. The empirical method is based on experimental data and dimensional analysis. Empirical models are used for water absorption process as well as single-layer drying process, which can adequately describe the drying kinetics. According to Kucuk et al. (2014), the number of constants for the models published in the literature varies between one constant and six constants. In Kucuk et al. (2014), a number of 67 drying curve models are reviewed and compared. Here, some of the most relevant models are briefly presented. Most of the drying curve equations have an exponential form being based on the first term solution for one-dimensional moisture transfer given by Eq. (5.15).

Let us introduce here the Midilli et al. (2002) model. In this model, four constants are considered and denoted as $a$, $b$, $n$. The parameters for Midilli et al. (2002) model are given in Table 5.7. These parameters are correlated with the temperature. The general form of this model gives the dimensionless moisture content changes in time. The model is defined by the following general equation:

$$\Phi(t) = a\exp(-kt^n) + bt \qquad (5.88)$$

Table 5.8 gives some selected models for thin-layer drying curve. These models have one up to six parameters fitted to the experimental data. The Lewis model is well suited to model continuous drying processes. The Page model is applicable form any types of dryers including rotary dryers. This model has been widely used to describe the drying behavior of a variety of biological materials. The drying of most biological materials is a diffusion-controlled process. For the case of okra, the present model represents convective drying adequately.

**Table 5.7**   Parameters for Midilli et al. (2002) model

| Constant | Mushrooms | Pollen | Shelled pistachio | Unshelled pistachio |
|---|---|---|---|---|
| $a$ | $0.9937 + 0.0003\ln T$ | $0.9975 + 0.0007\ln T$ | $0.9968 + 0.0007\ln T$ | $0.9968 + 0.0003\ln T$ |
| $k$ | $0.7039 + 0.0002\ln T$ | $1.0638 + 0.0006\ln T$ | $0.1439 + 0.0006\ln T$ | $0.1545 + 0.0002\ln T$ |
| $n$ | $0.8506 + 0.0005\ln T$ | $0.5658 + 0.0008\ln T$ | $0.9178 + 0.0008\ln T$ | $0.9247 + 0.0005\ln T$ |
| $b$ | $-0.0064 - 0.0004\ln T$ | $-0.0432 - 0.0001\ln T$ | $0.0501 + 0.0001\ln T$ | $0.0486 + 0.0004\ln T$ |

$T$ is temperature in degree Celsius; for these parameters, the time from Eq. (5.88) must be expressed in hours.

**Table 5.8** Semitheoretical and empirical models for thin-film drying curve

| Model name | Number of parameters | Model equation |
|---|---|---|
| Lewis | 1 | $\Phi = \exp(-kt)$ |
| Page | 2 | $\Phi = \exp(kt^n)$ |
| Modified Page | 2 | $\Phi = \exp[(kt)^n]$ |
| Henderson and Pabis | 2 | $\Phi = a\exp(kt)$ |
| Wang and Singh | | $\Phi = 1 + at + bt^2$ |
| Asymptotic model | 3 | $\Phi = a\exp(-kt) + c$ |
| Two-term model | 4 | $\Phi = a_1\exp(-k_1 t) + a_1\exp(-k_1 t)$ |
| Two-term exponential model | 4 | $\Phi = a\exp(-kt) + (1-a)\exp(-kat)$ |
| Modified Henderson and Pabis | 6 | $\Phi = a\exp(-kt) + b\exp(-gt) + c\exp(-ht)$ |

*Source*: Kucuk et al. (2014).

**Example 5.9**

Pollen is dried in air at 20° C. Determine the half drying time provided that the drying curve can be modeled based on Midilli et al. (2002) approach, Eq. (5.88):

- For 20° C, the constants for Midilli's model are calculated according to Table 5.7.
- One obtains $a = 0.9999, k = 1.0658, n = 0.5686, b = -0.0435$.
- The following data set is obtained from Midilli's model:

| Time (h) | 0 | 0.5 | 1 | 1.5 | 2 | 2.5 | 3 |
|---|---|---|---|---|---|---|---|
| $\Phi$ | 0.9999 | 0.4656 | 0.3009 | 0.196 | 0.118 | 0.05744 | 0.00811 |

- This data can be fit to a drying equation in the form from Eq. (5.22) as follows: $\Phi = 1.1282\exp(-1.33\,t)$.
- The lag factor and drying coefficient become $LF = 1.1282$ and $S = 1.33\,h^{-1}$.
- The half drying time results from Eq. (5.31)

$$t_{0.5} = -\frac{1}{S}\ln\left(\frac{1}{2\,LF}\right) = -\frac{1}{1.33}\ln\left(\frac{1}{2 \times 1.1282}\right) = 0.61\,h.$$

## 5.7 Conclusions

In this chapter, various aspects related to heat and moisture transfer are presented. Drying is a heat-driven process in which the temperature field influences the mass transfer process. In the same time, the moisture concentration influences thermal parameters such as density, specific heat, thermal conductivity, and thermal diffusivity. Therefore, the mechanisms of heat and moisture transfer are interdependent. The fundamental equations for heat and moisture transfer were introduced herein. Except for some very simplified cases, the equations must be solved simultaneously using numerical methods. However, several analytical solutions of time-dependent moisture diffusing for basic geometry exist and are useful for an approximate analysis of the processes during drying.

The chapter begins with introducing the transient moisture diffusion equation for basic geometries and its analytical solution. It is shown how the analytical solutions can be used to estimate the drying time for irregular and regular geometries. In this respect, the shape factors are introduced. Also it is demonstrated how the time-dependent analysis and its analytical solutions can be useful for moisture diffusion and moisture transfer coefficient determination.

The equations for simultaneous heat and moisture transfer are introduced for one-dimensional, two-dimensional, and three-dimensional geometries in Cartesian, cylindrical, and spherical coordinates are given. Some theoretical models for approximate solution of simultaneous heat and moisture transfer in drying applications are introduced. Here, the Luikov theory and sorption isotherm models are essential tools for modeling. In addition, some semitheoretical and empirical models for the drying curve are presented.


## 5.8   Study Problems

5.1   A moist material in the form of a slab of 20 mm thickness has the moisture diffusivity of $10^{-10}$ m$^2$/s. The moisture transfer coefficient during drying is determined to $h_m = 0.001$ m/s. The initial moisture content is $W_i = 0.5$ kg/kg, and the equilibrium moisture content is $W_e = 0.05$ kg/kg. Determine the moisture content in the middle plane when the surface reaches the equilibrium moisture content.

5.2   Determine the drying time of a slab product such that the final moisture content reduces 10 times with respect to the initial moisture. The mass transfer coefficient is 0.1 m/s. The diffusivity is 1E–9 m$^2$/s. The characteristic length is 1 mm.

5.3   Cucumbers modeled as cylinder of 20 cm long have a moisture diffusivity of 1E–9 m$^2$/s when the mass transfer coefficient is 1E–6 m/s. Determine the half drying time.

5.4   Grapes assumes sphere with the diameter of 8 mm have a moisture diffusivity of 1E–8 m$^2$/s and are dried in air with a moisture transfer coefficient of 1E–5 m/s. The equilibrium moisture content represents the 10th part of the moisture content at half drying time. Determine the time needed to dry the products until they reach the equilibrium moisture content.

5.5   A concrete beam has a square cross section of 900 cm$^2$. The moisture diffusivity of $10^{-11}$ m$^2$/s and the moisture transfer coefficient is $10^{-4}$ m/s. Determine the shape factor and the half drying time.

5.6   A wood slab of Douglas fir has the moisture diffusivity of 1E–9 m$^2$/s and is dried under a moisture transfer coefficient of 1E–7 m/s. Calculate the shape factor and the half drying time provided that the slab dimensions are given as 2 in. × 8 in. × 10 ft.

5.7   Okra modeled as long cylinder with $\beta_1 = 2$ and characteristic length of 3 mm has a moisture diffusivity of 6E–10 m$^2$/s and is dried in a medium where the drying coefficient is 1E–8 m/s. Calculate the shape factor and the half drying time.

5.8   Potato modeled as short cylinder with $\beta_1 = 2$ and characteristic length of 13 mm has a moisture diffusivity of 6E–11 m$^2$/s and is dried in a medium where the drying coefficient is 1E–6 m/s. Calculate the shape factor and the half drying time.

5.9   Potato modeled as prolate spheroid with $\beta = 2$ and characteristic length of 9 mm has a moisture diffusivity of 1E–6 m$^2$/s and is dried in a medium where the drying coefficient is 1E–4 m/s. Calculate the shape factor and the half drying time.

5.10   Consider the drying process of a moist solid that can be assimilated with an infinite cylinder. The drying data in terms of dimensionless moisture content is given in the table below. Determine the lag factor, the drying coefficient, the diffusivity, and the moisture transfer coefficient provided that the characteristic length (cylinder radius) is 7 mm.

| $t$ (min) | 0.02 | 28 | 55 | 83 | 111 | 139 | 167 | 194 | 222 | 250 |
|---|---|---|---|---|---|---|---|---|---|---|
| $\Phi_{exp}$ | 1.1 | 0.75 | 0.35 | 0.19 | 0.13 | 0.108 | 0.0629 | 0.062 | 0.055 | 0.05 |

5.11   Assume that a single thin layer of grain material of surface $2 \times 2 \, \text{m}^2$ is dried in forced convection air provided at the top side. The air velocity is 10 m/s, the layer thickness is 5 mm, the moisture diffusivity in air is 1E–7 $\text{m}^2$/s, the temperature of air is 60° C, and the humidity ratio is 0.021 kg/kg. Calculate the rate of drying in the first drying period.

5.12   During an experiment for Langmuir isotherm determination for a moist material, it is found that for water activity of 0.25, the equilibrium moisture content is 1.5 kg/kg. Also, for water activity of 0.025, the equilibrium moisture content is determined to 0.05 kg/kg. Determine the following: the monolayer moisture content, the BET constant $C$, and the equilibrium moisture content when water activity within the material is 0.15.

5.13   Mushrooms are dried in air at 35° C. Determine the half drying time provided that the drying curve can be modeled based on Midilli et al. (2002) approach, Eq. (5.88).

5.14   Shelled and unshelled pistachios are dried in air at 50° C. Determine the half drying time for both products provided that the drying curve can be modeled based on Midilli et al. (2002) approach, Eq. (5.88).

# References

Adebiyi G.A. 1995. A single expression for the solution of the one-dimensional transient conduction equation for the simple regular shaped solids. *Journal of Heat Transfer* 117:158–160.

Barbosa-Cánovas G.V., Vega-Mercado H. 1996. Dehydration of Foods. Springer Science+Business Media: Dordrecht.

Brunauer S., Emmett P.H., Teller E. 1938. Adsorption of gases in multimolecular layers. *Journal of the American Chemical Society* 60:309–319.

Chung D.S., Pfost H.B. 1967. Adsorption and desorption of water vapour by cereal grains and their products. I. Heat and free energy changes of adsorption and desorption. *Transactions of ASAE* 10:549–551.

Dincer I. 1996a. Modelling for heat and mass transfer parameters in deep-frying products. *Heat and Mass Transfer* 32:109–113.

Dincer I. 1996b. Development of a new number (the Dincer number) for forced-convection heat transfer in heating and cooling applications. *International Journal of Energy Research* 20:419–442.

Dincer I. 1998a. Moisture transfer analysis during drying of slab woods. *Heat and Mass Transfer* 34:317–320.

Dincer I. 1998b. Moisture loss from wood products during drying – Part II. Surface moisture content distributions. *Energy Sources* 20:77–83.

Dincer I., Dost S. 1995. Analytical model for moisture diffusion in solid objects during drying. *Drying Technology* 13:425–435.

Dincer I., Dost S. 1996. A modelling study for moisture diffusivities and moisture transfer coefficients in drying solid objects. *International Journal of Energy Research* 20:531–539.

Dincer I., Hussain M.M. 2002. Development of a new Bi–Di correlation for solids drying. *International Journal of Heat and Mass Transfer* 45:3065–3069.

Dincer I., Hussain M.M. 2004. Development of a new Biot number and lag factor correlation for drying applications. *International Journal of Heat and Mass Transfer* 47:653–658.

Dincer I., Hussain M.M., Sahin A.Z., Yilbas B.S. 2002. Development of a new moisture transfer (Bi-Re) correlation for food drying applications. *International Journal of Heat and Mass Transfer* 45:1749–1755.

EES. 2014. Engineering Equation Solver. F-Chart Software (developed by S.A. Klein), Middleton, WI.

Halsey G. 1948. Physical adsorption on non-uniform surfaces. *Journal of Chemical Physics* 16:931–937.

Henderson S. 1952. A basic concept of equilibrium moisture. *Agricultural Engineering* 33:29–32.

Hossain M.M., Cleland D.J., Cleland A.C. 1992. Prediction of freezing and thawing times for foods of regular multidimensional shape by using analytically derived geometric factor. *International Journal of Refrigeration* 15:227–234.

Kiranoudis C.T., Tsami E., Maroulis Z.B., Marinos-Kouris D. 2003. Drying of some fruits. *Drying Technology* 15:1399–1418.

Kucuk H., Midilli A., Kilic A., Dincer I. 2014. A review of thin-layer drying curve equations. *Drying Technology* 32:757–773.

Luikov A.Z. 1975. Systems of differential equations of heat and mass transfer in capillary-porous bodies (review). *International Journal of Heat and Mass Transfer* 18:1–14.

Midilli A., Kucuk H., Yapar Z. 2002. A new model for single-layer drying. *Drying Technology* 20:1503–1513.

Okos M.R., Narsimhan G., Singh R.K., Weitnaver A.C. 1992. Food dehydration. In: Handbook of Food Engineering. D.R. Heldman, D.B. Lund Eds., Marcel Dekker: New York.

Oswin C.R. 1946. The kinetics of package life. *III. Isotherm. Journal of the Society of Chemical Industry (London)* 65:419–421.

Sahin A.Z., Dincer I. 2004. Incorporation of the Dincer number into the moisture diffusion equation. *International Communications of Heat and Mass Transfer* 31:109–119.

Sahin A.Z., Dincer I., Yilbas B.S., Hussain M.M. 2002. Determination of drying time for regular multi-dimensional objects. *International Journal of Heat and Mass Transfer* 45:1757–1766.

Smith P.R. 1947. The sorption of water vapour by high polymers. *Journal of American Chemical Society* 69:646–651.

Treybal R.E. 1981. Mass Transfer Operations. 3rd ed. McGraw-Hill: London.

Van den Berg C. 1985. Water activity. In: Concentration and Drying of Foods. D. MacCarthy Ed., Elsevier: New York.

Yildiz M., Dincer I. 1995. Modelling and heat and mass transfer analysis during oil-frying of cylindrical sausages. *International Journal of Energy Research* 19:555–565.

# 6

# Numerical Heat and Moisture Transfer

## 6.1  Introduction

It is noted already from Chapter 5 that heat and moisture transfer processes during drying of moist materials are highly interlaced and often cannot be decoupled. The driving force of drying is the temperature gradient which continuously influences the time-dependent moisture transfer within the material. The solution to this type of complex process is normally obtained by numerical methods since no closed form can be found. Numerical analysis of heat and moisture transfer with a focus on drying is reviewed in Dincer (2010). The basic theory initially formulated by Luikov (1934) and Philip and De Vries (1957) suggest that a moisture transfer potential can be defined as an effective variable for moisture movement modeling within permeable materials. These theories are revisited by their authors in Luikov (1975) and De Vries (1987), respectively, and enhanced with schemes for numerical solutions of one-dimensional simultaneous heat and moisture transfer. An extension of the numerical solutions for two-dimensional problems is pioneered by Comini and Lewis (1976) under the assumption that the pressure effects on internal capillaries can be neglected.

Hussain and Dincer (2003a) performed a time-dependent numerical simulation of heat and moisture transfer during drying a rectangular material. The effect of temperature gradient on internal moisture movement has been neglected. However, the variation of moisture diffusivity with temperature and the variation of the thermal diffusivity with the moisture content are accounted for in the equations. The chosen numerical scheme has been of explicit type with uniform grid. A similar procedure for axisymmetric cylindrical geometry has been developed in Hussain and Dincer (2002, 2003b) and applied for various case studies of foodstuff drying. In many other applications of drying, spherically shaped materials are placed on trays, and drying air is blown from one of the sides. In such cases, the heat and moisture transfer is

*Drying Phenomena: Theory and Applications*, First Edition. İbrahim Dinçer and Calin Zamfirescu.
© 2016 John Wiley & Sons, Ltd. Published 2016 by John Wiley & Sons, Ltd.

symmetrical with respect to the azimuth angle (left–right), whereas with respect to the zenith angle (up–down), there is no symmetry. This type of problem is treated numerically in Hussain and Dincer (2003c) in two-dimensional spherical coordinates using the time-dependent explicit finite difference method.

Further extension of numerical schemes for three-dimensional drying problems can be found in Ranjan et al. (2002). Some secondary effects, such as material deformation during drying and local variation of thermophysical properties, are treated in Perré and May (2001) and Perré and Turner (2002).

The numerical solution of simultaneous heat and moisture transfer in drying is difficult due to the high nonlinearity, time dependence, and complexity of the system of partial differential equations (PDEs) which governs the process. In some cases, due to the change of thermophysical properties and material dimensions, the modeling equations change their type during numerical simulations: for example, a parabolic equation can become hyperbolic. Often, reasonable assumptions have to be made to reduce the system complexity. Thus, in many cases, the number of dimensions can be reduced due to symmetry. As seen in Chapter 5, the infinite slab, the infinite cylinder, and the sphere placed in a uniform drying agent can be modeled as one-dimensional processes. Also, a rectangular, cylindrical, or spherical material dried on a tray can be modeled in many situations as a two-dimensional object with left–right symmetry. The solution will determine the 2D temperature and moisture content distributions within the material in Cartesian $(x, y)$ or cylindrical or spherical $(r, \theta)$ coordinates. The three-dimensional modeling is generally used only in special cases with complicated geometry and material anisotropy.

In some specific cases, the equations can be partially decoupled which simplifies much the problem. Assume for instance that the thermal diffusivity variation with moisture content and temperature can be neglected. In many cases as such, the thermal diffusion equation has a closed form, with an analytical solution. Once the temperature variation in time and space is known, the moisture diffusivity can then be determined. Therefore, the moisture diffusion equation can be integrated numerically in this case.

Discretization in numerical methods allows for reduction of the order of the PDE. Thus, a PDE can be reduced to a system ordinary differential equation (ODE) with time being the sole variable. Moreover, the PDE can be even reduced to a system of algebraic equations in some numerical schemes.

The stability and grid independence of the solution are crucial in numerical schemes. In the simplest approach, the numerical scheme is explicit, in which the grid values for the next time step are determined based on the current values. This is how the solution marches in time. However, in numerical schemes with improved accuracy, the nodes' values for previous two or three time steps must be used to solve for the future time step. If this is the case, problems may arise in regards to setting the initial conditions at the grid nodes from the very vicinity of the boundary. Therefore, simpler methods may be required to predict the solution for the first few time steps. Yilbas et al. (2003) presented closed-form solution for heat and moisture transfer in the semi-infinite solid. This solution can be applied to predict the initial variation of moisture and temperature at the boundary region, since at the first time instant, any surface subjected to a temperature perturbation from the exterior behaves locally as a semi-infinite solid.

When more accurate solutions are required, the temperature and humidity distribution in the boundary layer surrounding the moist object must be known. It is interesting to note that in a continuous dryer where the drying agent is refreshed continuously, the heat, momentum, and

mass transfer can be modeled as time independent. However, inside the moist material subjected to drying, the process is always time dependent. Therefore, in such problems, two
domains must be considered: the external domain (or the drying agent/air) and the internal
domain (or the moist object). In the external domain, the main variables are the temperature,
velocity, and relative humidity which are time independent. In the internal domain, the main
parameters are density, permeability, porosity, sorption–desorption characteristics, thermal diffusivity, moisture diffusivity, specific heat, and surface tension, and the main variables (which
are time dependent) are the moisture content, the temperature, and, in some cases, the capillary
pressure and the void fraction. Depending on the case, the external flow equations can be solved
independently or simultaneously with the governing equations for the moist object.

Kaya et al. (2006, 2007, 2008a,b) developed a method for numerical heat and moisture transfer during drying in which the external flow governing steady-state equations are integrated
first to obtain the temperature profile around the moist object. Further, the local heat flux
and heat transfer coefficient at the material surface is determined based on the temperature gradient data. Thereafter, the local moisture transfer coefficient is determined based on heat and
mass transfer analogy. With these information, the convective boundary conditions for heat and
moisture transfer inside the materials can be set at the boundary. Then, the time-dependent heat
and moisture transfer equations can be solved to determine the moisture content and temperature variation in space and time. Further progress on the same topic is reported in Ozalp et al.
(2010a,b).

In this chapter, the main numerical methods for simultaneous heat and moisture transfer are
reviewed, with a focus on drying processes. In the first part, the differential equations (including PDE and ODE) are briefly introduced and classified. Then, the main numerical methods
including finite difference, finite element, and control volume are presented with discussions
on grid generation, convergence, and stability. Further, some relevant heat and moisture
transfer problems in one, two, and three dimensional space are formulated numerically, and
several examples are presented.

## 6.2   Numerical Methods for PDEs

As it is well known, the governing equations for heat, mass, and momentum transfer are of
partial differential type. These types of equations are represented by a mathematical expression
which includes partial derivatives of a multivariable function. In most of the cases concerned
here, the functions are that of temperature $T(x, y, z, t)$, moisture content $W(x, y, z, t)$, moisture
concentration $C_m(x, y, z, t)$, and velocity vector $\left[U_x, U_y, U_z\right] = U(x, y, z)$. The general representation of a PDE for $F(x, y, z, t)$ as the unknown function is as follows:

$$f\left(\frac{\partial F}{\partial t}, \frac{\partial F}{\partial x}, \frac{\partial^2 F}{\partial x^2}, \frac{\partial F}{\partial x \partial y}, \frac{\partial F}{\partial y}, \frac{\partial F}{\partial z}, \left(\frac{\partial F}{\partial x}\right)^2, \dots, F, x, y, z, t, \dots\right) = k \qquad (6.1)$$

where $k$ is a constant.

The PDE is said homogeneous provided that the constant $k = 0$, or, otherwise it is called nonhomogeneous. Moreover, if the partial derivatives are at the first power, then the PDE is said
linear and it has the property that $f(\text{const} \cdot F) = \text{const} \cdot f(F)$, where $F$ is the unknown function.
The highest derivative appearing in $f$ gives the order of the PDE. The equations generally

**Table 6.1** Classification of PDEs of two variables

| Equation type | Discriminant in Eq. (6.2) | Canonical form for 2D | Example | |
|---|---|---|---|---|
| Hyperbolic | $b^2 - 4ac > 0$ | $\dfrac{\partial F}{\partial\xi\partial\eta} = f\left(\xi,\eta,F,\dfrac{\partial F}{\partial\xi},\dfrac{\partial F}{\partial\eta}\right)$ | Wave equation: | $\dfrac{\partial^2 u}{\partial t^2} = u^2\left(\dfrac{\partial^2 u}{\partial x^2}\right)$ |
| Parabolic | $b^2 - 4ac = 0$ | $\dfrac{\partial^2 F}{\partial\xi^2} = f\left(\xi,\eta,F,\dfrac{\partial F}{\partial\xi},\dfrac{\partial F}{\partial\eta}\right)$ | Heat equation: | $\dfrac{\partial T}{\partial t} = a\left(\dfrac{\partial^2 T}{\partial x^2}\right)$ |
| Elliptic | $b^2 - 4ac < 0$ | $\dfrac{\partial^2 F}{\partial\xi^2} + \dfrac{\partial^2 F}{\partial\eta^2} = f\left(\xi,\eta,F,\dfrac{\partial F}{\partial\xi},\dfrac{\partial F}{\partial\eta}\right)$ | Laplace equation: | $\dfrac{\partial^2 T}{\partial x^2} + \dfrac{\partial^2 T}{\partial y^2} = 0$ |

*Note*: For parabolic equation, $\eta(x, y)$ must be chosen as independent on $\xi$; for elliptic equation in canonical form, the following change of variables apply $\xi \rightarrow 0.5(\xi+\eta)$ and $\eta \rightarrow 0.5(\xi-\eta)$.

involved in heat and moisture transfer are linear PDEs of first or second order. Using $F(x, y)$ as the unknown function, a linear second-order PDE can be written generally as follows:

$$a\frac{\partial^2 F}{\partial x^2} + b\frac{\partial^2 F}{\partial x\partial y} + c\frac{\partial^2 F}{\partial y^2} + d\frac{\partial F}{\partial x} + e\frac{\partial F}{\partial y} + fF + g = 0 \qquad (6.2)$$

where the coefficients $a, b, c, \ldots$ may be constant or variable.

Table 6.1 gives a classification of the linear PDEs as hyperbolic, parabolic, and elliptic. Note that for the scope of heat and moisture transfer modeling, the temporal derivative is never of higher order than one and mixed derivatives are not encountered. Thence, the equations of interest herein have $a = b = 0$. This means that the PDEs of most importance for time-dependent problems are the parabolic type. However, for time-independent problems, the elliptic and hyperbolic equations become important. Note that in problems where the coefficients are variable, the PDE can change the type depending on the domain region $(x, y, z, t)$. Any linear equation can be put in a "universal" canonical form as shown in Table 6.1 where the variables $(\xi, \eta)$ change the initial variable $(x, y)$ such that the equations $\xi_{xx} = \mu_1\xi_{yy}$ and $\eta_{xx} = \mu_2\eta_{yy}$ where $\mu_{1,2}$ are the roots of $a\mu^2 + b\mu + c = 0$.

Simultaneous heat and moisture transfer for drying problems involve a coupled system of at least two time-dependent PDEs: one for temperature field and one for the moisture content. In Chapter 5, these equations are given by Eqs. (5.69), (5.71), (5.76), (5.78)–(5.80), (5.83) and in the dimensionless form by Eq. (5.72). Also for capillary flow in porous materials, a set of three coupled equations is written for moisture, energy, and momentum conservation as in Eq. (5.85). The boundary and initial conditions are required for the particular solution to be obtained.

### 6.2.1 The Finite Difference Method

In the finite difference method, the PDEs are discretized based on Taylor expansions. The partial derivatives of first or higher order are in fact replaced with finite differences. To begin with, assume a continuous function $f(x)$ defined for $x \geq 0$ and having a continuous first

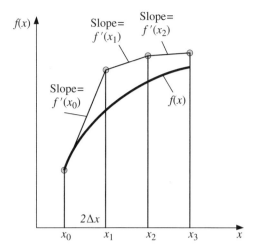

**Figure 6.1**   Approximation of a function based on Taylor expansion (Euler method)

derivative $f'(x)$ on that interval. The Taylor expansion $f(x_0+\Delta x)=f(x_0-\Delta x)+$ $2f'(x_0-\Delta x)\Delta x+2f''(x_0-\Delta x)(\Delta x)^2+4/3f'''(x_0-\Delta x)(\Delta x)^3+\cdots$ approximates the function in a point at $2\Delta x$ distance from $x_0-\Delta x$, where $\Delta x$ is a small increment. From here, a first-term approximation of the first derivative in $x_0$ can be obtained as follows:

$$f'(x_0)\cong f'(x_0-\Delta x)\cong \frac{f(x_0+\Delta x)-f(x_0-\Delta x)}{2\Delta x} \tag{6.3}$$

where the truncation error is of the order of $(\Delta x)^2$.

Consider now that the values of $f'$ are given at $n$ discrete nodes $x_i=x_0+2(i-1)\Delta x, i=1,\ldots,n$. In this case, the value of the function $f$ can be approximated at nodes $x_i$ using Eq. (6.3). This is known as Euler method for integration of ODE of the form $f'=F(x,f)$, known as initial value problems. Figure 6.1 shows how a continuous function $f(x)$ is approximated with a discontinuous solution with the help of Euler method. Assume that an initial value $f(x_0)=f_0$ is known. Then, the value of the derivative $f'(x_0)=F(x_0,f_0)$ can be calculated. If a sufficiently small $\Delta x$ increment is selected, then the value of the function at $x_0+2\Delta x$ is calculated from $f_0+2\Delta x f'(x_0)$. The process continues as shown in the figure.

Now, if the first derivative as approximated in Eq. (6.3) is introduced in the Taylor expansion $f(x_0+\Delta x)=f(x_0)+f'(x_0)\Delta x+0.5f''(x_0)(\Delta x)^2+\cdots$, then the following approximation is obtained for the second derivative:

$$f''(x_0)\cong \frac{f(x_0+\Delta x)-2f(x_0)+f(x_0-\Delta x)}{(\Delta x)^2} \tag{6.4}$$

where the truncation error is of the order of $(\Delta x)^2$.

Assume now a differential equations of the form $F(f'',f',f,x)=0$. Now, based on Eqs. (6.3) and (6.4), this equation can be converted to an algebraic system of the form

$F(f_{i-1}, f_i, f_{i+1}, x_i) = 0, i = 1, \ldots, n$ which can be solved with an adequate method. If the system is linear, then a coefficient matrix can be formed and inversed. If the system is nonlinear, specific iterative methods can be used to solve for $f_i$, $i = 1, \ldots, n$.

In some situations, it may be convenient that a time-dependent PDE is reduced to a system of ODEs by discretization of only the spatial derivatives. Assume for example the following form of PDE of a function $f(x, t)$:

$$\frac{\partial f}{\partial t} = F\left(\frac{\partial f}{\partial x}, \frac{\partial^2 f}{\partial x^2}, f, x\right)$$

If the partial derivatives of $f$ with respect to $x$ are discretized using finite differences in nodes $x_i, i = 1, \ldots, n$, then the following system of ODEs is obtained:

$$\frac{df_i}{dt} = F(f_{i-1}, f_i, f_{i+1}, x_i), \; i = 1, \ldots, n \tag{6.5}$$

where $f_i = f_i(t) = f(x_i, t)$ represents the unknown function value at node $x_i$.

There are well-established methods for solution of the ODEs of the form given in Eq. (6.5) starting with Euler, Runge–Kutta, Adams, and so on (see Faires and Burden, 2010). In these methods, the derivative of $f_i(t)$ with respect to variable $t$ (time) is discretized in a number of nodes $t_j = (j-1)\Delta t, j \in N, j \geq 1$, where $\Delta t$ is the time step which can be constant or sometimes adaptive. Using the first-term approximation, the Euler method discussed previously gives the following discrete approximation of $f(t)$ when the initial conditions are specified as $f_i(0) = f_0$

$$f_i(t_{j+1}) = f_i(t_j) + F(f_{i-1}(t_j), f_i(t_j), f_{i+1}(t_j), x_i)\Delta t, \; i = 1, \ldots, n, j \in N, j \geq 1 \tag{6.6}$$

where the step size $\Delta t$ must be sufficiently small such that the truncation error from Taylor expansion is less than a specified tolerance, denoted with $\varepsilon$. The time step must be chosen therefore as follows (see Faires and Burden, 2010):

$$\Delta t_j \leq \varepsilon, \text{ where } \varepsilon = \min\left\{\sqrt{\frac{2\varepsilon}{|f_i''(t_j)|}}, i = 1, \ldots, n\right\} \tag{6.7}$$

with the condition that $f_i''(t_j) \neq 0$.

Table 6.2 gives the most common numerical schemes for ODEs of first order. Among those methods, the Runge–Kutta method can be expanded for second-order PDE. This is done according to the so-called Runge–Kutta–Nyström method which approximates the solution of the following type of ODE: $f'' = F(f', f, t)$. This type of equation may be relevant for some types of second-degree PDE for drying processes which can be reduced to a set of ODE by finite difference or finite element discretization.

For example, a time-dependent diffusion problem can be modeled as follows: $f_{xx} = af_t$ where $f_{xx}$ is the second order partial derivative of $f(x, t)$ with respect to $x$ and $f_t$ is the partial derivative of $f$ with respect to $t$. Now, if time is discretized as $t_i = (i-1)\Delta t, i \in \{1, n\}$, then the PDE is converted to $f_{xx,i} = F(f(x, t_{j-1}), f(x, t_j), f(x, t_{j+1}))$ which is a set of $n$ second-order ODEs.

**Table 6.2** Numerical methods for systems of ODE of the form $f_i' = F_i(f_i(t), t), \ i = 1, \ldots, n, \ t \geq 0$

| Method | Main and auxiliary equations |
|---|---|
| Runge–Kutta | $f_i(t_{j+1}) = f_i(t_j) + \frac{1}{6}(k_{i,1} + 2k_{i,2} + 2k_{i,3} + k_{i,4})$ <br> $k_{i,1} = F_i(f_i, t_j)\Delta t$ <br> $k_{i,2} = F_i(f_i + k_{i,1}, t_j + 0.5\Delta t)\Delta t$ <br> $k_{i,3} = F_i(f_i + k_{i,2}, t_j + 0.5\Delta t)\Delta t$ <br> $k_{i,4} = F_i(f_i + k_{i,3}, t_j + \Delta t)\Delta t$ |
| Adams–Bashforth | $f_i(t_{j+1}) = f_i(t_j) + \dfrac{\Delta t}{24}k_i$ <br> $k_i = 55F_i(f_i(t_j), t_j) - 59F_i(f_i(t_{j-1}), t_{j-1}) + 37F_i(f_i(t_{j-2}), t_{j-2}) - 9F_i(f_i(t_{j-3}), t_{j-3})$ <br> where $f(t_j), j = 1,2,3$ can be obtained from Euler method, Eq. (6.6) |
| Adams–Moulton | $f_i(t_{j+1}) = f_i(t_j) + \dfrac{\Delta t}{24}k_i$ <br> $k_i = 9F_i(f_i^*(t_{j+1}), t_{j+1}) + 19F_i(f_i(t_{j-1}), t_{j-1}) - 5F_i(f_i(t_{j-2}), t_{j-2}) + F_i(f_i(t_{j-3}), t_{j-3})$ <br> $f_i^*(t_{j+1}) = f(t_j) + \dfrac{\Delta t}{24}k_i^*$ <br> $k_i^* = 55F_i(f_i(t_j), t_j) - 59F_i(f_i(t_{j-1}), t_{j-1}) + 37F_i(f_i(t_{j-2}), t_{j-2}) - 9F_i(f_i(t_{j-3}), t_{j-3})$ <br> where $f(t_j), j = 1,2,3$ can be obtained from Euler method, Eq. (6.6) |

**Table 6.3** Runge–Kutta–Nyström method for systems of second-order ODEs

| Equation of the form $f_i'' = F_i(f, x), i = \{1, n\}$ | Equations of the form $f_i'' = F_i(f, f', x), i = \{1, n\}$ |
|---|---|
| $f_i(x_{j+1}) = f_i(x_j) + \left[ f_i'(x_j) + \frac{1}{3}(k_{i,1} + 2k_{i,2}) \right]\Delta x$ | $f_i(x_{j+1}) = f_i(x_j) + \left[ f_i'(x_j) + \frac{1}{3}(k_{i,1} + k_{i,2} + k_{i,3}) \right]\Delta x$ |
| $f_i'(x_{j+1}) = f_i'(x_j) + \frac{1}{3}(k_{i,1} + 2k_{i,2} + k_{i,4})$ | $f_i'(x_{j+1}) = f_i'(x_j) + \frac{1}{3}(k_{i,1} + 2k_{i,2} + 2k_{i,3} + k_{i,4})$ |
| $k_{i,1} - \frac{3}{2}F_i(f(x_j), x_j)\Delta x$ | $k_{i,1} - \frac{\Delta x}{2}F_i(f(x_j), f'(x_j), x_j); K = \frac{\Delta x}{2}\left( f'(x_j) + \frac{1}{2}k_{i,1} \right)$ |
| $k_{i,2} = \frac{\Delta x}{2}F_i\left( f(x_j) + \frac{\Delta x}{2}\left( f_i'(x_j) + \frac{1}{2}k_{l,1} \right), x_j + \frac{\Delta x}{2} \right)$ | $k_{i,2} = \frac{\Delta x}{2}F_i\left( f(x_j) + K, f'(x_t) + k_{i,1}, x_j + \frac{\Delta x}{2} \right)$ |
| $k_{i,4} = \frac{\Delta x}{2}F_i(f(x_j) + \Delta x(f_i'(x_j) + k_{i,2}), x_j + \Delta x)$ | $k_{i,3} = \frac{\Delta x}{2}F_i\left( f(x_j) + K, f'(x_i) + k_{i,2}, x_j + \frac{\Delta x}{2} \right)$ |
| | $k_{i,4} = \frac{\Delta x}{2}F_i(f(x_j) + \Delta x(f_i'(x_j) + k_{i,3}), f'(x_i) + 2k_{i,3}, x_j + \Delta x)$ |

Furthermore, if $f(x, t)$ flattens at high $t$ such that $f(x, t_n) = f_n$, then one can impose initial and condition as $f(x, t_0) = f_0$ which closes the system of ODEs. Note that this situation is representative for the case of a drying process where the moisture content tends to the equilibrium value at all points within the material. The Runge–Kutta–Nyström method can be applied to determine $f_i(x_j)$ for $x_j = x_1 + (j-1)\Delta x$, where $j \in \{1, m\}$ and $(m-1)\Delta x = x_2 - x_1$ and $x \in [x_1, x_2]$. This method is given in Table 6.3 for two cases of second-order ODE systems.

### 6.2.1.1  Finite Difference Scheme for Hyperbolic PDEs

Let us consider now a hyperbolic PDE in one dimension with the unknown function $f(x, t)$ which can be reduced to $f_{tt} = f_{xx}$ with $x \in [0,1]$ and $t \geq 0$. Therefore, the coefficients in the general form given by Eq. (6.2) are $a = 1, b = 0, c = -1 \rightarrow b^2 - 4ac > 0$. The analysis presented here is representative for any other hyperbolic PDE. The finite difference discretization requires that the partial derivatives of second order are replaced with finite differences as given by Eq. (6.4). For discretization, one must consider a space step $\Delta x$ and a time step $\Delta t$. Then, one has $x_{j+1} = x_j + \Delta x$ and $t_{i+1} = t_i + \Delta t$. The discretized equation becomes

$$\frac{f(t_{i+1}, x_j) - 2f(t_i, x_j) + f(t_{i-1}, x_j)}{(\Delta t)^2} = \frac{f(t_i, x_{j+1}) - 2f(t_i, x_j) + f(t_i, x_{j-1})}{(\Delta x)^2} \tag{6.8}$$

It can be shown that the numerical stability condition for the scheme in Eq. (6.8) requires that

$$0 < \left(\frac{\Delta t}{\Delta x}\right)^2 \leq 1$$

In order to solve the PDE, two initial conditions are required, $f(x,0) = f_0(x)$ and $f'(x,0) = g_0(x)$, and two boundary condition which can be set as follows: $f(0,t) = f(1,t) = 0$. The scheme from Eq. (6.8) can be simplified when the time and space variables are such that for the domain of interest these steps are in the domain [0, 1]. In such cases, it is useful to take $\Delta t = \Delta x$ such that the finite difference scheme becomes explicit. This means that the unknown function for the next time step is determined based on the function values in the previous time steps. The scheme is given as follows:

$$f(x_j, t_{i+1}) = f(x_{i-1}, t_j) + f(x_{i+1}, t_j) - f(x_i, t_{j-1}) \tag{6.9}$$

with the initial conditions given in the discrete form as follows:

$$f(x_j, t_2) = \frac{1}{2}\left(f_0(x_{j-1}) + f_0(x_{j-1})\right) + g_0(x_j)\Delta t \tag{6.10}$$

### 6.2.1.2  Finite difference Scheme for Parabolic PDEs

Consider now a parabolic equation for which $b^2 - 4ac = 0$; the basic equation of this kind is the one-dimensional diffusion equation $f_t = f_{xx}$ for $x \in [0,1], t \geq 0$ with the initial condition $f(x,0) = f_0(x)$ and boundary conditions $f(0,t) = f(1,t) = 0$. There are explicit and implicit schemes for the discretization of these equations as it is shown in the following. The mesh grid for the explicit scheme is shown in Figure 6.2. The scheme relates the function value at time $t_{i+1}$ to the values of the function at the time $t_i$ for three locations: $x_{j-1}, x_{j1}, x_{j+1}$.

Consider that the ratio $r = \Delta t/(\Delta x)^2$ is chosen such that $r \leq 0.5$. In this case, the explicit scheme given as follows may be stable:

$$f(x_j, t_{i+1}) = (1 - 2r)f(x_j, t_i) + r\left(f(x_{j+1}, t_i) + f(x_{j-1}, t_i)\right) \tag{6.11}$$

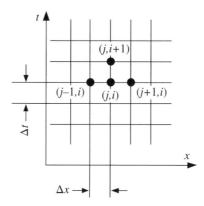

**Figure 6.2** Approximation of a function based on Taylor expansion (Euler method)

The explicit scheme has the disadvantage that it requires a very fine time space. This may be inconvenient in some situations. Therefore, a six-point implicit scheme can be used instead, known as the Crank–Nicolson scheme. The Crank–Nicolson implicit scheme expresses the function at a future time $t_{i+1}$ based on function values at time $t_i$ through an implicit equation which is as follows:

$$2(1+r)f\left(x_j, t_{i+1}\right) - r\left[f\left(x_{j+1}, t_{i+1}\right) + f\left(x_{j-1}, t_{i+1}\right)\right] = 2(1-r)f\left(x_j, t_i\right) + r\left[f\left(x_{j+1}, t_i\right) + f\left(x_{j-1}, t_i\right)\right]$$

$$(6.12)$$

Note that Eqs. (6.12) form a closed linear system of equations which can be solved by any dedicated method such as the direct Gaussian elimination or the iterative Gauss–Seidel method.

### 6.2.1.3  Finite Difference Scheme for Elliptic PDEs

Let us consider now the case of elliptic equations. Two types of those are representative: the homogeneous equation of Laplace $f_{xx} + f_{yy} = 0$ and the nonhomogeneous equation of Poisson $f_{xx} + f_{yy} = g(x, y)$, where $f - f(x, y)$. Appropriate transformations can be made so that the domain is specified as $x, y \in [0,1]$. Assuming equal mesh size for discretization in $x$ and $y$ directions $\delta = \Delta x = \Delta y$, then the finite difference equation for elliptic PDE is

$$f\left(x_{i+1}, y_j\right) + f\left(x_i, y_{j+1}\right) + f\left(x_{i-1}, y_j\right) + f\left(x_i, y_{j-1}\right) - 4f\left(x_i, y_j\right) = \mathcal{K} \qquad (6.13)$$

where $\mathcal{K} = 0$ for Laplace equation and $\mathcal{K} = g(x_i, y_j)\delta^2$ for Poisson equation and $x_i = x_0 + (i-1)\delta$, $i = 1, \ldots, n$ and $y_j = y_0 + (j-1)\delta$, $j = 1, \ldots, m$.

Again, the system of equation formed by Eq. (6.13) can be solved by Gauss or Gauss–Seidel methods. However, for large systems of equations, it is convenient to use the so-called the alternating direction implicit method (ADI method). Note that this method can be also used to solve parabolic and hyperbolic equations. The ADI method simplifies the numerical scheme because it is able to use a tridiagonal matrix algorithm which speeds up the numerical solution.

This method is proposed in Peaceman and Rachford (1955). In this case, Eq. (6.13) is rearranged as follows:

$$f(x_{i+1}, y_j) + f(x_{i-1}, y_j) - 4f(x_i, y_j) = \mathcal{K} - f(x_i, y_{j+1}) - f(x_i, y_{j-1}) \tag{6.14a}$$

$$f(x_i, y_{j+1}) + f(x_i, y_{j-1}) - 4f(x_i, y_j) = \mathcal{K} - f(x_{i+1}, y_j) - f(x_{i-1}, y_j) \tag{6.14b}$$

The iterative ADI scheme assumes initial values in each of the mesh points. In the first iteration, one uses Eq. (6.14a) and solves for LHS terms when the RHS are the initial values. The calculated solution at each point becomes initial value for the next iteration. Now, for the second iteration, Eq. (6.14b) is used with RHS calculated based on the initial values and LHS solved for. The iterations continues by using Eq. (6.14a,b) in an alternative fashion until convergence is achieved.

### 6.2.2   Weighted Residuals Methods: Finite Element, Finite Volume, Boundary Element

In the weighted residual methods, the domain is discretized by approximating it with geometrical blocks denoted, depending on the case, as: elements, volumes, or boundary elements. Depending on the actual implementation, the weighted residual methods can be of finite element type, finite volume type, boundary element type, or other less common types. In all these methods, the Green theorem is applied to reduce the problem order with one.

For each of the element, a PDE is written in integral form and eventually a system of ODEs or an algebraic system is formed depending on the problem type. The time-dependent PDEs are approximated with a set of ODEs. Once the ODE system or the algebraic system is solved, the discrete solutions are collected to approximate the entire domain. The great advantage of the weighted residual methods is that the boundary conditions are included as integrals in a functional which is to be minimized. In this way, the numerical procedure is independent of the boundary conditions.

The introduction to weighted residual methods presented in this section follows the main line from the work of Brio et al. (2010). Let us consider for exemplification the following PDE of advection–diffusion type in one dimension:

$$\frac{\partial f}{\partial t} + d\frac{\partial f}{\partial x} - a\frac{\partial^2 f}{\partial x^2} = 0, \, a > 0, d > 0, x \in [0,1], t \geq 0 \tag{6.15}$$

which is of parabolic type since $b^2 - 4ac = 0$. The following initial and boundary conditions are considered:

$$f(x,0) = x, f(0,t) = 0, f(1,t) = 1 \tag{6.16}$$

Now, one seeks an approximate solution of the following form:

$$f_n(x,t) = \sum_{i=0}^{n} k_i(t)\varphi_i(x) \cong f(x,t) \tag{6.17}$$

**Table 6.4** Weighted residual methods of finite element type

| Method | Test function |
|---|---|
| Finite element | $\psi_i(x)$ are piecewise polynomials |
| Galerkin finite element | $\psi_i(x) = \varphi_i(x)$, $i = 0, 1, \ldots, n$ |
| Boundary element | $\psi_i(x)$ based on Green theorem; problem degree reduced with one |
| Finite volume | $\psi_i(x)$ is either 0 or one for each volume element. The divergence theorem is applied |
| Spectral Galerkin | $\psi_i(x)$ are orthogonal polynomials |
| Petrov–Galerkin | $\psi_i(x) \neq \varphi_i(x)$, $i = 0, 1, \ldots, n$ |
| Collocation | $\psi_i(x) = \delta(x - x_i)$, $i = 0, 1, \ldots, n$ |

in which the functional form of the so-called trial functions $\varphi_i(x)$ can be conveniently specified and the variable coefficient $k_i(t)$ can be determined by minimization of the residual function $\Re_n$ in absolute value for all domain (spatial and temporal). Here, we write

$$\Re_n = \frac{\partial f_n}{\partial t} + d\frac{\partial f_n}{\partial x} - a\frac{\partial^2 f_n}{\partial x^2} \tag{6.18}$$

The minimization of $\Re_n$ requires necessarily and sufficiently the following integral to vanish

$$\int_0^1 \Re_n \psi_i(x) dx = \int_0^1 \left[ \frac{\partial f_n}{\partial t} \psi_i + d\psi_i \frac{\partial f_n}{\partial x} - a\left(\frac{\partial f_n}{\partial x}\right)\left(\frac{\partial \psi_i}{\partial x}\right) \right] dx = 0, i = 0, 1, \ldots, n \tag{6.19}$$

where $\psi_i(x), i = 0, 1, \ldots, n$ are denoted test functions and have the property to vanish only at the boundary; thence at $x = 0$ and $x = 1$, one has $\psi_i(x) = 0$, whereas $\psi_i(x) \neq 0$ for $x \in (0,1)$.

The domain is discretized into $n$ elements, in which each of the elements represents a segment for the one-dimensional problem. Therefore, the element can be represented as the segment between the nodes $x_i$ and $x_{i+1}, i = 0, 1, \ldots, n-1$. There are various methods for choosing the test functions as given in Table 6.4. Consider for a case study the use of Galerkin method in which we specify the trial function in form of a hat shape, namely,

$$\varphi_i(x) = \psi_i(x) = \begin{cases} \left|\frac{x - x_{i-1}}{x_i - x_{i-1}}\right| & \text{if } x \in [x_{i-1}, x_i] \\ \left| -\frac{x - x_{i+1}}{x_{i+1} - x_i} \right| & \text{if } x \in [x_i, x_{i+1}] \\ 0 & \text{otherwise} \end{cases} \tag{6.20}$$

Using Eq. (6.20) into Eq. (6.19) after some mathematical manipulation, one obtains the following set of ODEs:

$$\frac{2}{3}\delta\frac{dk_i}{dt} + \frac{\delta}{6}\left(\frac{dk_{i-1}}{dt} + \frac{dk_{i+1}}{dt}\right) + 0.5d(k_{i+1} - k_{i-1}) - \frac{a}{\delta}(k_{i+1} - 2k_i + k_{i-1}) - \left(\frac{a}{\delta} - \frac{d}{2}\right)\mathcal{F} = 0 \tag{6.21}$$

where $\mathcal{F}=0$ when $i=n-1$ and $\mathcal{F}=1$ otherwise, and an uniform grid is assumed with $\delta = x_{i+1} - x_i$.

The initial conditions required to solve the ODE system from Eq. (6.21) imposes that $k_i(0) = x_i$. A convenient scheme for numerical solution for the ODE system can be selected from Table 6.2.

## Example 6.1
A powder material with grains of 0.2 mm and moisture diffusivity of $10^{-7}$ m²/s is dried in a pneumatic dryer. The moisture content at entrance is 0.87 kg/kg, and the residence time in the dryer is 1 s. Determine the final moisture content knowing that the mass transfer coefficient is $10^{-4}$ m/s using two methods, namely, the finite element method and the lumped capacity method:

- The Biot number for mass transfer is calculated first $Bi_m = h_m L_c / D = 10^{-4} \times 10^{-4} / 10^{-7} = 0.1$.
- Lumped capacity model applies, namely,

$$-m_s \frac{dW}{dt} = h_m A (C_s - C_\infty)$$

where $m_s$ is the mass of the dry solid, $A$ is the exposed area, and $C_s, C_\infty$ are the moisture concentrations at the particle surface and in the bulk air.
- If one denotes $\rho_s$ the dry solid density, then $C_s = W \rho_s$ and $C_\infty = W_\infty \rho_s$, where $W$ is the moisture content in the particle.
- Since the typical humidity ratio in air is of the order of 10 g/kg and dry air density is ~1.3 kg/m³, it results that the order of magnitude of $C_\infty \cong 0.01$ kg moisture/m³. Assume a small dry bone density of the particle of 1000 kg/m³, then one has $W_\infty = 10^{-5}$ kg/kg which is negligible with respect to moisture content in the material. Therefore, one has

$$\frac{dW}{dt} + \frac{h_m A L_c \rho_s}{m_s L_c} W = 0$$

where $A L_c \rho_s \cong m_s$, therefore, with separation of the variables the equation becomes

$$\frac{dW}{W} = -\frac{h_m}{L_c} dt \rightarrow \frac{dW}{W} = -dt \rightarrow W = W_i \exp(-t) = 0.87 \exp(-t)$$

- After 1 s when the particle leaves the dryer, the moisture content is $W_f = 0.87 \exp(-1) = 0.32$ kg/kg.
- We will apply now the finite element method. Consider a grid with three nodes (forming three elements) at $t_i$ of 0, 0.5, 1 s. The Galerkin test and trial functions function are

$$\varphi_i(t) = \begin{cases} \left| \dfrac{t - t_{i-1}}{t_i - t_{i-1}} \right| & t_{i-1} \leq t \leq t_i \\[2mm] \left| \dfrac{t_{i+1} - t}{t_{i+1} - t_i} \right| & t_i < t \leq t_{i+1} \\[2mm] |0| & \text{otherwise} \end{cases}$$

- The approximate solution is $W \cong \mathbb{W} = \sum_{i=1}^{3} \mathcal{W}_i \varphi_i(t)$ and the residual and the integral are

$$\mathfrak{R}_3(t) = \frac{d\mathbb{W}}{dt} + \mathbb{W} \rightarrow \int_0^1 \mathfrak{R}_3(t)\varphi_j(t)dt = 0, \, j = 1, 2, 3$$

- Replacing the approximate solution in the integral, one eventually obtains

$$\sum_{i=1}^{3} \mathcal{W}_i \int_0^1 \left[ \varphi_j(t) \frac{d\varphi_i(t)}{dt} + \varphi_i(t)\varphi_j(t) \right] dt = 0, \, j = 1, 2, 3$$

- Denote the integral from the above with $I_{ij}$; then the following linear system is obtained $[I_{ij}][\mathcal{W}_i] = [0]$ where $\mathcal{W}_1 = 0.87$ and $\mathcal{W}_{2,3}$ are unknowns to be solved for.
- Note from the definition of $\varphi_i(t)$ that $\varphi_1(t) = 0$ for $t \in [0.5, 1]$ and $\varphi_3(t) = 0$ for $t \in [0, 0.5]$ and for the rest of the cases $\varphi_i(t) = 1$. This will give the following preferred linear system:

$$\begin{pmatrix} -\dfrac{5}{6} & \dfrac{2}{3} & \dfrac{7}{6} \\ 0 & -\dfrac{5}{6} & \dfrac{4}{3} \end{pmatrix} \begin{pmatrix} \mathcal{W}_2 \\ \mathcal{W}_3 \end{pmatrix} = \begin{pmatrix} 0 \\ 0 \end{pmatrix} \rightarrow \begin{pmatrix} \mathcal{W}_2 \\ \mathcal{W}_3 \end{pmatrix} = \begin{pmatrix} 0.519 \\ 0.325 \end{pmatrix}$$

- Remark the excellent agreement with the solution obtained by the lumped capacity model since $W_f \cong \mathcal{W}_3 = 0.325 \, \mathrm{kg/kg}$.

## 6.3 One-Dimensional Problems

Let us consider here a moist object subjected to drying in which a one-dimensional heat and moisture transfer process occurs. In this case, the equations for simultaneous heat and moisture transfer are given by Eq. (5.69) for the infinite slab, infinite cylinder, and spherical geometry. More advanced formulations for capillary porous media are that of Luikov given by Eqs. (5.83) and (5.85). In this section, we discuss some relevant cases of numerical equations for integration of the PDEs for time-dependent heat and moisture transfer in one dimension.

### 6.3.1 Decoupled Equations with Nonuniform Initial Conditions and Variable Boundary Conditions

In some drying problems, the moisture and thermal diffusivities can be assumed constant and uniform or taken at average values. However, these problems have closed-form solution of time-dependent one-dimensional thermal and moisture diffusion equations only for the case when the boundary conditions are constant in time and the initial conditions are uniform (i.e., the material has a uniform temperature and moisture content, initially). Equation (5.71) in Cartesian coordinates becomes

$$\text{For heat transfer:} \, \frac{\partial \theta}{\partial Fo} = \frac{\partial^2 \theta}{\partial \zeta^2} \tag{6.22a}$$

$$\text{For moisture transfer:} \left(\frac{D}{a}\right)\frac{\partial \phi}{\partial Fo} = \frac{\partial^2 \phi}{\partial \zeta^2} \tag{6.22b}$$

where $Fo = at/L^2$, $\zeta = z/L$, $L$ is the characteristic length (infinite slab domain half thickness), and $\theta$ and $\phi$ are the dimensionless temperature and the dimensionless moisture content given by, respectively,

$$\theta = \frac{T - T_\infty}{T_i - T_\infty} \tag{6.23a}$$

$$\phi = \frac{W - W_\infty}{W_i - W_\infty} \cong \frac{W}{W_\infty} \tag{6.23b}$$

with $T_i$, $W_i$ being the initial temperature and moisture content at the surface (boundary), respectively and $W_\infty$ and $W$ are defined in Chapter 5 referring to Eq. (5.71).

The equations are decoupled because the diffusivities are constant and there is assumed no influence of the temperature gradient on moisture movement through the porous material. The initial conditions are nonuniform, that is, neither temperature nor the moisture content is uniform throughout the slab at the initial moment. Therefore, the initial conditions are set as follows:

$$\theta(\zeta, 0) = \theta_0(\zeta) \tag{6.24a}$$

$$\phi(\zeta, 0) = \phi_0(\zeta) \tag{6.24b}$$

where $\theta_0(\zeta)$ and $\phi_0(\zeta)$ are specified functions of the length coordinate $\zeta$.

The symmetry boundary conditions are written for the slab midplane as follows:

$$\frac{\partial \theta(0, Fo)}{\partial Fo} = 0 \tag{6.25a}$$

$$\frac{\partial \phi(0, Fo)}{\partial Fo} = 0 \tag{6.25b}$$

whereas the convective boundary conditions at the surface are given as follows:

$$\frac{\partial \theta(1, Fo)}{\partial Fo} = -Bi\theta \tag{6.26a}$$

$$\frac{\partial \phi(1, Fo)}{\partial Fo} = \left(\frac{a}{D}\right)Bi_m \phi \tag{6.26b}$$

where $Bi = h_{tr}L/k$ and $Bi_m = h_m L/D$ are Biot numbers for heat transfer and mass transfer, respectively.

The reason why the factor $(a/D)$ is in the front of the RHS of Eq. (6.52b) is due to the fact that the dimensionless time variable is defined uniquely in terms of Fourier number for heat transfer $Fo$. Here, both Biot numbers, namely Biot number for heat transfer and Biot number for mass transfer, are allowed to vary with time. Because Eq. (6.22) are decoupled, they can be solved separately. However, numerical solution is required since the boundary conditions vary in time and the initial conditions are not uniform.

It is shown in Abbas et al. (2006) that the fully implicit scheme gives best results for equations such as Eq. (6.22) with initial and boundary conditions by Eqs. (6.24)–(6.26), respectively. However, the computational time is the longest for the fully implicit scheme. The general scheme recommended by Abbas et al. (2006) extended for heat and moisture transfer is

$$\theta_i^{j+1} = \theta_i^j + \frac{b}{r}\left(\theta_{i-1}^{j-1} - 2\theta_i^{j+1} + \theta_{i-1}^{j+1}\right) + \frac{1-b}{r}\left(\theta_{i-1}^j - 2\theta_i^j + \theta_{i+1}^j\right) \tag{6.27a}$$

$$\phi_i^{j+1} = \phi_i^j + \left(\frac{a}{D}\right)\left[\frac{b}{r}\left(\phi_{i-1}^{j-1} - 2\phi_i^{j+1} + \phi_{i-1}^{j+1}\right) + \frac{1-b}{r}\left(\phi_{i-1}^j - 2\phi_i^j + \phi_{i+1}^j\right)\right] \tag{6.27b}$$

where $r = \Delta Fo / (\Delta\zeta)^2$, $b \in [0,1]$, and $i \in \{0,1\}, j \in N, j \geq 0$.

Note that if the parameter $b = 0$, then the scheme becomes an the explicit finite difference scheme in three points, $(i-1,j)$, $(i,j)$, and $(i+1,j)$, where index $i$ refers to the discretized spatial variable $\zeta_{i+1} = \zeta_i + \Delta\zeta$ and $j$ refers to the discretized time variable $Fo_{j+1} = Fo_j + r(\Delta\zeta)^2$. Moreover, if $b = 1$, then the scheme becomes fully implicit in six points: $(i-1,j-1)$, $(i-1,j)$, $(i-1, j+1)$, $(i,j)$, $(i,j+1)$, and $(i+1,j)$. For any $b \in (0,1)$, the scheme is implicit according to the value of $b$ which plays the role of a weighting factor. The recommended value for $b$ given in Abbas et al. (2006) is 0.167.

The discretized initial conditions in a $n$ point spatial grid are $\theta_i^0 = \theta_0(\zeta_i)$, $\phi_i^0 = \phi_0(\zeta_i), i \in \{0, n\}$. For improved accuracy, the boundary conditions are discretized based on four-point formulae as follows:

$$-11\theta_0^{j+1} + 18\theta_1^{j+1} - 9\theta_2^{j+1} + 2\theta_3^{j+1} = 0 \tag{6.28a}$$

$$-11\phi_0^{j+1} + 18\phi_1^{j+1} - 9\phi_2^{j+1} + 2\phi_3^{j+1} = 0 \tag{6.28b}$$

$$-11\theta_{n-3}^{j+1} + 18\theta_{n-2}^{j+1} - 9\theta_{n-1}^{j+1} + 2\theta_n^{j+1} = 6(\Delta x)Bi\theta_n^j \tag{6.28c}$$

$$-11\phi_{n-3}^{j+1} + 18\phi_{n-2}^{j+1} - 9\phi_{n-1}^{j+1} + 2\phi_n^{j+1} = 6\left(\frac{a\Delta x}{D}\right)Bi_m\phi_n^j \tag{6.28d}$$

## Example 6.2

A moist material is dried in a continuous belt conveyor dryer. The system configuration is such that infinite slab geometry can be applied. The thermal and moisture diffusivities are constant but the thermal diffusivity is two times higher than moisture diffusivity. Furthermore, due to advancement of the material in the dryer, the boundary conditions change. The heat transfer coefficient is five times higher at the entrance than at the exit, and its decrease is linear. At the exit, Fourier number is given to be equal to 1. Also the entrance section $Bi$ is 1 and the dimensionless temperature and moisture distributions are linear with the core values of

$\theta_0(0) = 0.95$ and $\phi_0(0) = 1.1$. It is also given that Biot number for mass transfer is one-tenth of Biot number for heat transfer. Determine the dimensionless temperature and moisture content distribution at one fourth of the dryer length for $r = 1$. Note that $r$ is defined at Eq. (6.27).

- One starts by writing the governing equations for heat and moisture transfer in dimensionless format accounting for the fact that $(a/D = 2)$:

$$\frac{\partial \theta}{\partial Fo} = \frac{\partial^2 \theta}{\partial \zeta^2}; \frac{\partial \phi}{\partial Fo} = 2\frac{\partial^2 \phi}{\partial \zeta^2}$$

- The initial conditions for temperature for linear distribution becomes $\theta(\zeta) = 0.95 + 0.05\zeta$. Assume an uniform spatial discretization in nodes $i \in \{0, n\}$; therefore, the spatial step $\Delta\zeta = 1/n \to \zeta_i = i/n$ and $\theta_i^0 = 0.95 + 0.05i/n$.
- Similarly, the initial conditions for moisture content are $\phi_i^0 = 1.1 - 0.1i/n, n \in \{0, 1\}$.
- The symmetry boundary condition in the middle plane $(\zeta = 0)$ are as in Eq. (6.28a,b).
- The Biot number for heat transfer is correlated to Fourier number as it results from the problem enounce. At the entrance, when $Fo = 0 \to Bi = 1$. At the exit, when $Fo = 1 \to Bi = 0.2$. Therefore, $Bi = 1 - 0.8Fo$.
- The heat transfer boundary condition from Eq.(6.28c) becomes

$$-11\theta_{n-3}^{j+1} + 18\theta_{n-2}^{j+1} - 9\theta_{n-1}^{j+1} + 2\theta_n^{j+1} = \frac{6}{n}\left(1 - \frac{0.8jr}{n^2}\right)\theta_n^j$$

where it is noted that $Fo_j = j\Delta Fo = jr/n^2, j \in [0, j_{max}]$ with $j_{max} = \text{int}(n^2/r + 1)$.
- Similarly for moisture transfer, the discretized boundary condition is

$$-11\phi_{n-3}^{j+1} + 18\phi_{n-2}^{j+1} - 9\phi_{n-1}^{j+1} + 2\phi_n^{j+1} = -\frac{0.06}{n}\left(1 - \frac{0.8jr}{n^2}\right)\phi_n^j$$

- The results for the second time step are obtained as follows, assuming that $b = 1$ (fully implicit). The number of grid segments is taken to be 4. Then the number of time increments for all dryer length is 17. At one fourth, one has $j = 4$. The table below gives the calculated solution.
- It appears that the dimensionless moisture content at the center reduces from 1.1 to 1.015; whereas at the surface is 0.41. The average moisture content at one fourth of the dryer length is 0.76.

| $\zeta$ | $\phi_1^0$ | $\phi_1^1$ | $\phi_1^2$ | $\phi_1^3$ | $\phi_1^4$ | $\phi_1^5$ | $\theta_1^0$ | $\theta_1^1$ | $\theta_1^2$ | $\theta_1^3$ | $\theta_1^4$ | $\theta_1^5$ |
|---|---|---|---|---|---|---|---|---|---|---|---|---|
| 0 | 1.1 | 1.086 | 1.064 | 1.048 | 1.035 | 1.015 | 0.95 | 0.9568 | 0.9568 | 1.004 | 1.048 | 1.104 |
| 0.25 | 1.075 | 1.075 | 1.061 | 1.053 | 1.032 | 0.9737 | 0.9625 | 0.9625 | 0.9625 | 0.9648 | 0.9655 | 0.9814 |
| 0.5 | 1.05 | 1.05 | 1.05 | 1.036 | 0.9701 | 0.836 | 0.975 | 0.975 | 0.975 | 0.975 | 0.975 | 1.055 |
| 0.75 | 1.025 | 1.025 | 1.025 | 0.9538 | 0.7743 | 0.5799 | 0.9875 | 0.9875 | 0.9875 | 1.225 | 1.461 | 1.984 |
| 1 | 1 | 1 | 0.9288 | 0.6917 | 0.3841 | 0.4183 | 1 | 1.713 | 2.183 | 3.517 | 5.352 | 8.044 |

**Table 6.5** Equations and boundary conditions for simultaneous heat and moisture transfer in semi-infinite solid

|  | Heat transfer | Moisture transfer |
|---|---|---|
| Equations for $T(x,t)$ and $W(x,t)$ | $\dfrac{\partial^2 T}{\partial x^2} = \dfrac{1}{a}\left(\dfrac{\partial T}{\partial t}\right)$ | $\dfrac{\partial^2 W}{\partial x^2} = \dfrac{1}{D}\left(\dfrac{\partial W}{\partial t}\right)$ |
| Initial condition | $T(x,0) = 0$ | $W(x,0) = W_0$ |
| Surface conditions | $-\dfrac{\partial T(0,t)}{\partial x} = \dfrac{h_{\mathrm{tr}}}{k}[T_\infty - T(0,t)]$ | $W(0,t) = W_{\mathrm{s}}$ |
| Conditions at $x = \infty$ | $T(x=\infty,t) = T_i$ | $W(x=\infty,t) = W_0$ |

*Note*: $T_\infty$ is the bulk temperature of the drying agent.

## 6.3.2 Partially Coupled Equations

Let us consider here as a representative case for one-dimensional geometry, the semi-infinite solid. The particular form of Eq. (5.69) in Cartesian coordinates for the one-dimensional problem are given in Table 6.5. Here, we assume that the semi-infinite solid is subjected to convective boundary condition at the surface, and it has fixed temperature and moisture content at depth. At the surface, the moisture content is assumed constant, $W_{\mathrm{s}}$. Provided that the thermal and moisture diffusivity are constant, this problem has closed solutions as given in Table 1.16. For heat equation, the solution is written as

$$T(x,t) = T_i + (T_\infty - T_i)\mathrm{erfc}\left(\frac{0.5x}{\sqrt{at}}\right) - \exp\left[\frac{hx}{k} + \frac{h^2 at}{k^2}\right]\mathrm{erfc}\left[\frac{h\sqrt{at}}{k} + \frac{0.5x}{\sqrt{at}}\right] \tag{6.29}$$

The analytical solution for the moisture transfer equation is given in Yilbas et al. (2003) as follows:

$$\frac{W - W_0}{W_{\mathrm{s}} - W_0} = 1 - \mathrm{erf}\left(\frac{x}{2\sqrt{Dt}}\right) \tag{6.30}$$

Assume that the moisture diffusivity is dependent on temperature, while the thermal diffusivity is constant. This means that the analytical solution Eq. (6.29) is valid, whereas the solution for moisture content Eq. (6.30) does not apply. Therefore, the moisture transfer equation must be solved numerically. If one uses Eq. (5.66) for $D$, then the moisture transfer PDE becomes

$$\frac{\partial W(x,t)}{\partial t} = D_0 \exp\left(-\frac{E_{\mathrm{a}}}{\mathcal{R}T(x,t)}\right)\frac{\partial^2 W(x,t)}{\partial x^2} \tag{6.31}$$

where $T(x,t)$ is given by Eq. (6.29).

Let us apply the finite difference method for the spatial coordinate to solve Eq. (6.30) numerically. Consider a one-dimensional grid defined by $x_i = i\Delta x, i \in [0, n]$. In this case, a set of ODEs can be written as follows:

$$\frac{dW(x_i, t)}{dt} = D_0 \exp\left(-\frac{E_a}{\mathcal{R}T(x, t)}\right) \frac{W(x_i + \Delta x, t) - 2W(x_i, t) + W(x_i - \Delta x, t)}{(\Delta x)^2}, i \in [0, n] \quad (6.32)$$

## Example 6.3

This is a numerical example illustrating the integration of partially coupled time-dependent heat and moisture transfer equation in a drying process as discussed previously. The moist object is a parallelepiped with the thickness of 20 mm which is placed on a tray. For the time frame of interest, the material can be modeled as a semi-infinite solid having the base in contact with the tray which acts as a thermal insulator. The initial moisture content is $W_0 = 0.87\,kg/kg$, and the initial temperature of the material is $T_0 = 303\,K$. The surface moisture content is kept constant at 0.1 kg/kg. The drying air is kept at constant parameters at a temperature of $T_\infty = 354\,K$. The thermal diffusivity is $a = 1.6E{-}7\,m^2/s$, the thermal conductivity is $k = 0.577\,W/mK$, and the heat transfer coefficient is $h_{tr} = 200\,W/m^2\,K$.

Discretize the spatial domain in 50 segments. Determine the numerical solution based on the discretization scheme given in Eq. (6.32) for a time up to 500 s and compare the numerical solution with the analytical solution for moisture content at $x = 4\,mm$ obtained when the moisture diffusivity is taken constant at the initial value. Assume that the far field ($x = \infty$) is approximated at $x \geq 6\,mm$.

- Since the numerical grid has 50 nodes and the material thickness is 20 mm, one determines an uniform space step of $\delta = 0.02/50 = 0.4\,mm$. The grid is defined by $x_i = (i-1)\delta, i = 1, 2, \ldots$.
- The factor $\beta = h_{tr}/k = 346.7\,m^{-1}$.
- At $x_1$, one sets the surface condition $W_1 = W_s = 0.1\,kg/kg$.
- At $x_{16} = 6.4\,mm$, one sets the far-field condition that $W_{16} = 0.87\,kg/kg$.
- The solution is calculated for the range $x_1$ until $x_{16}$; therefore, 14 ODEs are formed as a closed system.
- The ODE system is solved with EES software; for the first run, the moisture diffusivity is calculated using the initial temperature and a value of $1.889E{-}9\,m^2/s$ is obtained.
- Figure 6.3 shows a comparison of the numerical and analytical solutions, where the analytical solution assumes a constant diffusivity. For the numerical case, the moisture diffusivity increased in time due to the temperature rise. The drying is more intense because the moisture diffusivity increases. The error caused by the constant diffusivity assumption is visible.
- Figure 6.4 shows the moisture content variation at four locations within the material. The variation at the surface layer is fastest, whereas in depth, the moisture content varies slowly.
- At the end of 600 s of numerical simulation, the moisture content and temperature distributions are represented as shown in Figure 6.5. The temperature varies quickly but with a constant gradient. The moisture content is low at the very superficial layer but increases rapidly and then it stabilizes.

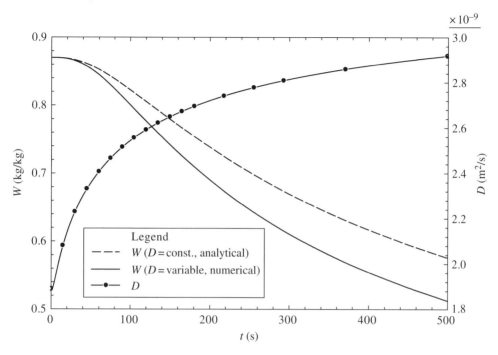

**Figure 6.3** The moisture content and the moisture diffusivity variation at a depth of 2 mm in the semi-infinite moist material subjected to drying. The analytical solution for $W$ at $D=$ constant as given by Eq. (6.30) is compared to the numerical solution for which $D$ varies with the temperature

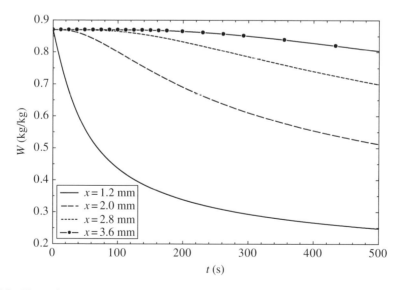

**Figure 6.4** The moisture content in the semi-infinite moist material subjected to drying at various depths

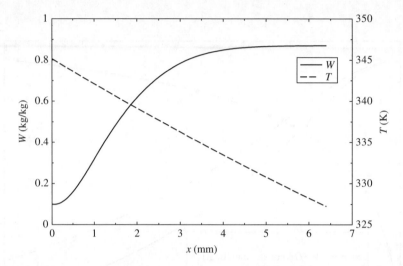

**Figure 6.5** Moisture content and temperature in the semi-infinite moist material subjected to drying at $t = 500\,\mathrm{s}$

### 6.3.3 Fully Coupled Equations

Let us consider a porous material modeled with the coupled Luikov equations for time-dependent heat and moisture transfer in one dimension. This equation includes both Soret and Dufour effects. The moisture movement process inside the porous medium caused by the temperature gradient is denoted as Soret effect. The heat flux carried by moisture movement affects weakly the temperature distribution and this is called Dufour effect. Assume that for a given case the pressure effect can be neglected. This situation is valid in many practical problems. The governing equations are derived from Luikov Eq. (5.85) by renouncing to pressure components as follows:

$$\frac{\partial}{\partial t}\begin{pmatrix} W \\ T \end{pmatrix} - \frac{\partial^2}{\partial x^2}\begin{pmatrix} k_{11} & k_{12} \\ k_{21} & k_{22} \end{pmatrix}\begin{pmatrix} W \\ T \end{pmatrix} = \begin{pmatrix} 0 \\ 0 \end{pmatrix} \tag{6.33}$$

The heat and moisture transfer equations given by Eq. (6.33) are fully coupled because the heat transfer equation comprises a term having a second derivative of moisture content and, furthermore, the moisture transfer equation includes Laplacian of temperature. The initial conditions are $W(x,0) = W_0$ and $T(x,0) = T_0$. Constant boundary conditions are written at $x = L$, namely, $W(L,t) = W_0$ and $T(L,t) = T_0$, and convective flux conditions at the exposed surface where $x = 0$, namely,

$$-\begin{pmatrix} D & 0 \\ 0 & k \end{pmatrix}\frac{\partial}{\partial x}\begin{pmatrix} W \\ T \end{pmatrix} = \begin{pmatrix} j \\ q \end{pmatrix} = \begin{pmatrix} h_{\mathrm{m}}\rho_{\mathrm{s}}(W - W_\infty) \\ h_{\mathrm{tr}}(T_\infty - T) \end{pmatrix} \tag{6.34}$$

where $\rho_{\mathrm{s}}$ is the density of the dry material.

In Eqs. (6.33) and (6.34), the Luikov coefficients $k_{ij}$ and $k, D, h_{tr}, h_m$ can be constant or variable. For the finite element scheme, we consider Galerkin approach and discretize the spatial domain in $n$ elements. The solutions are then approximated as follows:

$$\begin{pmatrix} W \\ T \end{pmatrix} \cong \begin{pmatrix} W_n \\ T_n \end{pmatrix} = \sum_{i=0}^{n} \begin{pmatrix} \varphi_i(x) W_i(t) \\ \varphi_i(x) T_i(t) \end{pmatrix} \tag{6.35}$$

where $\varphi_i(x)$ are the trial functions, and $W_i(t)$ and $T_i(t)$ are nodal functions to be determined.

The Galerkin method requires the trial functions to be identical to test functions and given by Eq. (6.20). The residual is constructed using the solution approximations from Eq. (6.35) and the test functions from Eq. (6.20). The finite element method requires that the following weighted residual integrals given below vanish:

$$\int_0^L \left\{ \sum_{i=0}^{n} \left[ \varphi_i(x)\varphi_j(x)\frac{d}{dt}\begin{pmatrix} W_i(t) \\ T_i(t) \end{pmatrix} - \frac{d^2\varphi_i(x)}{dx^2}\varphi_j(x)\begin{pmatrix} k_{11} & k_{12} \\ k_{21} & k_{22} \end{pmatrix}\begin{pmatrix} W_i(t) \\ T_i(t) \end{pmatrix} \right] \right\} dx = 0, j = 0, 1, \ldots, n \tag{6.36}$$

where it is assumed for simplicity that $k_{ij}$ are constant. This simplifies the integrals as follows:

$$\sum_{i=0}^{n} \left[ \frac{d}{dt}\begin{pmatrix} W_i(t) \\ T_i(t) \end{pmatrix}\int_0^L \varphi_i(x)\varphi_j(x)dx - \begin{pmatrix} k_{11} & k_{12} \\ k_{21} & k_{22} \end{pmatrix}\int_0^L \varphi_j(x)\frac{d^2\varphi_i(x)}{dx^2}dx\begin{pmatrix} W_i(t) \\ T_i(t) \end{pmatrix} \right] = 0, j = 0, 1, \ldots, n \tag{6.37}$$

Therefore, Eq. (6.37) together with the boundary conditions Eq. (6.34) form a closed system of ODEs which can be solved for $W_j(t)$ and $T_j(t)$ by typical methods starting from the initial conditions. Green theorem can be applied to Eq. (6.37) to relate the discretized functions with the boundary conditions and simplify the solution. Following Comini and Lewis (1976), the ODE system of $2(n+1)$ equations can be expressed as follows:

$$C_j \begin{pmatrix} \dot{W}_j \\ \dot{T}_j \end{pmatrix} + [K_j]\begin{pmatrix} W_j \\ T_j \end{pmatrix} + \begin{pmatrix} J_m \\ J_q \end{pmatrix} = 0, j = 0, 1, \ldots, n \tag{6.38}$$

where $C_j$ represents the factors for temporal derivative vector from Eq. (6.37), namely,

$$C_j = \int_0^L \varphi_i(x)\varphi_j(x)dx, j = 0, 1, \ldots, n$$

and $[K_j]$ are $n+1$ matrices of dimension $2 \times 2$ resulting from the second term of Eq. (6.37) when the integral is solved by parts, namely,

$$[K_j] = \begin{pmatrix} k_{11} & k_{12} \\ k_{21} & k_{22} \end{pmatrix}\int_0^L \left(\frac{d\varphi_i(x)}{dx}\right)\left(\frac{d\varphi_j(x)}{dx}\right)dx, j = 0, 1, \ldots, n$$

and the parameters $\mathcal{J}_m$ and $\mathcal{J}_q$ are boundary integrals derived from Green theorem and correspond to boundary conditions for moisture transfer and heat transfer, respectively. Here, $\mathcal{J}_m$ has the unit s$^{-1}$ and is defined by

$$\mathcal{J}_m = \sum_{i=1}^{n} \varphi_j(0) \frac{d\varphi_i(0)}{dx} \left(\frac{j}{\rho}\right)_{x=0} + \varphi_j(L) \frac{d\varphi_i(L)}{dx} \left(\frac{\rho_s D}{\rho}\right)_{x=L} \left(\frac{W_0 - W_{n-1}}{x_n - x_{n-1}}\right), j = 0, 1, \ldots, n$$

where $\rho$ is the moisture density, $\rho_s$ is the dry material density, and $D$ is the moisture diffusivity and

$$\mathcal{J}_q = \sum_{i=1}^{n} \varphi_j(0) \frac{d\varphi_i(0)}{dx} \left(\frac{j}{\rho_m C_{p,m}}\right)_{x=0} + \varphi_j(L) \frac{d\varphi_i(L)}{dx} \left(\frac{k}{\rho_m C_{p,m}}\right)_{x=0} \left(\frac{T_0 - T_{n-1}}{x_n - x_{n-1}}\right), j = 0, 1, \ldots, n$$

Comini and Lewis (1976) propose an iterative scheme for numerical solution of the ODE system from Eq. (6.38). The scheme starts from the initial conditions which allows for determination of $W_j, T_j, j = 0, 1, \ldots, n$ from Eq. (6.38). Next, the parameters $C_j, [K_j], \mathcal{J}_m, \mathcal{J}_q$ are calculated. Euler method can be applied to determine $W_{j,1,2}$ and $T_{j,1,2}$ at times $\Delta t$ and $2\Delta t$, respectively. Then, the following numerical scheme starts to be applied to determine $W_{j,m}, T_{j,m}, j = 0, 1, \ldots, n$ at times $m\Delta t$, for $m = 3, 4, 5, \ldots$, namely,

$$\left(\begin{matrix} W_j \\ T_j \end{matrix}\right)_{m+1} = -\left[\frac{1}{3}[K_j]_m + \frac{C_j I_2}{2\Delta t}\right]^{-1} \left[\left(\frac{1}{3}[K_j]_m - \frac{C_j I_2}{2\Delta t}\right)\left(\begin{matrix} W_j \\ T_j \end{matrix}\right)_m + \frac{1}{3}[K_j]_m \left(\begin{matrix} W_{j*} \\ T_{j*} \end{matrix}\right) + \left(\begin{matrix} \mathcal{J}_m \\ \mathcal{J}_q \end{matrix}\right)\right],$$

$$j = 0, 1, \ldots, n \tag{6.39}$$

where $I_2$ is the identity matrix of order 2, $\Delta t$ is the time step, the notation $[\ ]^{-1}$ signifies the inverse of a matrix and $W_{j*}$ and $T_{j*}$ are calculated as follows:

$$\left(\begin{matrix} W_{j*} \\ T_{j*} \end{matrix}\right) = \frac{1}{3}\left[\left(\begin{matrix} W_j \\ T_j \end{matrix}\right)_m + \left(\begin{matrix} W_j \\ T_j \end{matrix}\right)_{m-1} + \left(\begin{matrix} W_j \\ T_j \end{matrix}\right)_{m-1}\right]$$

Once the approximations $(W_{j,m}, T_{j,m}), j = 0, 1, 2, \ldots, n$ are obtained, they are plugged into Eq. (6.35) to obtain the final solution represented in the form $W_n(x_i, t_m)$ and $T_n(x_i, t_m)$, where index $n$ represents the total number of elements for discretization, $x_i, i = 0, 1, 2, \ldots, n$ is the spatial node, and $t_m = m\Delta t$ is the discretized time variable.

The coupled system from Eq. (6.33) can be also solved by finite difference method. One simple possibility is to express the second derivatives with central differences according to Eq. (6.4). Then, the following scheme is obtained:

$$\frac{d}{dt}\left(\begin{matrix} W_i \\ T_i \end{matrix}\right) = \frac{1}{(\Delta x)^2}\left(\begin{matrix} k_{11} & k_{12} \\ k_{21} & k_{22} \end{matrix}\right)\left(\begin{matrix} W_{i-1} - 2W_i + W_{i+1} \\ T_{i-1} - 2T_i + T_{i+1} \end{matrix}\right), i = \{1, n-1\} \tag{6.40}$$

where $\mathcal{W}_i$ and $\mathcal{T}_i$ are unknown functions of time and the slab thickness is discretized as $x_i = Li/n$. The initial conditions are given for $i = \{1, n-1\}$ in form $\mathcal{W}_i = W_{0,i}$ and $\mathcal{T}_i = T_{0,i}$. The boundary conditions are used to obtain the function values at nodes 0 and $n$. The discretized boundary conditions at $i = n$ require $\mathcal{W}_n = W_0$ and $\mathcal{T}_n = T_0$. The discretized boundary conditions at $i = 0$ are

$$-\frac{1}{2\Delta x}\begin{pmatrix} D & 0 \\ 0 & k \end{pmatrix}\begin{pmatrix} \mathcal{W}_2 - \mathcal{W}_0 \\ \mathcal{T}_2 - \mathcal{T}_0 \end{pmatrix} = \begin{pmatrix} h_m \rho_s (W - W_\infty) \\ h_{tr}(T_\infty - T) \end{pmatrix} \tag{6.41}$$

The resulting ODE system can be now solved using Runge–Kutta or other similar methods as given in Section 6.2.1. Example 6.4 clarifies the use of finite difference method for solution of the fully coupled equations given by Eq. (6.33) for simultaneous heat and mass transfer.

## Example 6.4

A drying process is governed by the coupled equation (6.33) that model the simultaneous heat and mass transfer in a slab material. The equations are made dimensionless in the domain $\zeta = [0,1]$ and $Fo \geq 0$, where $Fo$ is the dimensionless time given in the form of a Fourier number. The equations are given here in a dimensionless format in terms of the moisture content $\phi$ and temperature $\theta$. The equations are as follows:

$$\frac{\partial}{\partial Fo}\begin{pmatrix} \phi \\ \theta \end{pmatrix} = \begin{pmatrix} 0.1 & 1 \\ 1 & 0.1 \end{pmatrix}\frac{\partial^2}{\partial x^2}\begin{pmatrix} \phi \\ \theta \end{pmatrix}$$

with the initial conditions $\phi(\zeta,0) = 1.1 - 0.1\zeta$ and $\theta(\zeta,0) = 1$. The boundary conditions at $\zeta = 1$ are $\phi(1,Fo) = 1$ and $\theta(1,Fo) = 1$, and at $\zeta = 0$, they are as follows:

$$-\begin{pmatrix} 10^{-5} & 0 \\ 0 & 1 \end{pmatrix}\frac{\partial}{\partial \zeta}\begin{pmatrix} \phi \\ \theta \end{pmatrix} = \begin{pmatrix} \phi(0,Fo) \\ 2 - \theta(0,Fo) \end{pmatrix}$$

- We start with the discretization of $\zeta$ in $n+1$ nodes, $\zeta_i = i/n, i \in \{0,n\}$.
- The initial conditions are given for $i \in \{1, n-1\} \rightarrow \phi_i = 1.1 - 0.1i/n; \theta_i = 1$.
- The discretized boundary conditions at $\zeta = 0$ require cf. Eq. (6.41)

$$-\frac{n}{2}\begin{pmatrix} 10^{-5} & 0 \\ 0 & 1 \end{pmatrix}\begin{pmatrix} \phi_2 - \phi_0 \\ \theta_2 - \theta_0 \end{pmatrix} = \begin{pmatrix} \phi_0 \\ 2 - \theta_0 \end{pmatrix}.$$

- The discretized boundary conditions at $\zeta = 1$ require $\phi_n = 1$ and $\theta_n = 1$.
- The discretized equations for $i = \{1, n-1\}$ are

$$\frac{d}{dFo}\begin{pmatrix} \phi_i \\ \theta_i \end{pmatrix} = n^2\begin{pmatrix} 0.1 & 1 \\ 1 & 0.1 \end{pmatrix}\begin{pmatrix} \phi_{i-1} - 2\phi_i + \phi_{i+1} \\ \theta_{i-1} - 2\theta_i + \theta_{i+1} \end{pmatrix}.$$

- The integration of the defined ODE system has been done for $n = 10$ using the EES software with the results given in the table below for $Fo = 1$:

| $\zeta$ | 0 | 0.1 | 0.2 | 0.3 | 0.4 | 0.5 | 0.6 | 0.7 | 0.8 | 0.9 | 1 |
|---|---|---|---|---|---|---|---|---|---|---|---|
| $\phi$ | N/A | 1.93E–09 | 0.1885 | 0.3668 | 0.5263 | 0.6613 | 0.7696 | 0.8527 | 0.9145 | 0.9611 | 1 |
| $\theta$ | 1.845 | 1.683 | 1.534 | 1.403 | 1.293 | 1.205 | 1.138 | 1.087 | 1.051 | 1.023 | 1 |

- The distribution of $\phi$ and $\theta$ for $Fo = 1$ is shown in Figure 6.6. The dimensionless moisture content variation with Fourier number is shown in Figure 6.7 for three locations

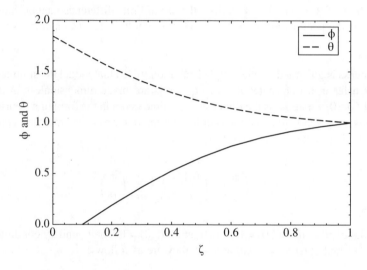

**Figure 6.6**  Distribution of the dimensionless moisture content and temperature at $Fo = 1$ for Example 6.4

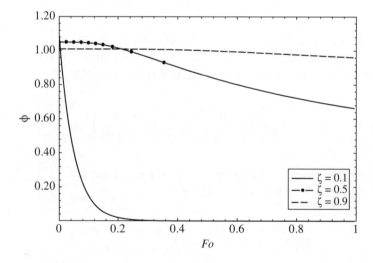

**Figure 6.7**  Dimensionless moisture content variation at three locations for Example 6.4

## 6.4 Two-Dimensional Problems

In this section, numerical solutions of two-dimensional time-dependent heat and moisture transfer in Cartesian, cylindrical, and spherical coordinates are presented with a focus on drying processes. The overwhelming majority of two-dimensional numerical analyses in heat and moisture transfer are made for two-dimensional problems. The discretization and numerical scheme are given and some case studies are presented.

### 6.4.1 Cartesian Coordinates

Let us consider here a rectangular permeable domain subjected to heat and moisture transfer with convective boundary conditions. We assume that the thermal diffusivity depends on moisture content and the moisture diffusivity depends on temperature according to Eq. (5.66). There is no heat generation inside the object and no deformation occurs during the process. Furthermore, the drying medium is air with constant parameters at any location around the domain such that the heat and mass transfer coefficients can be considered as uniform and constant at the boundary. In this case, the problem variables are the temperature and moisture content at any location inside the object, that is, $T(x, y, t)$ and $W(x, y, t)$, where $x \in [0, L]$ and $y \in [0, H]$. With these considerations, the following governing equations for heat and moisture transfer can be written:

$$\frac{1}{a(T, W)} \frac{\partial T}{\partial t} = \frac{\partial^2 T}{\partial x^2} + \frac{\partial^2 T}{\partial y^2} \qquad (6.42a)$$

$$\frac{1}{D(T)} \frac{\partial W}{\partial t} = \frac{\partial^2 W}{\partial x^2} + \frac{\partial^2 W}{\partial y^2} \qquad (6.42b)$$

The initial conditions are $T(x, y, 0) = T_i$ and $W(x, y, 0) = W_i$, and the boundary conditions are given exactly as in Table 5.4 (Chapter 5). The numerical scheme is constructed on a rectangular grid with step sizes $\Delta x$ and $\Delta y$, respectively. The grid is uniform, with $n + 1$ nodes in $x$ direction and $m + 1$ nodes in $y$ direction, namely, $x_i = i\Delta x$ and $y_i = j\Delta y$, where $i = 0, 1, 2, \ldots, n$ and $j = 0, 1, 2, \ldots, m$. At each node $(i, j)$, an approximate solution is determined for every time step in the form $(T_{i,j,k}, W_{i,j,k})$ where the index $k$ represent a discrete time step, $t_k = k\Delta t, k = 0, 1, 2, \ldots$. The numerical grid is shown in Figure 6.8

Table 6.6 gives the numerical scheme with explicit finite difference for the rectangular domain. The heat transfer and moisture transfer coefficients as well as the drying air temperature $T_d$ and the moisture content in the drying air $W_d$ must be specified as input parameters for the numerical scheme. The grid independence tests must be conducted to ensure that grid-independent results are obtained. Stability analysis must be performed in order to investigate the boundedness of the exact solution of the finite difference equations using the von Neumann method. The method introduces an initial line of errors as represented by a finite-Fourier series and applies in a theoretical sense to initial value problem. The stability criterion obtained for the previous difference equations is written as

$$\Delta t \leq \min \left\{ \frac{(\Delta x)^2 (\Delta y)^2}{2a \left[ (\Delta x)^2 + (\Delta y)^2 \right]}, \frac{(\Delta x)^2 (\Delta y)^2}{2D \left[ (\Delta x)^2 + (\Delta y)^2 \right]} \right\} \qquad (6.43)$$

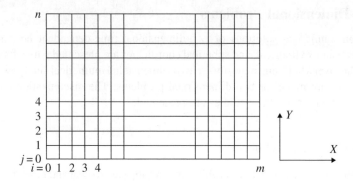

**Figure 6.8**  Two-dimensional numerical grid for the time-dependent finite difference scheme

**Table 6.6**  Explicit finite difference numerical scheme for 2D time-dependent heat and moisture transfer

| Heat transfer | | Moisture transfer | |
|---|---|---|---|

$$T_{i,j,k+1} = r(\Delta Y)^2 T_{i+1,j,k} + \left[1 - 2r\left((\Delta X)^2 + (\Delta Y)^2\right)\right] T_{i,j,k}$$
$$+ r(\Delta Y)^2 T_{i-1,j,k} + r(\Delta X)^2 T_{i,j+1,k} + r(\Delta X)^2 T_{i,j-1,k}$$

$$M_{i,j,k+1} = r_\mathrm{d}(\Delta Y)^2 M_{i+1,j,k} + \left[1 - 2r_\mathrm{d}\left((\Delta X)^2 + (\Delta Y)^2\right)\right] M_{i,j,k}$$
$$+ r_\mathrm{d}(\Delta Y)^2 M_{i-1,j,k} + r_\mathrm{d}(\Delta X)^2 M_{i,j+1,k} + r_\mathrm{d}(\Delta X)^2 M_{i,j-1,k}$$

| Heat transfer | | Moisture transfer | |
|---|---|---|---|
| Discretization parameter: | $r = \dfrac{a\Delta t}{(\Delta X)^2 (\Delta Y)^2}$ | Discretization parameter: | $r_\mathrm{d} = \dfrac{D\Delta t}{(\Delta X)^2 (\Delta Y)^2}$ |
| Initial condition: $T_{i,j,0} = T_i$ | | Initial condition: $W_{i,j,0} = W_i$ | |
| Boundary conditions on $x$: | Boundary conditions on $y$: | Boundary conditions on $x$: | Boundary conditions on $y$: |
| $T_{0,j,k} = \dfrac{T_{1,j,k} + C_1 \times T_\mathrm{d}}{1 + C_1}$ | $T_{i,0,k} = \dfrac{T_{i,1,k} + C_2 \times T_\mathrm{d}}{1 + C_2}$ | $W_{0,j,k} = \dfrac{W_{1,j,k} + C_3 \times W_\mathrm{d}}{1 + C_3}$ | $W_{i,0,k} = \dfrac{W_{i,1,k} + C_4 \times W_\mathrm{d}}{1 + C_4}$ |
| $T_{m,j,k} = \dfrac{T_{m-1,j,k} + C_1 \times T_\mathrm{d}}{1 + C_1}$ | $T_{i,n,k} = \dfrac{T_{i,n-1,k} + C_2 \times T_\mathrm{d}}{1 + C_2}$ | $W_{m,j,k} = \dfrac{W_{m-1,j,k} + C_3 \times W_\mathrm{d}}{1 + C_3}$ | $W_{i,n,k} = \dfrac{W_{i,n-1,k} + C_4 W_\mathrm{d}}{1 + C_4}$ |
| $C_1 = \dfrac{h_\mathrm{tr} \times \Delta X}{k}$ | $C_2 = \dfrac{h_\mathrm{tr} \times \Delta Y}{k}$ | $C_3 = \dfrac{h_\mathrm{m} \times \Delta x}{D}$ | $C_4 = \dfrac{h_\mathrm{m} \times \Delta y}{D}$ |

The numerical solutions approximate a set of $\{T, W\}$ surfaces in the domain $(x, y)$ with one pair of such surfaces for each time step. Figure 6.9 shows a qualitative representation of the solution at a particular time step. Figure 6.9a shows the momentary temperature surface where is visible that the temperature front propagates from the boundary toward the domain core. This indicates the direction of heat transfer which is toward the inside of the material (drying process). Figure 6.9b shows the moisture content surface which shows that the moisture progresses from the domain core toward the boundary that is in an opposed direction with respect to heat transfer.

Here, the use of the previous numerical scheme for three types of foodstuff cut in slices is exemplified. The materials are cut in rectangular shapes with the aspect ratios of 1:2,

(a)                                                    (b)

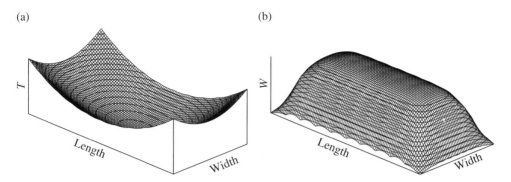

**Figure 6.9**   Qualitative representation of momentarily temperature and moisture content distributions in a rectangular material subjected to drying

**Table 6.7**   Drying parameters of some fruits slices

| Parameter | Value | | |
| Material | | | |
| | Fig (*Ficus carica*) | Apple | Peach |
| --- | --- | --- | --- |
| Size (cm) | $10 \times 20$ | $30 \times 20$ | $20 \times 20$ |
| Aspect ratio | 1:2 | 1:1.5 | 1:1 |
| $T_i$ (K) | 303 | 298 | 298 |
| $T_d$ (K) | 354 | 323 | 323 |
| $\omega$ (kg/kg) | 0.53 | 0.53 | — |
| $W_i$ (kg/kg db) | 0.87 | 5.25 | 8.09 |
| $k$ (W/m K) | 0.31 | 0.219 | 0.361 |
| $\rho$ (kg/m$^3$) | 1241 | 856 | 1259 |
| $C_p$ (J/kg K) | 850 | 851 | 851 |
| $h_{tr}$ (W/m$^2$ K) | 25 | 25 | 25 |
| $h_m$ (m/s) | 0.0001 | 0.0001 | 0.0001 |

1:1.5, and 1:1, respectively. Table 6.7 gives the product characteristics for each case. The foodstuff are dried in air with fixed temperature ($T_d$) and humidity ratio ($\omega$). The rectangular domain has been discretized and the explicit finite difference scheme applied to determine the temperature field and the moisture contour distribution at each time step. The results are shown in Figure 6.10 in form of contour plots for *Ficus carica* (figs) with aspect ratio of 1:2. Temperature in the slab increases as the time period progresses. This is because of the higher temperature of the drying air than the slab temperature. The temperature contours in the slab are elliptic profiles due to the rectangular shape of the drying product. Moreover, temperature distribution inside the slab is nonuniform, giving an indication that the temperature-dependent moisture diffusivity varies in the slab, which in turn affects the rate of moisture diffusion in the slab.

The contours in the figure are represented for three time moments at 60, 180, and 300 s. It is visible how the front of temperature advances toward inward whereas the moisture

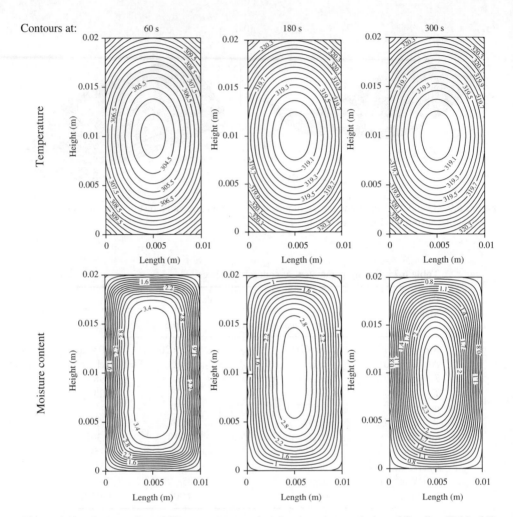

**Figure 6.10**   Contour plots of 2D temperature and moisture content at drying of fig slice (Table 6.7)

migrates toward periphery. The moisture in the slab reduces as the time period progresses. The initial moisture content reduces to nearly half of its initial value at the center of the slab after the time period of 300 s. Since the slab is rectangular, the resulting contours are in elliptic shape. The moisture distribution inside the slab is not uniform; therefore, the time required to reduce the moisture content to half of its initial value varies at each location in the rectangular slab.

The other option to represent the results of numerical analysis is by using surface plots. The surfaces from Figure 6.11 show the temperature and moisture distribution in the fig slab with aspect ratio 1:2, subjected to drying. Here is better visible how the temperature is higher at the slab boundary and decreases as one progresses toward the center. From the moisture content, the surface becomes visible that the product core is wet while the margins become dry. The minimum moisture content is found at the material corners.

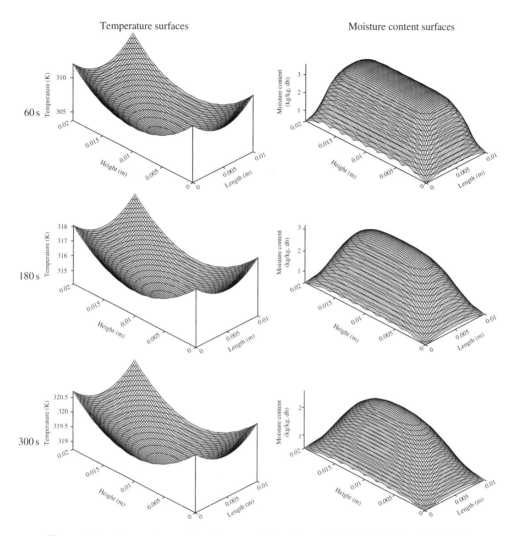

**Figure 6.11**  Temperature and moisture content surfaces at drying of fig slice (Table 6.7)

The contours and surfaces for temperature and moisture content are further shown in Figures 6.12–6.15 for the apple and peach slices subjected to drying. For the apple the simulations are reported for time instances of 60, 180, 300, and 600 s. It clearly appears that the aspect ratio influences quantitatively the results.

When compared with the fig drying, the 3D surfaces for temperature and moisture content for apple slice drying are more flatten. This is an effect of the aspect ratio of the slab, which is better remarked on the contour plots. At the time of 300 s, the temperature at the central region of the apple slab (1:1.5 aspect ratio) flattens at 315 K, whereas for the fig slab (1:2 aspect ratio), it flattens at 319 K. This means that heating through heat conduction is fasted when the aspect ratio progresses from rectangle toward a squared shape. The temperature reaches 321 K at the apple slice center after 600 s. The moisture content at the center decreases 1.4 kg/kg for fig at

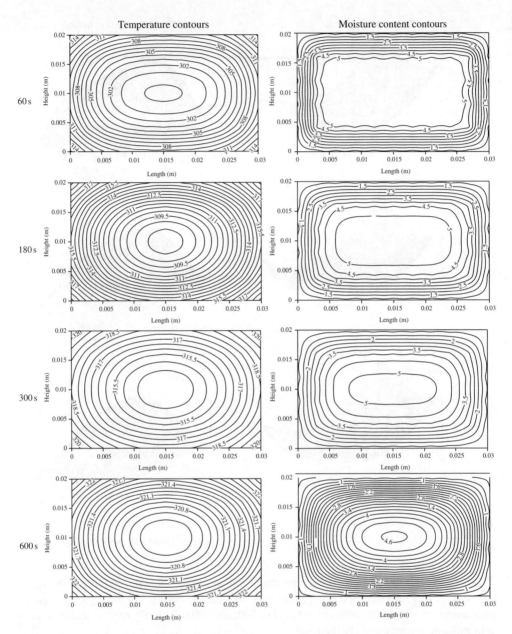

**Figure 6.12** Contour plots of 2D temperature and moisture content at drying of apple slice (Table 6.7)

300 s, whereas for apple, the moisture content at the central region remains constant at 5 kg/kg for the first 300 s; however, at 600 s, it becomes 4.6 kg/kg. This fact is a result of the lag factor, of which effect is represented by a delay of the drying process at the beginning. The apple surface dries from 1.5 to 1 kg/kg obtained after 600 s of drying.

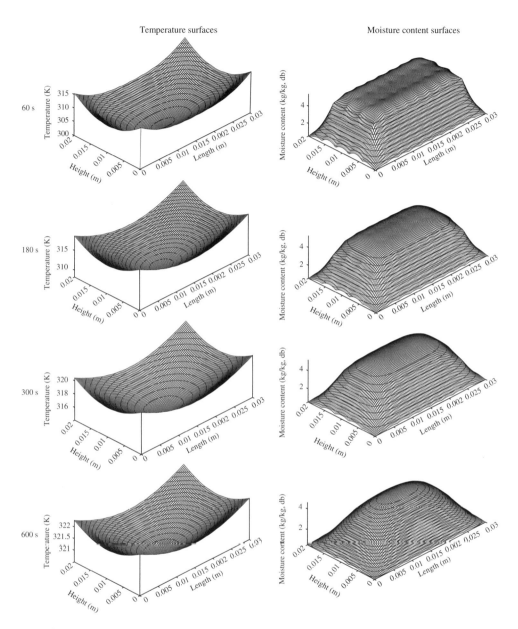

**Figure 6.13**   Temperature and moisture content surfaces at drying of apple slice (Table 6.7)

The moisture content at the corners is the lowest, having values below 1 kg/kg. The average moisture content can be estimated by integration over the domain; this is approximately 2 kg/kg at the end of the drying process. At the beginning of the drying, the average moisture content is approx. 4 kg/kg; therefore, the half drying time is ~600 s. Of course, the simulation results show in this case only the preliminary phase of drying as in an actual setting the drying must be conducted for longer times.

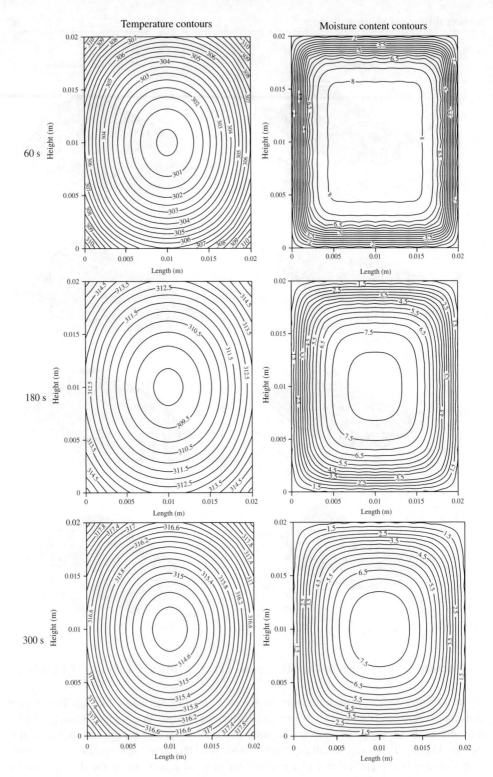

**Figure 6.14** Contour plots of 2D temperature and moisture content at drying of peach slice (Table 6.7)

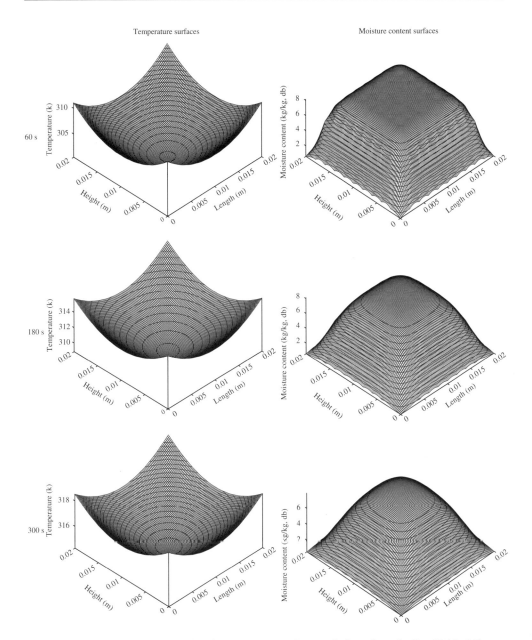

**Figure 6.15**   Temperature and moisture content surfaces at drying of peach slice (Table 6.7)

The influence of the effect of heat transfer coefficient on the temperature and moisture content distribution inside the slab is important. Obviously, the center temperature and surface temperature increases with the increase in heat transfer coefficient. The time required to reach the maximum limit reduces when the heat transfer coefficient is augmented because when the heat transfer coefficient is higher, then the heat transfer rate enhances. Of course, the value of the heat transfer coefficient influences the drying rate. However, often the heat transfer coefficient is limited due to material size, the surface properties, and air (gas) velocity. Forced convection drying shows high heat transfer coefficients and lower drying times, but this is obtained on the expense of power which needs to be consumed to drive gas blowers. The optimal heat transfer coefficient in the drying application depends on the nature of drying, initial temperature of the product to be dried, type of the product, and its moisture content.

The heat transfer rate is related to the temperature gradient. The temperature gradient at material center increases initially and then decreases regardless the value of heat transfer coefficient. The initial rise is due to sudden increase of temperature after steady behavior for a certain period of time. This phase is referred as the warming up of the solid. The surface temperature gradient decreases with time and the maximum heat transfer coefficient has the highest gradient all the time because the highest heat transfer rate occurs for the maximum heat transfer coefficient. Note that at air drying of slab products, the heat transfer coefficient is generally in the range of 25–50 W/m$^2$ K.

The rate of drying has an interesting variation in time. As observed from the contour and surface plots, the drying rate exhibits the general trend in which rate of drying takes place in the two well-known periods.

Firstly, it is constant for some period of time representing the constant rate period and then starts decreasing as the time period progresses expressing the falling-rate period. Note that all the above simulations compare well with experimental data. The mean percentage error between the measured and calculated values of temperature and moisture distribution is found to be 0.17 °C and 2.31 kg/kg, respectively.

### Example 6.5

Let us illustrate with a case study the application of the explicit finite difference scheme for a drying process of a rectangular-shaped apple slice having the thermal diffusivity dependent on the moisture content. The case study is expanded from Hussain and Dincer (2003a). The case study data is specified as given in Table 6.8. The scheme from Table 6.6 has been applied using grid and time steps that satisfy the stability criteria. The results of the case study are given in Figure 6.16 in form of contour plots at three moments for temperature and moisture content.

As it is known in drying problems, the variation of the center moisture content and temperature is essential. The numerical analysis provides good information about these parameters. Here, in Figure 6.17, the variation of temperature and moisture content at the center of the rectangular domain is shown for the case study. From the figure, it becomes clear that the moisture content variation and temperature variation inside the material have opposed trends. The temperature increases due to the propagation of temperature front under the external (heating) conditions. The moisture content decreases due to moisture migration toward the domain boundaries which act as sinks eliminating the product moisture into the drying agent.

**Table 6.8**  Input data for the apple slab of $4.8 \times 4.9 \times 2.0$ cm

| Parameter | Value | Parameter | Value | Parameter | Value |
| --- | --- | --- | --- | --- | --- |
| $T_i$ (K) | 303 | $k$ (W/m K) | $0.148 + 0.493W$ | $P$ (kg/m$^3$) | 856 |
| $T_d$ (K) | 354 | $C_p$ (kJ/kg K) | $1.4 + 3.22W$ | | |
| $W_i$ (kg/kg) | 0.87 | | | | |

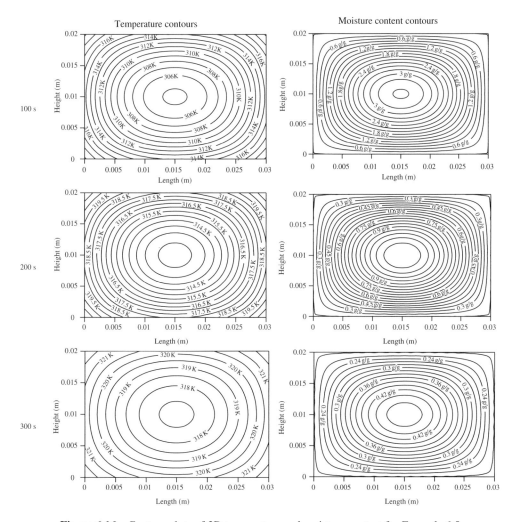

**Figure 6.16**   Contour plots of 2D temperature and moisture content for Example 6.5

## 6.4.2   *Cylindrical Coordinates with Axial Symmetry*

Here, a two-dimensional numerical analysis of heat and moisture transfer during drying of a cylindrical material with axial symmetry is carried out using an explicit finite difference approach. The main goal of the present study is to simulate the drying process in an axisymmetric cylindrical moist object by using the fundamental models of heat conduction and moisture diffusion. Similar assumptions as for the rectangular problem presented previously are made.

In the cylindrical domain geometry, the problem variables are the temperature and moisture content at any radius $r$ and axial position $z$. The solution is assumed independent on the polar angle $\theta$. The temperature and moisture content inside the object are the unknown functions

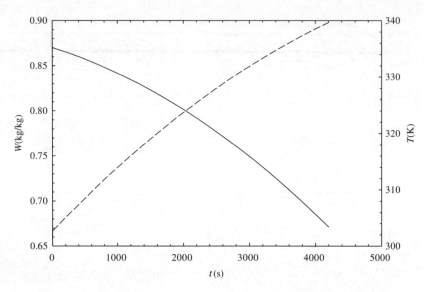

**Figure 6.17** Moisture content and temperature variation at the center of the apple slab for Example 6.5

**Table 6.9** Boundary conditions for simultaneous heat and mass transfer in cylindrical 2D domain with axial symmetry

| Boundary | Thermal boundary condition | Moisture conservation boundary condition |
|---|---|---|
| Axis: $r=0, z \in [0, L]$ | $\dfrac{\partial T(0,z,t)}{\partial r} = 0$ | $\dfrac{\partial W(0,z,t)}{\partial r} = 0$ |
| Periphery: $r = R, z \in [0, L]$ | $-k\dfrac{\partial T(R,z,t)}{\partial r} = h_{tr}[T(R,z,t) - T_\infty]$ | $-D\dfrac{\partial W(R,z,t)}{\partial r} = h_m[T(R,z,t) - T_\infty]$ |
| Bottom : $z = 0, r \in [0, R]$ | $-k\dfrac{\partial T(r,0,t)}{\partial z} = h_{tr}[T(r,0,t) - T_\infty]$ | $-D\dfrac{\partial W(r,0,t)}{\partial z} = h_m[T(r,0,t) - T_\infty]$ |
| Top : $z = L, r \in [0, R]$ | $-k\dfrac{\partial T(r,L,t)}{\partial z} = h_{tr}[T(r,L,t) - T_\infty]$ | $-D\dfrac{\partial W(r,L,t)}{\partial z} = h_m[T(r,L,t) - T_\infty]$ |

defined by $T(r, z, t)$ and $W(r, z, t)$, where $r \in [0, R]$ and $z \in [0, L]$. With these considerations, the following governing equations for heat and moisture transfer can be written:

$$\frac{1}{a(W)}\frac{\partial T}{\partial t} = \frac{1}{r}\frac{\partial}{\partial r}\left(r\frac{\partial T}{\partial r}\right) + \frac{\partial^2 T}{\partial z^2} \tag{6.44a}$$

$$\frac{1}{D(T)}\frac{\partial W}{\partial t} = \frac{1}{r}\frac{\partial}{\partial r}\left(r\frac{\partial W}{\partial r}\right) + \frac{\partial^2 W}{\partial z^2} \tag{6.44b}$$

The initial conditions are $T(r, z, 0) = T_i$ and $W(r, z, 0) = W_i$, and the boundary conditions are given in Table 6.9. It is assumed that the domain has axial symmetry. The numerical scheme is

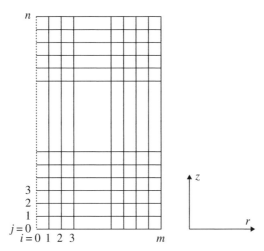

**Figure 6.18**   Axisymmetric numerical grid for the time-dependent finite difference scheme for heat and moisture transfer in cylindrical coordinates

constructed on a rectangular grid with step sizes $\Delta r$ and $\Delta z$, respectively. The grid is uniform, with $n+1$ nodes in $r$ direction and $m+1$ nodes in $z$ direction, namely, $x_i = i\Delta x$ and $y_i = j\Delta y$, where $i = 0, 1, 2, \ldots, n$ and $j = 0, 1, 2, \ldots, m$. Similarly as for the rectangular domain (above) at each node $(i, j)$, an approximate solution is determined for every time step in the form ($T_{i,j,k}$, $W_{i,j,k}$) where the index $k$ represent a discrete time step, $t_k = k\Delta t, k = 0, 1, 2, \ldots$. The numerical grid is shown in Figure 6.18.

Similarly as aforementioned, the heat transfer and moisture transfer coefficients as well as the drying air temperature $T_d$ and the moisture content in the drying air $W_d$ must be specified as input parameters for the numerical scheme. The grid independence tests must be conducted to ensure that grid-independent results are obtained. Stability analysis must be performed in order to investigate the boundedness of the exact solution of the finite difference equations as given by Eq. (6.43) where $\Delta x$ is replaced with $\Delta r$ and $\Delta y$ is replaced with $\Delta z$.

The explicit numerical scheme with finite differences for cylindrical coordinates is given in Table 6.10. This scheme marches in time starting from the initial values for temperature and moisture content at each point. The solution at each time step on the cylindrical axisymmetric domain is represented by temperature and moisture content surfaces.

The solution can be represented as 3D surfaces of temperature and moisture at any time step. Figure 6.19 shows a qualitative representation of temperature and moisture content surfaces for a cylindrical geometry with axial symmetry. The surfaces are somehow similar to those obtained for the rectangular domain: temperature is higher at the boundary and minimum in the geometrical center, whereas the moisture content is maximum at the geometrical center and has lower values at the boundary.

Figure 6.20 gives an example of graphical representation of numerical solutions at banana drying, assuming axial symmetry in cylindrical coordinates. The numerical simulation is given for a banana cylinder with length of 30 mm and radius of 10 mm. It is assumed that the thermal diffusivity varies with the moisture content. It is assumed that the banana density and thermal diffusivity remain constant during the time frame of the simulation. Their values are estimated

**Table 6.10** Explicit time-dependent finite difference numerical scheme for axisymmetric cylindrical coordinates

| Heat transfer | | Moisture transfer | |
|---|---|---|---|
| $T_{i,j,k+1} = AT_{i+1,j,k} + (1-2[A+B])T_{i,j,k}$ $+ AT_{i-1,j,k} + B(1+0.5j)T_{i,j+1,k} + B(1-0.5j)T_{i,j-1,k}$ | | $W_{i,j,k+1} = A_m W_{i+1,j,k} + (1-2[A_m+B_m])W_{i,j,k}$ $+ A_m W_{i-1,j,k} + B_m(1+0.5j)W_{i,j+1,k}$ $+ B_m(1-0.5j)W_{i,j-1,k}$ | |
| Discretization parameters: | $A = \dfrac{a\Delta t}{(\Delta z)^2}; B = \dfrac{a\Delta t}{(\Delta r)^2}$ | Discretization parameters: | $A_m = \dfrac{D\Delta t}{(\Delta z)^2}; B_m = \dfrac{D\Delta t}{(\Delta r)^2}$ |
| Initial condition: $T_{i,j,0} = T_i$ | | Initial condition: $W_{i,j,0} = W_i$ | |
| Boundary conditions on $r$: | Boundary conditions on $z$: | Boundary conditions on $r$: | Boundary conditions on $y$: |
| $T_{i,0,k} = T_{i,1,k}$ | $T_{0,j,k} = \dfrac{T_{1,j,k}+C_6 T_d}{1+C_6}$ | $W_{i,0,k} = W_{i,1,k}$ | $W_{0,j,k} = \dfrac{W_{1,j,k}+C_8 \times W_d}{1+C_8}$ |
| $T_{i,n,k} = \dfrac{T_{i,n-1,k}+C_5 \times T_d}{1+C_5}$ | $T_{m,j,k} = \dfrac{T_{m-1,j,k}+C_6 T_d}{1+C_6}$ | $W_{i,n,k} = \dfrac{W_{i,n-1,k}+C_7 W_d}{1+C_7}$ | $W_{m,j,k} = \dfrac{W_{m-1,j,k}+C_8 \times W_d}{1+C_8}$ |
| $C_5 = \dfrac{h_{tr}\Delta r}{k}$ | $C_6 = \dfrac{h_{tr}\Delta z}{k}$ | $C_7 = \dfrac{h_m\Delta r}{D}$ | $C_8 = \dfrac{h_m\Delta z}{D}$ |

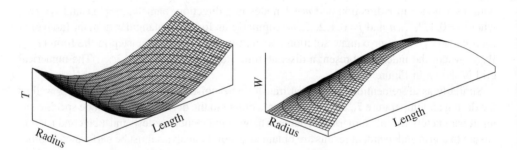

**Figure 6.19** Representation of numerical solutions for temperature and moisture content in cylindrical domain

to $k = 0.4592\,\text{W/m K}$ and $\rho = 810\,\text{kg/m}^3$. The heat capacity varies with the moisture content due to water loss. The specific heat is approximated with the equation $C_p = 837 + 1256W$ (J/kg K). The thermal diffusivity coefficient varies with moisture content as follows: $a(W) = k/\rho C_p = 5.5 \times 10^{-4}/(837 + 1256W)$.

The initial moisture content is taken 3.16 kg/kg. The initial temperature is 298 K and the drying air temperature is maintained to 323 K. In this context, the heat transfer coefficient is estimated to $h_{tr} = 25\,\text{W/m}^2\,\text{K}$, and the moisture transfer coefficient is $h_m = 0.0001\,\text{m/s}$. The moisture content inside the cylinder reduces as the time period increases.

The reduction rate of moisture content in the surface region is higher as compared to the interior of the object. The rapid drop of moisture content in the early heating period is because of the high moisture gradient in this region, which in turn derives considerable diffusion rates from inside to the surface. Temperature in the cylinder increases as the time period progresses.

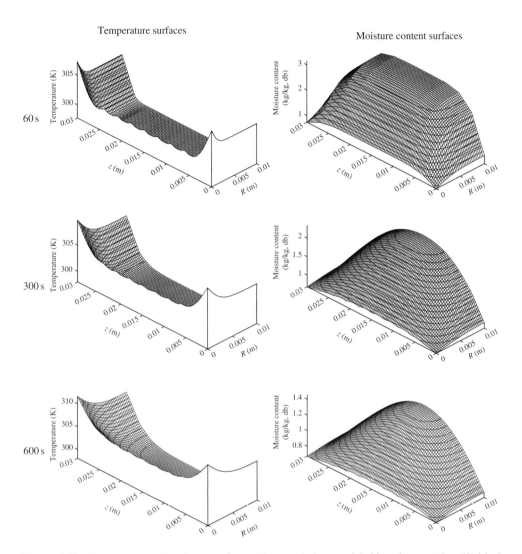

**Figure 6.20** Temperature and moisture surfaces at banana drying, modeled in axisymmetric cylindrical coordinates

This is because of the higher ambient temperature than the temperature of the object. Moreover, temperature in the cylinder is nonuniform, maximum at the surfaces and minimum at the center.

### Example 6.6
Here, the numerical integration results for a case study of drying of a cylindrically shaped broccoli with a diameter of 0.007 m and a length of 0.02 m is illustrated. The problem is assumed axisymmetric. The data for the case study is given in Table 6.11. The temperature and moisture are expressed in a dimensionless format as follows:

$$\theta = \frac{T - T_d}{T_i - T_d}; \phi = \frac{W - W_d}{W_i - W_d}$$

**Table 6.11**  Input data for broccoli drying, Example 6.6

| Parameter | Value | Parameter | Value | Parameter | Value |
|-----------|-------|-----------|-------|-----------|-------|
| $T_i$ (K) | 298 | $W_i$ (g/g (db)) | 9.57 | $\rho$ (kg/m$^3$) | 2195.27 |
| $T_d$ (K) | 333 | $K$ (W/m K) | $0.148 + 0.493W$ | $C_p$ (kJ/kg K) | $0.837 + 1.256W$ |

The dimensionless temperature distributions inside the axisymmetric cylindrical moist object for different drying times are given in Figure 6.21 as 3D surfaces. As seen in the figures, the temperature inside the moist object increases as the drying time progresses, which is due to the higher drying air temperature than the temperature of the object. Moreover, temperature distributions inside the moist object appear to be nonuniform and time dependent, and this clearly gives an indication that the temperature-dependent moisture diffusivity varies inside the object, which in turn affects the rate of moisture diffusion in the object.

The variations of reduced temperature at the center are shown in Figure 6.22 for five values of the heat transfer coefficient. The reduced center temperature increases with increasing heat transfer coefficient. It attains a maximum limit and flattens after a period of time. Note that the reduced surface temperature rises sharply during the first time step and then increases in a parabolic fashion. The rapid rise of the temperature in the surface vicinity of the object is because of the internal energy gain in this region, which is due to convective heating of the surface.

The variation of reduced center moisture content and reduced surface moisture content with time is shown in Figure 6.23. The reduction of moisture content is high in the early heating period, and as the heating period progresses, it reduces gradually until equilibrium moisture content is attained.

The attainment of high moisture gradient in the early period is because of the diffusion process. The reduction of moisture content is more pronounced at the surface, and as the heating progresses, it becomes constant. It is clear from the figure that drying takes place in two periods; first during the constant rate period in which rate of moisture reduction is constant and then the rate of moisture gradually reduces exhibiting falling-rate period.

### 6.4.3  Polar Coordinates

Let us consider here a cylindrical material exposed to drying for which there is no axial symmetry of the process. However, assume that the temperature and moisture content values do not vary with respect to $z$ axis. Then the only variations are with respect to the radius $r$ and the polar angle $\theta$. Polar coordinates are useful in this case to express the temperature and moisture content fields as $T(r, \theta, t)$ and $W(r, \theta, t)$. The heat and moisture transfer equations are as follows:

$$\frac{1}{a(T, W)} \frac{\partial T}{\partial t} = \frac{1}{r} \frac{\partial T}{\partial r} + \frac{\partial^2 T}{\partial r^2} + \frac{1}{r^2} \frac{\partial^2 T}{\partial \theta^2} \tag{6.45a}$$

$$\frac{1}{D(T)} \frac{\partial W}{\partial t} = \frac{1}{r} \frac{\partial W}{\partial r} + \frac{\partial^2 W}{\partial r^2} + \frac{1}{r^2} \frac{\partial^2 W}{\partial \theta^2} \tag{6.45b}$$

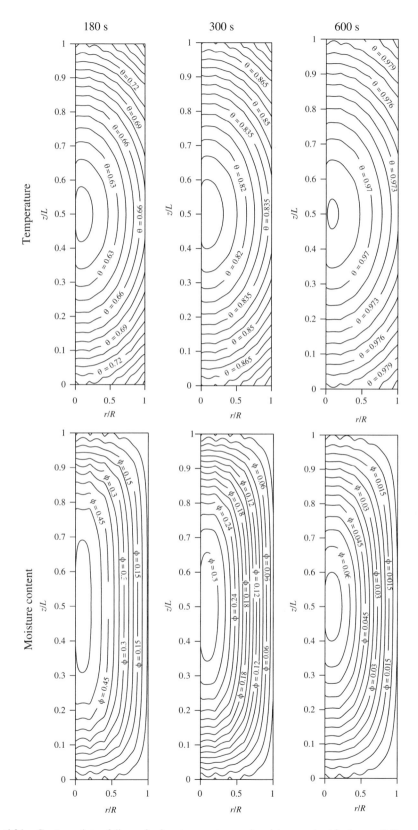

**Figure 6.21** Contour plots of dimensionless temperature and moisture content for broccoli, Example 6.6

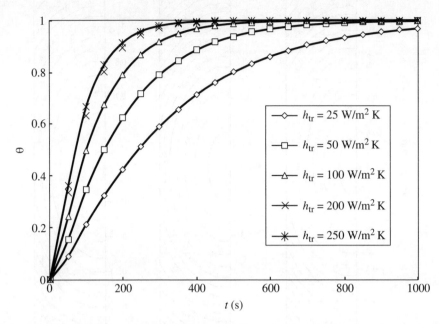

**Figure 6.22** Dimensionless temperature at the center of the cylindrical object as function of time for five values of the heat transfer coefficient, for Example 6.6

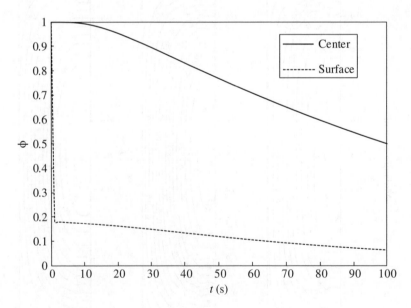

**Figure 6.23** Dimensionless moisture content at the center and at the surface of the cylindrical object variation in time, for Example 6.6

Here, the thermal diffusivity is generally dependent on the temperature and moisture content according to an equation such as Eq. (5.68). Furthermore, the moisture diffusivity depends on temperature according to Eq. (5.66). The initial conditions can be set by $T(r, \theta, 0) = T_i(r, \theta)$ and $W(r, \theta, 0) = W_i(r, \theta)$. The symmetry boundary conditions are applied at the center stating that at any polar angle there is no variation of temperature and moisture content with respect to radius as follows:

$$\frac{\partial T(0, \theta, t)}{\partial r} = \frac{\partial W(0, \theta, t)}{\partial r} = 0 \qquad (6.46)$$

The boundary conditions at periphery $(r = R)$ impose convective heat and moisture fluxes as follows:

$$-a(T, W)\frac{\partial T(R, \theta, t)}{\partial r} = h_{tr}(T_\infty - T) \qquad (6.47a)$$

$$-D(T)\frac{\partial W(R, \theta, t)}{\partial r} = h_m \rho_s (W - W_\infty) \qquad (6.47b)$$

There are also boundary conditions with respect to polar angle; these are denotes as angular boundary conditions. Since $\theta \in [0, 2\pi]$, the angular boundary conditions must be set at $\theta = 0$ and at $\theta = 2\pi$. The angular boundary condition requires two things. The first is periodicity of the solution, that is,

$$T(r, 0, t) = T(r, 2\pi, t) \qquad (6.48a)$$

$$W(r, 0, t) = W(r, 2\pi, t) \qquad (6.48b)$$

and the second angular boundary condition requires the continuity of the fluxes at the periodic boundary:

$$\frac{\partial T(r, 0, t)}{\partial \theta} = \frac{\partial T(r, 2\pi, t)}{\partial \theta} \qquad (6.49a)$$

$$\frac{\partial W(r, 0, t)}{\partial \theta} = \frac{\partial W(r, 2\pi, t)}{\partial \theta} \qquad (6.49b)$$

Alternating direction implicit scheme with finite differences can be used to solve the heat and moisture problem numerically. The ADI scheme given in Eq. (6.14) can be adapted to polar coordinates to solve Eq. (6.45) with the boundary conditions expressed by Eqs. (6.46)–(6.49). The scheme considers a mesh parameter $r$ defined as follows:

$$r = \min\{D, a\}\frac{\Delta t}{(\Delta r)^2 + (\Delta r)^2 (\Delta \theta)^2} \qquad (6.50)$$

According to the studies from the studies from Kaya et al. (2007) for drying application the recommended value of mesh parameter is $r = 0.5$. An interesting aspect refers to the variations

**Table 6.12** ADI numerical scheme for heat and moisture transfer on polar coordinates

| Direction | Discretization scheme for heat transfer equation |
|---|---|

$r$

$$T_{i,j}^{n+1}\left(\frac{2}{D\Delta t}+\frac{2}{\Delta r^2}\right)-T_{i+1,j}^{n+1}\left(\frac{1}{2r\Delta r}+\frac{1}{\Delta r^2}\right)+T_{i-1,j}^{n+1}\left(\frac{1}{2r\Delta r}-\frac{1}{\Delta r^2}\right)=$$
$$T_{i,j}^{n+\frac{1}{2}}\left(\frac{2}{D\Delta t}-\frac{2}{r^2\Delta\theta^2}\right)+T_{i,j+1}^{n+\frac{1}{2}}\left(\frac{1}{r^2\Delta\theta^2}\right)+T_{i,j-1}^{n+\frac{1}{2}}\left(\frac{1}{r^2\Delta\theta^2}\right)$$

$\theta$

$$T_{i,j}^{n+\frac{1}{2}}\left(\frac{2}{D\Delta t}+\frac{2}{r^2\Delta\theta^2}\right)-T_{i,j+1}^{n+\frac{1}{2}}\left(\frac{1}{r^2\Delta\theta^2}\right)-T_{i,j-1}^{n+\frac{1}{2}}\left(\frac{1}{r^2\Delta\theta^2}\right)=$$
$$T_{i,j}^{n}\left(\frac{2}{D\Delta t}-\frac{2}{\Delta r^2}\right)+T_{i+1,j}^{n}\left(\frac{1}{2r\Delta r}+\frac{1}{\Delta r^2}\right)+T_{i-1,j}^{n}\left(\frac{1}{\Delta r^2}-\frac{1}{2r\Delta r}\right)$$

| | Discretization scheme for moisture transfer equation |
|---|---|

$r$

$$W_{i,j}^{n+1}\left(\frac{2}{D\Delta t}+\frac{2}{\Delta r^2}\right)-W_{i+1,j}^{n+1}\left(\frac{1}{2r\Delta r}+\frac{1}{\Delta r^2}\right)+W_{i-1,j}^{n+1}\left(\frac{1}{2r\Delta r}-\frac{1}{\Delta r^2}\right)=$$
$$W_{i,j}^{n+\frac{1}{2}}\left(\frac{2}{D\Delta t}-\frac{2}{r^2\Delta\theta^2}\right)+W_{i,j+1}^{n+\frac{1}{2}}\left(\frac{1}{r^2\Delta\theta^2}\right)+W_{i,j-1}^{n+\frac{1}{2}}\left(\frac{1}{r^2\Delta\theta^2}\right)$$

$\theta$

$$W_{i,j}^{n+\frac{1}{2}}\left(\frac{2}{D\Delta t}+\frac{2}{r^2\Delta\theta^2}\right)-W_{i,j+1}^{n+\frac{1}{2}}\left(\frac{1}{r^2\Delta\theta^2}\right)-W_{i,j-1}^{n+\frac{1}{2}}\left(\frac{1}{r^2\Delta\theta^2}\right)=$$
$$W_{i,j}^{n}\left(\frac{2}{D\Delta t}-\frac{2}{\Delta r^2}\right)+W_{i+1,j}^{n}\left(\frac{1}{2r\Delta r}+\frac{1}{\Delta r^2}\right)+W_{i-1,j}^{n}\left(\frac{1}{\Delta r^2}-\frac{1}{2r\Delta r}\right)$$

of moisture and thermal diffusivity during numerical simulation. If this variation occurs, then adjustment of the tie step must be performed automatically such that one complies with a proposed set value of the mesh parameter. The ADI scheme splits the finite difference implicit formulae into two as given in Table 6.12. The first discretization is with respect to radial coordinate $r$ while the second one is given with respect to angular coordinate $\theta$. The discretization of the domain is shown in Figure 6.24 and consists on discrete elements encompassing angle $\Delta\theta$ and having a with corresponding to $\Delta r$. For an uniform grid the discretization on the radial direction is $r_i = i\Delta r, i \in \{0, n\}$ and in angular direction is $\theta_j = j\Delta\theta, j \in \{0, m\}$.

## 6.4.4 Spherical Coordinates

In many drying applications moist materials with spherical geometry are involved. If the size of the material is sufficiently large, full symmetry cannot be invoked such that the solutions are dependent not only on the radial coordinate, but also on the azimuth and zenith angles. However, as mentioned previously in this chapter, the symmetry with respect to one of the angles (say zenith angle) can be often observed. Assume that the solutions in spherical coordinates are independent on the zenith angle. In this case, the solution of the simultaneous time-dependent heat and moisture transfer is expressed in the form of spherical temperature field $T(r, \varphi, t)$ and moisture content $W(r, \varphi, t)$.

(a)                                          (b)

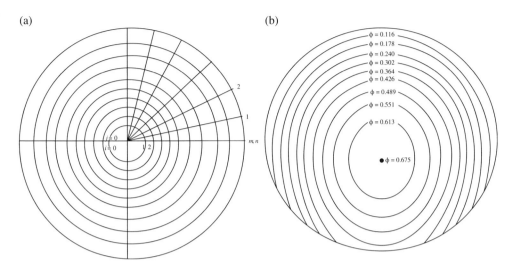

**Figure 6.24** Domain discretization for polar coordinates (a) and contours of dimensionless moisture content obtained for a porous material with moisture $D = 10^{-5}\,\mathrm{m^2/s}$ exposed to drying (b) (data from Ozalp and Dincer (2010b))

The problem requires the set of initial conditions according to $T(r,\varphi,0) = T_i(r,\varphi)$ and $W(r,\varphi,0) = W_i(r,\varphi)$, where $T_i$ and $W_i$ are functions expressing the initial conditions with respect to the radial and azimuthal coordinates. The solution is given in the domain $r \in [0,R]$ and $\varphi = [0,2\pi]$, where $R$ is the sphere radius. The governing equations for heat and moisture diffusion in 2D spherical coordinates are as follows:

$$\frac{1}{a(T,W)}\frac{\partial T}{\partial \theta} = \frac{1}{r^2}\frac{\partial}{\partial r}\left(r^2\frac{\partial T}{\partial r}\right) + \frac{1}{r^2\sin^2\varphi}\left(\frac{\partial^2 T}{\partial\varphi^2}\right) \tag{6.51a}$$

$$\frac{1}{D(T)}\frac{\partial W}{\partial \theta} = \frac{1}{r^2}\frac{\partial}{\partial r}\left(r^2\frac{\partial W}{\partial r}\right) + \frac{1}{r^2\sin^2\varphi}\left(\frac{\partial^2 W}{\partial\varphi^2}\right) \tag{6.51b}$$

Here, again the thermal diffusivity is generally dependent on the temperature and moisture content according to an equation such as Eq. (5.68). Furthermore, the moisture diffusivity depends on temperature according to Eq. (5.66). The angular boundary conditions state that the solution is periodic with $2\pi$ period with respect to azimuth angle and the continuity is assured at angular domain limits. Mathematically, these conditions are written as follows:

$$T(r,0,t) = T(r,2\pi,t); W(r,0,t) = W(r,2\pi,t) \tag{6.52a}$$

$$\frac{\partial T(r,0,t)}{\partial\varphi} = \frac{\partial T(r,2\pi,t)}{\partial\varphi}; \frac{\partial W(r,0,t)}{\partial\varphi} = \frac{\partial W(r,2\pi,t)}{\partial\varphi} \tag{6.52b}$$

The radial boundary conditions are set at $r = 0$ for symmetry, and $r = R$ for convective fluxes. The radial symmetry boundary conditions are as follows:

$$\frac{\partial T(0, \varphi, t)}{\partial r} = \frac{\partial W(0, \varphi, t)}{\partial r} = 0 \tag{6.53}$$

The convective boundary conditions are as follows:

$$-a(W)\frac{\partial T(R, \varphi, t)}{\partial r} = h_{tr}(T_\infty - T) \tag{6.54a}$$

$$-D(T)\frac{\partial W(R, \varphi, t)}{\partial r} = h_m(W - W_\infty) \tag{6.54b}$$

For the numerical solution a radial–angular grid is constructed; if the grid is uniform, then it can be described by $r_i = i\Delta r, i \in \{0, n\}$ and $\varphi_j = j\Delta\varphi, j \in \{0, m\}$. The grid looks similar as that shown in Figure 6.24a. As specified in Hussain and Dincer (2003c) in order to assure the stability of the finite difference numerical scheme, the time step must be chosen such that the following condition is satisfied:

$$\Delta t \leq \min\left\{\frac{(\Delta r)^2 (R\Delta\varphi)^2}{2a\left[(\Delta r)^2 + (R\Delta\varphi)^2\right]}, \frac{(\Delta r)^2 (R\Delta\varphi)^2}{2D\left[(\Delta r)^2 + (R\Delta\varphi)^2\right]}\right\} \tag{6.55}$$

The explicit numerical scheme with finite differences is given in Table 6.13. Numerical analysis of foodstuff drying in 2D spherical coordinates has been pursued in Hussain and Dincer (2003c) where the simulation results were compared with experimental data. The available experimental data considered potato drying, where the potato is assimilated with a sphere with 40 mm diameter. The considered thermophysical properties and process parameters are given

**Table 6.13** Explicit time-dependent finite difference numerical scheme for 2D spherical coordinates

| Heat transfer | | Moisture transfer | |
|---|---|---|---|
| $T_{i,j,k+1} = A(1+j)T_{i+1,j,k} + [1 - 2(A + B)]T_{i,j,k} +$ $A(1-j)T_{i-1,j,k} + BT_{i,j+1,k} + BT_{i,j-1,k}$ | | $M_{i,j,k+1} = A_m(1+j)M_{i+1,j,k} + [1 - 2(A_m + B_m)]M_{i,j,k}$ $+ A_m(1-j)M_{i-1,j,k} + B_m M_{i,j+1,k} + B_m M_{i,j-1,k}$ | |
| Discretization parameters: | $A = \dfrac{a\Delta t}{(\Delta r)^2}; B = \dfrac{a\Delta t}{R^2\sin^2\varphi(\Delta\varphi)^2}$ | Discretization parameters: | $A_m = \dfrac{D\Delta t}{(\Delta r)^2}; B_m = \dfrac{D\Delta t}{R^2\sin^2\varphi(\Delta\varphi)^2}$ |
| Initial condition: $T_{i,j,0} = T_i$ | | Initial condition: $W_{i,j,0} = W_i$ | |
| Boundary conditions on $r$: | Boundary conditions on $\varphi$: | Boundary conditions on $r$: | Boundary conditions on $\varphi$: |
| $T_{0,j,k} = T_{1,j,k}$ | $T_{i,0,k} = T_{i,n,k}$ | $W_{0,j,k} = W_{1,j,k}$ | $W_{i,0,k} = W_{i,n,k}$ |
| $T_{m,j,k} = \dfrac{T_{m-1,j,k} + C_9 \times T_d}{1 + C_9}$ | $T_{i,0,k} = 0.5[T_{i,1,k} + T_{i,n-1,k}]$ | $W_{m,j,k} = \dfrac{W_{m-1,j,k} + C_{10}W_d}{1 + C_{10}}$ | $W_{i,0,k} = 0.5[W_{i,1,k} + W_{i,n-1,k}]$ |
| $C_9 = \dfrac{h_{tr}\Delta r}{k}$ | | $C_{10} = \dfrac{h_m\Delta r}{D}$ | |

**Table 6.14** Process and thermophysical parameters considered for potato drying

| Parameter | Value | Parameter | Value |
|---|---|---|---|
| $a$ (m$^2$/s) | $1.31 \times 10^{-7}$ | $T_i$ (K) | 296 |
| $h_{tr}$ (W/m$^2$ K) | 25–250 | $T_d$ (K) | 333 |
| $h_m$ (m/s) | 0.0001 | $W_i$ (kg/kg (db)) | 5.25 |

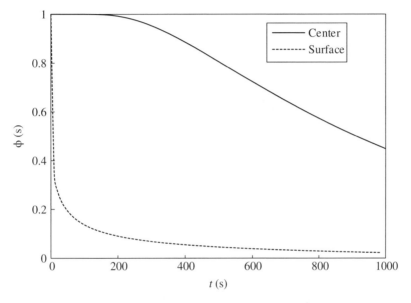

**Figure 6.25** Variation of dimensionless moisture content at center and surface during drying of a spherical potato (data from Hussain and Dincer (2003c); process description in Table 6.14)

in Table 6.14. In the numerical study, the heat transfer coefficient is varied in the range 25–250 W/m$^2$ K.

For better numerical results with reduced error the problem is solved using dimensionless formulation for the moisture content and temperature. An important result is represented by the moisture content variation at sphere center and at the surface. This can be seen in the plot shown in Figure 6.25. As observed, the moisture content at the center has a long lag time of ~150 s, during which no change can be remarked. The moisture content reduces more rapidly in the vicinity of the surface where a high moisture gradient manifests. This produces considerable diffusion rates from bulk of the substrate to the surface.

The rate of drying with time is shown in Figure 6.26. It is clear from the figure that during early heating period, the rate of drying is constant, thus exhibiting constant rate period. As the time period progresses, the rate of drying continuously decreases representing falling-rate period. The moisture content decreases both at the center and at the surface as the time period progresses. The time required for the reduced moisture content to drop its half value at the

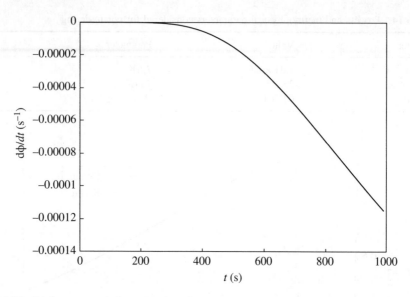

**Figure 6.26** Drying rate variation with time for a spherical potato (data from Hussain and Dincer (2003c); process description in Table 6.14)

center is about 1000 s. It is to be noted that the reduced moisture content is nonuniform in the sphere; consequently, half-time requirement varies at each location. Numerical model shows much higher agreement with the experimental values than the analytical model. The maximum difference between numerical results and experimental data is less than ±5.0% for temperature distribution and less than ±10.0% for moisture content distribution while these comparisons are higher than these values.

## 6.5   Three-Dimensional Problems

Although numerical heat and moisture transfer can be solved in 2D for majority of practical situations, there are some significant numbers of cases when a three-dimensional simulation is required. Because the pressure effect in drying applications (except for very special conditions) can be neglected, the 3D modeling of coupled heat and moisture transfer must include six equations: three for heat transfer and three for mass transfer. Cartesian coordinates are the most relevant because 3D numerical simulations are justified only for irregular geometries.

Some progress in 3D numerical analysis of drying processes is reported in Perré and May (2001) and Perré and Turner (2002) which use a finite difference scheme. These scheme is however difficult to apply for 3D irregular materials because the boundary conditions formulation becomes very complicated. In this respect, the finite volume method is more facile, and in addition, it closes energy and mass balance equations for each of the control volume used for the 3D domain discretization and the overall domain. Perré and Turner (1999) developed control volume schemes for numerical solution of Luikov coupled equation for heat and moisture transfer in 3D Cartesian coordinates. The mass and energy conservation equations are discretized for the 3D domain and solved in the discrete form such that for each control volume, the mass and energy are conserved.

Here, we expand the Luikov equations given in Eq. (6.33). The Luikov-type energy conservation (related to heat transfer process) in 3D domain can be written, regardless the kind of the coordinate system, as follows:

$$\rho C_p \frac{\partial T}{\partial t} = \mathrm{div}\left[(k + \delta \varepsilon \rho_w D \Delta h_{lv}) \nabla T + (\varepsilon \rho_w D \Delta h_{lv}) \nabla W\right] \tag{6.56}$$

where $\delta$ is the thermogradient coefficient ($\mathrm{K}^{-1}$), $\varepsilon$ is the volume of vapor per unit of total volume ($\mathrm{m^3_w/m^3}$), $D$ is the moisture diffusion coefficient, $\Delta h_{lv}$ is the latent heat of moisture evaporation, $\rho$ and $C_p$ are the density and specific heat of the moist material, "div" represents the divergence operator, and $\nabla \bullet$ is the gradient operator. The moisture transfer equation (which expresses the mass balance equation) is given as follows:

$$\rho_s \frac{\partial W}{\partial t} = \mathrm{div}\left[(\delta \varepsilon \rho_w D) \nabla T + (\varepsilon \rho_w D) \nabla W\right] \tag{6.57}$$

where $\rho_s$ is the density of the dry material.

The boundary condition for the heat transfer equation expresses the continuity of heat fluxes at the faces of the control volume. Therefore, this boundary condition is given in the units of $\mathrm{W/m^2}$ as follows:

$$-k \frac{\partial T}{\partial n} + q_n = h_{tr}(T_\infty - T) + h_m \rho_s (1 - \varepsilon) \Delta h_{lv}(W - W_\infty) \tag{6.58}$$

where $n$ represents the normal direction to the surface and $q_n$ represents the external heat flux produced by any kind of heat source and entering the surface.

The boundary condition for mass transfer equation expresses the mass continuity at the surface (no mass accumulation). This equation is expressed in units of mass flux of moisture, $\mathrm{kg/m^2\ s}$, as follows:

$$-\delta \varepsilon \rho_w D \frac{\partial T}{\partial n} - \rho_s D \frac{\partial W}{\partial n} + j_n = h_m \rho_s (W - W_\infty) \tag{6.59}$$

where the first term represents the moisture pushed outside the control volume by the temperature gradient at the boundary, the second term represents the moisture diffusion in the drying gas, $j_n$ represents a source mass flux entering the control volume in normal direction, and the term RHS represents the mass convection at the surface.

Assuming constant properties, the coefficients for the Luikov equation can be determined as given in Table 6.15. Now, using a matrix notation for more compactness, Eqs. (6.56) and (6.57) become

$$\frac{\partial}{\partial t} \begin{pmatrix} W \\ T \end{pmatrix} = \mathrm{div}\left[\begin{pmatrix} k_{11} & k_{12} \\ k_{21} & k_{22} \end{pmatrix} \nabla \begin{pmatrix} W \\ T \end{pmatrix}\right] \tag{6.60}$$

**Table 6.15**  Luikov coefficients for simultaneous heat and moisture transfer with constant thermophysical properties

| Equation | Temperature term | Moisture content term |
|---|---|---|
| Moisture transfer | $k_{11} = \dfrac{\delta \varepsilon \rho_{\mathrm{w}} D}{\rho_{\mathrm{s}}}$ | $k_{22} = \dfrac{\varepsilon \rho_{\mathrm{w}} D}{\rho_{\mathrm{s}}}$ |
| Heat transfer | $k_{21} = \dfrac{k + \delta \varepsilon \rho_{\mathrm{w}} D \Delta h_{\mathrm{lv}}}{\rho C_p}$ | $k_{22} = \dfrac{\varepsilon \rho_{\mathrm{w}} D \Delta h_{\mathrm{lv}}}{\rho C_p}$ |

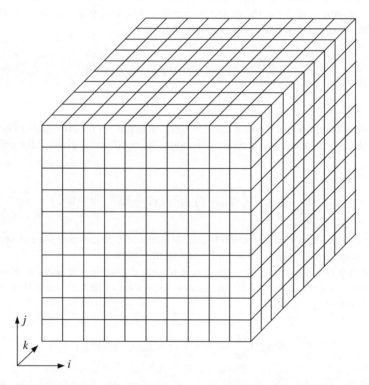

**Figure 6.27**  Discretization of a 3D domain in Cartesian coordinates for the control volume method

The volume discretization with finite volumes can be made in form of cuboidal elements with the edges $\Delta x, \Delta y, \Delta z$. The discretization elements become a mesh with nodes $(x_i, x_i, x_k)$, $i \in \{0, n\}, j \in \{0, m\}, k \in \{0, p\}$ as suggested in Figure 6.27. The volume elements do not need to be uniform. Each volume element has a volume of $v_{i,j,k} = \Delta x_i \Delta y_j \Delta z_k$, where $\Delta x_i = x_{i+1} - x_i, i \in \{0, n-1\}, \Delta y_j = y_{j+1} - y_j, j \in \{0, m-1\}, \Delta z_k = z_{k+1} - z_k, k \in \{0, p-1\}$. Thus, the total number of elements is $mnp$. The surface area of the element $(i, j, k)$ is

$\Delta S_{i,j,k} = 2\left(\Delta x_i \Delta y_j + \Delta x_i \Delta z_k + \Delta y_j \Delta z_k\right)$. Applying now the divergence theorem to Eq. (6.60) for each volume $v_{i,j,k}$, the following equations are obtained:

$$\frac{d}{dt}\begin{pmatrix} \mathcal{W}_{i,j,k} \\ \mathcal{T}_{i,j,k} \end{pmatrix} = \int_{v_{i,j,k}} \frac{\partial}{\partial t}\begin{pmatrix} W \\ T \end{pmatrix} dv = \int_{\Delta S_{i,j,k}} \left[\begin{pmatrix} k_{11} & k_{12} \\ k_{21} & k_{22} \end{pmatrix} \nabla \begin{pmatrix} W \\ T \end{pmatrix}\right] \cdot n \, dS \tag{6.61}$$

where $i \in \{0, n-1\}, j \in \{0, m-1\}, k \in \{0, p-1\}$, $n$ is the normal vector on the surface pointing inward, and $\mathcal{W}_{i,j,k}$ and $\mathcal{T}_{i,j,k}$ are functions of time representing the moisture content and temperature for each volume element.

The elements at the faces form surfaces positioned at left $(i=0)$, right $(i=n)$, front $(j=0)$, back $(j=m)$, bottom $(k=0)$, and top $(k=p)$. Therefore, the surface areas of boundary elements are $\Delta S_{0,j,k} = \Delta S_{n,j,k} = \Delta y_j \Delta z_k$, $\Delta S_{i,0,k} = \Delta S_{i,m,k} = \Delta x_i \Delta z_k$, and $\Delta S_{i,j,0} = \Delta S_{i,j,p} = \Delta x_i \Delta y_i$. In Eq. (6.61), only first-order derivative appears. The spatial derivative can be now discretized using finite difference scheme.

Assume that the discretization grid is uniform. Then, the discretization of space coordinates for a uniform mesh becomes $x_i = i \Delta x, i \in \{0, m\}, y_j = j \Delta y, j \in \{0, n\}, z_\ell = \ell \Delta z, \ell \in \{0, p\}$. The discretization scheme of Eq. (6.61) can be obtained if the spatial derivatives are replaced with finite differences. The following discretization scheme is obtained:

$$\frac{d}{dt}\begin{pmatrix} \mathcal{W}_{i,j,k} \\ \mathcal{T}_{i,j,k} \end{pmatrix} = \begin{pmatrix} k_{11} & k_{12} \\ k_{21} & k_{22} \end{pmatrix} \mathcal{A}_{i,j,k} \begin{pmatrix} \Delta y \Delta z \\ \Delta x \Delta z \\ \Delta x \Delta y \end{pmatrix} \tag{6.62}$$

where $i \in \{0, n-1\}, j \in \{0, m-1\}, k \in \{0, p-1\}$ and the $\mathcal{A}_{i,j,k}$ is the following matrix:

$$\mathcal{A}_{i,j,k} = \begin{pmatrix} \dfrac{2\mathcal{W}_{i+1,j,k} - \mathcal{W}_{i,j,k} - \mathcal{W}_{i+2,j,k}}{\Delta x} & \dfrac{2\mathcal{W}_{i,j,k} - \mathcal{W}_{i,j+1,k} - \mathcal{W}_{i,j+2,k}}{\Delta y} & \dfrac{2\mathcal{W}_{i,j,k} - \mathcal{W}_{i,j,k+1} - \mathcal{W}_{i,j,k+2}}{\Delta z} \\ \dfrac{2\mathcal{T}_{i+1,j,k} - \mathcal{T}_{i,j,k} - \mathcal{T}_{i+2,j,k}}{\Delta x} & \dfrac{2\mathcal{T}_{i,j,k} - \mathcal{T}_{i,j+1,k} - \mathcal{T}_{i,j+2,k}}{\Delta y} & \dfrac{2\mathcal{T}_{i,j,k} - \mathcal{T}_{i,j,k+1} - \mathcal{T}_{i,j,k+2}}{\Delta z} \end{pmatrix} \tag{6.63}$$

The boundary conditions at the external faces can be also discretized to introduce additional equations which close the scheme. The resulting ODE system will have $2mnp$ equations which can be solved with a selected method starting from initial values for $\mathcal{W}_{i,j,k}$ and $\mathcal{T}_{i,j,k}$.

Let us now derive the finite difference scheme for Eq. (6.60). The finite difference approach can be the preferred scheme in some situations. Note that Eq. (6.60) with constant coefficient in Cartesian coordinates becomes

$$\frac{\partial}{\partial t}\begin{pmatrix} W \\ T \end{pmatrix} = \begin{pmatrix} k_{11} & k_{12} \\ k_{21} & k_{22} \end{pmatrix} \begin{pmatrix} \dfrac{\partial^2 W}{\partial x^2} + \dfrac{\partial^2 W}{\partial y^2} + \dfrac{\partial^2 W}{\partial z^2} \\ \dfrac{\partial^2 T}{\partial x^2} + \dfrac{\partial^2 T}{\partial y^2} + \dfrac{\partial^2 T}{\partial z^2} \end{pmatrix} \tag{6.64}$$

The time variable is discretized as follows: $t_q = q\Delta t, q = 0, 1, 2, \dots$ . Then, the implicit finite difference scheme for 3D Cartesian coordinates is given as follows:

$$\frac{1}{\Delta t}\begin{pmatrix} W_{i,j,\ell}^{q+1} - W_{i,j,\ell}^{q} \\ T_{i,j,\ell}^{q+1} - T_{i,j,\ell}^{q} \end{pmatrix} = \begin{pmatrix} k_{11} & k_{12} \\ k_{21} & k_{22} \end{pmatrix} A_{i,j,k}^{q} \tag{6.65}$$

where $i \in \{1, n-1\}, j \in \{1, m-1\}, k \in \{1, p-1\}$ and

$$A_{i,j,k}^{q} = \begin{pmatrix} \dfrac{W_{i+1,j,k}^{q+1} - 2W_{i,j,k}^{q+1} + W_{i-1,j,k}^{q+1}}{(\Delta x)^2} + \dfrac{W_{i,j+1,k}^{q+1} - 2W_{i,j,k}^{q+1} + W_{i,j-1,k}^{q+1}}{(\Delta y)^2} + \dfrac{W_{i,j,k+1}^{q+1} - 2W_{i,j,k}^{q+1} + W_{i,j,k-1}^{q+1}}{(\Delta z)^2} \\[4mm] \dfrac{T_{i+1,j,k}^{q+1} - 2T_{i,j,k}^{q+1} + T_{i+1,j,k}^{q+1}}{(\Delta x)^2} + \dfrac{T_{i,j+1,k}^{q+1} - 2T_{i,j,k}^{q+1} + T_{i,j-1,k}^{q+1}}{(\Delta y)^2} + \dfrac{T_{i,j,k+1}^{q+1} - 2T_{i,j,k}^{q+1} + T_{i,j,k-1}^{q+1}}{(\Delta z)^2} \end{pmatrix} \tag{6.66}$$

The discretized boundary conditions are added to Eq. (6.65) to form a closed system of $(n+1) \times (m+1) \times (p+1)$ equations. This system can be solved for $(W_{i,j,k}^{q+1}, T_{i,j,k}^{q+1})$ at each time step $q$ starting from the initial conditions set at $q = 0$ as follows: $\left( W_{i,j,k}^{0}, T_{i,j,k}^{0} \right)$, $i \in \{0, n\}, j \in \{0, m\}, k \in \{0, p\}$.

## 6.6 Influence of the External Flow Field on Heat and Moisture Transfer

The vast majority of heat and moisture transfer studies assume constant convective heat and mass transfer coefficients in the analysis. This fact may not reflect the reality because the heat and mass transfer coefficients may vary along the material surface due to boundary layer effects of the external flow field. Kaya et al. (2006, 2007, 2008a,b) and Ozalp and Dincer (2010a,b) studied the effect of the external flow field on heat and moisture transfer coefficients along the moist object surface and their impact on moisture content and temperature distribution inside the material. This type of study must focus in six essential parts: (i) analysis of the external flow and temperature fields by a CFD package, (ii) determination of the spatial variations of the convective heat transfer coefficients, (iii) calculation of the spatial variations of the convective mass transfer coefficients using the analogy between the thermal and concentration boundary layers, (iv) computation of the temperature and moisture distributions inside the moist material using a finite difference-based implicit numerical method, and (v) repetition of the aforementioned studies by changing the aspect ratio.

The external flow field stabilizes quickly and it makes sense to assume that it is at steady state with constant thermophysical properties. The governing equations are of Navier–Stokes type. Here, these equations are given in Table 6.16 for a rectangular domain.

A proper CFD package or numerical scheme can be selected to solve the Navier–Stokes equations. Once the external flow field is solved numerically, the local convection heat transfer coefficient can be determined from the temperature field, according to the following equation:

$$-k_g \frac{\partial T}{\partial n} = h_{tr}(T - T_\infty) \tag{6.67}$$

**Table 6.16**  Governing equations for the external flow field

| Type | Equation |
| --- | --- |
| Continuity (mass conservation) | $\dfrac{\partial u}{\partial x} + \dfrac{\partial v}{\partial y} = 0$ |
| Momentum conservation | $\rho\left(u\dfrac{\partial u}{\partial x} + v\dfrac{\partial u}{\partial y}\right) = -\dfrac{\partial p}{\partial x} + \mu\left(\dfrac{\partial^2 u}{\partial x^2} + \dfrac{\partial^2 u}{\partial y^2}\right)$ |
|  | $\rho\left(u\dfrac{\partial v}{\partial x} + v\dfrac{\partial v}{\partial y}\right) = -\dfrac{\partial p}{\partial y} + \mu\left(\dfrac{\partial^2 v}{\partial x^2} + \dfrac{\partial^2 v}{\partial y^2}\right)$ |
| Energy conservation | $u\dfrac{\partial T}{\partial x} + v\dfrac{\partial T}{\partial y} = a\left(\dfrac{\partial^2 T}{\partial x^2} + \dfrac{\partial^2 T}{\partial y^2}\right)$ |

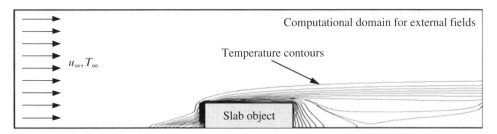

**Figure 6.28**  External flow field and computational domain around a moist slab material (adapted from Kaya et al. (2008a))

where $k_g$ is the thermal conductivity of the drying gas (air).

After the convective heat transfer is determined, the convective mass transfer coefficient can be calculated using the analogy between the thermal and concentration boundary layers according to the following equation:

$$h_m = h_{tr}\left(\frac{D_w Le^{\frac{1}{3}}}{k_g}\right) \tag{6.68}$$

where $D_w$ is the diffusivity of water vapor in drying agent (air).

When the heat and moisture transfer coefficients are provided, the problem of simultaneous heat and moisture transfer inside the moist material can be solved numerically based on the methods discussed in this chapter. For exemplification, Figure 6.28 shows the temperature contours of a slab object placed in a flow field with forced convection. From the contours profiles, it becomes visible that the temperature gradients along the surface of the object are not uniform which will result in nonuniform convective heat transfer coefficient distributions along it.

The variation of the local heat transfer coefficient at the surface of a slab object placed in a flow field is shown in Figure 6.29. The slab object is represented in 2D coordinates. The coordinate $l$ in the figure is taken along the surface. Thence, $l$ varies from 0 to $a + b + c$, where $a = c$

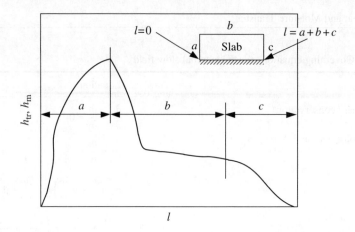

**Figure 6.29** Qualitative representation of local heat and moisture transfer coefficients for a slab object in an external flow field

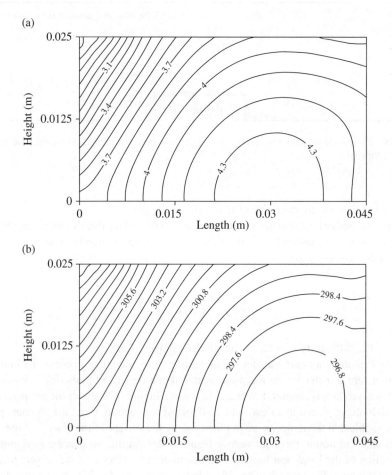

**Figure 6.30** Numerical results showing the moisture content (a) and temperature (b) contours in a kiwi fruit slice after 30 min drying in air at 50 °C with a velocity of 0.3 m/s (data from Kaya et al. (2008a))

are the lengths of the vertical edges and $b$ is the length of the horizontal edge. Now, from Eq. (6.63), it can be inferred that the mass transfer coefficient has, qualitatively, the same variation as the heat transfer coefficient. Figure 6.30 shows the moisture content (a) and temperature (b) contours obtained for a slab material during drying, when the influence of the external flow field is accounted for. The data refer to a kiwi fruit slice dried with air at 50 °C, when air velocity was 0.3 m/s and the drying time 30 min.

From the figure, it is clear that the contours are not symmetric. The drying is more pronounced in front where the moisture coefficient is higher, thence the moisture content reduced faster. The temperature is higher in region where the heat transfer coefficient is higher. This result can be compared with the contours from Figure 6.10 where constant heat and moisture transfer coefficients are considered and the therefore symmetrical contours were predicted.

## 6.7  Conclusions

This chapter introduces and discusses the main numerical methods for solving problems of heat and moisture transfer with a focus on drying processes. Then, the main numerical methods including finite difference, finite element, and control volume (or finite volume) are presented with discussions on grid generation, convergence, and stability. The numerical schemes are given for one-, two-, and three-dimensional coordinates. Further, some relevant heat and moisture transfer problems with practical drying applications and data are formulated numerically with various approaches and several examples are presented. It is shown that uniform heat and mass transfer assumption is reasonable for many practical applications. When this assumption is made, then the predicted contours of moisture content and temperature are symmetric with respect to domain center, provided that the material is subjected to uniform external conditions. However, when the external conditions are studied in detail, variable heat and moisture transfer coefficients can be considered. Consequently, the predicted moisture content and temperature distribution inside the material are more accurate and do not show symmetry with respect to the domain center.

## 6.8  Study Problems

6.1  Consider one-dimensional moisture transfer process in Cartesian coordinates. Following the steps from Example 6.1, solve the equation using the finite element method in three nodes.

6.2  Solve the one-dimensional moisture transfer equation in dimensionless format using the finite element method with convective condition at the surface for $Bi_m = 1$.

6.3  Rework the Example 6.3 for another set of data of your choice.

6.4  The Luikov equation is used to model a heat and moisture transfer process in dimensionless format. The coefficient matrix is $\begin{pmatrix} 0.02 & 1 \\ 1 & 0.02 \end{pmatrix}$ and initial conditions are $\phi(\zeta,0) = 1.2 - 0.2\zeta$ and $\theta(\zeta,0) = 1$. The boundary conditions at $\zeta = 1$ are $\phi(1,Fo) = 1$ and $\theta(1,Fo) = 2$, and at $\zeta = 0$, they are as follows: $-\begin{pmatrix} 10^{-4} & 0 \\ 0 & 1 \end{pmatrix} \dfrac{\partial}{\partial\zeta}\begin{pmatrix} \phi \\ \theta \end{pmatrix} = \begin{pmatrix} \phi(0,Fo) \\ 2 - \theta(0,Fo) \end{pmatrix}$.

Solve the problem using the finite difference method.

6.5 Consider an apple slice of square shape of $2 \times 2 \text{ cm}^2$ with the properties and process conditions given in Table 6.7. Determine the temperature and moisture content surfaces for 10, 100, and 1000 s during drying.

6.6 Rework Example 6.11 provided that $T_d = 350 \text{ K}$ and the cylindrical broccoli diameter of 10 and 30 mm length.

6.7 Potato of 50 mm diameter is dried in a process described by data given in Table 6.14. Determine the profiles and moisture content after 10, 100, and 1000 s.

# References

Abbas K.A., Ahmad M.M.H.M., Sapuan S.M., Wan M.A., Jamilah B., Dincer I. 2006. Numerical analysis of heat transfer in cooling of fish packages. *International Communications in Heat and Mass Transfer* 33:889–897.

Brio M., Zakharian A.R., Webb G.M. 2010. Numerical Time-dependent Partial Differential Equations for Scientists and Engineers. Elsevier: Burlington, MA.

Comini G., Lewis R.W. 1976. A numerical solution of two-dimensional problems involving heat and mass transfer. *International Journal of Heat and Mass Transfer* 19:1387–1392.

De Vries D.A. 1987. The theory of heat and moisture transfer in porous media revisited. *International Journal of Heat and Mass Transfer* 30:1343–1350.

Dincer I. 2010. Heat and mass transfer during food drying. In: Mathematical Modeling of Food Processing, Volume in the Contemporary Food Engineering Series, Chapter 10, Taylor and Francis/CRC Press: New York, pp. 253–300.

Faires J.D., Burden R.L. 2010. Numerical Analysis 9th ed. Brooks/Cole: Boston.

Hussain M.M., Dincer I. 2002. Numerical analysis of two-dimensional heat and mass transfer during drying of cylindrical moist solids. *The Arabian Journal for Science and Engineering* 27:137–148.

Hussain M.M., Dincer I. 2003a. Numerical simulation of two-dimensional heat and moisture transfer during drying a rectangular object. *Numerical Heat Transfer, Part A* 43:867–878.

Hussain M.M., Dincer I. 2003b. Two-dimensional heat and moisture transfer analysis of a cylindrical moist object subjected to drying: a finite-difference approach. *International Journal of Heat and Mass Transfer* 46:4033–4039.

Hussain M.M., Dincer I. 2003c. Analysis of two-dimensional heat and moisture transfer during drying of spherical objects. *International Journal of Energy Research* 27:703–713.

Kaya A., Aydin O., Dincer I. 2006. Numerical modeling of heat and mass transfer during forced convection drying of rectangular moist objects. *International Journal of Heat and Mass Transfer* 49:3094–3103.

Kaya A., Aydin O., Dincer I. 2007. Numerical modeling of forced-convection drying of cylindrical moist objects. *Numerical Heat Transfer, Part A* 51:843–854.

Kaya A., Aydin O., Dincer I. 2008a. Experimental and numerical investigation of heat and mass transfer during drying of Hayward kiwi fruits (Actinidia Deliciosa Planch). *Journal of Food Engineering* 88:323–330.

Kaya A., Aydin O., Dincer I. 2008b. Heat and mass transfer modelling of recirculating flows during air drying of moist objects for various dryer configurations. *Numerical Heat Transfer, Part A* 53:18–34.

Luikov A.V. 1934. On thermal diffusion of moisture. *Russian Journal of Applied Chemistry* 8:1354–1364.

Luikov A.V. 1975. Systems of differential equations of heat and mass transfer in capillary-porous bodies (review). *International Journal of Heat and Mass Transfer* 18:1–14.

Ozalp A.A., Dincer I. 2010a. Hydrodynamic-thermal boundary layer development and mass transfer characteristics of a circular cylinder in confined flow. *International Journal of Thermal Sciences* 49: 1799–1812.

Ozalp A.A., Dincer I. 2010b. Laminar boundary layer development around a circular cylinder: fluid flow and heat-mass transfer characteristics. *Journal of Heat Transfer* 132:121703 (1–17).

Peaceman D.W., Rachford Jr. H.H. 1955. The numerical solution of parabolic and elliptic differential equations. *Journal of Society for Industrial and Applied Mathematics* 3:28–41.

Perré P., May B.K. 2001. A numerical drying model that accounts for the coupling between transfers and solid mechanics. Case of highly deformable products. *Drying Technology* 19:1629–1643.

Perré P., Turner I.W. 1999. A 3-D version of TransPore: a comprehensive heat and mass transfer computational model for simulating the drying of porous media. *International Journal of Heat and Mass Transfer* 42:4501–4521.

Perré P., Turner I.W. 2002. A heterogeneous wood drying computational model that accounts for material property variation across growth rings. *Chemical Engineering Journal* 86:117–131.

Philip J.R., De Vries D.A. 1957. Moisture movement in porous materials under temperature gradients. *Transactions of the American Geophysics Union* 46:222–232.

Ranjan R., Irudayaraj J., Mahaffy J. 2002. Modeling simultaneous heat and mass transfer using the control-volume method. *Numerical Heat Transfer, Part B* 41:463–476.

Yilbas B.S., Hussain M.M., Dincer I. 2003. Heat and moisture diffusion in slab products due to convective boundary condition. *Heat and Mass Transfer* 39:471–476.

# 7

# Drying Parameters and Correlations

## 7.1 Introduction

The analysis and modeling of drying require a clear identification of relevant parameters which influence the process. Because the drying process has several stages, the heat and moisture transfer mechanisms can change during the process, and other phenomena can manifest. For example, in the early stages, the capillarity accounts for the moisture movement, whereas the diffusional mechanism applies latter at lower moisture content. Material shrinkage, texture change, and other processes may occur concurrently. As discussed earlier in Chapters 5 and 6, the heat and moisture transfer processes are highly interdependent. All these facts make the accurate mathematical modeling of drying very difficult and very much dependent on experimental data for validation. Simple correlations for quick, firsthand estimations of drying processes are therefore required. Besides being used in preliminary analyses which guide the design process in early stages, the correlations may be also useful for detailed mathematical modeling of drying.

There are many correlations available in the drying literature for determination of various moisture transfer parameters. These types of correlations are often required for mathematical modeling and used in the design and analysis of drying applications. The correlations can simplify the process analysis and the mathematical modeling of drying which is generally based on a set of time-dependent differential equations with boundary and initial conditions. When the modeling equations are solved, the moisture content and temperature profiles at any moment within the moist material are determined.

The evolution of the temperature and moisture content profiles in time gives the drying dynamics, whereas the average moisture content within the material and the average temperature are connected to drying kinetics. Physical–chemical properties of the material and the operational parameters (temperature, humidity ratio, pressure, velocity, moisture transfer

*Drying Phenomena: Theory and Applications*, First Edition. İbrahim Dinçer and Calin Zamfirescu.
© 2016 John Wiley & Sons, Ltd. Published 2016 by John Wiley & Sons, Ltd.

coefficient, and the type of the drying agent) control the processes at the material surface and the boundary layer. Moisture diffusivity, porosity, sinuosity, tortuosity, and other capillarity parameters influence the moisture transfer within the permeable material.

Therefore, drying correlations are essential within the larger panoply of theoretical and experimental methods for modeling and analysis of these types of energy-intensive operations. Drying is one of the most widely used unit operation in the industry. No other unit operation exists capable of handling such a large variety of materials. Development of drying correlations has become one of the prime concerns of the researchers who aim to find the means of attaining optimum process conditions for good quality drying processes, which leads to energy savings. Toward this goal, accurate determination of moisture transfer parameters for the drying operation is essential. Accurate determination of moisture transfer parameters is important in order to obtain good quality dried products leading to energy savings. Moreover, decreasing the energy consumption in the drying process will decrease environment impact in terms of pollutants and hence protect the environment.

Numerous experimental and theoretical studies have been carried out on the determination of drying profiles for various food products in correlation with the operational parameters. Jia et al. (2000) elaborated a mathematical model to determine the moisture fields within a grain kernel during drying. In Ruiz-Cabrera et al. (1997), the effective diffusivity in permeable moist materials has been correlated with the characteristics of the diffusion path. Akiyama et al. (1997) developed a model that correlates the heat and moisture transfer rates with the process of crack formation in viscoelastic materials subjected to drying. A method of identification of the receding evaporation front during foodstuff drying is proposed in Chou et al. (1997). El-Naas et al. (2000) developed a correlation for the mass transfer coefficient in a spouted bed dryer based on the experimental results.

Dincer and Hussain (2002) developed a correlation between Biot number for mass transfer and Dincer number for moisture transfer. This correlation allows ultimately for determination of the moisture transfer parameters of a moist material subjected to drying, provided that the drying gas velocity and the drying coefficient are known. A further correlation has been developed by Dincer and Hussain (2004) in the form of Biot number for mass transfer versus lag factor. Again, this correlation is useful for moisture transfer parameter determination and prediction of the drying time.

In this chapter, the most important drying parameters are introduced. Thereafter, several drying correlations which are useful for quick, firsthand estimation of drying parameters are presented.

## 7.2   Drying Parameters

### 7.2.1   Moisture Transfer Parameters

Probably, the most important drying process parameters are the moisture transfer coefficient $h_m$ (m/s) and moisture diffusivity $D$ (m²/s), both being related to the Biot number for mass transfer ($Bi_m$). These parameters were defined elsewhere in this book. The moisture diffusivity appears in Fick's law of diffusion and expresses the proportionality between the moisture flux and the moisture concentration gradient. Therefore, moisture diffusivity accounts for the moisture movement within permeable materials. According to its basic definition, the moisture diffusivity is introduced by

$$D = -\frac{J_m}{\nabla C_m} = -\frac{J_m}{\rho_s \nabla W} \tag{7.1}$$

where $J_m$ is the mass flux (kg/m$^2$ s), $\nabla C_m$ is the moisture concentration gradient (kg/m$^4$), $\nabla W$ is the moisture content gradient (kg/kg), and $\rho_s$ is the dry material density (kg/m$^3$).

In capillary-porous material, the effective moisture diffusivity is smaller than the moisture diffusivity of the same material having no pores. For the porous material, an effective moisture diffusivity related to the void fraction and tortuosity is introduced as indicated by Eq. (1.76).

In a mass transfer process at the interface between a material and a surrounding gas phase, it is important to compare the moisture concentration gradients at the two adjacent sides. This is done with the help of Biot number for mass transfer introduced previously in Eq. (1.77a). The Biot number is also related to the characteristic length for the diffusion path within the material. For a slab, typically, the characteristic length is the half width; for a cylinder or a sphere, the characteristic length is typically equal to their radius. The moisture diffusivity divided to the characteristic length has the same units (m/s) as the moisture transfer coefficient, and therefore, the ratio defining the Biot number for mass transfer is dimensionless. Therefore, one has

$$Bi_m = \frac{h_m}{D/L_c} = \frac{h_m L_c}{D} \tag{7.2}$$

The moisture transfer coefficient is defined for an interface between a material and a surrounding gas based on Eq. (1.76). The moisture transfer coefficient represents the ratio of the moisture flux exiting the material divided to the moisture concentration difference across the mass transfer boundary layer. If one denotes $C_{m,s}$ the moisture concentration in the surrounding gas (air) at the interface and $C_{m,\infty}$ the moisture concentration in the bulk gas, then the moisture transfer coefficient becomes

$$h_m = \frac{J_m}{C_{m,s} - C_{m,\infty}} = \frac{J_m}{\rho_s (W_s - W_\infty)} \tag{7.3}$$

where $W_s$ is the moisture content at the material surface and $W_\infty$ is defined through the product $C_{m,\infty} = \rho_a W_\infty$.

Besides being correlated with the moisture diffusivity $D$ through the Biot number for mass transfer, the moisture transfer coefficient can be correlated to the binary diffusivity of moisture in air (gas) $D_{ma}$ according to Sherwood number. Indeed, taking drying as an example, at the surface of the moist material, the moisture vapor which leaves the surface diffuses into the surrounding gas. The mass flux of the diffusing vapor depends on the moisture concentration gradient in the gas at the interface and the binary diffusion coefficient. According to its definition, the Sherwood number becomes

$$Sh = \frac{h_m L_c}{D_{ma}} \tag{7.4}$$

where $L_c$ represents the characteristic length, often taken differently than that for the definition of Biot number.

As known, the binary diffusion coefficient of water vapor in air is approximately 0.25 mm²/s at standard conditions and increases with the second power of temperature and with pressure according to (see Chapter 1)

$$D_{ma} \cong 2 \times 10^{-10} T^2 P \qquad (7.5)$$

with $P$ in atm, $T$ in K, and $D_{ma}$ in m²/s.

Kneule (1970) proposed an alternative correlation for the binary diffusivity coefficient of water vapor in air, which is given as follows:

$$D_{ma} = 0.3172 \frac{T^{1.81}}{P} \qquad (7.6)$$

where pressure is given in Pa and temperature in °C and the $D_{ma}$ results in m²/s.

Recall that Biot number for mass transfer is $Bi_m = h_m L_{c,Bi}/D$, where $D$ is the moisture diffusivity within the permeable material. Thence, the ratio between Sherwood and Biot numbers is $Sh/Bi_m = (D_{ma}/D)(L_{c,Sh}/L_{c,Bi})$, where $L_{c,Sh}$ stands for the characteristic dimension for Sherwood number definition and $L_{c,Bi}$ is the characteristic dimension for mass transfer Biot number definition. Sherwood number can be correlated with Reynolds number according to a mass and momentum analogy. Various correlations for Sherwood number are already given in Table 1.17.

The drying time is an important parameter in drying analysis. In addition, the duration of each of the drying period must often be determined. The time during a drying process can be conveniently expressed in the dimensionless form of Fourier number for mass transfer, namely,

$$Fo_m = \frac{Dt}{L_c} \qquad (7.7)$$

The drying time can be correlated with the dimensionless moisture content at the moist material center. Indeed, taking the first term approximation of the general analytical solution given by Eq. (5.10) for one-dimensional time-dependent moisture transfer, it results that the dimensionless moisture content at the material center is defined as

$$\Phi = A_1(\mu_1, Bi_m) \exp(-\mu_1^2 Fo_m) \qquad (7.8)$$

The factor $A_1(\mu_1, Bi_m)$ from Eq. (7.8) represents the lag factor LF (see Chapter 5) which is generally higher than one, but close to one. If in Eq. (7.8) one denotes $\mu_1^2 Fo_m = St$, where $S$ is the drying coefficient and $t$ is the time, then accounting for $Fo_m$ definition by Eq. (7.7) the following mathematical expression (also discussed in Chapter 5) is obtained from moisture diffusivity:

$$D = \frac{S L_c^2}{\mu_1^2} \qquad (7.9)$$

Now, from the definition of Biot number for mass transfer, combining Eq. (7.2) with Eq. (7.9), the following expression results for the moisture transfer coefficient:

$$h_m = Bi_m \mu_1^2 S L_c \qquad (7.10)$$

## Example 7.1

A permeable material in a form of a flat plate having thickness of 10 mm and length of 100 mm is placed on a tray. Air is blown at 2 m/s and 40 °C. Determine the moisture transfer coefficient and moisture diffusivity provided that the Biot number for mass transfer is 0.2.

- The material is in a form of plate in contact with air in forced convection; therefore, the characteristic dimension is the plate length $L_{plate} = 0.1$ m.
- For air at 40 °C and an assumed pressure of 1 atm, the kinematic viscosity is $\nu = 1.7\text{E}{-}5\,\text{m}^2/\text{s}$.
- The Reynolds number becomes

$$Re_{plate} = \frac{U L_{plate}}{\nu} = 11{,}763$$

- The binary diffusion coefficient of water vapor in air is given by Eq. (7.5): $D_{ma} = 6.26\text{E}{-}8\,\text{m}^2/\text{s}$.
- Schmidt number for air can be now calculated with Eq. (1.78): $Sc = \nu/D_{ma} = 271.5$.
- The following correlation for Sherwood number from Table 1.17 is used: $Sh = 0.664\,Re_{plate}^{0.5} Sc^{1/3} = 466$.
- The moisture transfer coefficient results from Eq. (7.4): $h_m = 292\text{E}{-}3\,\text{m/s}$.
- The moisture diffusivity results from Eq. (7.2) where the moisture transfer coefficient and the Biot number for mass transfer have known values; the equation is solved for $D$ resulting in

$$D = \frac{h_m L_{slab}}{Bi_m}$$

where $L_{slab}$ is taken as the thickness of the plate $L_{slab} = 0.01$ m because the plate bottom sitting on the tray can be considered impermeable. Therefore, one has $D = 1.46\text{E}{-}5\,\text{m}^2/\text{s}$.

## 7.2.2 Drying Time Parameters

From Eq. (7.8), it can be inferred that the dimensionless moisture content at the very beginning of the drying process can reach values higher than the unity. In fact, this first term approximation of the solution equation is not valid for the very first stage of drying. In the actual process, there is a lag time period in which the moisture content at the product center does not vary. This is of course a transient behavior; the long-term trend is a decrease of the moisture content at any location within the material.

This feature is well modeled qualitatively and quantitatively by Eq. (7.8) which reveals that a time interval is required until the moisture content at the material centre starts decreasing: $\Phi$ becomes

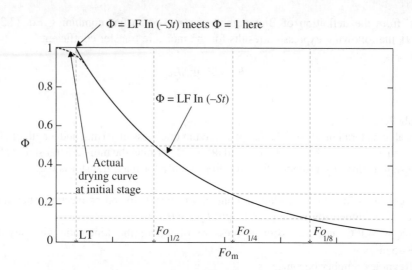

**Figure 7.1** Drying curve expressing the exponential variation of the dimensionless moisture content at material center ($\Phi$) with respect to the dimensionless time ($Fo_m$)

smaller than one. The time duration from the initial moment of drying process ($t = 0$, $Fo_m = 0$) until the moment when the dimensionless moisture content predicted by Eq. (7.8) at material center start decrease below the unity ($\Phi < 1$) is referred as the *"lag time."*

The dimensionless lag time is expressed in terms of Fourier number and simply denoted with LT. From Eq. (7.8), accounting for the fact that $A_1(\mu_1, Bi_m) = $ LF, the lag time equation results as follows:

$$LT = \frac{\ln(\text{LF})}{\mu_1^2} \tag{7.11}$$

Figure 7.1 illustrates the drying curve of a typical drying process (regardless of the geometry of the moist material). This curve is given in the form of dimensionless moisture content at the material center versus the dimensionless time. Here, the dimensionless time defined by Eq. (7.7) can be also expressed based on Eq. (7.8) from which the following equation for $Fo_m$ results:

$$Fo_m = \frac{1}{\mu_1^2} \ln\left(\frac{\text{LF}}{\Phi}\right) \tag{7.12}$$

The lag period is clearly illustrated in the figure and covers the moments between $Fo_m = 0$ and $Fo_m = $ LT. At times $Fo_m > $ LT, one can talk about the moisture content halving times. The time needed for the moisture content at material center to decrease by 50% ($\Phi = 50\%$), also introduced in Chapter 5, is referred to as the half drying time. The half drying time includes the initial lag time and its dimensionless form given as follows:

$$Fo_{1/2} = \frac{1}{\mu_1^2} \ln(2\text{LF}) \tag{7.13}$$

When the lag time is subtracted from the half drying time, the remaining corresponds to the so-called halving time. On the other hand, the consecutive halving times that follow can be expressed based on $Fo_{1/4} = \mu_1^{-2} \ln(4LF)$, $Fo_{1/8} = \mu_1^{-2} \ln(8LF)$, and so on. Therefore, the halving time of centerline moisture content, denoted here with HT, becomes

$$\text{HT} = Fo_{1/2} - \text{LT} = Fo_{1/2} - Fo_{1/4} = Fo_{1/4} - Fo_{1/8} = \frac{\ln(2)}{\mu_1^2} \tag{7.14}$$

The drying time can also be expressed in terms of the number of halving times beyond the lag time. The number of halving times is denoted with NH. Assume that the desired final moisture content at the centerline is $\Phi_f$. Then, the number of halving times can be defined as follows:

$$\text{NH} = \frac{Fo_m - \text{LT}}{\text{HT}} = -\frac{\ln(\Phi_f)}{\ln(2)} \tag{7.15}$$

The dimensionless drying time results from the summation of the lag time and the number of halving times required to reduce the moisture content at the centerline to the desired value. Mathematically, this is written as follows:

$$Fo_{m,f} = \text{LT} + (\text{NH} \times \text{HT}) \tag{7.16}$$

**Example 7.2**
The lag factor determined during drying of a cylindrical moist material is 1.15. Determine the dimensionless lag time, the halving time, and the number of halving times to reduce the dimensionless moisture content to $\Phi_f = 10\%$, provided that Biot number for mass transfer is 0.5.

- Use Eq. (5.51) to calculated the eigenvalue $\mu_1 = [0.72 \ln(6.796 \, Bi_m + 1)]^{1/1.4} = 1.047$.
- Use Eq. (7.8) to determine the dimensionless lag time LT = 0.1355.
- Use Eq. (7.14) to determine the halving time HT = 0.662.
- Use Eq. (7.15) to determine the number of halving times NH $= -\ln(\Phi_f)/\ln(2) = 3.322$.
- From Eq. (7.16), the dimensionless drying time is $Fo_{mf} = 3.349$.

## 7.3   Drying Correlations

### 7.3.1   Moisture Diffusivity Correlation with Temperature and Moisture Content

In Equation (5.79), the moisture diffusivity is correlated with the temperature according to an Arrhenius-type law. In fact, for better accuracy, the moisture content influence on the activation energy must be taken into the account. In addition, it is shown in Mourad et al. (1996) that the preexponential factor must be related to the temperature and moisture content. Moreover, in some cases, the preexponential factor is influenced by the initial moisture content. The general form of moisture diffusivity correlation is as follows:

$$D = D_0(T, W) \exp\left(-\frac{E_a(T, W)}{RT}\right) \tag{7.17}$$

Note that for the correlation of Zielinska and Markowski (2007), the activation energy can be expressed as follows:

$$E_a = E_{a0} + cTW \tag{7.18}$$

where $c$ and $E_{a0}$ are constants; the $E_{a0}$ is the activation energy for moisture movement in the dry material.

If the activation energy expressed by Eq. (7.17) is introduced into Eq. (5.79), the following equation which correlates the moisture diffusivity with the material temperature and the moisture content is obtained:

$$D = D_0(T, W) \exp\left(-\frac{E_{a0} + cTW}{\mathcal{R}T}\right) \tag{7.19}$$

where $D_0$ is the preexponential factor, $T$ is the temperature in K, $W$ is the moisture content, and $\mathcal{R}$ is the universal gas constant.

For the preexponential factor, a general form of the correlation can be derived from Mourad et al. (1996), namely, $D_{0(T,W)} = \mathcal{D}W_0 \exp[(c_1 T - c_2)W]$, where $c_{1,2}$ are positive constants. Zielinska and Markowski (2007) developed a correlation of the form given by Eq. (7.19) for the moisture diffusivity where the preexponential factor is taken as a constant. The correlation has been developed based on experimental data obtained from carrot drying. In the experiments, the carrots were cut in cubical pieces and dried in air at temperature in the range of 60–90 °C. The dryer used for this purpose has been a spouted bed dryer. It has been determined that:

- The preexponential factor is $D_0 = 6.32\text{E}{-}7\,\text{m}^2/\text{s}$.
- The activation energy for the dry material is $E_{a,0} = 2371\mathcal{R}$.
- The constant $\perp$ is $c = 0.25\mathcal{R}$.

In conclusion, the correlation of Zielinska and Markowski (2007) becomes (see Table 7.1)

$$D = 6.32 \times 10^{-7} \exp\left(-\frac{2371 + 0.25TW}{T}\right) \tag{7.20}$$

The correlation coefficient obtained for Eq. (7.19) with respect to the experimental data from Zielinska and Markowski (2007) is $r^2 = 0.929$. The experimental data covered the range of moisture diffusivities of 3.52E–9 m²/s until 6.92E–9 m²/s. The experimental moisture content has been in the range of 0–9 kg/kg.

**Table 7.1** Parameters for moisture diffusivity correlation (Eq. (7.17)) for various foodstuff subjected to drying

| Reference | Material | $D_0$ (m²/s) | $E_a$ (kJ/mol) |
|---|---|---|---|
| Pabis and Henderson (1961) | Maize | 3.493E–4 | 31.228 |
| Abid (1983) | Maize | 1.946E–6 | 25.5 |
| Cobbinah    Stage I et al. (1987)    Stage II | Maize Maize | 5.671E–10 2.464E–10 | 3.082 3.261 |
| Patil (1989) | Maize | $1.454 \times 10^{-5} W$ | 2.226 |
| Chu and Hustrulid (1968) | Maize | $4.204 \times 10^{-8} \ \exp[(0.045T - 5.485)W]$ | 20.908 |
| Mourad et al. (1996) | Corn | $1.9507 \times 10^{-7} W_0 \ \exp[(0.0183T - 2.37)W]$ | 17.875 |
| Zielinska and Markowski (2007) | Carrot | 6.32E–7 | $19.7 + 0.0021TW$ |

## Example 7.3

Let us assume that carrots cut in slab-like pieces with average thickness of 5 mm are dried on trays. Dry air at 60 °C is blown at 0.1 m/s in a perpendicular direction on the carrots' axes. Assume that the average temperature of the carrot is 10 K lower than the temperature of air. Determine the drying time provided that the initial moisture content of 8 kg/kg is reduced to 1 kg/kg.

- Because the carrot is placed on trays, the characteristic length of the carrot is the thickness $L = 0.005$ m.
- The Reynolds number can be calculated based on the characteristic length; the air kinematic viscosity at 60 °C is $\nu = 1.897 \text{E} - 5 \text{ m}^2/\text{s}$.
- The Reynolds number becomes $Re = UL/\nu = 26$.
- With Eq. (7.5), the binary diffusion coefficient of water vapor in air for 60 °C and 1 atm becomes $D_{ma} = 2.22 \text{E} - 5 \text{ m}^2/\text{s}$. Therefore, Schmidt number becomes $Sc = \nu/D_{ma} = 0.8547$.
- Using the Sherwood correlation for forced flow over a plate, given in Table 1.17, one has

$$Sh = 0.664 Re^{1/2} Sc^{1/3} = 3.118$$

- The moisture transfer coefficient results: $h_m = ShD_{ma}/L = 13.8 \text{E} - 3 \text{ m/s}$.
- The moisture diffusivity is determined with Eq. (7.19) where $T = 333 - 10 = 323$ K. Therefore, one has

$$D = 6.32 \times 10^{-7} \ \exp(-7.34 - 0.25W)$$

- During drying, the moisture content decreases from 8 to 1 kg/kg. Therefore, the moisture diffusivity varies from 0.55E–10 to 3.19E–10 m²/s. The average is $\bar{D} = 1.875 \text{E} - 10 \text{ m}^2/\text{s}$.
- Calculate the Biot number for mass transfer $Bi_m = h_m L/\bar{D} = 369,154$.
- From Eq. (5.11) for $Bi_m > 100 \rightarrow \mu_1 = \pi/2$.

- From Eq. (5.13) for $Bi_m > 100 \rightarrow LF \cong A_1 = 2/\mu_1 = 4/\pi$.
- Using $\Phi_f = 1/8$ in Eq. (7.12), the Fourier number at the end of drying becomes

$$Fo_{mf} = \frac{1}{\mu_1^2} \ln\left(\frac{LF}{\Phi_f}\right) = \frac{4}{\pi^2} \ln\left(\frac{32}{\pi}\right) = 0.941$$

- The drying time results from Fourier number $t_f = Fo_m L^2 / \bar{D} = 34.5\,h$.

## 7.3.2 Correlation for the Shrinkage Ratio

It is well known that drying of moist material is often accompanied by shrinkage. The shrinkage is mainly due to the removal of moisture which may lead to a change of the material volume. Assume that the initial volume of the moist material is $V_0$ and the volume of the dried material is $V_f$. In general, one has $V_f \leq V_0$. The shrinkage ratio parameter can be defined as the ratio of the final versus the initial volume of the material subjected to drying. Therefore, one has

$$SR = \frac{V_f}{V_0} = f\left(\frac{W_f}{W_0}\right) \tag{7.21}$$

where $W_0$ is the initial moisture content and $W_f$ is the final moisture content, respectively.

Besides the shrinking ratio, the characteristic length change during the drying is an important parameter to be estimated. The characteristic length is needed in the calculation of various dimensionless numbers related to heat and moisture transfer (Reynolds, Nusselt, Sherwood, Biot, Dincer). A simple and practical way to estimate the change of characteristic length during drying is by assuming that the material contracts or expands equally on all directions. This being the case, the length ratio of dried and undried material is

$$\frac{L_f}{L_0} = (SR)^{1/3} \tag{7.22}$$

where $L_0$ is the initial characteristic dimension and $L_f$ is the final characteristic dimension, respectively.

A linear dependence of shrinkage ratio with the moisture content ratio is often considered as follows:

$$SR = c\left(\frac{W_f}{W_0} - 1\right) + 1 \tag{7.23}$$

where $c \in (0,1)$ is a regression constant.

The form of Eq. (7.23) reflects the fact that if the moisture content is the same as the initial moisture content, then there is no shrinking ($SR = 1$). Using experimental data for carrot drying, Zielinska and Markowski (2007) developed a correlation of the form similar to that given by Eq. (7.23). The regression constant is $c \cong 0.957$ for raw carrots and $c \cong 0.925$ for blanch carrots.

Therefore, the variation of the characteristic length for carrots due to moisture loss during drying can be approximated using averaged regression coefficient as follows:

$$L_c = L_0 \left[ 0.941 \left( \frac{W_f}{W_0} - 1 \right) + 1 \right]^{1/3} \tag{7.24}$$

### 7.3.3  Biot Number–Reynolds Number Correlations

These types of correlations are constructed for forced convection drying application in which the drying agent (air) is blown forcedly over the moist product, generally placed on trays. In forced convection, the momentum and the thermal and the concentration boundary layer thicknesses depend on the fluid velocity, more specifically on Reynolds number. Recall typical correlations between the thermal boundary layer thickness and Reynolds number at a power of approximately negative 0.5. The higher the Reynolds number, the smaller the boundary layer thickness is. When the concentration boundary layer is small, the moisture transfer coefficient is high. Therefore, the moisture removal rate is high. This situation leads to a shorter drying time of a moist product.

On the other hand, the drying rate depends on the moisture diffusivity within the product. Furthermore, in an indirect manner, the drying rate depends on the thermal diffusivity within the moist product. Since Biot number correlates the diffusivity with the transfer coefficient and the transfer coefficient is correlated with the Reynolds number, it results that in a drying process, Biot number can be correlated with Reynolds number.

Development of $Bi_m$–$Re$ correlations is based on the experimental data. The moisture diffusivity and moisture transfer coefficient and thermal conductivity are determined based on the drying curve according to established methodologies. Section 5.4 discusses the methodology of moisture diffusivity and transfer coefficient determination. Once these parameters are known, the Biot number for mass transfer and the Reynolds number are determined and correlated based on regression methods.

The procedure for $Bi_m$–$Re$ correlation determination goes as follows:

- The drying curve giving the moisture content variation in time is determined based on experiment (literature experimental data can be used).
- The dimensionless moisture content is then determined and its variation in time.
- The data is regressed in the form of an exponential drying curve $\Phi = LF \exp(-St)$ as also described in Chapter 5. This will determine the lag factor LF and the drying coefficient $S$.
- The Biot number for mass transfer $(Bi_m = h_m L/D)$ can be determined based on Eq. (5.77) for the infinite slab and Eqs. (5.78a,b) for the infinite cylinder and sphere, respectively. The Biot number will depend only on the lab factor and material geometry as given in Table 7.2.
- The drying air (gas) velocity, the characteristic length, and the kinematic viscosity are used to determine the Reynolds number $Re = UL_c/\nu$.
- Sufficient experimental data is used to develop a $Bi_m$–$Re$ correlation of the form $Bi_m = aRe^{-b}$.

Once the $Bi$–$Re$ correlation is established, this can be used to predict the Biot number for heat transfer for an actual situation:

- Based on flow characteristics (kinematic viscosity and velocity) and the characteristic length, Reynolds number is determined for the considered geometry.

**Table 7.2** Moisture diffusivity correlated with lag factor and drying coefficient

| Geometry | Biot number based on lag factor | Lag factor based on Biot number |
|---|---|---|
| Infinite slab | $Bi_m = \dfrac{1 - 3.94813 \ln(\text{LF})}{5.1325 \ln(\text{LF})}$ | $\text{LF} = \exp\left(\dfrac{1}{3.94813 + 5.1325 Bi_m}\right)$ |
| Infinite cylinder | $Bi_m = \dfrac{1 - 1.974 \ln(\text{LF})}{3.3559 \ln(\text{LF})}$ | $\text{LF} = \exp\left(\dfrac{1}{1.974 + 3.3559 Bi_m}\right)$ |
| Sphere | $Bi_m = \dfrac{1 - 1.316 \ln(\text{LF})}{2.76396 \ln(\text{LF})}$ | $\text{LF} = \exp\left(\dfrac{1}{1.316 + 2.76396 Bi_m}\right)$ |

*Source*: Dincer and Dost (1996).

- Using the *Bi–Re* correlation, Biot number is determined starting from Reynolds number.
- Providing that the moisture diffusivity is known, then the moisture transfer coefficient can be determined from Eq. (5.75).
- Once Biot number for mass transfer is predicted, the lag factor of the drying curve can be determined according to the considered geometry. This is done by solving for the lag the corresponding $Bi_m$. The solution is given Table 7.2 in the form of lag factor versus Biot number correlation.
- The characteristic root results from Biot number and the moist material geometry, namely, Eq. (5.16) for infinite slab, Eq. (5.51) for infinite cylinder, and Eq. (5.60) for sphere.
- The drying coefficient is calculated from Eq. (5.74) which is solved for $S \to S = D\mu_1^2/L^2$.
- The dimensionless moisture content at material center is correlated with $\Phi = L\,F\exp(-St)$.

Dincer and Dost (1996) developed a $Bi_m$–*Re* correlation starting from experimental data for infinite slab, infinite cylinder, and spherical moist materials with the following type of foodstuff: (i) prune slices assimilated with an infinite slab, (ii) potato assimilated with an infinite cylinder, (iii) potato assimilated with a spherical material. The drying agent has been air in forced convection with order of magnitude of Reynolds number of 1000 and air temperature at approximately 50 °C. The characteristic length for test data has been in the range of 1–10 mm. The resulting correlation is as follows:

$$Bi_m = 22.55\,Re^{-0.59} \qquad (7.25)$$

where

$$Re = \frac{UL_c}{\nu}$$

The correlation from Eq. (7.25) has been found sufficiently accurate with respect to the test experimental data. For infinite slab, the average error in moisture content prediction is ±7.1%; for infinite cylinder, it is ±4.0%; and for the spherical geometry, it is ±9.8%. The correlation coefficient is 0.72. Example 7.4 clarifies the use of the correlation.

Kinematic viscosity of air (or drying gas) is required for Reynolds number calculation. For dry air, the kinematic viscosity is influenced by temperature. The kinematic viscosity can be calculated as the ratio between viscosity and air density $\nu = \mu/\rho$, both $\mu$ and $\rho$ being functions of temperature. The variation of air density and dynamic viscosity with temperature (in °C) can be estimated as proposed in Zielinska and Markowski (2007) as follows:

- For air density,

$$\rho \left(\text{kg/m}^3\right) = -3.51 \times 10^{-8} T^3 + 1.585 \times 10^{-5} T^2 - 4.699 \times 10^{-3} T + 1.292 \qquad (7.26)$$

- For air dynamic viscosity,

$$\mu \left(\text{Pas}\right) = 2.563 \times 10^{-4} \frac{T^{2/3}}{T + 114} \qquad (7.27)$$

**Example 7.4**
Prune of 10 mm thickness and with moisture diffusivity of 6.69E−8 m²/s are dried in air at 60 °C blown at 3.5 m/s. Predict the moisture transfer coefficient and the half drying time using the $Bi_m$–$Re$ correlation from Eq. (7.25).

- Assume that the prune slice can be assimilated to an infinite slab.
- For air at 60 °C, one finds the kinematic viscosity of $\nu = 1.895 \times 10^{-5} \text{m}^2/\text{s}$.
- The characteristic length of the infinite slab is $L = 5$ mm.
- The Reynolds number becomes $Re = UL/\nu = 3.5 \times 0.005/1.895 \times 10^5 = 922$.
- The Biot number for mass transfer becomes $Bi_m = 20.55 Re^{-0.59} = 0.366$.
- From Eq. (5.75), the moisture transfer coefficient results: $h_m = Bi_m D/L = 4.896 \times 10^{-6} \text{m/s}$.
- The half drying time is given by Eq. (5.44) which requires the lag factor and the drying coefficient.
- The lag factor results from Table 7.2 for the infinite slab: LF = 1.187.
- The eigenvalue results from Eq. (5.16) $\rightarrow \mu_1 = 2.01 \times 10^{-5}$.
- From Eq. (5.74), the drying coefficient becomes $\rightarrow S = D\mu_1^2/L^2 = 8.13 \times 10^{-4} \text{s}^{-1}$.
- The half drying time becomes

$$t_{0.5} = -\frac{1}{S}\ln\left(\frac{1}{2\text{LF}}\right) = 1063 \text{ s.}$$

### 7.3.4  Sherwood Number–Reynolds Number Correlations

Sherwood number – also referred to as the Nusselt number for mass transfer – is somehow similar to Biot number. This number is defined in Chapter 1 according to Eq. (1.78a) as the ratio of moisture transfer coefficient multiplied to characteristic length to the binary diffusion coefficient of moisture into the drying agent (gas). We denote here with $D_{ma}$ the binary diffusion coefficient of moisture in gas (e.g., water vapor diffusion in air).

Once Sherwood number is known, the moisture transfer coefficient can be calculated starting from the characteristic length and the binary diffusion coefficient, which, in turn, can be determined for any pair of gases (e.g., water vapor and air), the temperature, and the pressure. Furthermore, if the moisture transfer coefficient and Biot number for mass transfer are known, then the moisture diffusivity can be determined straightforwardly. In brief, by a successive application of a Sherwood–Reynolds correlation and a Biot–Reynolds correlation, all relevant drying parameters such as moisture diffusivity and moisture transfer coefficient, lag factor, drying coefficient, half drying time, and moisture content can be determined.

In general, Sherwood number is correlated with Reynolds number according to mass and momentum transfer analogy. Here, we introduce the Sherwood–Reynolds correlation of El-Naas et al. (2000) which is applicable to spouted bed dryers. Spouted beds are important in various unit operations including dryers for granular materials, pebbles, prills, pellets, coarse solids with particle size greater than 1 mm, and sludge. The drying agent is blown from underneath and penetrates a granular bed in the form of a spout as shown in Figure 7.2. There is a minimum velocity which can fluidize the particles from the spout which are pneumatically

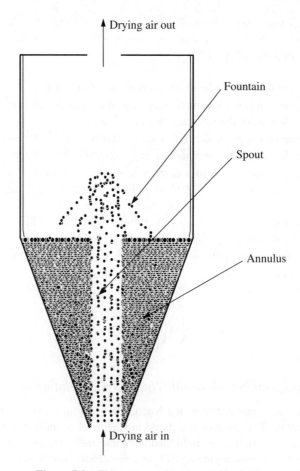

**Figure 7.2**  Flow patterns in a spouted bed dryer

conveyed upwardly and dried in the same time. While flowing in cocurrent with the drying agent, the particles subjected to drying release moisture. There is an intense moisture transfer coefficient due to forced convection effects at the particle surface.

El-Naas et al. (2000) developed their correlation based on experimental data from spouted beds operating with particles of 1.5–3.2 mm. The drying agent is air with temperature in the range of 333–363 K. The gas velocity relative to particle has been from 0.65 to 1.35 m/s corresponding to Reynolds number in the range of hundreds. For data reduction, the mass transfer equation has been integrated and averaged over the bed height. The correlation is of the type $Sh = aRe^b$, and it is given as follows:

$$Sh = 0.000258\, Re^{1.66} \tag{7.28}$$

Note that for a fluid bed or a spouted bed, Reynolds number is determined based on the relative velocity of the gas phase with respect to the solid. In addition, the void fraction $\varepsilon$ must be taken into the account. If the velocity based on a fixed reference system is denoted with $U_g$ for the air phase and $U_m$ for the moist material (solids), then the relative velocity to be used in Reynolds number determination is $U = \varepsilon(U_g - U_m)$.

The mass and momentum transfer analogy requires also the consideration of Schmidt number which is the equivalent of Prandtl number for heat and momentum analogy. Schmidt number represents the ratio of kinematic viscosity to the binary coefficient of diffusion. Often, in drying operation, the moist material can be assimilated with a sphere dried in a current of air. For this configuration, Bird et al. (2002) give the following Sherwood–Reynolds number correlation:

$$Sh = 2 + 0.6\, Sc^{1/3} Re^{1/2} \tag{7.29}$$

Here, the equivalent sphere diameter required to determine the Reynolds number is based on the volume ($V$) of sphere-like moist material: $R = \sqrt[3]{6V/\pi}$.

**Example 7.5**
Cylindrical-shaped granules for a moist material are dried in a spouted bed dryer where air at 60 °C is the drying agent. The air velocity relative to the granules is of 1 m/s, and the characteristic dimension of the granules assimilated with infinite cylinders is $R = 1.75$ mm. Determine the moisture diffusivity, moisture transfer coefficient, and half drying time.

- For air at 60 °C, one finds the kinematic viscosity of $\nu = 1.895 \times 10^{-5}\, \mathrm{m^2/s}$.
- With air velocity of 1 m/s, Reynolds number becomes $Re = UR/\nu = 92$.
- From Eq. (7.27), Sherwood number becomes $Sh = 0.471$.
- Assume air pressure of 1 atm; the binary diffusing coefficient of water vapor in air is given by Eq. (7.5) as follows: $D_{ma} = 22.5\mathrm{E}{-}6\, \mathrm{m^2/s}$.
- From Eq. (7.4), the moisture transfer coefficient results as follows: $h_m = Sh\, D_{ma}/R = 6.05\mathrm{E}{-}3\, \mathrm{m/s}$.
- We now apply $Bi$–$Re$ correlation from Eq. (7.25) $\rightarrow Bi_m = 1.424$.
- From $Bi_m$ and moisture coefficient, the moisture diffusivity results: $D = h_m R/Bi_m = 7.4\mathrm{E}{-}6\, \mathrm{m^2/s}$.

- From Table 7.2, for the infinite cylinder, the lag factor results: $LF = 1.16$.
- The eigenvalue from Eq. (5.51a) becomes $\mu_1 = 1.462$.
- From Eq. (5.74), the drying coefficient becomes

$$S = \frac{D\mu_1^2}{R^2} = 5.194\,s^{-1}.$$

- The half drying time becomes

$$t_{0.5} = -\frac{1}{S}\ln\left(\frac{1}{2\,LF}\right) = 0.162\,s.$$

### 7.3.5  Biot Number–Dincer Number Correlation

Dincer number for moisture transfer $(Di_m)$ relates the drying coefficient to the flow velocity. This dimensionless number is introduced by Eq. (5.23): $Di_m = U/SY$, where $U$ is the flow velocity, $S$ is the drying coefficient, and $Y$ is the characteristic dimension. A correlation between Biot number for mass transfer and Dincer number for moisture transfer is reported in Dincer and Hussain (2002). This type of correlation is helpful for determination of the moisture diffusivity and moisture transfer coefficient provided that the drying coefficient is known. The correlation can also be used to determine the drying coefficient, the lag factor, and the drying curve provided that the flow velocity is known.

A consistent set of experimental data from drying of foodstuff with planar, cylindrical, and spherical geometry has been employed to develop the $Bi_m$–$Di_m$ correlation. It is already known that both cooling and drying processes have similar nature due to the fact that in cooling the temperature of the product is reduced from a certain level to another and in drying the moisture content is reduced from a certain point to another. Based on this analogy, the Dincer number can be applied to a drying process, representing the influence of the flow velocity of the drying fluid on the drying coefficient of the product subjected to drying.

Furthermore, since Biot number for mass transfer is a function of both product and drying medium properties, therefore, it is connected to Dincer number for moisture transfer. The correlation developed in Dincer and Hussain (2002) is of the form $Bi_m = aDi_m^{-b}$ where $a$ and $b$ are nondimensional constants of the regression. In order to develop the correlation, experimental data is used to determine the moisture diffusivity and moisture transfer coefficient and the drying coefficient. Once those are known, the Dincer and Biot numbers for mass transfer are calculated and then correlated on a logarithmic chart representing $Bi_m$ versus $Di_m$. The correlation is expressed as follows (see Figure 7.3):

$$Bi_m = 24.848\,Di_m^{-3/8} \tag{7.30}$$

The use of $Bi_m$–$Di_m$ correlation to determine the moisture diffusivity and transfer coefficient for foodstuffs with slab, cylindrical, and spherical geometry is illustrated for three vegetables as given in Table 7.3. A typical calculation scheme for determination of moisture diffusivity and

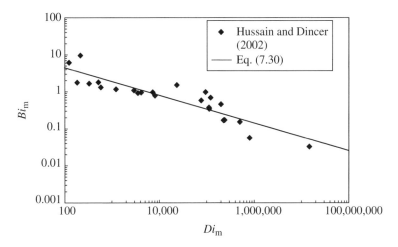

**Figure 7.3** Experimental (marks) and predicted (line) Biot number for mass transfer as a function of Dincer number for moisture transfer during various foodstuff drying

**Table 7.3** The use of $Bi_m$–$Di_m$ correlation to determine the moisture diffusivity and transfer coefficient for foodstuff

| Foodstuff | Prune | Okra | Potato |
|---|---|---|---|
| Shape | Slab | Cylinder | Sphere |
| Air temperature (°C) | 60 | 80 | 40 |
| Air velocity (m/s) | 3 | 1.2 | 1 |
| Characteristic dimension (m) | 0.0075 | 0.003 | 0.009 |
| Drying coefficient (s$^{-1}$) | 7E–5 | 1E–4 | 9E–4 |
| Lag factor | 1.00016 | 1.1981 | 1.0074 |
| Dincer number, $Di_m$ | 5,714,286 | 4,000,000 | 123,457 |
| Biot number, $Bi_m$ | 0.07 | 0.08 | 0.31 |
| Eigenvalue, $\mu_1$ | 0.1407 | 1.2593 | 0.2781 |
| Moisture diffusivity (m²/s) | 1.99E–7 | 5.67E–10 | 9.42E–7 |
| Moisture transfer coefficient (m/s) | 1.97E–6 | 1.61E–8 | 3.27E–5 |
| Half drying time (s) | 9,904 | 8,739 | 778 |

*Source*: Dincer and Hussain (2002).

moisture transfer coefficient assumes that the drying coefficient and air velocity are given as input data. Then, the calculation steps are as follows:

- Determine Dincer number from the characteristic dimension $L$, air velocity, and drying coefficient.
- Determine Biot number based on Dincer number from $Bi_m$–$Di_m$ correlation (Eq. (7.30)).
- Determine the eigenvalue $\mu_1$ as a function of $Bi_m$.
- Determine the moisture diffusivity from $\mu_1, S, L$ according to Eq. (5.74): $D = SL^2/\mu_1^2$.

- Determine the moisture transfer coefficient from Eq. (5.75): $h_m = Bi_m D/L$.
- Determine the lag factor from drying curve or using equations given in Table 7.2.
- Determine the half drying time from Eq. (5.44): $t_{0.5} = -\ln(1/2LF)/S$.

## Example 7.6

Spherically shaped potato with an average radius of 9 mm is dried in air at 40 °C, blown at 1 m/s. Use $Bi_m$–$Di_m$ and $Bi_m$–$Re$ correlation to determine the moisture diffusivity, moisture transfer coefficient, and half drying time.

- For air at 40 °C, one finds the kinematic viscosity of $\nu = 1.7 \times 10^{-5}\,\mathrm{m^2/s}$.
- With air velocity of 1 m/s and $R = 0.009$ m, Reynolds number becomes $Re = UR/\nu = 103$.
- From $Bi_m$–$Re$, Eq. (7.25), Biot number for mass transfer becomes $Bi_m = 1.335$.
- From $Bi_m$–$Di_m$ Eq. (7.30), Dincer number for moisture transfer becomes $Di_m = 2434$.
- Based on Dincer number definition, Eq. (5.23) $\rightarrow S = U/Di_m R = 0.235\,\mathrm{s^{-1}}$.
- From Eq. (5.60), the eigenvalue becomes $\mu_1 = [1.1223\ln(4.9\,Bi_m + 1)]^{1/1.4} = 1.795$.
- From Eq. (5.74) the moisture diffusivity becomes $\rightarrow D = SR^2/\mu_1^2 = 0.223\mathrm{E}{-}6\,\mathrm{m^2/s}$.
- From Eq. (5.75), the moisture transfer coefficient becomes $h_m = Bi_m D/R = 0.17\mathrm{E}{-}3\,\mathrm{m/s}$.
- From Table 7.2, for the sphere, the lag factor becomes $\rightarrow LF = 1.221$.
- The half drying time becomes

$$t_{0.5} = -\frac{1}{S}\ln\left(\frac{1}{2LF}\right) = 3.8\ \mathrm{s}.$$

## 7.3.6 Regression Correlations for $\mu_1$ Eigenvalues versus Lag Factor

For objects with regular geometry such as the infinite slab, the infinite cylinder, and the sphere subjected to drying, the eigenvalue of the time-dependent moisture diffusion equation is typically correlated with the Biot number for mass transfer. However, as given in Table 7.2, the Biot number for mass transfer is correlated with the lag factor. Therefore, the eigenvalue $\mu_1$ can be correlated with the lag factor directly.

The correlations of $\mu_1$ with the lag factor were developed in Dincer et al. (2000). As shown in Dincer et al. (2000, 2002a) and Dincer and Hussain (2002), these correlations are very useful for determination of moisture diffusivity and moisture transfer coefficient for moist materials which have a regular geometry. Table 7.4 gives the equations for the first eigenvalue for the three regular geometries.

## Example 7.7

Okra of cylindrical shape with average radius of 3 mm is dried in air at 80 °C blown with a velocity of 1.2 m/s. The drying curve is obtained by regressing the experimental data as follows: $\Phi = 1.2\ \exp(-0.0001t)$, where $\Phi$ is the dimensionless moisture content at the center. Determine the moisture diffusivity, moisture transfer coefficient, and half drying time.

**Table 7.4** Regression correlations for $\mu_1$ eigenvalues versus lag factor

| Geometry | Correlation equation |
|---|---|
| Infinite slab | $\mu_1 = -419.24LF^4 + 2013.8LF^3 - 3615.8LF^2 + 2880.3LF - 858.94$ |
| Infinite cylinder | $\mu_1 = -3.4775LF^4 + 25.285LF^3 - 68.43LF^2 + 82.468LF - 35.638$ |
| Sphere | $\mu_1 = -8.3256LF^4 + 54.842LF^3 - 134.01LF^2 + 145.83LF - 58.124$ |

*Source*: Dincer et al. (2000).

- For air at 40 °C, one finds the kinematic viscosity of $\nu = 1.7 \times 10^{-5}$ m$^2$/s.
- From the given drying curve equation, the drying coefficient results as $S = 1E-4\,s^{-1}$ and $LF = 1.2$.
- Dincer number can be calculated as follows: $Di_m = U/SR = 4E-6$.
- From $Bi_m$–$Di_m$ (Eq. (7.30)), Biot number for moisture transfer becomes $Bi_m = 0.083$.
- From Table 7.4, the eigenvalue for cylindrical shape is determined: $\mu_1 = 2.267$.
- From Eq. (5.74), the moisture diffusivity becomes $\rightarrow D = SR^2/ \rightarrow D = SR^2/\mu_1^2 = 0.175E-9\,m^2/s$.
- From Eq. (5.75), the moisture transfer coefficient becomes $h_m = Bi_m D/R = 4.85E-9\,m/s$.
- The half drying time becomes

$$t_{0.5} = -\frac{1}{S}\ln\left(\frac{1}{2\,LF}\right) = 8755\ s.$$

## 7.3.7 Biot Number–Drying Coefficient Correlation

In many practical cases, the drying coefficient and Biot number for mass transfer can be directly correlated. Because it offers a means for moisture transfer coefficient and moisture diffusivity determination, a Biot number–drying coefficient correlation is a helpful tool to the design engineers in drying industries. It allows for calculating the parameters affecting the drying process in a simple and accurate manner and to optimize the drying process. The intended correlation is of the form $Bi_m = aS^b$ where $a$ and $b$ are regression coefficients.

The correlation has been developed in the past work by Dincer et al. (2002b). The experimental data for planar, cylindrical, and spherical moist product has been processed to determine the drying coefficient, the lag factor, the moisture diffusivity, and the moisture transfer coefficient. The processed data has been correlated using typical regression methods to determine the coefficients $a$ and $b$ for the assumed power law equation for Biot number. The correlation is given as follows (see Figure 7.4):

$$Bi_m = 1.2388\,S^{1/3} \tag{7.31}$$

The experimental data used for the regression procedure covered various foodstuffs such as potato, apple, and yam dried in air with a drying coefficient of the order of $1E-3\ s^{-1}$ and $Bi_m \geq 0.1$. The obtained moisture content profiles are in good agreement with the experimental data

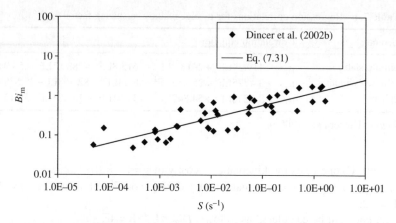

**Figure 7.4** Experimental (marks) and predicted (line) Biot number for mass transfer as a function of drying coefficient during various foodstuff drying

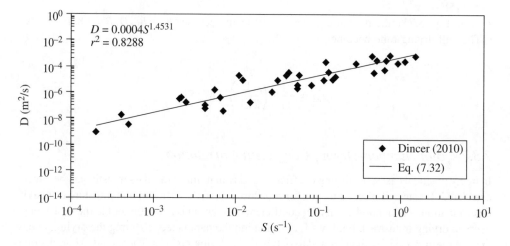

**Figure 7.5** Prediction validation of diffusion coefficient using the $D-S$ correlation

taken from the literature sources as shown in Figure 7.5. The maximum errors between predicted and measured moisture content values for the respective products are found to be ±17.32%, ±23.73%, and ±22.7% with respect to measured moisture content data, respectively.

The use of $Bi_m-S$ correlation to determine the moisture diffusivity and transfer coefficient is illustrated in the data given in Table 7.5. Here, the drying coefficient and the lag factor result from the drying curve expressed in terms of the dimensionless moisture content versus time. The correlation is used then to determine the Biot number for mass transfer starting from the known drying coefficient. Furthermore, based on the lag factor, the eigenvalue can be determined according to the equations given in Table 7.4. Subsequently, the moisture diffusivity and the moisture transfer coefficient are determined according to Eqs. (5.74) and (5.75), respectively.

**Table 7.5** The use of $Bi_m$–$S$ correlation to determine the moisture diffusivity and transfer coefficient for foodstuff

| Foodstuff | Potato | Apple | Yam |
|---|---|---|---|
| Shape | Slab | Cylinder | Sphere |
| Drying coefficient ($s^{-1}$) | 2.7E–3 | 2E–3 | 3.1E–3 |
| Lag factor | 1.262 | 1.3908 | 1.1486 |
| Biot number, $Bi_m$ | 0.181 | 0.641 | 0.1893 |
| Eigenvalue, $\mu_1$ | 0.4606 | 1.2593 | 0.8037 |
| Moisture diffusivity ($m^2/s$) | 2.15E–8 | 1.66E–6 | 2.79E–6 |
| Moisture transfer coefficient (m/s) | 1.12E–6 | 2.84E–5 | 7.60E–5 |

*Source*: Dincer et al. (2002b).

**Example 7.8**

A moist material of spherical shape with average radius of 10 mm is dried according to the drying curve given as follows: $\Phi = 1.15 \exp(-0.003t)$, where $\Phi$ is the dimensionless moisture content at the center. Determine the moisture diffusivity, moisture transfer coefficient, and half drying time.

- From the given drying curve equation, the drying coefficient results as $S = 3E-3\,s^{-1}$ and LF $= 1.15$.
- From $Bi_m$–$S$ Eq. (7.31), the Biot number for moisture transfer becomes $Bi_m = 0.179$.
- From Table 7.4 $\mu_1 = 1.199$ and from Eq. (5.74), $\rightarrow D = SR^2/\mu_1^2 = 0.21E-6\,m^2/s$.
- From Eq. (5.75), the moisture transfer coefficient becomes $h_m = Bi_m D/R = 3.73E-6\,m/s$.
- The half drying time becomes $t_{0.5} = 278$ s.

## 7.3.8   Moisture Diffusivity–Drying Coefficient Correlation

In many practical cases, the drying coefficient and Biot number for mass transfer can be directly correlated. Because it offers a means for moisture transfer coefficient and moisture diffusivity determination, a Biot number–drying coefficient correlation is a helpful tool to the design engineers in drying industries.

Such a Biot versus lag factor correlation allows for calculating the parameters affecting the drying process in a simple and accurate manner and to optimize the drying process. The intended correlation is of the form $Bi_m = aS^b$ where $a$ and $b$ are regression coefficients. The following moisture diffusivity versus drying coefficient correlation is presented in Dincer (2010):

$$D = 0.0004\, S^{1.4531} \tag{7.32}$$

where $S$ is given in $s^{-1}$ and the moisture coefficient results in $m^2/s$.

The correlation coefficient for the regression (Eq. (7.32)) is 0.82 (see Figure 7.5), and the correlating validity range is $\in \left[10^{-4}, 10^{1}\right]\,s^{-1}$. The predicted diffusivity is in the range of $D \in \left[10^{-10}, 10^{-3}\right]\,m^2/s$.

### 7.3.9  Biot Number–Lag Factor Correlation

In this section, a Biot number–lag factor ($Bi_m$–LF) correlation for drying applications is presented which is considered as a useful tool for practical drying applications. Development of this correlation is due to Dincer and Hussain (2004) which used experimental data acquired from various sources for determination of the regression coefficients. With this correlation, moisture transfer parameters such as moisture diffusivity and moisture transfer coefficient for three regular shaped objects, for example, slab, cylinder, and sphere, can be predicted. The results showed an appreciably high agreement between the measured and predicted moisture content values from the correlation.

The correlation is given by the following equation:

$$Bi_m = 0.0576\,LF^{26.7} \tag{7.33}$$

from which the correlation coefficient has been found to be 0.9181 as shown in Figure 7.6.

The following procedure is employed to determine the moisture transfer parameters and dimensionless moisture distribution in a drying process:

- The drying curve is determined based on experiments and represented as dimensionless moisture content versus time; from the drying curve of usual form $\Phi = LF\exp(-St)$, the lag factor and drying coefficient are determined.
- Based on the lag factor, the Biot number for mass transfer is calculated.
- The eigenvalue is determined based on Biot number.
- The moisture diffusivity is determined by Eq. (5.74) and the moisture transfer coefficient by Eq. (5.75).

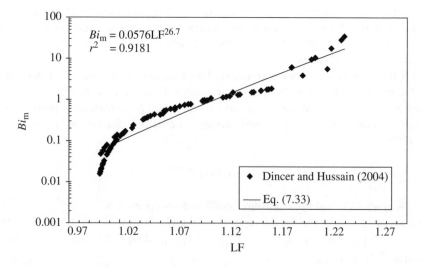

**Figure 7.6**  Prediction validation of Biot number for mass transfer using the $Bi_m$–LF correlation

### 7.3.10  Graphical Determination of Moisture Transfer Parameters in Drying

Here, a graphical method is introduced for determination of moisture transfer characteristics of a moist product during drying. The method assumes that the lag factor and drying coefficient are previously determined. A chart is constructed such as the one shown in Figure 7.7 which corresponds to a slab product. Similar charts can be constructed for other geometries. The chart consists of four plots, namely, (i) lag time versus lag factor, (ii) eigenvalue versus lag factor, (iii) eigenvalue versus Biot number, and (iv) halving time versus Biot number. The procedure for estimating the drying process parameters from Figure 7.7 is as follows:

- Starting from subplot in Figure 7.7a, LT is obtained since LF is assumed as known.
- Moving through subplots in Figure 7.7b and c, the eigenvalue and $Bi_m$ number are obtained, respectively.
- For the same value of $Bi$ number, halving time is determined from the subplot in Figure 7.6d.
- Moisture diffusivity $D$ is calculated using Eq. (5.74).
- Moisture transfer coefficient $h_m$ is calculated using Eq. (5.75).

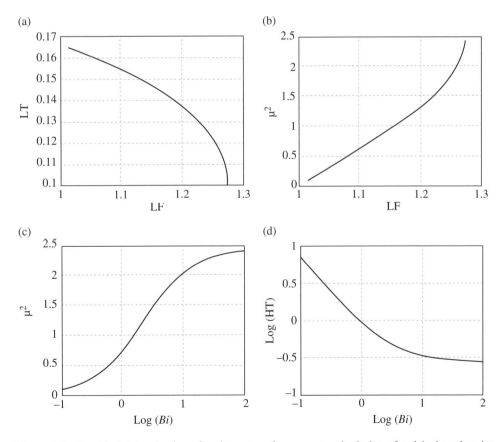

**Figure 7.7**  Graphical determination of moisture transfer parameters in drying of a slab-shaped moist material (data from Sahin and Dincer (2002))

Figure 7.7 can also be used to determine the values of lag factor and drying coefficient for various solids subject to drying provided that the moisture diffusivity $D$ and moisture transfer coefficient $h_m$ are given. Because $Bi_m$ number is available, the procedure in this case is as follows: one starts from the subplot in Figure 7.7d and proceed backward through subplots in Figure 7.7c, b, and a to determine HT, $\mu_1$, LF, and LT, respectively.

**Example 7.9**

A carrot slab of 1 cm thickness is to be dried to a centerline moisture content of 20% of the initial moisture content. The moisture diffusivity $D$ and the moisture transfer coefficient $h_m$ are given to be $5.189 \times 10^{-9}$ m²/s and 6.6084E−7 m/s, respectively. Experimentally, the drying time is found to be 16,200 s as obtained for drying profile (e.g., moisture content change with time). Let us estimate the drying time graphically.

- The log of Biot number in this case is $\log(Bi_m) = \log\left(\dfrac{h_m L}{D}\right) = \log$
$\left(\dfrac{6.6084 \times 10^{-7} \times 0.005}{5.189 \times 10^{-9}}\right) = -0.196.$
- Referring to Figure 7.7 and starting from zone (d), the following parameters are obtained as we come back through zones (c), (b), and (a) in order: $\log(\text{HT}) = 0.15$ or $\text{HT} = 1.4125$, $\mu^2 = 0.5$, $\text{LF} = 1.08$, $\text{LT} = 1.025$.
- On the other hand, $\text{NH} = -\dfrac{\ln(\Phi)}{\ln(2)} = -\dfrac{\ln(0.2)}{\ln(2)} = 2.32.$
- Therefore, $Fo_m = \text{LT} + \text{NH} \times \text{HT} = 0.125 + 2.32 \times 1.4125 = 3.402.$
- Based on $Fo_m$, the drying time is calculated as follows: $t = \dfrac{Fo_m L^2}{D} = \dfrac{3.402 \times 0.005^2}{5.189 \times 10^{-9}} = 16,390$ s.
- The drying coefficient is $S = \dfrac{\mu^2 D}{L^2} = \dfrac{0.5 \times 5.189 \times 10^{-9}}{0.005^2} = 1.038 \times 10^{-4}\,\text{s}^{-1}.$

## 7.3.11 Moisture Transfer Coefficient

Moisture transfer coefficient generally depends on the flow regime around the moist material. The hydrodynamics of the flow of drying gas around the moist material is in high connection with the type of dryer and the geometry. It depends on the geometry of the moist material and the flow conditions expressed by velocity, binary diffusion coefficient, boundary layer thickness, humidity ratio in the drying air, and so on. In this section, several dryer specific correlations for the moisture transfer coefficient are presented.

### 7.3.11.1 Rotary Drum Dryer

The rotary drum drier is the same as a cloth dryer. It consists of a rotating drum in which moist material in sheets (e.g., cloth) or particles is placed and rotated. Hot dry air is injected in the drum which removes the moisture. A mass transfer coefficient describes the moisture transfer process between the moist material surface and the drying gas. In a rotary drum dryer, the moisture movement in liquid phase is influenced simultaneously by the thermal gradients and by the centrifugal forces. Here, the correlation for Sherwood number developed by Bassily and Colver (2003) is introduced as follows:

$$Sh = 0.02888 \, m^{0.5802} \dot{N}_{d}^{0.0633} Gu^{-1.3811} Re^{0.3567} Sc^{1.2272} \tag{7.34}$$

where $m$ is the mass of material in kg, $\dot{N}_{d}$ is the rotation speed of the drum, $Gu$ is the Gukhman number (also introduced in Chapter 4), $Re$ is the Reynolds number with respect to the drum diameter, and $Sc$ is the Schmidt number.

The characteristic dimension used in Eq. (7.34) for $Sh$ and $Re$ numbers is the drum diameter $d_{c}$. The Gukhman number is a dimensionless temperature based on the dry-bulb temperature $T_{a}$ of the drying air and the wet-bulb temperature $T_{wb}$; the Gukhman number is defined as follows:

$$Gu = \frac{T_{a} - T_{wb}}{T_{a}} \tag{7.35}$$

The velocity of the drying air (gas) required for Reynolds number determination is calculated based on the volumetric flow rate of air which enters in the drying drum and the cross-sectional flow area. Here, the cross-sectional flow area is determined as follows:

$$A_{flow} = \frac{\pi d_{c}^{2}}{4} - \frac{m}{\rho L_{d}} \tag{7.36}$$

where $L_{d}$ is the length of the drum and $\rho$ is the density of the moist material. The Reynolds number becomes

$$Re = \frac{\dot{V}_{g} d_{d}}{A_{flow} \nu_{g}} \tag{7.37}$$

where $\dot{V}_{g}$ is the volumetric flow rate of the drying gas and $\nu_{g}$ is the kinematic viscosity.

The correlation is constructed using experimental data covering a rotation speed of the drum dryer of 35–50 rpm and a drum load of 2–4 kg. The correlation coefficient for Eq. (7.34) obtained when experimental data is regressed to determine the coefficients is found to be 0.9941.

### 7.3.11.2 Permeable Thin Bed with through Flow Drying

In many drying applications, permeable moist materials are in the form of thin beds or sheets. Abundant applications of permeable sheet drying are in the textile industry. In this case, air as drying agent is blown through the permeable sheet material where an intense heat and moisture transfer occur; the humidity is further carried by the air.

The flow velocity through the permeable material is correlated with the pressure gradient through the porous layer according to the Darcy law as follows:

$$U = -\frac{k}{\mu}\left(\frac{dP}{dz}\right) \tag{7.38}$$

where $\underset{\sim}{k}$ is the permeability of the porous (permeable) layer measured in m$^2$, μ is the dynamic viscosity, $P$ is the pressure, and $z$ is the spatial coordinate in the direction of the layer thickness.

Darcy's law offers the opportunity to define a Reynolds number for the capillaries. This number is based on the flow velocity through capillaries and flow kinematic viscosity. The characteristic length is normally taken to be the square root of permeability. Therefore, Reynolds number definition is

$$Re = \frac{U\underset{\sim}{k}^{0.5}}{v_g} \tag{7.39}$$

Bassily and Colver (2003) developed a Sherwood number correlation for Sherwood number for the porous material. This correlation is given as follows:

$$Sh = 0.2085 \, Re^{1.6396} Sc^{0.9872} \tag{7.40}$$

The coefficient of correlation for Eq. (7.40) is 0.9971. Once Sherwood number is determined, the moisture transfer coefficient can be calculated based on characteristic dimension and the binary diffusion coefficient.

## 7.4   Conclusions

In this chapter, the main parameters for drying process analysis and modeling are described. These parameters are categorized in two kinds: moisture transfer parameters and drying time parameters. The moisture transfer parameters are represented which are the moisture diffusivity and moisture transfer coefficients, as well as the two drying curve parameters which are the lag factor and the drying coefficient. The drying time parameters include the lag time, halving time, number of halving times, and dimensionless drying time.

The modeling and analyses are well facilitated by the development of correlations between drying parameters and other parameters characterizing the process such as dimensionless numbers: Reynolds, Biot, Sherwood, Dincer, and so on. A number of correlations are presented in the chapter, and examples are given for determination of drying parameters with the help of correlations.

## 7.5   Study Problems

7.1   A moist material in the form of a slab of 30 mm thickness and with moisture diffusivity of 1E−8 m$^2$/s is dried in air at 50 °C which is blown at 5 m/s. Predict the moisture transfer coefficient and the half drying time using the $Bi_m$–$Re$ correlation from Eq. (7.1).

7.2   Spherically shaped grapes of 5 mm diameter are dried in a spouted bed dryer where air at 40 °C is the drying agent. The air velocity relative to the granules is 2 m/s. Determine the moisture diffusivity, moisture transfer coefficient, and half drying time.

7.3   A moist material in the form of a cylinder of 50 mm diameter and with moisture diffusivity of 1E−9 m$^2$/s is dried in air at 30 °C which is blown at 0.5 m/s. Predict the moisture transfer coefficient and the half drying time using the $Bi_m$–$Re$ correlation from Eq. (7.1).

7.4   Prunes assimilated with a slab geometry with an average width of 10 mm is dried in air at 50 °C, blown at 1.2 m/s. Use $Bi_m$–$Di_m$ and $Bi_m$–$Re$ correlation to determine the moisture diffusivity, moisture transfer coefficient, and half drying time.

7.5 Potato of spherical shape with average radius of 15 mm is dried in air at 50 °C, blown with a velocity of 0.6 m/s. The drying curve is obtained by regressing the experimental data as follows: $\Phi = 1.025 \exp(-0.0005\, t)$, where $\Phi$ is the dimensionless moisture content at the center. Determine the moisture diffusivity, moisture transfer coefficient, and half drying time.

7.6 A moist material of cylindrical shape with average radius of 20 mm is dried according to the drying curve given as follows: $\Phi = 1.25 \exp(-0.0001\, t)$, where $\Phi$ is the dimensionless moisture content at the center. Determine the moisture diffusivity, moisture transfer coefficient, and half drying time.

7.7 Consider a prune slab of 1 cm thickness to be dried. Experimental observation shows that the lag factor and the drying coefficient are 1.0086 and $7{:}91389 \times 10^{-5}\ \mathrm{s}^{-1}$, respectively. It is also observed that the experimental drying time for a 10% centerline moisture content of the initial value is 28,800 s. Use the graphical method to estimate the moisture diffusivity and the moisture transfer coefficient of this prune slab.

# References

Abid M.S. 1983. Etude du séchage en couche fluidisée à flottation. Etude des mécanismes et acquisition des données. PhD Thesis. Toulouse National Polytechnic Institute, Toulouse: France, pp. 120.

Akiyama T., Liu H., Hayakawa K. 1997. Hygrostress multi-crack formation and propagation in cylindrical viscoelastic food undergoing heat and moisture transfer processes. *International Journal of Heat and Mass Transfer* 40:1601–1609.

Bassily A.M., Colver G.M. 2003. Correlation of the area-mass transfer coefficient inside the drum of a clothes dryer. *Drying Technology* 21:919–944.

Bird R.B., Steward W.E., Lightfood E.N. 2002. Transport Phenomena. John Wiley and Sons Inc.: New York.

Chou S.K., Hawlader M.N.A., Chua K.J. 1997. Identification of the receding evaporation front in convective food drying. *Drying Technology* 15:1353–1367.

Cobbinah S., Abid M., Laguerie C., Gibert H. 1987. Application of the non-linear diffusion model to the drying of cereals: a step function approximation of the variable diffusivity. In: Drying, A.S. Mujumdar Eds., Hemisphere Publishing Corporation: New York, pp. 24–28.

Dincer I. 2010. Heat and mass transfer during food drying. In: Mathematical Modelling of Food Processing (a volume in the contemporary Food Engineering series), Chapter 10, Taylor and Francis/CRC Press: New York, pp. 253–300.

Dincer I., Dost S. 1996. A modelling study for moisture diffusivities and moisture transfer coefficients in drying of solid objects. *International Journal of Energy Research* 20:531–539.

Dincer I., Hussain M.M. 2002. Development of a new Bi–Di correlation for solids drying. *International Journal of Heat and Mass Transfer* 45:3065–3069.

Dincer I., Hussain M.M. 2004. Development a new Biot number and lag factor correlation for drying applications. *International Journal of Heat and Mass Transfer* 47:653–658.

Dincer I., Hussain M.M., Yilbas B.S., Sahin A.Z. 2002. Development of a new drying correlation for practical applications. *International Journal of Energy Research* 26:245–251.

Dincer I., Sahin A.Z., Yilbas B.S., Al-Farayedhi A.A., Hussain M.M. 2000. Exergy and energy analysis of food drying systems. Progress Report 2, KFUPM Project # ME/ENERGY/203.

El-Naas M.H., Rognon S., Legros R., Mayer R.C. 2000. Hydrodynamics and mass transfer in a spouted bed dryer. *Drying Technology* 18:323–340.

Jia C.-C., Sun D.-W., Cao C.-W. 2000. Mathematical simulation of temperature and moisture fields within a grain kernel during drying. *Drying Technology* 18:1305–1325.

Kneule F. 1970. Drying. Arkady: Warsaw.

Mourad M., Hemati M., Laguerie G. 1996. A new correlation for estimation of moisture diffusivity in corn kernels from drying kinetics. *Drying Technology* 14:873–894.

Pabis S., Henderson S.M. 1961. Grain drying theory II. A critical analysis of the drying curve for shelled corn. *Journal of Agricultural Engineering Research* 6:272–277.

Ruiz-Cabrera M.A., Salgado-Cervantes M.A., Waliszewski-Kubiak K.N., Garcia-Alvarado M.A. 1997. The effect of path diffusion on the effective diffusivity in carrot slabs. *Drying Technology* 15:169–181.

Sahin A.Z., Dincer I. 2002. Graphical determination of drying process and moisture transfer parameters for solids drying. *International Journal of Heat and Mass Transfer* 45:3267–3273.

Zielinska M., Markowski M. 2007. Drying behavior of carrots dried in a spout-fluidized bed dryer. *Drying Technology* 25:261–270.

# 8

# Exergoeconomic and Exergoenvironmental Analyses of Drying Processes and Systems

## 8.1 Introduction

Regardless of the type of drying application in various sectors, ranging from food industry to petrochemical and from lumber industry to industrial, the drying commodities, such as food-stuff, wood, paper, cloth, and air, need large quantities of thermal energy in order to achieve the drying processes. The dryer must supply sufficient heat to the moist material such that the vapor pressure of the moisture is sufficiently increased to promote migration from the core to the surface. This process of heat supply is highly irreversible by its nature because it involves creating and maintaining of a sufficient temperature gradient. Furthermore, a dry gas (generally dry air) must be supplied to the process. The humidity in the supply air must be maintained low so as to increase its moisture carrying capability and to ensure a sufficiently low equilibrium moisture content within the moist material. Often, air dryers or heat pumps are used to produce a sufficiently dry air. Of course, this operation requires additional energy supply and generates more irreversibility.

Therefore, drying is a process characterized by high energy consumption and high exergy destruction. This is why drying system assessment and design must be performed using multiple types of analyses, namely, thermodynamic analysis through energy and exergy, economic analysis, and environmental impact analysis. In Chapter 4, the energy and exergy methods are applied for drying systems and processes as the most fundamental ones. These are methods of thermodynamic analysis and optimization and are used in a unified manner as expressed by the mass, energy, entropy, and exergy balance equations. A drying process or system is better when it consumes less energy and when it destroys less exergy. Furthermore, as stated before, exergy

*Drying Phenomena: Theory and Applications*, First Edition. İbrahim Dinçer and Calin Zamfirescu.
© 2016 John Wiley & Sons, Ltd. Published 2016 by John Wiley & Sons, Ltd.

analysis allows for clear identification of irreversibilities, system weak points, and pathways of performance improvement. Exergy analysis also related to the system economics and to the environmental impact of the system or its environmental benefits, whatever is the case.

Both energy consumption and exergy destruction by the systems bring an operational cost. If the design is better, the operational costs are obviously lower. However, constructing a better system requires development of a better design and the use of better and expensive materials, better components, and therefore more investment to cover a higher system cost.

In the real world, the economic factor has one of the most important influences in selecting any technical design. Ultimately, a business case must be presented in terms of investment cost and profitability, and this guides the system development. Noteworthy, the cost is a very volatile factor; therefore, any technical-economic analysis is not absolute, but subjected to regional and temporal economic constraints.

The common technical-economic analysis of thermal systems is referred to as thermoeconomic analysis which combines the method of energy analysis with economic analysis. Basically, in thermoeconomics, the ratio between energy loss and capital cost is determined and guides the path toward determining the best design. Although the thermoeconomics (energy/cost analysis) gives a valuable answer for the design and selection of processes and systems in engineering, this method is much connected to the economy fluctuation. Exergoeconomic analysis instead provides a fairer comparative assessment of competing designs because this is based on exergy, which, as known, represents the maximum potential of a system to do work with respect to a reference environment. The exergoeconomic analysis is a combination of exergy and economic analyses in which a unique cost for exergy is rationally defined.

Let us consider a drying system. The system must be engineered such that it consumes the least exergy for one unit of dried product. In fact, the system works in an environment, and the minimum exergy required to run must be equal to the reversible work for moisture separation at the reference state (temperature, pressure, composition of the reference environment). Then the reversible work is compared with the actual exergy required to drive the system, and exergy efficiency is obtained to assess the system with respect to an absolute basis. Furthermore, the exergy consumed by the system comes at a cost.

Rosen (1986) emphasizes the existence of a correlation between capital cost of devices and irreversibilities. But how do we assess the exergy cost? As pointed out by Wall (1993), both cost and exergy should reflect values. Here, some possible ways of assigning a cost value to exergy are presented. One can assume that the cost of exergy is equal to the average cost of electric power, since exergy is the value of a work. Another choice is to consider the exergetic cost of a fuel (dollar per megajoule) obtained by dividing the fuel price (dollar per kilogram) to the specific chemical exergy (megajoule per kilogram). Other choices are possible, including methods based on sectorial exergy utilization in countries or regions (see Dincer and Zamfirescu, 2011), but ultimately, a unique figure can be established in dollar per megajoule for the specific cost of exergy. More detailed exergoeconomic analysis can be pursued by accounting not only on system performance but also considering the exergetic cost required for system construction.

The detailed exergoeconomic analysis must consider a time frame during which the system operates and after which a salvage value is obtained for the system. Once the time frame is set, the total dried product can be determined and the total exergy consumed as well. Additional exergy is required for system construction, and some exergy is recovered from system salvage.

Thinking in this way, figures of merit can be adopted for system assessment for the time frame, given in terms of exergy or in terms of cost or even in terms of cost rate.

Bejan (1982) connected the second law-based thermodynamic analysis with economic analysis of thermal systems. It is seen that thermal system optimization can be effectively pursued for minimizing the thermodynamic irreversibility under economic constraints. In a further work, Bejan et al. (1996) propose the term of exergoeconomics. The concepts of utility and exergy tax become now specific to macroeconomics. The application of exergoeconomic analysis and optimization methodology to thermal systems is later detailed in Bejan (1996, 2001, 2006).

The methodology of exergoeconomic analysis with a focus on system modeling based on balance equations including mass, energy, entropy, exergy, and cost balances is presented in Dincer and Rosen (2013). Many publications on exergoeconomics have been published in recent years. Tsatsaronis (1987) identifies exergy-economic cost accounting, exergy-economic calculus analysis, exergy-economic similarity number, and product/cost efficiency diagrams as the main methodologies for thermal system exergoeconomic analyses. Further developments in exergoeconomics are reported in Szargut et al. (1988). The theory of exergetic cost is developed in Lozano and Valero (1993) to cover various aspects such as assessment of alternative energy savings within the system, cost allocation, operation optimization, single- to multiobjective optimization, and local and partial optimization of subsystems. Kotas (1995) develops interesting exergoeconomic analyses for power plants. Rosen and Dincer (2003a) investigate power plants using various fuels based on the exergoeconomic methods. Moran et al. (2011) give an illustrative example of exergoeconomic analysis for a cogeneration plant. Exergoeconomic optimization of a dual pressure combined cycle power plant with a supplementary firing unit is described in Ahmadi and Dincer (2011a). With regard to drying systems, one notes the work by Ganjehsarabi et al. (2014) which studies a heat pump tumbler dryer.

Exergy analysis offers a means to develop quantification methods for environmental impact of anthropogenic activity. This aspect is commented for the case of drying systems in Dincer (2002a) and for hydrogen energy systems in Dincer (2002b). In fact, the exergoeconomic analysis can be further completed with exergoenvironmental analysis to obtain a more complete picture about the studied system. One study as such, which tackles both the exergoeconomic and exergoenvironmental analyses of a solid oxide fuel cell plant, is presented by Meyer et al. (2009). One original method of application of exergoeconomic analysis has been developed by Rosen and Dincer (2003b) in the form of exergy–cost–energy–mass (EXCEM) which helps clarify the relationship between capital cost, exergy, and environmental impact of systems and processes.

Connelly and Koshland (2001a,b) propose a method to quantify the environmental impact of anthropic activity through exergy. Ahmadi and Dincer (2010) used a genetic algorithm for multiobjective optimization to determine the best operational parameters of a cogeneration plant when exergy efficiency and environmental impact factor are the objective functions. The analysis method is extended for gas turbine power plant in Ahmadi and Dincer (2011b), for combined cycle power plants in Ahmadi et al. (2011a), and for trigeneration systems in Ahmadi et al. (2011b). The exergoenvironmental analysis of heat recovery steam generator of combined cycle power plant is pursued in Kaviri et al. (2013). A desalination plant with reverse osmosis is analyzed in Blanco-Marigorta et al. (2014) using the exergoenvironmental method. The exergy analysis and environmental impact of a solar-driven heat pump drying system are studied in Ozcan and Dincer (2013).

In this chapter, exergoeconomic and exergoenvironmental analyses of thermal systems are introduced with a focus on dryers. The economic value of exergy is emphasized. Two representative exergoeconomic methods are detailed, namely, the EXCEM methods and the specific exergy cost (SPECO) method. Two case studies are presented for drying process analysis from combined exergetic, economic, and environmental standpoints.

## 8.2   The Economic Value of Exergy

In this section, we dwell into an exergy-based theory of value which may form the foundation for exergoeconomic analysis. In economics, the theory of value attempts to explain the correlation between value and price of traded goods and services. In today's economy, the trades are normally made by paying a price in money as a standardized currency for payments of goods, services, and debts. The price must reflect the value of the trade which is related to costs and profitability. The theory of value offers an ideological basis for quantification of the benefit from a traded good or service. This helps assign a price to a value.

Three theories of value received much attention in the last century. The first is the "power theory of value" which states that political power and the economy (which is constrained by laws of trade) are so highly interlaced that prices are established based on an internal hierarchy of values of the society rather than on a production and demand balance. The second is the "labor theory of value" which states that the value is determined by the labor developed to produce the good or the service including the labor spent to accumulate any required capital for the production process.

The third is the "utility theory of value" which quantifies the value of tradable goods and services based on their utility. There is no direct way of measuring the utility as a representation of the preferences for trading various services or goods because this depends on subjective factors of human individuals such as wishes and wants. However, the utility can be observed indirectly through the price that is established by trading activity. The price is determined by the balance between marginal utility and marginal cost. Here, the term "marginal" means an infinitesimal change. Let us assume that a quantity $Q$ of products is traded; if one denotes $U$ a quantified utility of the product, then the utility marginality is the derivative $dU/dQ$; also, if the production cost is $C$, then the marginal cost would be $dC/dQ$. The price according to the utility theory of value is spontaneously established such that

$$\frac{dU}{dQ} = \frac{dC}{dQ} \tag{8.1}$$

which means that the marginal utility must be equal to the marginal cost.

The marginal utility theory explains, for example, why gold price is much higher than that of water. The marginality of gold cost (the change in cost for an infinitesimal change of quantity) is obviously much higher than the marginality of water cost. The utility of water is much higher than the utility of gold, but the scarcity of gold is high, whereas the availability of water is high. This makes the marginal utility of water much smaller than the marginal utility of gold, and therefore, the marginalities of the costs of water and gold must be on the same relationship.

As Georgescu-Roegen (1986) points out, if a theory quantifies value through a conservable quantity, then it falls in fallacy. For example, if when applying the theory of labor one assumes that any type of labor can be valued through the amount of mechanical work (or energy) deployed to do it, then this leads to misappropriations (which are in fact common) because energy does conserve and value does not. Value degrades or grows. Another mistake is to assume that the utility is expressed in terms of mass (amounts, quantities) of a precious metal (gold). This is a fallacy because mass is conserved (we do not consider here any nuclear reactions) and therefore it cannot quantify value.

The true values that humans and also any other living organisms appreciate are the sources of low entropy or high exergy. These quantities do not conserve. When used, a low-entropy source is converted in a high-entropy waste by living organism, which in the meantime performs their activity. Equivalently, one says that the high-exergy sources are degraded by earth systems (living species, natural cycles, etc.): exergy degrades from the source to waste. The net exergy absorbed by the earth, consequently, is gradually destroyed, but during this destruction, it manages to drive the earth's water, wind, and other natural systems, as well as life on earth. Because it does not conserve, a source of low entropy can be used only once and never reused. The same stands for exergy: once destroyed, it cannot be reused. Exergy is also related to the surrounding environment as it accounts for its temperature, pressure, and species concentration. Therefore, due to these attributes, exergy can be used in establishing a theory of value. In fact, exergy represents the part of energy which is useful to society, and therefore, it has economic value.

Furthermore, once the economic value of exergy is expressed in terms of currency, then it can effectively be used for exergoeconomic and exergoenvironmental analyses. Various methods can be approached to price exergy for analysis purpose. Recent reviews on this topic are presented in Rosen (2008) and Dincer and Rosen (2013). It is important to determine sound methods to set the prices and the costs in relation to exergy content. This in fact requires formulation of a theory of costs based on exergy. It has been suggested that when analyzing a thermal system is reasonable to distribute costs in relation to outputs and accumulations of exergy. With regard to the prices of physical resources (fuels, materials), these also must be set in a tight relation with the resource exergy content such that to foster resource saving and effective technology.

Let us do a simple attempt to quantify exergy in terms of monetary currency. In doing this, one considers the main fuels in a society. The energy content is taken as the lower heating value (LHV) of the fuel, and the exergy content is the chemical exergy of each fuel. Table 8.1 gives the specific energy and exergy content of fuels considered in this brief analysis. The price per unit of mass of each fuel is also given. When the price is divided to chemical exergy, then the exergy specific price $C_{cx}$ is obtained. Table 8.1 is constructed for Canada; however, the methodology presented subsequently is general.

In our approach, a country or region must be considered first. Then, the primary energy sources are inventoried. For Canada, the following primary energy sources can be considered: coals, refined natural gas, natural gas liquids, crude oil derivate, hydro, nuclear, and biomass derivate (here, wind and solar are neglected as not highly represented). Further, the method for exergy price estimation is described as follows:

- Cost of each fuel type is obtained from the market and expressed in dollar per kilogram. For the case of hydro and nuclear, this step is skipped.
- Based on the available statistics, the consumed energy fraction (CEF) for each type of fuel is determined. In Table 8.1, the CEF is obtained from the previous work by Dincer and Zamfirescu (2011, Chapter 17). The CEF represents the fraction of specified primary energy source from the total energy consumed from primary sources.

**Table 8.1** Calculation table for exergy price based on primary energy sources for Canada (data for year 2008)

| Fuel | | Data input | | | | | | | Calculated item | | | | |
|---|---|---|---|---|---|---|---|---|---|---|---|---|---|
| | | $\overline{\text{LHV}}$ (MJ/kg) | $ex^{ch}$ (MJ/kg) | $C_f$ ($/kg) | CEF (%) | $\overline{\text{LHV}}$ (MJ/kg) | $\overline{ex}^{ch}$ (MJ/kg) | $\overline{C}_f$ ($/kg) | $\gamma$ | $C_{ex}$ ($/GJ) | CEF×$\gamma$ | CExF (%) | CExF×$C_{ex}$ ($/GJ) |
| Coals | | 24.0 | 15.0 | 0.15 | 8.3 | 24.0 | 15.0 | 0.15 | 0.62 | 10.0 | 0.052 | 5.1 | 0.51 |
| Refined natural gas | | 50.7 | 52.4 | 0.18 | 30.8 | 50.7 | 52.4 | 0.18 | 1.03 | 3.43 | 0.318 | 31.3 | 1.07 |
| Natural gas liquids | LPG | 46.0 | 54.9 | 1.43 | 2.9 | 31.6 | 35.6 | 0.93 | 1.13 | 26.1 | 0.033 | 3.3 | 0.85 |
| | Methanol | 19.9 | 22.4 | 0.47 | | | | | | | | | |
| | Ethanol | 28.8 | 29.5 | 0.88 | | | | | | | | | |
| Crude oil derivate | Gasoline | 43.5 | 47.7 | 1.74 | 45.1 | 42.4 | 45.5 | 1.37 | 1.07 | 30.1 | 0.484 | 47.7 | 14.3 |
| | Diesel | 42.8 | 44.2 | 1.78 | | | | | | | | | |
| | Kerosene | 43.1 | 49.1 | 0.94 | | | | | | | | | |
| | Fuel oil | 40.1 | 41.1 | 1.03 | | | | | | | | | |
| Hydro | | N/A | N/A | N/A | 7.4 | N/A | N/A | N/A | 1 | 34.7 | 0.074 | 7.3 | 0.16 |
| Nuclear | | N/A | N/A | N/A | 5.0 | N/A | N/A | N/A | 1 | 89.6 | 0.050 | 4.9 | 6.6 |
| Biomass derivate | Whole tree | 19.7 | 22.1 | 0.2 | 0.5 | 15.6 | 17.8 | 0.05 | 1.14 | 2.8 | 0.005 | 0.4 | 0.14 |
| | Wood pellets | 14.6 | 18.5 | 0.2 | | | | | | | | | |
| | Wood chips | 10.0 | 11.0 | 0.2 | | | | | | | | | |
| | Pinewood | 18.9 | 24.8 | 0.5 | | | | | | | | | |
| | Sawdust | 8.0 | 8.5 | 0.2 | | | | | | | | | |
| | Straw | 14.5 | 16.5 | 0.3 | | | | | | | | | |
| | Rice straw | 14.1 | 15.9 | 0.8 | | | | | | | | | |
| | Waste paper | 17.7 | 20.1 | 0.1 | | | | | | | | | |
| | Biogas | 22.5 | 23.2 | 2.0 | | | | | | | | | |

Equations used:

$$C_{ex} = \frac{\overline{C}_f}{\overline{ex}^{ch}}$$

$$\gamma = \frac{\overline{ex}^{ch}}{\overline{\text{LHV}}}$$

$$\text{CExF} = \frac{\text{CEF}\gamma}{\sum(\text{CEF}\gamma)}$$

Average price of exergy: 8.4¢/kW h

Average price of electricity: 11.0¢/kW h

N/A, not applicable.

- The LHV and specific chemical exergy for categories of fuels are averaged. For example, the mean LHV for natural gas liquids is an average of the LHVs for LPG, methanol, and ethanol.
- The specific price of fuel is averaged for each fuel category. For example, the mean fuel price $\bar{C}_f$ for fuels obtained from crude oil (crude oil derivate) is the average of prices of gasoline, diesel, kerosene, and fuel oil.
- The quality factor for each category of fuel is determined as indicated in the table, that is, by the ratio between average specific chemical exergy and LHV. The quality factor for hydropower and nuclear energy is one.
- The exergetic price $C_{ex}$ for each fuel category is determined as shown in the table by the ratio between the average fuel cost and specific chemical exergy. The exergetic price of hydropower results from the specific price of electric power divided by 0.8, as it is fair to assume that the exergy efficiency of the hydro power plant is 80%. The exergetic price of nuclear energy is determined from the price of electric power divided to 0.31 based on the fact that, as shown in Dincer and Zamfirescu (2014), the exergy efficiency of CANDU power plants is 31.3%, in average. The cost of electric power in Canada is taken at an average of 11¢/kW h.
- The consumed exergy factor for each fuel category is calculated with $CExF_i = CEF_i\gamma_i / \sum(CEF_i\gamma_i)$, where $i$ is an index representing each type of the fuel. These represent the weighting factors.
- The average price of exergy results as a weighted average $C_{ex} = \sum CExF_i \times C_{ex,i}$.
- The results show that the exergy price for Canada, based on the exergy of the primary resources, is 8.4¢/kW h. This compares well with the average electricity price of 11¢/kW h.

Once a price of exergy is determined, further models can be created to establish a costing scheme for other items of interest in an exergoeconomic analysis. Nonenergetic cost such as labor, material supply, environment remediation expenditure, incidental expenditures, and so on can be priced using exergy content as a basis for cost accounting. The economic value of system outputs can be also allocated based on exergy.

## 8.3 EXCEM Method

Among other methods of exergoeconomic analysis, the so-called EXCEM method is distinguished by the fact that it represents an extension of the typical exergy analysis which is based on four balance equations: mass, energy, entropy, and exergy balance. In EXCEM, there is one more balance equation to be written, namely, the cost balance equation. Note that in EXCEM the balance equations are written for quantities that do not conserve such as entropy, exergy, and costs. For such quantities, the generation and consumption terms must be considered in order to obtain a balance equation. The EXCEM method originated from the work of Rosen (1986) and has been further developed in the subsequent works by Rosen and Dincer (2003a,b) and Dincer and Rosen (2013).

The basic rationale underlying an EXCEM analysis is that an understanding of the performance of a system requires an examination of the flows of each of the quantities represented by EXCEM into, out of, and at all points within a system. Figure 8.1 illustrates the application of the EXCEM method. The equations for thermodynamic balances, which are part of the EXCEM method, are also discussed in Chapter 4, where exergetic analysis of the drying

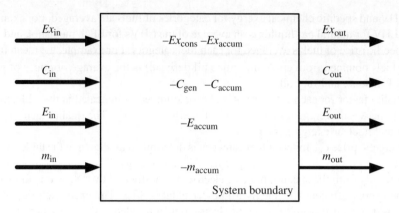

**Figure 8.1**   System modeling sketch to the application of the EXCEM method

systems is discussed. For completeness, we give the general form of the thermodynamic balances here, in terms of mass, energy, entropy, and exergy balance, respectively, as follows:

$$\sum \dot{m}_{in} - \sum \dot{m}_{out} = \frac{dm}{dt} \tag{8.2a}$$

$$\sum \dot{E}_{in} - \sum \dot{E}_{out} = \frac{dE}{dt} \tag{8.2b}$$

$$\sum \dot{S}_{in} - \sum \dot{S}_{out} + \dot{S}_{gen} = \frac{dS}{dt} \tag{8.2c}$$

$$\sum \dot{E}x_{in} - \sum \dot{E}x_{out} - \sum \dot{E}x_{d} = \frac{d\dot{E}x}{dt} \tag{8.2d}$$

Note that in Eq. (8.2d) the rate of exergy destruction $Ex_d$ represents a consumption of exergy due to internal irreversibilities. The exergy output includes the exergy in products and the exergy lost at system interaction with the surroundings: $\sum \dot{E}x_{out} = \sum \dot{E}x_{prod} + \dot{E}x_{lost}$. It is remarked from Eqs. (8.2) that the general form of a balance equation in rate format is as follows:

Input rate + generation rate − output rate − consumption rate = accumulation        (8.3)

The balance equations are sometimes written in terms of amounts as follows:

Input amount + generated amount − output amount − consumed amount = amount accumulated

(8.4)

The cost balance is essentially different than the thermodynamic balances, although it has the same general form as given by Eqs. (8.3) and (8.4). In cost balance equation, only the cost input and cost generation are defined, whereas in thermodynamic balances, all terms are defined with the help of scientific relationships. For cost accumulation and cost outputs, there is no strict formulation; rather, these types of costs are subjectively allocated, depending on the type and purpose of the system and other economic considerations. For example, costs may be distributed proportionally to all outputs and accumulations of a quantity (such as mass, energy, or exergy) or all nonwaste outputs and accumulations of a quantity. The cost balance equation in rate and amount form is written, respectively, as follows:

$$\dot{C}_{in} + \dot{C}_{gen} - \dot{C}_{out} = \frac{dC}{dt} \tag{8.5a}$$

$$C_{in} + C_{gen} - C_{out} = \Delta C \tag{8.5b}$$

The cost inputs are generally defined by the knowledge of the prices for each input and the rate of supply. For example, the fuel rate supplied to the system is generally known; also known is the specific cost of the fuel. Then, determination of the cost rate due to fuel input is just obtained by multiplication of the rate of fuel consumption and the specific cost of fuel. Cost generation corresponds to the appropriate capital and other costs associated with the creation and maintenance of a system. That is,

$$C_{gen} = C_{cap} + C_{o\&m} + C_{occ} \tag{8.6}$$

where index gen refers to the capital costs, o&m refers to operation and maintenance cost, and occ refers to other cost creation due to various reasons.

The capital costs are often the most significant component of the total cost generation. Hence, the consideration of only capital costs closely approximates the results when cost generation is considered. The other reasons for cost creations may be, depending on the case, interest, insurance, and so on. The "cost generation rate" term in a differential cost balance represents the total cost generation levelized over the operating life of the system. The cost term denoting the "amount of cost generated" in an integral cost balance represents the portion of the total cost generation accounted for in the time interval under consideration. Cost generation components other than capital costs often are proportional to capital costs. Hence, the trends described are in qualitative agreement with those identified when the entire cost generation term is considered.

The system assessment in EXCEM is made based on indicators expressing the thermodynamic loss versus the capital cost. The energetic loss ratio is defined therefore as follows:

$$R_{en} = \frac{E_{loss}}{C_{cap}} \tag{8.7}$$

where $E_{loss} = \sum E_{in} - \sum E_{prod} - \Delta E_{accum}$ and $\sum E_{prod}$ represents the energy retrieved in the form of products; this is a part of the energy output: $\sum E_{out} = \sum E_{prod} + E_{loss}$.

Equation (8.7) is written in general in terms of amounts, where $E_{loss}$ represents the total energy lost during the system lifetime and $C_{cap}$ is the present worth of capital discounted for the entire lifetime. Equation (8.7) can be also given in terms of rates of lost energy and rate of capital cost. The exergetic loss ratio is defined as follows:

$$R_{ex} = \frac{Ex_{loss}}{C_{cap}} \tag{8.8}$$

where $Ex_{loss} = \sum Ex_{in} - \sum Ex_{prod} - \sum Ex_d - \Delta Ex_{accum}$ represents the exergy wasted into the surroundings.

Equation (8.8) is written in general in terms of amounts, where $Ex_{loss}$ represents the total energy lost during the system lifetime. If analysis requires, Eq. (8.8) can be given in terms of rates of lost exergy divided to the rate of capital cost. In this case, the capital discounted capital in present worth is levelized for the entire lifetime.

Although there may be some variations in the economic engineering analysis of the system, here we describe a simple approach which leads to determination of the discounted capital in present worth and does not consider loans, but only invested capital. The economic analysis requires first to establish a set of parameters as given in Table 8.2.

The cost generation and cost outputs can be determined based on a set of economic calculations for the cash flow, as given in Table 8.3. Note that the accumulated cost represents a generated capital. Assume that an equity capital $C_{cap}$ is invested at the beginning of the system lifetime. Due to the generated capital at the end of the system lifetime, the total available capital is increased to $GC + C_{cap}$, where GC is the generated capital. Therefore, a capital productivity can be defined as shown in the table as the ratio of total available capital at the end of the lifetime to the invested capital at the beginning of the lifetime.

**Table 8.2** Parameters required for an economic analysis

| Parameter | Symbol | Description |
|---|---|---|
| System lifetime | $n$ | Number of years (or discount periods) of the system under investigation |
| Real discount rate | $r$ | Rate of return for the discounted cash flow to determine value |
| Inflation rate | $i$ | This is a subunitary factor accounting for inflation rate |
| Tax on credit due to renewable energy | $t_c$ | When renewable energy is used, a tax on credit may apply, $t_c \in [0,1)$ |
| Tax on income | $t_i$ | Tax applied on any income of money coming from output costs, $t_i \in [0,1)$ |
| Tax on property | $t_p$ | Tax applied to the property value, $t_p \in [0,1)$ |
| Tax on salvage | $t_s$ | Tax applied on the salvage value of the system, $t_s \in [0,1)$ |
| Capital salvage factor | CSF | At the end of lifetime, a fraction CSF of the invested capital is recovered |
| Levelized product price | LPP | The price in dollar with which the unit of product is sold (in present worth) |
| Annual production | AP | Number of units of product for 1 year (or one discount period) |
| Levelized cost of consumables | LCC | The sum of all input costs (fuels, supplies) levelized for present worth |
| Annual consumption | AC | The annual consumption of consumables |
| Operation and maintenance cost factor | OMCF | A subunitary fraction of either the net income or the invested capital |
| Invested capital | $C_{cap}$ | A monetary value (in dollar) paid upfront (in equity) to initiate the business |

**Table 8.3**  Equations for economic analysis

| Parameter | Equation | Description |
|-----------|----------|-------------|
| Market discount rate | $r_m = (r+1)(i+1)-1$ | The discount rate taking into account the inflation rate |
| Present value factor | $PVF = (1+r_m)^{-N}$ | A factor used to determine the present value of cash flow |
| Capital recovery factor | $CRF = \dfrac{r_m}{1-PVF}$ | The ratio of a constant annuity to the present value of receiving that annuity for a given number of discount periods |
| Present worth factor | $PWF = CRF^{-1}$ | A factor used to determine the present value of a series of values |
| Present worth income ($) | $PWI = LPP\ AP\ PWF$ | The income from a business expressed in the present worth of money. This is equal to cost output from product selling |
| Present worth costs ($) | $C_{in} = LCC\ AC\ PWF$ | This is a cost input due to the cost of fuels and supplies |
| Cost of o&m | $C_{o\&m} = PWI\ OMCF$ | Here, the o&m cost are taken as a fraction of income from product sales |
| Net income ($) | $NI = PWI - C_{o\&m} - C_{in}$ | Income from a business expressed in present worth |
| Tax credit deduction ($) | $TCD = t_c\ C_{cap}$ | Tax deduction due to investment in renewable energy |
| Taxable income ($) | $TI = NI - TCD$ | The amount of income subjected to taxation |
| Tax on income ($) | $TOI = t_I\ TI$ | The tax applied to the income |
| Tax on property ($) | $TOP = t_p\ C_{cap}(1-t_i)$ | Tax applied on the property |
| Other cost creation | $C_{occ} = TOI + TOP$ | The cost creation to support taxation on income and property |
| Generated costs ($) | $C_{gen} = C_{cap} + C_{o\&m} + C_{occ}$ | This is the total costs generated by the system during lifetime |
| Salvage value ($) | $SV = CSF\ C_{cap}\ PVF(1-t_s)$ | Salvage value expressed in present worth |
| Generated capital (M$) | $GC = NI + SV - C_{cap} - TOI - TOP$ | Generated capital is equal to the accumulated cost within the system, according to the cost balance equation (8.5b) |
| Capital productivity | $CP = (C_{cap} + GC)/C_{cap}$ | This factor compares the amount of available cash at the end of the lifetime $(C_{cap} + GC)$ to the invested equity capital |

The capital productivity factor as well as the exergetic loss ratio or energetic loss ratio may help select a design option of the (drying) system among many possibilities. In general, a higher invested capital may lead to better systems with less exergetic and energetic losses. However, the loss ratio can sometimes show a maximum.

If a maximum of loss ratio exits, then this is an indication for the best system selection or the optimum capital investment. In addition, the capital productivity factor may be affected by the invested capital. In some conditions a maximum capital productivity can be observed for a given investment cost. However, the system design showing the maximum capital productivity is generally different than the design showing the lowest exergetic loss ratio and different than the system showing the lowest energetic loss ratio. These facts may lead to trade-off selection problems for the best system design. Here, the application of the EXCEM method is illustrated for drying systems with the help of Example 8.1.

## Example 8.1

A small-scale solar dryer that is used to dry sultana grapes on trays is set for a time frame of 15 years. The system is described as shown in Figure 8.2. It includes photovoltaic thermal (PV/T) arrays, an air blower, and a drying tunnel, where the products are placed on trays. The tunnel operates discontinuously. The system operates only during daytime for a period of 3 months/year during the harvesting season. Determine the energetic and exergetic loss ratios and the capital productivity.

- Problem data and assumptions
  - The initial moisture content is 2.28 kg/kg which is reduced to 0.16 kg/kg.
  - The lag factor is $LF = 1.24$ and the drying coefficient is $S = 1E-5\,s^{-1}$; the average characteristic dimension for grapes assimilated as cylinders is $R = 5\,mm$.
  - The dimensionless moisture content must be reduced to $\Phi_f = 7\%$.
  - The load of grapes is $16\,kg/m^2$ of tray.
  - The total tray area is $10\,m^2$; five trays of $2\,m^2$ are used.
  - The heat loss factor is 10% from the heat required to evaporate the moisture.
  - The temperature drop of air in the dryer is $\Delta T_{air} = 5\,°C$.
  - The width of the trays is 1 m.
  - Air temperature at inlet is $T_0 = 25\,°C$ with relative humidity of 50%.

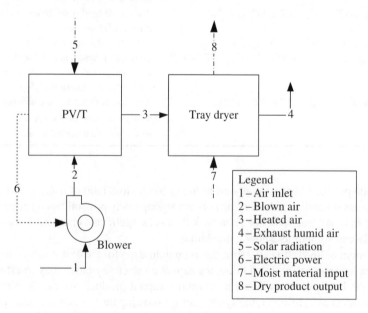

**Figure 8.2**   Solar dryer system for Example 8.1

- ○ Specific heat of dry grapes is 2000 J/kg K.
- ○ The average solar radiation intensity is 500 W/m$^2$; PV conversion efficiency is 15%.
- Determination of the drying time
  - ○ The drying time is determined from $\Phi_f = LF \ln(-St_f) \rightarrow t_f = -S^{-1} \exp(\Phi_f/LF) = 78.84$h.
  - ○ The total load of the dryer is 160 kg grapes for each 3.2 days; assume that a charge is for 4 days.
  - ○ In total, 22 charges are used per season; therefore, 3520 kg grapes are dried per year or 52.8 t/lifetime.
  - ○ The operational time for all lifetime is therefore $t_{op} = 22 \times 4 \times 24 \times 15 = 31,680$h.
- Determination of the heat duty for drying
  - ○ In order to determine the flow velocity, apply $Bi_m$–LF correlation from Table 7.2 $\rightarrow Bi_m = 0.797$.
  - ○ Find the eigenvalue from Table 7.4 $\rightarrow \mu_1 = 1.392$.
  - ○ Find moisture diffusivity from Eq. (5.74) $\rightarrow D = 1.29E-10$m$^2$/s.
  - ○ From Eq. (7.25) $\rightarrow$ the Reynolds number with respect to grape radius is $Re_R = 289$.
  - ○ Kinematic diffusivity of air at 40 °C is $\nu = 1.7E-5$m$^2$/s.
  - ○ From $Re_R$ definition $\rightarrow$ air velocity $U = 0.98$m/s.
  - ○ The average rate of moisture removal is $j = (2.28-0.16)/t_f \times 16 = 0.123$mg/m$^2$s.
  - ○ The heat flux required to evaporate the moisture is $q = j\Delta h_{lv}(@30\ °C) = 300$W/m$^2$.
  - ○ The average heat transfer coefficient from air to the tray is $h_{tr} = 30$W/m$^2$K.
  - ○ The minimal heat duty for drying is $\dot{Q} = 3000$W.
  - ○ The actual heat duty accounting for heat losses is $\dot{Q}_{in} = 3300$W.
- Mass balance equation
  - ○ Mass balance equation in integral form for all lifetime is $m_1 + m_7 = m_4 + m_8$.
  - ○ The mass flow rate of air for $\Delta T_{air} = 5\ °C$ is $\dot{m}_{air} = 65$g/s $\rightarrow \dot{V}_{air} = 602$l/s.
  - ○ The total mass of dry air circulated for lifetime is $m_{air} = \dot{m}_{air}t_{op} = 7457$t.
  - ○ The humidity ratio at the air entrance is $\omega_1(T_0, rh = 50\%) = 9.9$g/kg.
  - ○ The total mass of humid air in state 1 is $m_1 = (1+\omega_1)m_{air} = 75,310$t.
  - ○ The total mass of evaporated water is $m_w = jAt_{op} = 140.507$t.
  - ○ The mass flow rate of air at exit is $m_4 = m_1 + m_w = 75,450$t.
  - ○ The grape masses at inlet and outlet are $m_7 = m_7 + m_w = 193.807$t, $m_8 = 52.8$t.
- Energy balance equation
  - ○ The energy balance for an integral form for all lifetime is shown in Figure 8.2; accumulation is neglected: $E_1 + E_5 + E_7 = E_{loss}$.
  - ○ Here, the energy in input air and moist grapes is zero: $E_1 = E_7 = 0$ and $E_{loss} = E_{Q,loss} + E_4 + E_8$.
  - ○ The lifetime heat loss is $E_{Q,loss} = 300 \times 31,680 \times 3600 = 34.21$GJ.
  - ○ The heat loss with the exhaust humid air is $E_4 = m_{air}(h_4 - h_0)$; here, the enthalpy of air in state 4 is determined as follows:
    - Total moisture in exit air $m_{w,4} = \omega_1 m_{air} + m_w = 877.388$t.
    - The humidity ratio in exit air $\omega_4 = m_{w,4}/m_{air} = 11.77$g/kg.
    - The enthalpy in exit air becomes $h_4 = h(\omega_4, T_4 = 40\ °C) = 70.57$kJ/kg.
    - The enthalpy in surrounding air $h_0 = h(rh = 50\%, T_4 = 25\ °C) = 50.34$kJ/kg.
    - The lost energy with exit air becomes $E_4 = 1509$GJ.
  - ○ The heat loss with the product is $E_8 = m_8 C_p (T_{prod} - T_0) = 0.528$GJ.

- Total energy loss $E_{loss} = 1543\,GJ$.
- From the energy balance for lifetime results $E_5 = E_{loss} = 1543\,GJ$.
- Exergy balance equation
  - For the system lifetime, one has $Ex_1 + Ex_5 + Ex_7 = Ex_4 + Ex_8 + Ex_d$.
  - One neglects $Ex_1 = Ex_7 \cong 0$, and one uses 0.95 for exergy of solar radiation $Ex_5 = 0.95 E_5 = 1466\,GJ$.
  - Use exergy definition $(Ex = H - H_0 - T_0(S - S_0)) \rightarrow Ex_4 = 72.3\,GJ$, $Ex_8 = 8.71\,MJ$.
  - The exergy balance determines $Ex_d = 1394\,GJ$.
  - The total exergy loss is $Ex_{loss} = Ex_4 + Ex_8 + Ex_d = 1466\,GJ$.
- Estimation of the invested capital
  - The drying system comprises the PV/T array, the blower, and the dryer tunnel; for the dryer tunnel, the capital cost is proportional to the surface area of the trays and of the duct. There is $10\,m^2$ surface area for the trays; assume that spare trays are included $\rightarrow 20\,m^2$.
  - The free space between the loaded trays of 1 m width is ~2 cm; tunnel height becomes 60 cm.
  - Tunnel lateral surface area is $2 \times (0.6 + 1) \times 2 \cong 7\,m^2$.
  - Assume that the total area of metal sheets for making ducts and trays is roughly $40\,m^2$; take the price of galvanized metal sheet including machining and labor of $200/m^2 \rightarrow $8000 tunnel cost.
  - For the blower, assume $\Delta P = 10mm\,H_2O$ with 50% efficiency $\rightarrow \dot{W}_{blower} = 118\,W$; assume cost of the blower is $300.
  - The minimum area for PV array comes from $\dot{W}_{blower} = 500 A_{PV}/0.15 \rightarrow A_{PV} = 1.57\,m^2$.
  - The cost of PV at $500/m^2$ is $787.
  - The solar thermal air heater area assuming 60% efficiency is $A_{SAH} = E_5/t_{op}/0.6/500 = 45\,m^2$.
  - Assume the price of solar air heater of $100/m^2 \rightarrow $4500.
  - Total estimated capital is $C_{cap} = 8000 + 787 + 300 + 4500 = 13,587$.
  - The energetic loss ratio is $R_{en} = 1543/13,587 = 113.6\,MJ/\$$.
  - The exergetic loss ratio is $R_{ex} = 1466/13,587 = 107.9\,MJ/\$$.
- Cost balance equation
  - For lifetime, the cost balance is $C_1 + C_5 + C_7 + C_{cap} + C_{o\&m} + C_{occ} = \Delta C + C_8 + C_4$.
  - Here, air in and out and solar radiation are at no cost $C_1 = C_5 = C_7 = 0$.
  - For $r = 0.1$, $i = 0.01$, $n = 15 \rightarrow r_m = 0.111$, $PVF = 0.2394$, $CRF = 0.1459$, $PWF = 6.852$.
  - Take the levelized cost of dried grapes at $15/kg; the annual production is $AP = m_8/n = 3.52t$.
  - Present worth income is $PWI = $361,803$.
  - Take the levelized cost of grapes at $3/kg; the annual consumption is $AC = m_7/n = 12.887t$.
  - Present worth cost input is $C_{in} = $264,921$, assume $OMCF = 0.1 \rightarrow C_{o\&m} = $18,090$.
  - Net income is $NI = $78,792$, assume solar energy tax $t_c = 0.4 \rightarrow TCD = t_c C_{cap} = $5435$.
  - Taxable income is $TI = $73,537$; assume tax on income $t_i = 0.35 \rightarrow TOI = $25,657$.
  - Assume tax on property $t_p = 0.05 \rightarrow TOP = $442$, $C_{occ} = $26,117$, $C_{gen} = $57,794$.
  - Assume capital salvage factor $CSF = 0.3$ and tax on salvage $t_s = 0.15 \rightarrow SV = $839$.
  - The generated capital is $GC = $39,928$ and capital productivity $CP = 3.9$.

## 8.4   SPECO Method

Another method to perform exergoeconomic analysis is the SPECO method. The method has been applied to assess and improve various devices, such as power plants. The SPECO method has been described by various authors (Tsatsaronis and Moran, 1997; Bejan et al., 1996). Cost accounting with SPECO methods uses cost balances and is generally concerned with determining the actual cost of goods or services, providing a rational basis for pricing them. In the same time, the method provides a means for allocating and controlling expenditures and determines useful assessment parameters to assist in creating and evaluating design and operating decisions.

For a system operating at steady state, there may be a number of entering and exiting material streams as well as heat and work interactions with the surroundings. Since exergy measures the true thermodynamic value of such effects and cost should only be assigned to commodities of value, it is meaningful to use exergy as a basis for assigning costs in energy systems. Such "exergy costing" provides a rational basis for assigning costs to the interactions that a thermal system experiences with its surroundings and to the sources of inefficiencies within it. In this method, the costs are determined based on the exergy content. Therefore, a monetary value of exergy can be used as determined in Section 8.2.

In a SPECO economic analysis, a cost balance is usually formulated in a rate form (cost rates $\dot{C}$ given in dollar per hour) and applied to the overall system operating at steady state. Let us consider a general drying system. Heat is applied to the drying system as input, together with other types of inputs including an invested capital. Each physical (heat, work, material) stream entering and exiting the system is expressed through its exergy content. To each stream $i$, a SPECO is allocated $C_{ex,i}$. Figure 8.3 shows a drying system representation for the SPECO method application. The following streams can be remarked:

- $\dot{C}_Q = C_{ex,Q}\dot{Ex}^Q$, representing the rate of cost input due to the supplied heat for the dryer, where $C_{ex,Q}$ is a specific cost for exergy provided in the form of heat and $\dot{Ex}^Q$ is the exergy rate provided to the dryer in the form of heat.

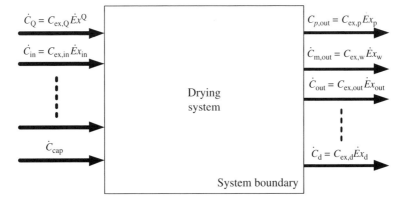

**Figure 8.3**   SPECO method application to a drying system showing the input and output cost stream

- $\dot{C}_{in} = C_{ex,in}\dot{Ex}_{in}$, representing the cost rate in physical inputs; all types of inputs are considered (energy inputs, material inputs, e.g., drying air); for each input, the exergy rate is given $\dot{Ex}_{in}$ and the specific cost of the respective type of exergy.
- $\dot{C}_{cap}$, representing the rate of capital cost, which will be explained later.
- $\dot{C}_{p,out} = C_{ex,p}\dot{Ex}_p$, representing the rate of cost in dried product, where $\dot{Ex}_p$ is the exergy rate in dried product and $C_{ex,p}$ is the allocated cost to the exergy of the dried product.
- $\dot{C}_{m,out} = C_{ex,w}\dot{Ex}_w$, representing the cost of moisture generated by the dryer with $\dot{Ex}_w$ being the exergy rate of moisture and $C_{ex,w}$ the specific cost of exergy allocated for the moisture.
- $\dot{C}_{out} = C_{out}\dot{Ex}_{out}$, representing the cost rate of any other type of output.
- $\dot{C}_d = C_{ex,d}\dot{Ex}_d$, representing the cost rate of exergy destruction which is related to the allocated specific cost of exergy destruction $C_{ex,d}$ and the rate of exergy destruction $\dot{Ex}_d$.

The cost balance for the drying system is written as follows:

$$\dot{C}_Q + \sum \dot{C}_{in} + \dot{C}_{cap} = \dot{C}_{p,out} + \dot{C}_{m,out} + \sum \dot{C}_{out} \qquad (8.9)$$

The cost balance equation (8.9) states that the total cost of the exiting exergy streams (including the cost in exergy destruction) equals the total expenditure to obtain them, namely, the cost of the entering exergy streams plus the capital and other costs. The capital investment rate $\dot{C}_{cap}$ expressed in dollar per hour is determined based on the invested capital expressed in present worth value $C_{cap}$, the capital recovery factor CRF (of which equation is given in Table 8.3), and operation and maintenance cost factor (OMCF) which is here defined with respect to the invested capital. A typical value of around 1.06 can be assumed for OMCF for SPECO analysis, if not specified otherwise. Therefore, the capital investment rate becomes

$$\dot{C}_{cap} = \frac{C_{cap}\text{CRF OMCF}}{n_{hour}} \qquad (8.10)$$

where $n_{hour}$ represents the number of operating hours per year.

All cost data used in an economic analysis must be brought to the same reference year: the year used as a basis for the cost calculations. For the cost data based on conditions at a different time, a normalization is performed with the aid of an appropriate cost index.

Cost balances as given by Eq. (8.9) can be written for each of the system components and for the overall system. When the cost balances are solved simultaneously, the cost rate of exergy destructions can be estimated. Exergoeconomic factors can be devised to account for the contribution of nonexergy-related costs. The system assessment can be done with respect to the cost rate of exergy destruction divided to the cost rate for the capital or as the ratio of the difference between product and fuel cost and the fuel cost.

## Example 8.2

In this example, the exergoeconomic analysis of a heat pump tumbler dryer for cotton fabric is demonstrated according to the SPECO method. The data for the exergy analysis of the dryer is taken from the case study presented in Ganjehsarabi et al. (2014). Based on these data and cost assumption, the SPECO analysis is conducted herein.

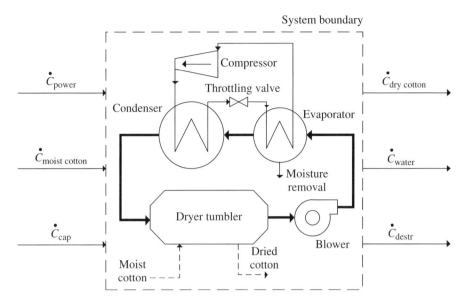

**Figure 8.4** SPECO method applied for Example 8.2 (heat pump tumbler dryer)

The heat pump tumbler dryer diagram is shown in Figure 8.4. The system inputs are as follows: power to drive the compressor, blower and drum motor, and moist cotton fabric to be dried. The system outputs are as follows: dry cotton fabric as product and moisture to be removed. As shown in the figure, the SPECO method requires modeling the system in terms of specific cost rates for exergy (dollar per hour) at steady state. In this respect, the system input are as follows: cash rate to cover the costs of consumed power, cost rate to cover the moist cotton fabric fed to the dryer, and cost rate of capital invested into the system. The output cost rates are for dried product, removed water, and exergy destruction:

- The cost rate balance equation for the SPECO analysis of the dryer becomes

$$\dot{C}_w + \dot{C}_{mm} + \dot{C}_{cap} = \dot{C}_{prod} + \dot{C}_{mr} + \dot{C}_{ex,d}$$

  where subscript w means power, mm means moist material, and mr is moisture removal.
- The SPECO starts after the exergy analysis is performed; here, the following exergy data is retained from Ganjehsarabi et al. (2014):
  ○ Exergy efficiency of the system is $\psi = 0.11$.
  ○ Exergy destruction rate is $\dot{Ex}_d = 2140\,\text{W}$.
  ○ Moisture removal rate is $\dot{m}_{mr} = 2.4\,\text{kg/h}$.
  ○ Product exit rate $\dot{m}_{prod} = 0.94\,\text{kg/h}$.
  ○ Capital cost for the compressor is $2245, condenser $336, evaporator $241, valve $31, dryer $555, and blower $150.
- Power input determined based on exergy balance is $\dot{Ex}_w = 2404\,\text{W}$.
- The total cost for capital is estimated to $C_{cap} = 2245 + 336 + 241 + 31 + 555 + 150 = \$3558$.

- The system lifetime is assumed $n = 15$ years with $t_{op} = 1000 \text{h/year}$.
- With a discount rate of $i = 0.04 \rightarrow \text{CRF} = 0.08994$.
- Assume OMCF $= 1.06$; from Eq. (8.10) $\rightarrow \dot{C}_{cap} = 0.3392 \$/\text{h}$.
- Assume specific cost of exergy is $C_{ex} = 8\text{¢/kW h} \rightarrow$ cost rate of power $\dot{C}_w = C_{ex}\dot{E}x_w = 0.1924 \$/\text{h}$.
- Cost of exergy destruction is $\dot{C}_{ex,d} = C_{ex}\dot{E}x_d = 0.1712 \$/\text{h}$.
- The removed water is generally wasted; assume that its cost rate can be quantified by exergy content.
- Chemical exergy of water is $0.75 \text{ kJ/mol} \rightarrow \dot{E}x_{mr} = 27.83 \text{ W} \rightarrow \dot{C}_{mr} = 6.18\text{E} - 17 \$/\text{h (negligible)}$.
- The cost balance equation now becomes

$$0.1924 + \dot{C}_{mm} + 0.3392 = \dot{C}_{prod} + 0 + 0.1712 \rightarrow \Delta\dot{C} = \dot{C}_{prod} - \dot{C}_{mm} = 0.3604 \ \$/\text{h}$$

where $\Delta\dot{C}$ can be used to determine the minimum price for the drying service.
- The exergetic loss ratio in rate form is $R_{ex} = \dot{E}x_d/\dot{C}_{cap} = 22.7\text{E}7 \text{MJ/\$}$.

## 8.5   Exergoenvironmental Analysis

Exergy analysis is expanded into exergoeconomic analysis. As shown in Dincer and Rosen (2013), when considering exergy in environmental impact analysis, the inventory analysis phase has to account more carefully for mass and energy flows into, out of, and through all the stages of the life cycle; next, the energy flow is associated with an exergy flow; eventually, environmental impact indicators based on exergy can be developed and studied.

Drying is a thermal separation process. It requires exergy for doing the separation work. The exergy can be supplied directly as heat/fuel combustion or solar energy. However, dryers can be driven by electricity only when electrical heating or heat pump heating and dehumidification are used. In fact, the exergy supplied to the system will cover all irreversibilities and losses in addition to providing the separation work to extract moisture.

When the environmental impact is accounted for, one must consider the impact of the system effluents to the environment but also the pollution and resource depletion related to the drying system construction. Therefore, the system boundary for the analysis is set at larger limits than that of the actual system. In fact, one can model a system construction process in terms of required exergy for construction and the environmental impact. Also, one can elaborate a model for the environmental impact due to system effluents. Therefore, the system itself is expanded with two subsystems: upstream (the system construction process) and downstream (effluent-environmental and decommissioning processes). All three subsystems – upstream process, actual dryer, and downstream process – must be supplied with exergy, or otherwise no subsystem can run. For the dryer itself, the exergy supply in the form of any kind of fuels is referred as direct exergy. For the upstream and downstream subsystems, the supplied exergy is referred to as the indirect exergy. Therefore, the total exergy supply to be considered in an extended exergoenvironmental analysis is

$$\dot{E}x_{supply} = \dot{E}x_{direct} + \dot{E}x_{indirect}^{upstream} + \dot{E}x_{indirect}^{downstream} \tag{8.11}$$

A similar equation can be also written for the environmental impact, namely,

$$EI_{total} = EI_{direct} + EI_{indirect} \tag{8.12}$$

where EI is a general environmental impact indicator; $EI_{direct}$ represents the downstream environmental impact, that is, the environmental impact caused by the system due to its operation; and $EI_{indirect}$ represents the environmental impact associated with system construction.

The general representation of a drying system, illustrating the environmental pollution through the effluents, is shown in Figure 8.5. Here, only the direct environmental impact is represented in the figure. In most of the exergoenvironmental analyses, only the direct impact is accounted for. However, when the exergoenvironmental analysis is expanded to a more general exergetic life cycle analysis, all direct and indirect impacts are considered.

In many practical situations, the energy that drives a process is electricity or a petrochemical fuel generated elsewhere (power plant or refinery, respectively). It is generally known in a certain economy what is the environmental impact and exergy efficiency for electric power or consumer fuels production. Once the rate of fuel consumption is specified, the indirect exergy consumption and environmental impact associated with fuel supply can be easily determined.

Other types of environmental impacts may be considered in the analysis. There are several kinds of impact categories, like depletion of resources (abiotic or biotic), land use (competition, biodiversity loss, life support function loss), climate change, desiccation, stratospheric ozone depletion, human toxicity and ecotoxicity, acidification, photooxidant formation, eutrophication, waste heat generation, noise and odor (in water and air), and ionizing radiation. The toxicity regards aquatic (marine, freshwater) and terrestrial ecosystems. Eutrophication represents the overincrease of chemical nutrients in the ecosystem which causes productivity augmentation. This creates an unbalance in the ecosystem because other biological species may occur and experience an increase in number with respect to other species that can deplete.

Table 8.4 presents the main quantifiable environmental impact categories and their descriptions. The first quantifiable impact is called abiotic resource depletion which is expressed with

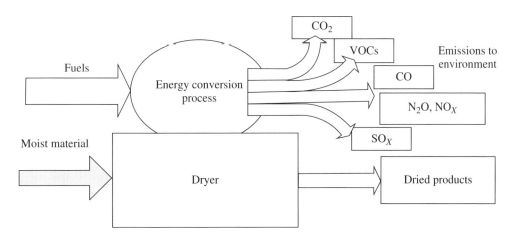

**Figure 8.5**   Environmental impact of a drying system

**Table 8.4**  Quantification of environmental impact

| Impact category | Unit | Definition |
|---|---|---|
| Depletion of abiotic resources | Kilogram antimony equivalent | Abiotic resources include all nonliving resources (coal, oil, iron ore, renewable energies, etc.) |
| Climate change | GDP ($CO_2$ equivalent) | Kilogram of $CO_2$ equivalent. See other chapters of the book for more detailed definition |
| Ozone depletion | ODP (CFC11 equivalent) | Ozone depletion potential (ODP) is a time-dependent parameter characterizing the potential of a substance to deplete the stratospheric ozone layer. Represents the amount of a substance that depletes the ozone layer relative to CFC11 (trichlorofluoromethane) |
| Photooxidant formation | Kilogram of ethylene equivalent | Depends on photochemical ozone creation potential expressed with respect to the reference substance, ethylene |
| Toxicity | Kilogram DCB equivalent | Human toxicity and ecotoxicity produced in air, freshwater, seawater, or terrestrial. It is quantified with respect to the toxicity of DCB |
| Acidification | Kilogram $SO_2$ equivalent | Expressed in $SO_2$ emitted in Switzerland equivalent. There are three acidification factors: ammonia, $NO_x$, and $SO_2$ |
| Eutrophication | Kilogram $PO_4$ equivalent | Covers all potential impacts of high levels of nutrients, the most important of which are nitrogen (N) and phosphorus (P) |

DCB, 1,4-dichlorobenzene.

reference to antimony resource. Each kind of resource is assigned an abiotic depletion potential (ADP) which is expressed in kilogram of antimony equivalent to kilogram of resource.

A primary objective of exergoenvironmental analysis is the identification and quantification of direct and/or indirect environmental impact in correlation with exergy destroyed by the system. Another objective is minimizing the environmental impact. Often, this is done by finding means of increasing the energy efficiency of the system. Another approach to environmental impact minimization is the reduction of the polluting effluents.

In recent years, particular emphasis has been placed on releases of carbon dioxide, since it is the main greenhouse gas, and optimization of thermal systems based on this parameter has received much attention. A focus of many studies is to consider emissions of the following types of atmospheric pollutants: CO, $NO_x$, $SO_2$, $CO_2$, and $CH_4$. In the case of drying systems which are fueled by a combustion process, the atmospheric pollution through flue gas emissions is very relevant.

The inventory analysis is the core of the environmental impact analysis as it determines the streams of materials and their impact on the environment. All flows of matter exchanged with the environment and their interrelationship within the analyzed system must be inventoried. The impact categories are then determined according to the definitions given in Table 8.4. Figure 8.6 shows the relative importance of impact factors for solar-driven dryers; this information can be used as a guideline for environmental analyses of dryers.

The SPECO and EXCEM analyses can be extended for exergoenvironmental analysis. Regardless of the actual method used to conduct an exergoenvironmental analysis, the environmental impact by the system and all its components must be determined first. This is a purely environmental impact analysis phase.

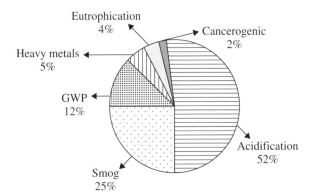

**Figure 8.6**  Relative environmental impact factor for a solar-driven drying system (Data from Koroneos and Nanaki (2012))

As suggested in Meyer et al. (2009), for each exergy stream entering or exiting the system, a compounded environmental impact rate can be assigned. In order to give an example, let us consider a combustion process. The stream of flue gases injects into the surrounding atmosphere an exergy at a rate that can be easily determined from the exergy analysis. The same flow impacts the environment at a rate that can be determined based on the specific method of life cycle environmental impact assessment. The ratio of the rate of environmental impact to the rate of exergy can be then determined for any exergy stream, including for streams of exergy destruction. Therefore, a balance equation for the environmental impact of a drying system can be formulated by analogy to Eq. (8.9):

$$\dot{E}I_Q + \sum \dot{E}I_{in} + \dot{E}I_{sys} = \dot{E}I_{p,out} + \dot{E}I_{m,out} + \sum \dot{E}I_{out} \tag{8.13}$$

where $\dot{E}I_Q$ is the environmental impact rate due to heating source for the drying process; $\sum \dot{E}I_{in}$ is related to any input stream of energy or matter, including the moist product; $\dot{E}I_{sys}$ represents the environmental impact of the system (or its subsystems when the components are analyzed); $\dot{E}I_{p,out}$ is due to the release in the environment of the dried product themselves; $\dot{E}I_{m,out}$ refers to moisture release; and $\sum \dot{E}I_{out}$ refers to other outputs and exergy destruction.

As mentioned previously, one of the most important environmental impacts of drying systems is related to combustion subprocess. Many dryers are driven directly by fuel combustion. For electrically driven dryers, the combustion process can be inferred indirectly at power generation phase (i.e., outside the system).

Carbon dioxide emission from fuel combustion can be relatively easy correlated with fuel combustion rate based on general combustion parameters such as stoichiometry, excess air, and so on. More difficult to determine through modeling are the rates of CO and $NO_x$, since detailed kinetic modeling for combustion process would be required. Some simplified models can be used as an alternative, such those by Toffolo and Lazzaretto (2004) who relate the mass flow rate of CO and $NO_x$ emissions with the fuel (natural gas) consumption:

$$\frac{\dot{m}_{CO}}{\dot{m}_f} = 179,000 \frac{\exp\left(\dfrac{7800}{T}\right)}{P^2 \tau \left(\dfrac{\Delta P}{P}\right)^{0.5}} \qquad (8.14a)$$

$$\frac{\dot{m}_{NO_x}}{\dot{m}_f} = 150,000 \frac{\tau^{0.5} \exp\left(-\dfrac{71,100}{T}\right)}{P^{0.05} \tau \left(\dfrac{\Delta P}{P}\right)^{0.5}} \qquad (8.14b)$$

where $T$ is the adiabatic flame temperature, $P$ is combustion pressure, $\Delta P$ is the pressure drop across the combustion chamber, and $\tau$ is the residence time in the combustion zone (approximately 2 ms).

When a system with higher exergy efficiency is constructed, the associated environmental impact will be low. This aspect is illustrated with the help of Figure 8.7 which shows the correlation between environmental impact and exergy efficiency. The data for Figure 8.7 is taken from a heat pump dryer (HPD) case study published by Ozcan and Dincer (2013). A parametric study shows that the environmental impact decreases if a system with higher exergy efficiency is devised. For the environmental impact quantification, only the carbon dioxide emissions (direct and indirect) are considered.

Two types of energy sources are studied in Ozcan and Dincer (2013): solar photovoltaics (PV) and the supply from the Canadian grid. For both cases, the associated emissions for power

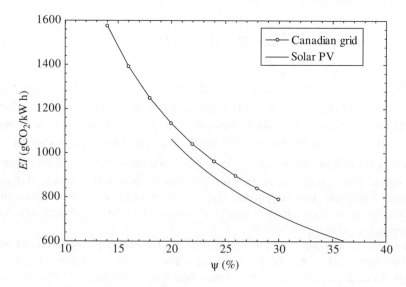

**Figure 8.7** Environmental impact and exergy efficiency for a heat pump dryer (Data from Ozcan and Dincer (2013))

generation are determined and correlated with the power consumption by the HPD. When the dryer is more efficient, it consumes less power for the same duty, therefore emitting less carbon dioxide pollutant.

Furthermore, if solar PV is the power source, the emissions are much reduced and efficiency increased. Noteworthy is that a design with better efficiency is definitely more expensive. If the environmental impact would be plotted against the system cost, then the curves will be similar as shown in the figure; however, the objectives will be contradictory: when the system is more expensive, it pollutes less and vice versa. This situation leads to Pareto trade-off problem in which a rational selection must be made for two divergent objective functions in a multiobjective optimization.

## 8.6   Conclusions

The present chapter concerns exergoeconomic and exergoenvironmental analyses of the thermal system in general and with a specific focus on drying systems. Two potential exergoeconomic methods, such as EXCEM and SPECO, are considered and treated for drying systems and applications. Some detailed illustrative examples are presented to illustrate the use of these methods for practical drying applications. A method for exergoenvironmental analysis is introduced with a focus on the exergetic life cycle assessment and environmental impact assessment. The environmental impact in correlation with exergy efficiency is commented for a case study data referring to an HPD. The choice of the energy source to drive the dryer affects the analysis as shown. These methods are of practical importance where high amount of energy is used and potential environmental consequences are evident.

## 8.7   Study Problems

8.1   Based on the methodology given in Section 8.2, determine the average price of exergy in the United States; use public statistical data.

8.2   Rework Example 8.1 for similar data but double lifetime.

8.3   Consider a dryer of a tunnel foodstuff of your choice; perform an EXCEM analysis assuming that the heat input is derived from natural gas combustion.

8.4   Perform a detailed exergy analysis for the system presented in Figure 8.4. Apply thereafter the SPECO method.

8.5   Expand problem 4 with a parametric study to observe the influence of investment cost on exergy efficiency.

8.6   Consider a solar-driven heat pump dryer such as that from Ozcan and Dincer (2013). Perform detailed exergoeconomic and exergoenvironmental analyses to assess the system.

8.7   What is exergoeconomics? What is the difference between exergoeconomics and thermoeconomics?

8.8   What are the advantages of costing based on exergy compared to costing based on energy?

8.9   Explain how the exergy-based price of various fuels can be determined. Provide examples.

8.10  Explain the correlation between capital cost, exergy efficiency, and environmental impact.

# References

Ahmadi P., Dincer I. 2010. Exergoenvironmental analysis and optimization of a cogeneration plant system using multimodal genetic algorithm (MGA). *Energy* 35:5161–5172.

Ahmadi P., Dincer I. 2011a. Thermodynamic analysis and thermoeconomic optimization of a dual pressure combined cycle power plant with a supplementary firing unit. *Energy Conversion and Management* 52:2296–2308.

Ahmadi P., Dincer I. 2011b. Thermodynamic and exergoenvironmental analyses, and multi-objective optimization of a gas turbine power plant. *Applied Thermal Engineering* 31:2529–2540.

Ahmadi P., Dincer I., Rosen M.A. 2011a. Exergy, exergoeconomic and environmental analyses and evolutionary algorithm based multi-objective optimization of combined cycle power plants. *Energy* 36:5886–5898.

Ahmadi P., Rosen M.A., Dincer I. 2011b. Greenhouse gas emission and exergo-environmental analyses of a trigeneration energy system. *International Journal of Greenhouse Gas Control* 5:1540–1549.

Bejan A. 1982. Entropy Generation through Heat and Fluid Flow. Wiley: Toronto.

Bejan A. 1996. Entropy Generation Minimization. CRC Press: Florida.

Bejan A. 2001. New century, new methods. *Exergy, An International Journal* 1:2.

Bejan A. 2006. Advanced Engineering Thermodynamics. 3rd ed. John Wiley & Sons: New York.

Bejan A., Tsatsaronis G., Moran M. 1996. Thermal Design and Optimization. Wiley: New York.

Blanco-Marigorta A.M., Masi M., Manfrida G. 2014. Exergo-environmental analysis of a reverse osmosis desalination plant. *Energy* 76:223–232.

Connelly L., Koshland C.P. 2001a. Exergy and industrial ecology. Part 1: An exergy-based definition of consumption and a thermodynamic interpretation of ecosystem evolution. *Exergy, An International Journal* 3:146–165.

Connelly L., Koshland C.P. 2001b. Exergy and industrial ecology. Part 2: A non-dimensional analysis of means to reduce resource depletion. *Exergy, An International Journal* 1:234–255.

Dincer I. 2002a. Exergy and sustainability. Proceedings of SET-2002: First International Conference on Sustainable Energy Technologies, 1–10.

Dincer I. 2002b. On energetic, exergetic and environmental aspects of drying systems. *International Journal of Energy Research* 26:717–727.

Dincer I., Rosen M.A. 2013. Exergy: Energy, Environment, and Sustainable Development. Elsevier: Oxford, UK.

Dincer I., Zamfirescu C. 2011. Sustainable Energy Systems and Applications. Springer: New York.

Dincer I., Zamfirescu C. 2014. Advanced Power Generation Systems. Elsevier: New York.

Ganjehsarabi H., Dincer I., Gungor A. 2014. Exergoeconomic analysis of a heat pump tumbler dryer. *Drying Technology* 32:352–360.

Georgescu-Roegen N. 1986. The entropy law and the economic process. *Eastern Economic Journal* 12:3–25.

Kaviri A.G., Mohd Jaafar M.N., Mat Lazim T., Barzegaravval H. 2013. Exergoenvironmental optimization of heat recovery steam generators in combined cycle power plant through energy and exergy analysis. *Energy Conversion and Management* 67:27–33.

Koroneos C.J., Nanaki E.A. 2012. Life cycle environmental impact assessment of a solar water heater. *Journal of Cleaner Production* 37:154–161.

Kotas T.J. 1995. The Exergy Method of Thermal Plant Analysis. Krieger: Malabar, Florida.

Lozano M.A., Valero A. 1993. Theory of exergetic cost. *Energy* 18:939–960.

Meyer L., Castillo R., Buchgeister J., Tsatsaronis G. 2009. Application of exergoeconomic and exergoenvironmental analysis to an SOFC system with allothermal biomass gasifier. *International Journal of Thermophysics* 12:177–186.

Moran M., Shapiro H.N., Boettner D.D., Bailey M.B. 2011. Fundamentals of Engineering Thermodynamics. 7th ed. John Wiley and Sons Inc: New York.

Ozcan H., Dincer I. 2013. Exergy analysis and environmental impact assessment of solar-driven heat pump drying system. In: Causes, Impacts and Solutions to Global Warming. I. Dincer, C.O. Colpan, F. Kadioglu Eds., Springer: New York.

Rosen M.A. 1986. The development and application of a process analysis methodology and code based on exergy, cost, energy and mass. Ph.D. Thesis. Department of Mechanical Engineering, University of Toronto, Toronto.

Rosen M.A., Dincer I. 2003a. Exergoeconomic analysis of power plants operating on various fuels. *Applied Thermal Engineering* 23:643–658.

Rosen M.A., Dincer I. 2003b. Exergy–cost–energy–mass analysis of thermal systems and processes. *Energy Conversion and Management* 44:1633–1651.

Szargut J., Morris D.R., Steward F.R. 1988. Exergy Analysis of Thermal, Chemical, and Metallurgical Processes. Hemisphere: New York.

Toffolo A., Lazzaretto A. 2004. Energy, economy and environment as objectives in multi-criteria optimization of thermal system design. *Energy* 29:1139–1157.

Tsatsaronis G. 1987. A review of exergoeconomic methodologies. In: Second Law Analysis of Thermal Systems. American Society of Mechanical Engineers: New York, pp. 81–87.

Tsatsaronis G., Moran M.J. 1997. Exergy-aided cost minimization. *Energy Conversion and Management* 38:1535–1542.

Wall G. 1993. Exergy, ecology and democracy-concepts of a vital society. *ENSEC'93:* International Conference on Energy Systems and Ecology, 5–9 July, Cracow, Poland, 111–121.

Rosen M.A. 2008. A concise review of exergy-based economic methods. 3rd IASME/WSEAS Int. Conf. on Energy & Environment, University of Cambridge, UK, February 23–25.

# 9

# Optimization of Drying Processes and Systems

## 9.1  Introduction

From the perspective of today's society, it is no longer adequate to develop a drying system only for performing its desired task with high efficiency, but rather other colateral factors must be accounted for. The increasingly concerns regarding safety, environmental impact, energy and material resource minimization, and product quality place an expanded pressure of the design process for drying systems.

Designing a better drying system is an obvious engineering goal. But what means better and how to make a better design? These two basic questions are addressed here. The general method of "searching for the best" is known as optimization. The drying system can be designed for better efficiency, better cost-effectiveness, better utilization of energy source, better environment, better sustainability, better product security, and so on.

A clear specification of what "the best" means is required for a fully deterministic search for optimum. But sometimes, optimization problems are fuzzy and involve subjective choices. This is because often a trade-off selection among designs is required. For example, in many cases, the design showing the lowest environmental impact has also the highest cost, whereas the low cost designs have generally higher environmental impact.

Special mathematical methods for vector optimization or multiobjective optimization can be found useful in solving trade-off problems. However, the core of drying systems optimization is the minimization of exergy destruction (or entropy generation). Drying is a thermal separation process. Therefore, drying is directly governed by the laws of thermodynamics. An ideal drying process will proceed reversibly. Thermodynamics offers the possibility to determine the minimum required work for moisture separation. This is the reversible work for the process. Any actual drying process will require more than reversible work. In fact, since drying is driven thermally, there is required even more amount of thermal energy to compensate for the

*Drying Phenomena: Theory and Applications*, First Edition. İbrahim Dinçer and Calin Zamfirescu.
© 2016 John Wiley & Sons, Ltd. Published 2016 by John Wiley & Sons, Ltd.

heat-to-work conversion factor. These facts help establishing a scale of values which says that the drying system which approaches the most reversible process is the best one.

Thermodynamic optimization facilitates the design process not only in the sense of minimizing the irreversibilities, but it helps reducing the environmental impact and improving the system economics. Indeed, as shown in Chapter 8, thermodynamic analysis through exergy method can be expanded into exergoenvironmental and exergoeconomic analyses. Furthermore, thermodynamic optimization, through minimization of irreversibility, can be expanded to cover a trade-off optimization for lowering the environmental impact and increasing the economic profitability. It is possible to go even beyond exergoeconomic–environmental optimization by properly considering other objectives such as product quality. Dried products do not need to be only dried, but also obtaining certain quality for the texture can be specified. The product deformation must be restricted to certain values as a design requirement. When the optimization problem becomes complex, its proper formulation is increasingly important. Mathematical modeling and optimization plays a crucial role, but more importantly is the proper problem formulation. The objective functions and optimization criteria must be rigorously described together with the assumptions, the mathematical model, setting of boundary conditions, and identification of design variables. The mathematical method for optimization must be carefully selected.

The use of exergy for drying systems optimization and design is discussed in Dincer (2011). It is shown that dryer optimization for maximization exergy efficiency leads to determination of the optimum drying conditions when the design configuration is known. On the other hand, the economic resources must be properly allocated for the design and operating parameters optimization by a trade-off problem of economic profitability and exergy efficiency minimization. Moreover, since a drying system has many components, a holistic optimization is required to optimize the parts (components) and the whole (the system). When doing this, the optimal configuration and optimal operating conditions of the drying system are discovered.

Optimization of an engineered drying system is a significant process that must be implemented accordingly. Fundamentally, this optimization should start from exergy destruction (or entropy generation) minimization. The discussion of this method in the framework of moisture transfer through porous material (such as in moist materials subjected to drying) is presented in Bejan et al. (2004). The method has been applied in terms of exergy efficiency maximization of drying systems as shown in Dincer and Rosen (2013). Exergy efficiency is an effective indicator of exergy destructions of a system relative to the exergy input. Indeed, as discussed in Chapter 1 and discussed further subsequently in this chapter, the total exergy destruction represents the sum of exergy destroyed within the system plus the exergy destroyed at the interaction of the system with its surroundings.

The relative exergy destruction represents the ratio between the total exergy destruction and the exergy input into the system. The second law of thermodynamics requires that the relative exergy destruction in a system is always higher than zero and smaller than one. In a design process, relative exergy destruction plays the role of an objective function that must be minimized. However, most often, the exergy efficiency is maximized, where exergy efficiency is the difference between the unity and the relative exergy destruction. An ideal system, according to the second law of thermodynamics, will have an exergy efficiency equal to unity and zero exergy destruction. The exergy efficiency indicates the best system performance in relative terms with respect to the exergy consumption (input).

This chapter reviews the main assessment parameters for the drying systems and processes. These parameters can be used to define objective functions for system optimization. Further in the chapter, a brief review of mathematical optimization methods is provided including vector and multiobjective optimization and numerical methods. Some applications are presented for drying systems optimization from single and multiple criteria.

## 9.2   Objective Functions for Drying Systems Optimization

Assessment criteria for drying processes and systems are very important for analyses and design. An assessment criterion must be able to assign a figure of merit to the drying system such that from a particular point of view, the system under the focus can be compared with other competing system. In addition, through assessment indicators, the effects of the system in economy, environment, and so on can be quantified.

We discuss first the drying system assessment based on thermodynamics, because this is considered the most fundamental. Performance assessment for drying systems is discussed in Chapter 4. We expand here that section by including additional criteria. Here, the focus is of formulating objective functions that can eventually be used for drying systems optimization.

### 9.2.1   Technical Objective Functions

Energy and exergy efficiency are the most representative parameters to be used as objective functions for drying systems optimization. Here, the energy and exergy efficiencies presented in Section 4.2.1 are revisited and reformulated in a more general way. An example is given at the end of this section, and some insight on the dying process optimization briefly mentioned. In this respect, let us consider a moist material in a surrounding environment at temperature $T_0$. Assume that the moist material is in thermodynamic equilibrium with the environment. Figure 9.1, which is a variation of Figure 4.1, shows a model of the process considered here, which is an ideal, reversible one.

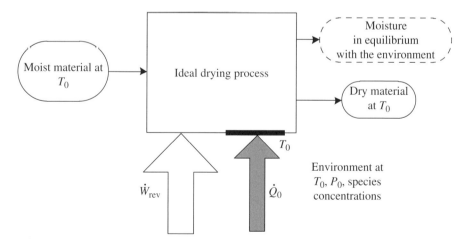

**Figure 9.1**   Thermodynamic representation of an ideal (reversible) drying process

The dryer represented in the figure process is the moist material coming at state #1 and delivers the dried material at state #2 and moisture at state #3 in equilibrium with the environment. The process of moisture removal is not spontaneous. In order to work, it draws heat from the environment at $T_0$ and it is supplied with work from exterior, equal to the required reversible work. The reversible work and heat input can be calculated from the mass, energy, entropy, and exergy balance equations, provided that sufficient thermodynamic information exists for estimating the enthalpy and entropy of the moist product and of the moisture. In addition, the pressure $P_0$ and the chemical composition of the surrounding environment must be specified.

Notice that the total energy used to drive the ideal drying process can be written as $E_{ideal} = W_{rev} + Q_0$. However, since $Q_0$ is supplied spontaneously by the surrounding environment, the ideal drying process required only the reversible work as an external supply. This work has to be compared to the energy or exergy input required for drying in an actual system. The representation from Figure 9.2 shows an actual drying system.

The actual dryer is a finite size device operating during a finite time interval. For any heat input, a temperature gradient is required; therefore, the system boundary is at higher temperature than the process boundary. For any mass transfer, a concentration gradient is required. Therefore, a drying agent is required to produce a concentration gradient. The total energy required to run the actual dryer becomes

$$E_{in} = W_{in} + Q_{in} + E_{da} \tag{9.1}$$

where $W_{in}$ is the amount of work supplied directly, $Q_{in}$ is the amount of heat supplied by heat transfer at system boundary, and $E_{da}$ is the amount of energy carried by the drying agent, with respect to a reference state.

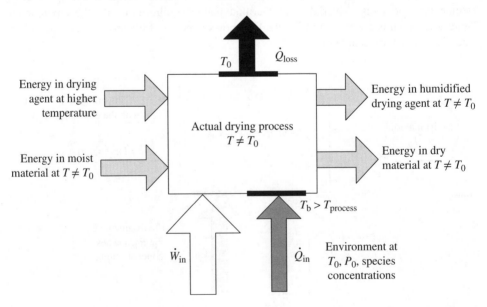

**Figure 9.2**  Energy balances for an actual (irreversible) drying process (finite time/finite size process)

There are three types of losses of the system: the heat loss through the system boundary (it includes the work losses), the energy lost with the products which exit the dryer at a temperature higher than the environment, and the energy lost with the humidified air (when this energy is not recovered). In heat pump dryers, the humidified air is recycled; therefore, the energy lost is quantified in terms of energy in the extracted moisture. Since the drying agent is recycled and the heat input is provided by mechanical mean, the only energy input in the heat pump dryer is in form of work $W_{in}$; the terms $Q_{in}$ and $E_{da}$ are zero in Eq. (9.1) for heat pump dryers.

### 9.2.1.1　Determination of Reversible Work for a Drying Process

Let us analyze the determination of reversible work for a drying (moisture separation) process. The mass, energy, and entropy balance equations for the ideal (reversible) system from Figure 9.1 are, respectively, as follows:

$$\dot{m}_p + \dot{m}_p W_{in} = \dot{m}_r + \dot{m}_p + \dot{m}_p W_{out} \tag{9.2a}$$

$$\dot{m}_p h_p + \dot{m}_p W_{in} h_{bw} + \dot{W}_{rev} + \dot{Q}_0 = \dot{m}_r \omega_0 h_{w,0} + \dot{m}_p h_{p,0} + \dot{m}_p W_{out} h_{bw} \tag{9.2b}$$

$$\dot{m}_p s_p + \dot{m}_p W_{in} s_{bw} + \frac{\dot{Q}_0}{T_0} = \dot{m}_r \omega_0 s_{w,0} + \dot{m}_p s_{p,0} + \dot{m}_p W_{out} s_{bw} \tag{9.2c}$$

Here, $W$ is the moisture content, $\omega$ is the inlet humidity ratio, and subscript r means the removed moisture.

The reversible work results from Eq. (9.2) which can be simultaneously solved to determine $\dot{m}_r$, $\dot{Q}_0$, $\dot{W}_{rev}$. The enthalpy of the bounded moisture $h_{bw}$ results from Eq. (4.12) and the bounded moisture entropy $s_{bw}$ is given by Eq. (4.13). From Eq. (9.2a), the following expression for the removed moisture rate results:

$$\dot{m}_r = \dot{m}_p \Delta W \tag{9.3}$$

where $\Delta W = (W_{in} - W_{out})$.

The heat flux of entropy input results from Eq. (9.2b) as follows:

$$\frac{\dot{Q}_0}{T_0} = \dot{m}_p \left[ \Delta W \omega_0 s_{w,0} + \Delta s_p \right] \tag{9.4}$$

where $\omega_0$ is the humidity ratio in the surrounding environment, $s_{w,0}$ is the specific moisture entropy in the surrounding environment, $\Delta s_p = s_{m,out} - s_{m,in}$, and $s_m$ is the specific entropy of the moist material.

Here, for a capillary porous material, the specific entropy results from Eq. (9.2b) as a superposition of the dry bone material entropy and the moisture entropy assumed all in vapor phase in the capillaries. Therefore, the specific entropy for moist capillary porous material is $s_m = s_p + W s_{bw}$, where $s_p$ is the specific entropy of the dry bone material and $s_{bw}$ is the specific entropy of the moisture in the capillaries.

On the other hand, for the reversible work, the following equation results, after introducing Eqs. (9.3) and (9.4) in Eq. (9.2c):

$$\dot{W}_{rev} = \dot{m}_p \left( \omega_0 \Delta W ex_{w,0} + \Delta ex_p \right) \tag{9.5}$$

where the specific exergy of moisture in the surrounding environment is $ex_{w,0} = h_{w,0} - T_0 s_{w,0}$ and $\Delta ex_p$ represents the specific exergy change of the moist material on dry basis:

$$\Delta ex_p = \Delta h_p - T_0 \Delta s_p \tag{9.6}$$

where $\Delta h_p = h_{m,out} - h_{m,in}$ and $h_m$ is the specific enthalpy of the moist material, which for a capillary porous material can be approximated with $h_m = h_p + W h_{bw}$, where $h_p$ is the specific entropy of the dry bone material and $h_{bw}$ is the specific entropy of the moisture in the capillaries.

### 9.2.1.2 Determination of the Actual Energy and Exergy Input for a Drying System

It is obvious that the actual exergy and energy required to drive a drying process depend on system configuration and operational parameters. Without a system configuration clearly specified, energy and exergy requirements cannot be estimated. In fact, the simplified conceptual model for a dryer shown in Figure 9.2 is very helpful for deriving the general features of any drying system in terms of energy and exergy demand. In this case, the balance equations for mass, energy, and exergy are as follows:

$$\dot{m}_a(1 + \omega_{in}) + \dot{m}_p + \dot{m}_p W_{in} = \dot{m}_a(1 + \omega_{out}) + \dot{m}_p + \dot{m}_p W_{out} \tag{9.7a}$$

$$\dot{m}_a(h_a + \omega h_w)_{in} + \dot{m}_p h_{m,in} + \dot{W}_{in} + \dot{Q}_{in} = \dot{m}_a(h_a + \omega h_w)_{out} + \dot{m}_p h_{m,out} + \dot{Q}_{loss} \tag{9.7b}$$

$$\dot{m}_a ex_{ha,in} + \dot{m}_p ex_{m,in} + \dot{W}_{in} + \dot{Q}_{in}\left(1 - \frac{T_0}{T_b}\right) = \dot{m}_a ex_{ha,out} + \dot{m}_p ex_{m,out} + \dot{E}x_d \tag{9.7c}$$

where $\dot{E}x_d$ is the exergy destruction due to internal irreversibilities and $ex_{ha}$ is the specific exergy of the humid air.

From Eq. (9.7a), the humidity ratio in the humid drying agent results as follows:

$$\omega_{out} = \omega_{in} + \frac{\dot{m}_r}{\dot{m}_a} \tag{9.8a}$$

where $\dot{m}_r = \dot{m}_p \Delta W$.

If the drying agent is recycled, such as in the heat pump dryers, then Eq. (9.8a) is not meaningful for the process, and in Eqs. (9.7), $\dot{m}_a$ must be set to zero. From Eq. (9.7b), the following equation for the actual energy demand for the drying process results:

$$\dot{E}_{in} = \dot{W}_{in} + \dot{Q}_{in} + \dot{m}_a(h_a + \omega h_w)_{in} = \dot{m}_a(h_a + \omega h_w)_{out} + \dot{m}_p \Delta h_p + (h_{tr}A)_1(T - T_0) \tag{9.8b}$$

where $(h_{tr}A)_l$ is the heat transfer resistance between the system and the surroundings, $T$ is the average drying process temperature, and $T_0$ is the surroundings temperature, and $\dot{Q}_{loss} = (h_{tr}A)_l(T-T_0)$.

If air is recirculated, then $\dot{m}_a$ must be set to zero in Eq. (9.8b). Since the heat in the expelled air and the generated dry product is not recovered, the total exergy destruction is $Ex_{d,tot} = \dot{m}_a ex_{a,out} + \dot{m}_p ex_{m,out} + \dot{Ex}_d$. If a heat pump dryer is used with recycling of the air, then $Ex_{d,tot} = \dot{m}_p ex_{m,out} + \dot{Ex}_d$. According to Eq. (9.7c), the total exergy destruction becomes

$$\dot{Ex}_{d,tot} = \dot{m}_a ex_{a,in} + \dot{m}_p ex_{m,in} + \dot{W}_{in} + (h_{tr}A)_h(T_b-T)\left(1-\frac{T_0}{T_b}\right) \tag{9.9a}$$

$$\dot{Ex}_{d,tot} = \dot{m}_a\left(ex_{a,in} - ex_{a,out}\right) + \dot{m}_p ex_{m,in} + \dot{W}_{in} + (h_{tr}A)_h(T_b-T)\left(1-\frac{T_0}{T_b}\right) \tag{9.9b}$$

where Eq. (9.9a) applies for a dryer which expels the humidified drying agent, whereas Eq. (9.9b) is used when the expelled drying agent is recycled.

In Eqs. (9.9), the parameter $(h_{tr}A)_h$ is the heat transfer resistance between the heat source and the system, $T_b$ is the average temperature of the boundary between the system and the heat source, and $\dot{Q}_{in} = (h_{tr}A)_h(T_b-T)$.

### 9.2.1.3 Efficiency-Based Objective Functions

The energy efficiency of the drying system is defined as follows:

$$\eta = \frac{E_{ideal}}{E_{in}} \tag{9.10}$$

and represents the ratio between the ideal (minimum) energy for moisture removal ($E_{ideal}$) and the actual energy input into the system.

In some approaches, $E_{ideal}$ is calculated for the actual operating conditions of the dryer, but assuming an operation without any losses. This is the case for which Eq. (4.19) is developed which is rewritten here as follows:

$$\eta = \left(\frac{\dot{m}_p}{\dot{m}_a}\Delta W\right)\frac{h_{w,out} - h_{bw,in}}{\Delta h_{ha} + w_{in}} \tag{9.11}$$

where $h_{w,out}$ represents the specific enthalpy of the moisture in the humid air at exit, $h_{bw,in}$ is the specific enthalpy of the bounded moisture in the moist material and inlet as given by Eq. (4.12), $\Delta h_{ha} = \Delta h_a + \Delta(\omega h_w)$ is the change of specific enthalpy of the humid air (inlet–outlet), and $w_{in} = \dot{W}_{in}/\dot{m}_a$ is the work per unit of mass flow rate of dry air required by the system as input.

When the energy efficiency given by Eq. (9.11) is used as an objective function for the drying systems optimization, one needs to account for the fact that $\Delta W$ and $h_{bw,in}$ are fixed quantities. Indeed, the change of moisture content is imposed by technological demand for the design. Also, the moist material is clearly defined and so is the enthalpy of its bounded moisture. Therefore, the energy efficiency objective function to be maximized becomes

$$\text{EEOF}_1 = m\frac{h_{\text{w,out}} - h_{\text{bw,in}}}{\Delta h_{\text{ha}} + w_{\text{in}}} \tag{9.12}$$

where $\perp$ is a short notation for $\dot{m}_p/\dot{m}_a$.

Another fair approach for the energy efficiency objective function is to take $E_{\text{ideal}} = W_{\text{rev}}$ in Eq. (9.10) which sets the system boundary as in Figure 9.1, assuming that the moist material enters at $T_0$ and leaves as dry material also at $T_0$. This gives a more absolute assessment parameter of a drying process in terms of energy efficiency. Therefore, for energy efficiency, one has

$$\eta = \frac{\dot{m}_p\left(\omega_0\Delta W\text{ex}_{\text{w,0}} + \Delta\text{ex}_p\right)}{\dot{m}_a\left(h_a + \omega h_w\right)_{\text{out}} + \dot{m}_p\Delta h_p + \left(h_{\text{tr}}A\right)_1\left(T - T_0\right)} \tag{9.13}$$

Accounting for the fact that the reversible work is fixed (the numerator), an energy input objective function representing the denominator of Eq. (9.13) can be set for minimization. This function is:

$$E_{\text{in}}\text{OF} = \dot{m}_a\left(h_a + \omega h_w\right)_{\text{out}} + \dot{m}_p\Delta h_p + \left(h_{\text{tr}}A\right)_1\left(T - T_0\right) \tag{9.14}$$

However, it is important to note that often in numerical optimization normalized objective functions are preferred for better effectiveness of numerical algorithms. Therefore, it is meaningful and effective to use an energy efficiency objective function to be maximized as follows:

$$\text{EEOF}_2 = \frac{m\left(\omega_0\Delta W\text{ex}_{\text{w,0}} + \Delta\text{ex}_p\right)}{\left(h_a + \omega h_w\right)_{\text{out}} + m\Delta h_p + \left(h_{\text{tr}}A\right)_1\left(T - T_0\right)/\dot{m}_a} \tag{9.15}$$

Regarding the exergy efficiency of the drying system, this is the ratio between the exergy used by the ideal system and the exergy input, written as follows:

$$\psi = \frac{\text{Ex}_{\text{ideal}}}{\text{Ex}_{\text{in}}} \tag{9.16}$$

where $\text{Ex}_{\text{ideal}}$ represents the ideal exergy required to remove the moisture.

If $\text{Ex}_{\text{ideal}}$ is estimated with respect to the actual operating conditions for the process, then Eq. (4.21) can be adopted for exergy efficiency. In this approach, $\text{Ex}_{\text{ideal}} = \dot{m}_r\left(\text{ex}_{\text{w,out}} - \text{ex}_{\text{bw,in}}\right)$, where $\dot{m}_r$ is the removed moisture mass flow rate and $\text{ex}_{\text{w,out}}$ is the specific entropy of the moisture in the exhaust stream. Furthermore, the exergy of the exhausted air and the exergy of the dried product are lost exergies; therefore, the exergy input is taken as follows: $\dot{E}x_{\text{in}} = \dot{m}_a\left(\text{ex}_{\text{ha,in}} + w_{\text{in}}\right)$, where $\text{ex}_{\text{ha,in}}$ is the specific exergy of the humid air at input. However, for a heat pump dryer, the exergy input will be $\dot{E}x_{\text{in}} = \dot{m}_a\left(\text{ex}_{\text{ha,in}} - \text{ex}_{\text{ha,out}} + w_{\text{in}}\right)$. We therefore write the exergy efficiency as follows:

$$\psi = \Delta W\left(m\frac{\text{ex}_{\text{w,out}} - \text{ex}_{\text{bw,in}}}{\Delta\text{ex}_{\text{ha}} + w_{\text{in}}}\right) \tag{9.17}$$

where for a heat pump dryer $\Delta\text{ex}_{\text{ha}} = \text{ex}_{\text{ha,in}} - \text{ex}_{\text{ha,out}}$ and for a dryer which wastes the humidified drying agent $\Delta\text{ex}_{\text{ha}} = \text{ex}_{\text{ha,in}}$.

The exergy efficiency objective function to be maximized becomes

$$\text{ExEOF}_1 = m \frac{\text{ex}_{\text{w,out}} - \text{ex}_{\text{bw,in}}}{\Delta\text{ex}_{\text{ha}} + w_{\text{in}}} \tag{9.18}$$

Based on similar consideration as for energy, it is fair to take the reversible work as ideal exergy input for drying. Furthermore, as it results from Eqs. (9.7) and (9.9), if the exergy in the dry product is not considered useful, then the system fully destroys the input exergy, regardless its configuration, $\dot{\text{Ex}}_{\text{in}} = \dot{\text{Ex}}_{\text{d,tot}}$. However, if the dried product is a fuel (e.g., dried biomass), then the exergy in the product is a useful output. The exergy balance equation becomes $\dot{\text{Ex}}_{\text{in}} = \dot{m}_{\text{p}}\text{ex}_{\text{m,out}} + \dot{\text{Ex}}_{\text{d,tot}}$. Therefore, the exergy efficiency becomes

$$\psi = \frac{\dot{W}_{\text{rev}}}{\dot{\text{Ex}}_{\text{in}}} = \frac{\dot{m}_{\text{p}}\left(\omega_0 \Delta \mathcal{W}\text{ex}_{\text{w,0}} + \Delta\text{ex}_{\text{p}}\right)}{\dot{m}_{\text{a}}\left(\text{ex}_{\text{a,in}} - \text{ex}_{\text{a,out}}\right) + \dot{m}_{\text{p}}\left(\text{ex}_{\text{m,in}} + \text{ex}_{\text{m,out}}\right) + \dot{W}_{\text{in}} + \left(h_{\text{tr}}A\right)_{\text{h}}\left(T_{\text{b}} - T\right)\left(1 - \dfrac{T_0}{T_{\text{b}}}\right)} \tag{9.19}$$

where $\text{ex}_{\text{a,out}}$ is included only for a heat pump dryer with fully recycling of the exhaust air, or otherwise $\text{ex}_{\text{a,out}}$ is set to zero; also if the product is not a fuel, then $\text{ex}_{\text{m,out}}$ is set to zero.

The exergy efficiency given by Eq. (9.20) gives two opportunities to formulate objective functions as follows:

• Total exergy destroyed objective function to be minimized (dimensional)

$$\text{TExDOF} = \dot{m}_{\text{a}}\left(\text{ex}_{\text{a,in}} - \text{ex}_{\text{a,out}}\right) + \dot{m}_{\text{p}}\text{ex}_{\text{m,in}} + \dot{W}_{\text{in}} + \left(h_{\text{tr}}A\right)_{\text{h}}\left(T_{\text{b}} - T\right)\left(1 - \frac{T_0}{T_{\text{b}}}\right) \tag{9.20}$$

• Exergy efficiency objective function to be maximized (dimensionless)

$$\text{ExEOF}_2 = \frac{m\left(\omega_0 \Delta \mathcal{W}\text{ex}_{\text{w,0}} + \Delta\text{ex}_{\text{p}}\right)}{\text{ex}_{\text{a,in}} - \text{ex}_{\text{a,out}} + m\text{ex}_{\text{m,in}} + \dot{W}_{\text{in}} + \left(h_{\text{tr}}A\right)_{\text{h}}\left(T_{\text{b}} - T\right)\left(1 - \dfrac{T_0}{T_{\text{b}}}\right)/\dot{m}_{\text{a}}} \tag{9.21}$$

It is worth to remark here one more time the importance of exergy efficiency in thermodynamic optimization. As known from the early work of Bejan (1982), thermodynamic optimization must be formulated in terms of total entropy generation minimization. Alternatively, the optimization problem can be formulated in terms of total exergy destruction minimization because $\dot{\text{Ex}}_{\text{d,tot}} = T_0\dot{S}_{\text{g,tot}}$. Here, both $\dot{S}_{\text{g,tot}}$ and $\dot{\text{Ex}}_{\text{d,tot}}$ are dimensional quantities. It is better for optimization to proceed versus a dimensionless quantity expressed in relative terms to the problem scale. For a drying process, Eq. (9.19) expresses the magnitude of the reversible work relative to the total exergy destruction plus the exergy output (in situations when the product is a fuel). In fact, as given by Eq. (9.19), the exergy efficiency is a dimensionless form of total exergy destruction.

### 9.2.1.4 Other Technical Objective Functions and Constraints

Some other technical parameters can be used as objective functions or constraints in a drying systems optimization. An exergetic assessment parameter for drying systems is the sustainability index mentioned previously in Chapter 4. This parameter must be maximized, so according to Eqs. (4.26) and (9.17), a sustainability index objective function can be formulated for a drying system as follows:

$$SIOF = \frac{\Delta ex_{ha} + w_{in}}{\Delta ex_{ha} + w_{in} - m\Delta \mathcal{W}(ex_{w,out} - ex_{bw,in})} \qquad (9.22)$$

Drying time is also a technical parameter characterizing a drying process. One wants always to minimize the drying time for better productivity. In an optimization problem, a dimensionless drying time objective function can be formulated based on Eqs. (7.11)–(7.16). The dimensionless drying time objective function to be minimized becomes

$$DDTOF = \frac{1}{\mu_1^2} \ln\left(\frac{LF}{\Phi_f}\right) \qquad (9.23)$$

where $\mu_1$ is the eigenvalue, LF is the lag factor, and $\Phi_f$ is the final dimensionless moisture content. It is obvious that DDTOF is a function of constructive parameters and operating conditions of the drying system.

The drying quality (DQ) parameter can be used as a constraint in a drying optimization problem. DQ has been introduced previously by Eq. (4.28). A range of technically sound DQ can be specified to limit the optimization domain. Table 9.1 gives some possible technical constraints for the optimization. Besides DQ, these include the drying effectiveness, shrinking ratio, and half time. The shrinkage ratio is a technical parameter much related with the product quality. The acceptable range for the shrinkage ratio can be specified. Other technical constraints can be formulated depending on the case, such as the range of Reynolds number, the maximum admissible temperature difference inside the dryer, the admissible range of material temperature, and so on.

**Table 9.1** Parameters utilizable as constraints for drying systems optimization

| Parameter | Definition | Remarks |
|---|---|---|
| Drying quality | $DQ = \dfrac{\dot{m}_r}{\dot{m}_p(1 + W_{in})}$ | Represents the ratio between removed mixture rate and feeding rate of moist material |
| Drying effectiveness | $DE = \dfrac{\omega_{out}}{\omega_{in}}$ | Represents the ratio between the humidity ratios in the drying agent at the output with respect to input |
| Shrinkage ratio | $SR = c\left(\dfrac{W_{out}}{W_{in}} - 1\right) + 1$ | Represents the volume ratio of the dry versus moist material. It can be correlated with moisture content ratio as in Eq. (7.23) with $c$ a regression constant |
| Half time | $t_{0.5} = -\dfrac{1}{S}\ln\left(\dfrac{1}{2LF}\right)$ | Represents the time required for the moisture content to reach half of its initial value. $S$ is drying coefficient and LF is lag factor |

## 9.2.2 Environmental Objective Functions

Here, some relevant objective functions which account for the environmental impact of the drying system are introduced. The environmental emissions for a drying system can be determined based on the actual dryer type. The energy source type influences the kinds and magnitude of system emissions and effluents and therefore the environmental impact. As stated already in Chapter 8, the environmental impact is of two kinds: direct (due to direct pollution during system operation) and indirect (accounting for all environmental impact produced during system construction).

Before formulating objective functions for environmental impact, it is instructive to analyze some relevant cases of pollution produced by drying systems. The rate of pollution is to be determined in correlation with production rate of dry material. First of all, let us consider the environmental pollution associated with the national grid. As an example, Figure 9.3 shows the grid emission indicator (GEI) given in grams of greenhouse gas (GHG) emissions per kilowatt-hour of power generated in Canada and its provinces. The average GEI for Canada is 200 g/kW h.

Now, the GEI can be used to estimate the dryer emissions. The exergy efficiency of a heat pump drying system can be used to determine the required power demand as follows: $\dot{E}x_{in} = \dot{m}_p w_{rev}/\psi$, where $\dot{m}_p$ is the production rate of dry material and $w_{rev}$ is the specific reversible work defined by $\dot{W}_{rev} = \dot{m}_p w_{rev}$. The specific exergy input can be now defined as follows:

$$ex_{in} = \frac{w_{rev}}{\psi} \tag{9.24}$$

where $ex_{in}$ is measured in kW h/kg d.b. (dry basis).

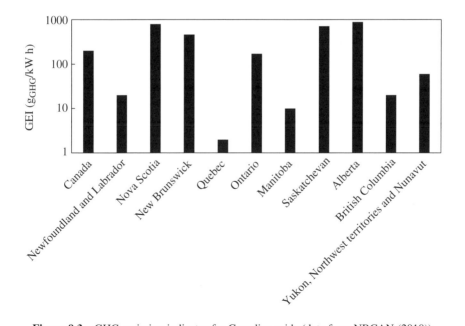

**Figure 9.3**   GHG emission indicator for Canadian grids (data from NRCAN (2010))

In this regard, the dryer emission indicator in terms of grams of GHG emitted per mass unit of product (dry basis) can be defined as follows:

$$DEI = ex_{in} GEI = \frac{w_{rev} GEI}{\psi} \tag{9.25}$$

In Eq. (9.25), one recognizes that for a specified drying process and a specified geographical region, the parameters GEI and $w_{rev}$ are fixed. Therefore, minimization of DEI required maximization of $\psi$. This means that $ExEOF_{1,2}$ can be used for optimizing the system both with respect to highest exergy efficiency and lowest dryer emission indicator.

If a combustion fuel is used for generating heat for a drying process, the emission indicator can be determined based on the analysis of the combustion process. Here, we show that the emission indicators for combustion-driven dryers can be correlated with dryer exergy efficiency for a defined region. Taking Canada as an example, the overall energy and exergy efficiencies considering all sectors are 40% and 37%, respectively, as shown in the study from Dincer and Zamfirescu (2011, Chapter 17). This number, together with the data presented in Dincer and Zamfirescu (2014, Chapter 2), leads to approximate determination of the life cycle GHG emissions for various energy consumption technologies, as shown in Figure 9.4. The figure then shows an exergetic life cycle emission indicator of GHG, which is measured in grams of GHG emission with respect to source exergy consumption.

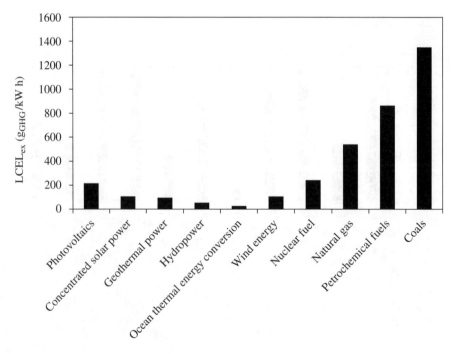

**Figure 9.4** Life cycle exergetic emission indicator for various energy sources usage (approximated for Canada)

Furthermore, the consumed exergy fraction for each type of fuel in Canada is given in Table 8.1 from where it results that the exergy fractions for fossil fuels are 5.1% for coals, 31.3% for natural gas, and 51% for petrochemical fuels, representing a total of 87.4% from Canada's exergy consumption. Therefore, among fossil fuels exergy, the share of coal is 5.8%, of natural gas 35.8%, and of petrochemical fuels is 58.4%. From Figure 9.4, $LCEI_{ex}$ is 600 g/kW h for natural gas, 900 g/kW h for petrochemical fuels, and 1350 g/kW h for coals. Henceforth, the average emission from fossil fuels in terms of $LCEI_{ex}$ is 786 g/kW h. This demonstrates that $LCEI_{ex}$ can be determined for a region of interest. When $LCEI_{ex}$ is used to quantify the exergetic specific emission of GHG, then the following equation results for DEI:

$$DEI = \frac{w_{rev}LCEI_{ex}}{\psi} \tag{9.26}$$

Besides the GHG emission, other atmospheric pollutants may be considered for more detailed environmental impact assessment. The atmospheric pollutants released by the combustion processes are given in Table 9.2. An air pollution indicator (API) can be devised in a similar manner as for the GHG emission indicator. The API is higher than GEI or $LCEI_{ex}$ and it can be generally correlated to it. For a first-hand estimation, in Table 9.3, a correlation between API and $LCEI_{ex}$ is provided. Using the API, the equation for DEI becomes

**Table 9.2** Atmospheric pollutants released by combustion systems

| Pollutant | Explanations |
| --- | --- |
| GHG | Greenhouse gases. These are gases that produce greenhouse effect. The main GHGs are $CO_2$, $CH_4$, $N_2O$. Greenhouse effect is the main cause of global warming |
| CO | Carbon monoxide. Arises mostly from the incomplete combustion of fuels and poses health and life of animals and humans at high risk upon inhalation. Intense emission of CO in open spaces may affect birds, whereas emissions in close spaces (residences, garages) may induce death of exposed persons |
| $SO_2$ | Sulfur dioxide. It is a corrosive gas which is hazardous to human health and harmful to the natural environment. It results from combustion of coal and fuel oil, smelting of nonferrous metal ores, oil refining, electricity generation, and pulp and paper manufacturing. It causes respiratory difficulties, damages green plants, and is a precursor of acid precipitation |
| $NO_x$ | Nitrogen oxides (NO and $NO_2$). It is produced by combustion of fuels at high temperature in all combustion facilities from large scale to small scale (including motor engines, furnaces, etc.). It can lead to respiratory problems of humans and animals. It can form acids in high-altitude atmosphere |
| VOCs | Volatile organic compounds. Are volatile organic particles resulting from hydrocarbon combustion in engines. Have harmful effects in the atmosphere; it impedes formation of stratospheric ozone |
| PM | Particulate matter. Particles in the air (fly ash, sea salt, dust, metals, liquid droplets, soot) come from a variety of natural and human-made sources. Particulates are emitted by factories, power plants, vehicles, and so on and are formed in the atmosphere by condensation or chemical transformation of emitted gases. PM cause of health and environmental effects include acid precipitation, damage to plant life and human structures, loss of visibility, toxic or mutagenic effects on people, and possibly nonaccidental deaths |

**Table 9.3** Rough estimation of life cycle air pollution versus GHG indicator for various dryer systems

| Dryer system description | API / LCEI$_{ex}$ |
|---|---|
| Conventional large-scale dryer based on fossil fuels | 2.4 |
| Medium-scale dryer using combustion heating and grid power for auxiliaries | 2.5 |
| Solar heat pump dryer using PV arrays | 3.72 |
| Grid-connected heat pump dryer | 2.1 |
| Conventional solar dryer | 1.8 |

$$DEI = \frac{w_{rev}API}{\psi} \tag{9.27}$$

Sustainability is much related to environmental impact. A greenization factor is introduced in Dincer and Zamfirescu (2013) to quantify the sustainability with the help of the pollutant emission indicator. Provided that a reference drying system is specified, having EI$^0$ as the reference emission indicator, the following equation is used to define the greenization factor:

$$GF = \frac{EI^0 - EI}{EI^0} \tag{9.28}$$

It is important to note that the greenization factor varies from 0 to 1. A greenization factor of zero indicates that the system is not greenized. If the system is fully greenized, then the greenization factor is 1. Fully greenized systems depend on sustainable energy sources that have zero or minimal environmental impact during the utilization stage (although some environmental impact is associated with system construction). Depending on the specific problem analyzed, various types of environmental impact factors may be formulated.

Although in general the maximization of exergy efficiency leads to improved environmental impact, some objective functions can be formulated in terms of minimization of emission indicator or maximization of greenization factor.

## 9.2.3 Economic Objective Functions

Various economic objective functions can be formulated for drying systems optimization. A very interesting option for drying systems design is the minimization of the levelized product price. Any drying process is used for generation of a product which is valued economically (it is to be sold). The levelized product price represents the selling price of the product levelized with respect to the present value and for a time horizon equal to the lifetime of the system. A way of formulating the levelized product price is as follows:

$$LPP = \frac{C_{cap} + C_{o\&m} + C_{oc}}{LT\,AP} \tag{9.29}$$

where AP is the annual production in kilogram of dry product, $C_{cap}$ is the capital cost, and $C_{o\&m}$ and $C_{oc}$ are the total operational and maintenance and other costs for the lifetime, respectively.

As a matter of fact, Eq. (9.29) can be expanded using present worth economic analysis. Therefore, a levelized dry product price objective function to be minimized can be derived as follows:

$$
\text{LPPOF} = \frac{r_m \mathcal{C}_{cap}(1-t_i)\left[2 + \text{CP} + t_p(1-t_i) + \dfrac{\text{LCC}(1-\text{PVF})\text{AC}}{r_m(1-t_i)\mathcal{C}_{cap}} - \text{CSF}(1-t_S)\text{PVF} - t_i t_c\right]}{\text{AP}(1+\text{PVF})(1-\text{OMCF})}
$$

$$(9.30)$$

where all the parameters are defined in Tables 8.2 and 8.3.

The levelized product price must be set to a competitive value which leads to setting a constraint to the optimization problem. Another economic objective function can be the payback period (PBP). The PBP is defined as the number of years required to return the investment cost $\mathcal{C}_{cap}$ taking in account the annual operational costs and annual income (AI). As defined, the PBP is given by the following equation:

$$
\text{PBP} = \frac{\mathcal{C}_{cap}}{\text{AI} - \mathcal{C}_{o\&m} + \mathcal{C}_{oc}}
$$

$$(9.31)$$

In some cases, it may be preferred to set the PBP as a constraint, whereas the levelized product price is taken as an objective function.

## 9.3 Single-Objective Optimization

This is the most basic optimization case when one single-objective function is formulated and must be minimized or maximized. The optimum exists if the objective function is concave or convex. These types of problems are often encountered in thermal system engineering when design constraints are applied to an objective function. The objective function is generally represented by a design assessment parameter such as those detailed in Section 9.2.

### 9.3.1 Trade-off Problems in Drying Systems

One typical design optimization opportunity for drying systems is represented by the maximization of the moisture transfer conductance subjected to real-world constraints such as finite size materials and finite time processes. This type of problem is illustrated in the diagram shown in Figure 9.5 where the variation of the moisture transfer conductance against a design parameter such as Biot number for mass transfer ($Bi_m$) is presented.

The magnitude of Biot number is considerably influenced by the size of the moist material and the moisture transfer coefficient, which in turn is influenced by flow conditions in the drying chamber (Reynolds number, Nusselt number, Dincer number, humidity ratio). There is a limited practical range for $Bi_m$ due to the system constraints. Therefore, the design can be based on one or more parameters which have to be optimized. When Biot number is small, the moisture transfer conductance due to diffusion through the moist material is high, whereas the moisture transfer through convection at the material surface has low conductance. When Biot

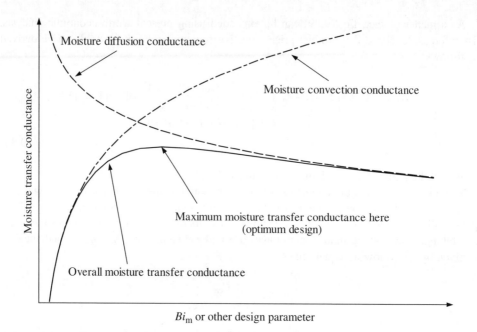

**Figure 9.5**  Drying process optimization as a trade-off problem of balancing between moisture diffusion through the material and convective moisture transfer at the surface

number is high, the conductance of moisture transfer is high in the material surface, whereas the moisture diffusion conductance through the material is small. The overall moisture transfer conductance represents a superposition of the moisture diffusion through the material and moisture convection at material surface. The superposition is made according to the following equation:

$$\bar{h}_m = \left( \frac{1}{h_m} + \frac{L_c}{D} \right)^{-1} \tag{9.32}$$

where $\bar{h}_m$ is the overall moisture transfer coefficient, $h_m$ is the moisture transfer coefficient, and $D$ is the moisture diffusivity.

The best (optimized) design is identified in Figure 9.5 by the point of overall maximum moisture transfer conductance. Mathematically, this point is determined by setting the first derivative of the objective function to zero. However, as pointed out in Bejan (2003), the first-hand approximation of the optimum can be found at the intersection of two asymptotes approximating the overall process at two extremes. In the example given here, the moisture diffusion controls the process at the extreme case represented by small Biot number. Also, the moisture convection controls the process at large Biot number (other extreme). As visible in the figure, the intersection of the asymptotes falls close to the optimum Biot number; thence, it represents a first-hand approximation for the optimization problem. Similar trade-off problems with concave objective functions are often found in thermal design optimization (see Figure 9.6).

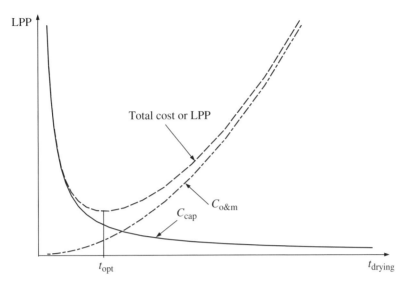

**Figure 9.6** Optimization of drying time for levelized product price or total cost minimization

When two opposing trade-offs are superimposed, a convex objective function can be formed. Here, we give an example of a drying time optimization problem having a convex objective function. Drying time depends on the actual operating condition which in turn depends on design parameters. If the drying time is short, then (potentially) the production rate is high. However, devising a dryer which features shorter drying time may be expensive, whence the capital cost is high. Furthermore, a high-quality dryer has higher efficiency, and therefore, its cost of operation and maintenance is small. In brief, for short drying time, the design features high capital cost and small operating and maintenance cost. At longer drying time, a system features lower capital cost, but higher operating and maintenance cost. Thenceforth, a trade-off exists between the capital cost and operation and maintenance cost. Since the total cost is a summation of the capital and operating and maintenance cost, namely,

$$C_{tot} = C_{cap} + C_{o\&m}$$ (9.33)

an optimum drying time can be found which minimizes the total costs.

Furthermore, the levelized product price (the price with which the dry product is sold) is in direct connection with the total cost. Therefore, the LPP shows a very similar trend with the total cost, and it has therefore a minimum. Finding the dryer design featuring the lowest selling price of the product is of major importance for the economic success of the system. Often, LPP shows also a minimum versus the production capacity.

Note that constraints play a major role in formulation of the optimization problem. As mentioned previously, essentially the constraints arise from limitations such as finite time for the processes, finite size of the system, finite material and financial resources, and so on. The constraints impose restrictions in the problem variables by fixing lower and upper bounds. For example, a highest safe temperature can be prescribed depending to the materials used in the design. Also, minimum values of the temperature may be specified for moist material affected by deformation. In a drying system, a minimum and a maximum humidity ratio can be set for the drying air. In addition, the maximum flow rate can be specified or the allowable pressure drop within the air ducts can be set.

Many constraints in thermal systems arise because of conservation laws, particularly those related to mass, momentum, and energy. For instance, under steady-state conditions, mass inflow to the system must equal mass outflow. This condition gives rise to an equation that must be satisfied by the relevant design variables, thus restricting the values that may be employed in the search for an optimum. Similarly, energy balance considerations are important in thermal systems and may limit the range of temperatures, heat fluxes, dimensions, and so on that may be used. Several such constraints are often satisfied during modeling and simulation because the governing equations are based on conservation principles. In this way, the objective function being optimized already considers these constraints. In such cases, only the additional limitations that define the boundaries of the design domain remain to be considered.

### 9.3.2 Mathematical Formulation and Optimization Methods

In a single-objective optimization problem, it is required that a single quantity is minimized or maximized by variation of a number of controlled parameters. Minimization or maximization is mathematically equivalent. The varied parameters bear the name of decision variables. To select these decision variables, it is important to (i) include all important variables that could affect the performance and cost-effectiveness of the system, (ii) not include variables with minor importance, and (iii) distinguish among independent variables whose values are amenable to change.

Let us consider a real function denoted as objective function $F_{obj}$ which is defined on a domain $\Omega \in \mathbb{R}^n$. Mathematically, one writes $F_{obj} : \Omega \to \mathbb{R}$. The optimization problem requires to find the set of optimum decision variables $\{X_{opt}\} \in \Omega$ for which $F_{obj}$ is minimum. The elements of $\{X_{opt}\}$ are denoted optimal solutions, and each element is a vector of $n$ elements $X_{opt} = (x_1, x_2, \ldots, x_n)_{opt}$. The problem is stated as follows:

$$\min\{F_{obj}(X)|F_{obj} : \Omega \to \mathbb{R}, \Omega \in \mathbb{R}^n, \text{constraints}\} \tag{9.34a}$$

$$\{X_{opt}\} = \underset{X \in \Omega}{\arg\min}\{F_{obj}(X)|F_{obj} : \Omega \to \mathbb{R}, \Omega \in \mathbb{R}^n, \text{constraints}\} \tag{9.34b}$$

The formulation given previously to the optimization problem is very general as it assumes that the problem has multiple solutions – that is, there are multiple minima (locally and globally) and all are considered solutions. Because in general many solutions exist, the optimization problem is denoted as multimodal. Of course, particular problems can have a unique solution in form of a global minimum (or maximum). The formulation from Eq. (9.34a) asks for the minimum value of the objective function. The formulation from Eq. (9.34b) asks for the set of optimal solutions.

The existence of the optimal solution is necessary if the objective function is continuous on a compact interval, as stated by the theorem of Weierstrass. The optimality conditions represent a set of necessary and sufficient conditions that can be used to identify the optimal solution. With respect to optimality, the optimization problems are of three categories, as follows:

1. Unconstrained Optimization
   - In this problem, no constraints are specified.
   - The necessary optimality condition is given by Fermat theorem which states that the optima are necessarily found in the stationary points or at the domain boundary. In a stationary point, the gradient of the objective function is zero; therefore, the necessary optimality condition is

$$\nabla F_{obj}\left(X_{opt}\right) = 0 \qquad (9.35)$$

   - The sufficient optimality condition for a twice differentiable objective function is determined by the sign of the second-order derivative or the Hessian (matrix of the second-order derivatives) which has to be positive for a minimum or negative for a maximum.
   - Note that if the objective function and its derivative are not continuous, the Fermat optimality conditions do not apply.
2. Equality-Constrained Optimization
   - The Lagrange multiplier method defines and finds the optimality.
3. Equality- and Inequality-Constrained Optimization
   - The Kuhn–Tucker conditions can be used to identify the optimum solution.

In the case when the objective function is not continuous, there are several numerical methods to apply for solving the optimization problem. A brief description of the methods is as follows:

- *Linear Programming*: Studies the case in which the objective function $f$ is linear and the set $A$, where $A$ is the design variable space, is specified using only linear equalities and inequalities
- *Integer Programming*: Studies linear programs in which some or all variables are constrained to take on integer values
- *Quadratic Programming*: Allows the objective function to have quadratic terms, while the set $A$ must be specified with linear equalities and inequalities
- *Nonlinear Programming*: Studies the general case in which the objective function or the constraints or both contain nonlinear parts
- *Stochastic Programming*: Studies the case in which some of the constraints depend on random variables
- *Dynamic Programming*: Studies the case in which the optimization strategy is based on splitting the problem into smaller subproblems
- *Combinatorial Optimization*: Concerns problems where the set of feasible solutions is discrete or can be reduced to a discrete one

Another class of optimization methods is by evolutionary algorithms. An evolutionary algorithm utilizes techniques inspired by biological evaluation reproduction, mutation, recombination, and selection. Candidate solutions to the optimization problem play the role of individuals in a population, and the fitness function determines the environment within which the solutions "live." Evolutionary algorithm methods include genetic algorithms (GAs), artificial neural networks, and fuzzy logic.

A GA is a search method used for obtaining an optimal solution which is based on evolutionary techniques that are similar to processes in evolutionary biology, including inheritance, learning, selection, and mutation. The process starts with a population of candidate solutions called individuals and progresses through generations, with the fitness of each individual being evaluated. Fitness is defined based on the objective function. Then, multiple individuals are selected from the current generation based on fitness and modified to form a new population. This new population is used in the next iteration, and the algorithm progresses toward the desired optimal point.

### Example 9.1

In this case study, an example of constrained optimization for a tray dryer is presented. The problem is an extension of the case study presented in Vargas et al. (2005). The chamber has the form of a duct with dimensions $L_1 \times L_2 \times L_3$ as shown in Figure 9.7. Dry air is blown in the duct using a blower which consumes electrical energy at the rate $\dot{W}_{\text{blower}}$. Determine the maximum moisture removal provided that the $\dot{W}_{\text{blower}}$ and the dimensions $L_2, L_3$ are fixed.

- This is an optimization problem under constraints in which the decision variable is $L_1$ (since it is stated that $L_2, L_3$ cannot change). The problem can be formulated mathematically as follows:

$$\max\left\{\dot{m}_r(L_1)\,|\,\dot{W}_{\text{blower}}, L_2, L_3 \text{ fixed}\right\}$$

where $\dot{m}_r$ is the moisture removal rate.
- We need to derive an analytical expression. The moisture removal rate is related to the heat transfer from the air to the moist material.
- Assume that the heat required for the moist preheating can be neglected; the heat transfer rate $\dot{Q}$ between the drying air and the moist material in the tray relates to the evaporation rate $\dot{m}_r$ as follows:

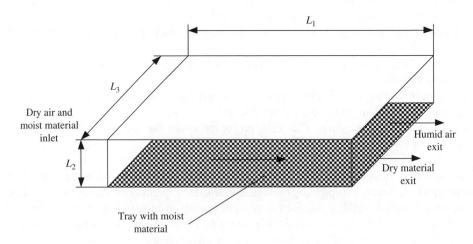

**Figure 9.7**  Tray dryer configuration for optimization Example 9.1

$$\dot{Q} = \dot{m}_r \Delta h_{lv}$$

where $\Delta h_{lv}$ is the latent heat of moisture evaporation.

- It is reasonable to assume that during the drying process the moist material temperature does not change (since the most predominant process is the evaporation of moisture surface).
- The rate of heat transfer between air and the moist product subjected to drying is given as follows:

$$\dot{Q} = h_{tr} A_{tr} \frac{T_{in} - T_{out}}{\ln \frac{T_{in} - T_p}{T_{out} - T_p}}$$

where $h_{tr}$ is the heat transfer coefficient between air and tray and $A_{tr}$ is the heat transfer area, which is the same as tray area $A_{tr} = L_3 L_1$.

- The energy balance for the air enclosed in the control volume $L_1 \times L_2 \times L_3$ can be written as follows:

$$\dot{Q} = \dot{m} C_p (T_{in} - T_{out})$$

where $\dot{m}$ is the mass flow rate of air.

- Combining the given two equations, the air temperature at the dryer outlet results as follows:

$$T_{out} = T_p + (T_{in} - T_p) \exp\left( -\frac{h_{tr} L_3}{\dot{m} C_p} L_1 \right)$$

- If the expression for $T_{out}$ is inserted into the energy balance equation for air, then the following expression is obtained:

$$\dot{Q} = \dot{m}_r \Delta h_{lv} = \dot{m} C_p (T_{in} - T_p) \left[ 1 - \exp\left( -\frac{h_{tr} L_3}{\dot{m} C_p} L_1 \right) \right]$$

where one recalls that $\dot{Q} = \dot{m}_r \Delta h_{lv}$.

- Therefore, the moisture removal rate becomes

$$\dot{m}_r = \left( \frac{C_p (T_{in} - T_p)}{\Delta h_{lv}} \right) \dot{m} \left[ 1 - \exp\left( -\frac{h_{tr} L_3}{\dot{m} C_p} L_1 \right) \right]$$

- Now, the $\dot{W}_{blower}$ constraint is used to relate the mass flow rate to the duct length $L_1$.
- The power consumed by the blower to circulate air above the trays is given by $\dot{W}_{blower} = \dot{m} \Delta P / \rho$, where $\Delta P$ is the pressure drop across the air duct and $\rho$ is the air density.
- The pressure drop across the drying chamber $\Delta P$ is related to the duct length and the hydraulic diameter $D_h$ of the duct as follows:

$$\Delta P = 2 f \rho \frac{L}{D_h} \left( \frac{4 \dot{m}}{\pi \rho D_h^2} \right)^2 = 32 f L \frac{\dot{m}^2}{\pi^2 \rho D_h^5}$$

where $f$ is the friction coefficient and $D_h = 2L_2L_3/(L_2 + L_3)$ is the hydraulic diameter of the duct.

- Further, the mass flow rate of air becomes

$$\dot{m} = \left(\frac{\pi^2 \rho^2 D_h^5 \dot{W}_{blower}}{32 f L_1}\right)^{\frac{1}{3}}$$

- When the given expression for mass flow rate is introduced in the equation for $\dot{m}_r$, the following results:

$$\dot{m}_r = \left(\frac{C_p(T_{in} - T_p)}{\Delta h_{lv}}\right)\left(\frac{\pi^2 \rho^2 D_h^5 \dot{W}_{blower}}{32 f L_1}\right)^{\frac{1}{3}}\left\{1 - \exp\left[-\frac{h_{tr}L_3}{C_p}\left(\frac{32 f L_1}{\pi^2 \rho^2 D_h^5 \dot{W}_{blower}}\right)^{\frac{1}{3}} L_1\right]\right\}$$

- In the given expression, all parameters are constant except $L_1$ which is the decision variable. Let us make the following notations:

$$K_1 = \left(\frac{C_p(T_{in} - T_p)}{\Delta h_{lv}}\right)\left(\frac{\pi^2 \rho^2 D_h^5 \dot{W}_{blower}}{32 f}\right)^{\frac{1}{3}}; \quad K_2 = \left[\frac{h_{tr}L_3}{C_p}\left(\frac{32 f}{\pi^2 \rho^2 D_h^5 \dot{W}_{blower}}\right)^{\frac{1}{3}}\right]^{\frac{3}{4}}$$

- The mass flow rate becomes

$$\dot{m}_r = K_1 \frac{1}{L_1^{\frac{1}{3}}}\left[1 - \exp\left(-K_2 L_1^{\frac{4}{3}}\right)\right]$$

- Let us make a change of variable as follows:

$$K_2 L_1^{\frac{4}{3}} = x^{\frac{4}{3}} \rightarrow L_1^{\frac{1}{3}} = \frac{1}{K_2^{\frac{1}{4}}} x^{\frac{1}{3}}$$

- The removed moisture rate becomes

$$\dot{m}_r(x) = \left(K_1 K_2^{\frac{1}{4}}\right)\frac{1}{x^{\frac{1}{3}}}\left[1 - \exp\left(-x^{\frac{4}{3}}\right)\right]$$

- Since $K_1$ and $K_2$ are constants, the objective function which maximizes the moisture removal rate is

$$F_{obj}(x) = \frac{1 - e^{-x^{\frac{4}{3}}}}{x^{\frac{1}{3}}}$$

- The constraint is already embedded in the analytical expression for $F_{obj}$. Therefore, the initial constrained optimization problem is converted into an unconstrained optimization expressed by $\max\{F_{obj}(x)\}$. Thence, Fermat theorem applies and the stationary point is obtained by setting the first derivative of the objective function to zero, as follows:

$$\frac{dF_{obj}}{dx} = \frac{4}{3}e^{-x^{\frac{4}{3}}} - \frac{1}{3}\frac{1-e^{-x^{\frac{4}{3}}}}{x^{\frac{4}{3}}} = 0 \rightarrow x_{opt} = 1.88993$$

- The second derivative of the objective function is determined as follows:

$$\frac{d^2F_{obj}}{dx^2} = -\frac{16}{9}x^{\frac{1}{3}}e^{-x^{\frac{4}{3}}} - \frac{4}{9}\frac{e^{-x^{\frac{4}{3}}}}{x} + \frac{4}{9}\frac{1-e^{-x^{\frac{4}{3}}}}{x^{\frac{7}{3}}}$$

- For $x = x_{opt} = 1.88993$, the value of the second derivative is calculated as $-0.1444 < 0$.
- Therefore, the objective function has a maximum at the stationary point $x_{opt} = 1.88993$ and the maximum is simply calculated as $\max\{F_{obj}\} = 8.411E-7$.
- The optimum $L_1$ becomes $L_{1,opt} = K_2^{-\frac{3}{4}}x_{opt} = 1.88993K_2^{-\frac{3}{4}}$.
- The maximum moisture removal rate becomes $\max\{\dot{m}_r\} = 8.411 \times 10^{-7}\left(K_1K_2^{\frac{1}{4}}\right)$.
- As a brief conclusion, this example shows that a fundamental optimum exists for the length of tray drying chambers, which maximized the moisture removal rate.

### 9.3.3 Parametric Single-Objective Optimization

Parametric optimization can be useful in several instances. This technique involves the optimization against a parameter. One of the uses is for finding global optimum of functions defined on compact intervals. The Fermat optimality criteria do not guarantee that the optimum is a global optimum for the specified domain. Multiple optima can be found on a compact domain, and one of them or several of them is the overall maximum or minimum. Therefore, a parametric search for the global optima can be devised in some cases deterministically. It is generally known that if the objective function is a polynomial, then its global minimization problem reduces to a convex problem. One technique of global parametric optimization is presented in Faísca et al. (2007). Further review on global optimization algorithms is given in Pardalos and Rosen (1986).

Other useful application of parametric optimization is the graphical search for optimum. Furthermore, trade-off problems with multiple assessment criteria can be reasonably solved by parametric single-objective optimization. In these problems, the effect that the perturbation of parameter produces on the optimum is quantified. In order to introduce the parametric optimization, let us consider an objective function $F(x,y) \in \mathbb{R}$ having real arguments. The optimization problem required

$$\min\{F(x,y)|(x,y) \in \Omega \in \mathbb{R}^2, \text{constraints}\} \tag{9.36}$$

The problem expressed by Eq. (9.36) is a single-objective two-variable optimization. One possible way (not the only one) to convert the problem from Eq. (9.36) into a parametric optimization is to firstly write

$$y(x_0) = \arg\min\{F(x,y)|(x,y) \in \Omega \in \mathbb{R}^2, \text{constraints}, x = x_0\} \qquad (9.37a)$$

$$\min\{F(x,y(x_0))|x \in [x_{\min}, x_{\max}], \text{constraints}\} \qquad (9.37b)$$

The problem expressed by Eq. (9.37a) is a single-objective single-variable optimization. This determines the optimum $y$ when $x_0$ is a parameter. The problem given in Eq. (9.37b) is again a single-objective, single-variable optimization with constraints. The problems in Eqs. (9.36) and (9.37) are equivalent. As mentioned, the problem from Eqs. (9.37) is often preferred in engineering applications because it can lead to global optimum and because it gives a better (sometimes graphical) representation of the trade-offs.

An example of a 2D convex function is shown in Figure 9.8, represented as a 3D surface. This function has a minimum. The parametric minimization process is represented graphically in Figure 9.9. Basically, three values of $x$ can be assumed as $x_1 < x_2 < x_3$ to cover the compact interval where $x$ is defined. Then, the objective functions $F_{\text{obj}}(x_1, y)$, $F_{\text{obj}}(x_2, y)$, $F_{\text{obj}}(x_3, y)$ are plotted for the $y$ domain. Since the surface is convex, minima are observed for each curve. In the middle, for $x = x_2$, the minimum minimorum is found. The search may require some trial and error. Eventually, a curve connecting the minima of each of the curves $F(x_i, y)$ can be drawn, and the minimum on this envelope gives the answer to the optimization problem.

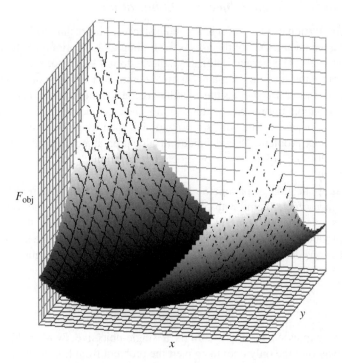

**Figure 9.8**  Example of a 2D convex function

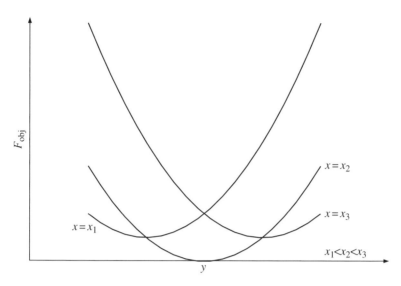

**Figure 9.9** Graphical representation of parametric minimization process of a 2D objective function

**Example 9.2**

Study the maximization of energy efficiency based on Eq. (9.11) for a tunnel dryer having a number $n$ of trays stacked vertically. The decision variables are $n$ and $L_1$ where $L_1$ is the length of the tray. The constraint is the total power of the blower. Use the results from Example 9.1. Make a graphical representation for the problem. It is given that $T_{in} = 60\,°C$, $T_p = 40\,°C$, $f = 0.01$, $L_2 = 0.2\,m$, $L_3 = 1\,m$, $h_{tr} = 50\,W/m^2\,K$.

- Since the total power is fixed and there are $n$ branches, it is reasonable to assume that the power divides equally such that $\dot{W}_{blower} = n\dot{W}_{branch}$.
- In Eq. (9.11), one notes that $\dot{m}_p \Delta W = \dot{m}_r$ and $\dot{m}_a \Delta h_a = \dot{Q}$. Therefore, the dryer energy efficiency becomes

$$\eta = \left(1 + \frac{\dot{W}_{blower}}{n\dot{m}_r \Delta h_{lv}}\right)^{-1} \frac{h_{w,out} - h_{bw,in}}{\Delta h_{lv}}$$

where $\dot{Q}$ is the heat transfer rate for one branch.
- The factor $K_1$ is $K_1 = 0.004377\left(\dot{W}_{blower}/n\right)^{1/3}$.
- The factor $K_2$ is $K_2 = 0.1711\left(\dot{W}_{blower}/n\right)^{1/4}$.
- The optimum $L_1$ is $L_{1,opt} = 1.88993K_2^{-\frac{3}{4}} = 7.104\left(\dot{W}_{blower}/n\right)^{-3/16}$.
- The moisture removal rate is $\dot{m}_{r,max} = 8.411 \times 10^{-7}\left(K_1 K_2^{\frac{1}{4}}\right) = 2.368 \times 10^{-9}\left(W_{blower}/n\right)^{19/48}$.
- The heat transfer rate is $\dot{Q}_{tot} = n\dot{Q} = n\dot{m}_{r,max}\Delta h_{lv} = 0.05697n\left(\dot{W}_{blower}/n\right)^{19/48}$.
- The optimum mass flow rate of air is

$$\dot{m}_{a,\,\text{opt}} = \left( \frac{\pi^2 \rho^2 D_h^5 \dot{W}_{\text{blower}}}{32 f n \times 7.104 \left( \dot{W}_{\text{blower}}/n \right)^{\frac{-3}{16}}} \right)^{\frac{1}{3}} = 0.2717 n^{\frac{19}{48}} \dot{W}_{\text{blower}}^{\frac{13}{48}}$$

- The temperature of air at outlet is

$$T_{\text{out}} = T_p + \left( T_{\text{in}} - T_p \right) \exp\left( -\left( \frac{h_{\text{tr}} L_3}{C_p} \right) \frac{L_{1,\text{opt}}}{\dot{m}_{a,\,\text{opt}}} \right)$$

which after replacing all known parameters it becomes

$$T_{\text{out}} = 40 + 20 \ \exp\left( -1.297 n^{-\frac{19}{48}} \dot{W}_{\text{blower}}^{-\frac{11}{24}} \right)$$

- Once $T_{\text{out}}$ is known, the enthalpy of water vapor in outlet air stream is determined.
- The results of the parametric optimization are presented graphically in Figure 9.10.
- The curve are obtained by varying first $L_1$ and calculating the energy efficiency for $n \in \{1,2,5,10,20\}$.
- Then, the optimum $L_1$ and the maxima of efficiency $\eta_{\text{max}}$ are determined and the curve superimposed.
- It is observed that the number of trays is small (under 4) and it is beneficially to install longer trays.
- Also if the number of trays is large (more than 4), then it is beneficially to shorten the tray length.

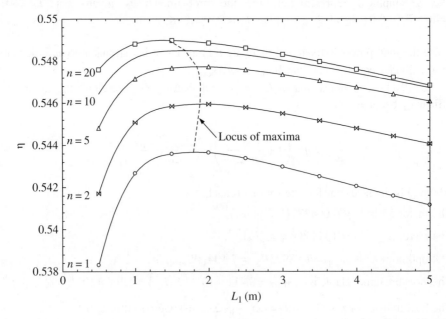

**Figure 9.10**  Graphical representation of parametric minimization process of a 2D objective function

- Overall, it is better to use many stacked trays with a length that decreases.
- The benefit returns however if the tray number is increased over 20 since the cost of trays increases much, but the increase in efficiency is marginal.

## 9.4  Multiobjective Optimization

In many practical situations, the objective function is not convex. This means that there are no maxima or minima since the usual optimality conditions do not hold. In order to illustrate the case, we take a closer look to Example 9.2. The results from Figure 9.10 show in fact a trade-off optimization (with no maximum). As seen, the curve indicated that "locus of maxima" increases continuously. However, after a while ($n > 20$), the increase rate is slow, which indicates the region of trade-off. In order to select a solution from the "locus of maxima," another parameter must be considered, beside the energy efficiency in Figure 9.10. This parameter can be the total area of the trays $A = nL_1L_2$ which increases much when the number of trays is higher.

As a matter of fact, the energy efficiency of the one-time optimized dryer can be plotted against the total tray area. If one does this, then the plot from Figure 9.11 is obtained. All discrete points indicated $n = 1 - 20$ on the figure represent optimal solutions. The dashed line is only drawn to illustrate the overall tendency, although the problem is discrete (discontinued). According to Pareto theory of optimality, the points $n$ in this optimization reached an equilibrium in the sense that they cannot be improved anymore. The dashed line is denoted as Pareto frontier and will be discussed subsequently. The problem of selection of the best is difficult and sometimes subjective, but the Pareto frontier helps with the decision.

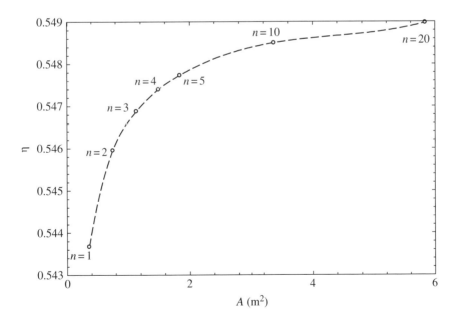

**Figure 9.11**  Pareto frontier for the parametric optimization problem from Example 9.2

According to the so-called Pareto law, in a system at equilibrium (not thermodynamic equilibrium but an equilibrium of design configuration meaning that it cannot be further improved), 20% of causes produces 80% of the effects. This allows one to select 20% of the best solutions as the most representative in the Pareto curve. This means only one point for our problem, that is, $n = 5$ as the preferred optimum.

With this simple introduction, we dwell further into multiobjective optimization. In thermal and energy systems design, efficiency (energy and/or exergy), production rate, output, quality, and heat transfer rate are common quantities that are to be maximized, while cost, input, environmental impact, and pressure are quantities to be minimized. Any of these can be chosen as the objective function for a problem, but it is usually more meaningful and useful to consider more than one objective function. When more than one objective function is considered in the optimization, we refer to the procedure as multiobjective optimization.

One of the common approaches for dealing with multiple-objective functions is to combine them into a single-objective function that is to be minimized or maximized. For example, in the design of heat exchangers and cooling systems for electronic equipment, it is desirable to maximize the heat transfer rate. However, this often comes at the cost of increased fluid flow rates and corresponding frictional pressure losses.

Another approach which has attracted much attention in recent years is the multiobjective optimization. With this approach, two or more objective functions that are of interest in a given problem are considered, and a strategy is developed to balance or trade off each objective function relative to the others.

Referring to drying system, the engineering goal is to find design configurations and operating conditions that maximize some performance indicators such as energy and exergy efficiencies, energy input, sustainability index, and greenization factor or minimize other indicators such as drying time, exergy destruction, dryer emission indicators, levelized product price, PBP, and total costs. The performance indicators (or assessment criteria) constitute themselves in objective functions. In general, it is impossible to optimize all objective functions because they vary in opposite directions. For example, efficiency increases lead to an expensive system. Hence, the best design selection must be made based on a trade-off analysis.

The multiobjective optimization requires the determination of the Pareto front which indicates the best solutions. Here, one needs to recognize optimization variables, which are the parameters that influence the system performance. In common multiobjective optimization of engineering systems with fixed configuration, the parameters to be optimized are the design (e.g., the duty of the heat exchangers) and operational parameters (e.g., mass flow rates). Figure 9.12 shows how the design degree of freedom may improve the design performance.

When the degree of freedom of the design is higher, the system performs better and the successive Pareto fronts approach the target point for optimization. Analogous Pareto fronts of multigeneration systems are shown in Figure 9.12a with incrementally higher performance as the number of outputs increases. A three-dimensional representation of optima is more useful. This is shown in Figure 9.12b. Two surface cuts are shown. The first represents the optimized system with fixed economic or environmental impact indicators (the vertical line hatched surface). Only exergy efficiency is optimized along this cut. The curve suggests that exergy destructions reduce when the number of outputs increases. Another cut is at fixed exergy destruction (or fixed exergy efficiency). Again, the increase in number of outputs enhances the revenue and reduces the environmental impact. These two cuts describe a 3D Pareto surface.

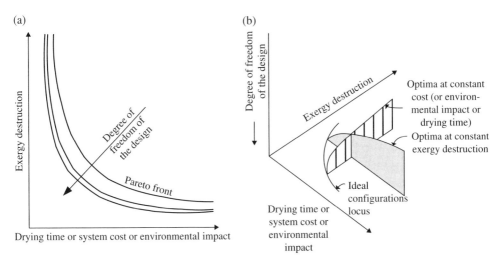

**Figure 9.12** Multiobjective optimization of multigeneration systems (modified from Bejan and Lorente (2006)): (a) two-dimensional Pareto fronts and (b) three-dimensional optimal representation

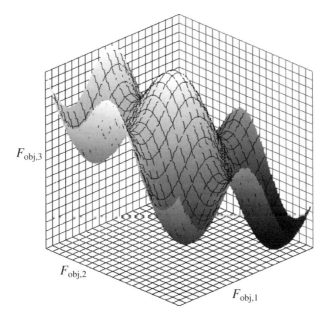

**Figure 9.13** Example of an irregular Pareto frontier of a three-objective optimization problem

In general, multiobjective optimization problems are solved using GAs which are capable to tackle with discrete problems and find global optima. Deb (2001) describes in detail the genetic multiobjective optimization algorithms as a class of evolutionary algorithms. Figure 9.13 shows a 3D Pareto frontier which is not smooth: it has multiple valleys and peaks. The GA

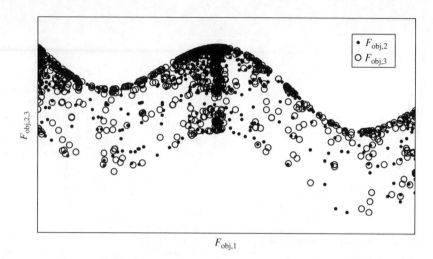

**Figure 9.14** Example of an irregular Pareto frontier of a three-objective optimization problem

proceeds by an educated search until the solution equilibrium is found at the Pareto frontier. Figure 9.14 shows the solution search by the GA. The points agglomerate at the Pareto frontier.

### Example 9.3

This example continues Example 9.3. In this case, the optimization is made with respect to three objectives, namely, the production rate of dry material, the total area of the trays, and the drying time. It is given that the tray load is of 100 kg dry basis per linear meter and that the change of moisture content in the dryer is $\Delta W = 0.001\,\text{kg/kg}$.

- First objective function is the mass flow rate of product. This is determined based on the equations presented in Example 9.3 which give an expression for $\dot{m}_r$. Further, the dry basis mass flow rate of product results from the equation:

$$\dot{m}_r = \dot{m}_p \Delta W$$

- The residence time in the dryer is calculated based on the tray length $L_1$ and production rate:

$$t_{\text{drying}} = \frac{L_1}{\dot{m}_p m_{\text{load}}}$$

where $m_{\text{load}}$ represents the material load on dry basis.
- The results are shown in Figure 9.15 and represent the Pareto frontier search using a genetic algorithm. The genetic algorithm has been set such that the number of individuals in a generation is 32, the number of generation is 64, and the maximum mutation rate is 0.2625.
- As observed, the shortest drying time corresponds to the largest product generation rate (the capacity or the scale of the dryer), but also, it corresponds with the largest tray area, that is, with the highest investment cost.

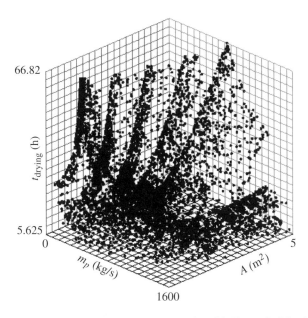

**Figure 9.15** Three-objective optimization of the dryer analyzed in Example 9.2 with respect to drying time ($t_{drying}$), mass flow rate of dry product ($\dot{m}_p$), and total tray area (m$^2$)

## 9.5 Conclusions

In this chapter, key aspects on optimization of drying processes and systems are presented and discussed. Optimization problem is in general formulated with the help of objective functions. Here, the objective functions are classified in three categories: technical, environmental, and economic. Since exergy and energy efficiencies play a major role for technical assessment of the drying system, several energy and exergy formulations for drying systems and the associated objective functions are included in the process. The reversible work of a drying system is important for the quantification of the departure of the actual system from the ideal reversible one. Some other technical objective functions include sustainability index and drying time. The selection of constraints is important and required for a true optimization of drying processes and systems. Some dryer emission indicators are presented which are useful for formulating environmental objective functions. For economic objective function, the levelized product price and the PBP are described as one of the most representative for dryers. Single- and multiple-objective optimizations of dryers are discussed with some illustrative examples to really highlight the importance of optimization. Dryer systems and processes optimization plays a crucial role in drying technologies.

## 9.6 Study Problems

9.1   Consider a through type solar dryer. Perform an optimization having the energy and exergy efficiencies as objective functions.

9.2     Identify the source of exergy destruction in a drying process and explain how this loss can be reduced. Then, optimize the dryer using appropriate decision variables and consider the total exergy destruction as your objective function.

9.3     When an engineer or designer is deciding on the preferred optimal point on a Pareto frontier, what point will be selected and why?

9.4     Does the Pareto frontier change with changing fuel costs? Explain.

9.5     For a tunnel dryer with moist material preheating section, evaluate the exergy and energy efficiencies, the Pareto frontier, and the total exergy destruction and compare these with the values for the dryer without the preheater.

9.6     Rework Example 9.1 for a different type of dryer. Also, optimize the system and draw a Pareto frontier for exergy efficiency and levelized product price.

9.7     Optimize a dryer based on exergy efficiency objective function, dryer emission indicator, and levelized product price objective function. Use a genetic algorithm.

9.8     Rework Example 9.2 for different data set and perform an optimization by considering exergy efficiency and total cost rate as two objective functions for multiobjective optimization and discuss the Pareto frontier?

9.9     For a dryer, minimize the levelized product price and dryer emission indicator assuming that the decision variables are the dry product flow rate, the drying time, and the aspect ratio of drying chamber.

9.10    Rework Example 9.3 for a different input data.

# References

Bejan A. 1982. Entropy Generation Minimization Through Heat and Fluid Flow. Wiley: New York.

Bejan A. 2003. Optimal internal structure of volumes cooled by single-phase forced and natural convection. *Journal of Electronic Packages* 125:200–207.

Bejan A., Dincer I., Lorente S., Miguel A.F., Reis A.H. 2004. Porous and Complex Flow Structures in Modern Technologies. Springer: New York.

Bejan A., Lorente S. 2006. Constructal theory of generation of configuration in nature and engineering. *Journal of Applied Physics* 100:041301.

Deb K. 2001. Multi-objective Optimization Using Evolutionary Algorithms. John Wiley and Sons Ltd: New York.

Dincer I. 2011. Exergy as a potential tool for sustainable drying systems. *Sustainable Cities and Society* 1:91–96.

Dincer I., Rosen M.A. 2013. Exergy: Energy, Environment, and Sustainable Development. Elsevier: Oxford, UK.

Dincer I., Zamfirescu C. 2011. Sustainable Energy Systems and Applications. Springer: New York.

Dincer I., Zamfirescu C. 2012. Potential Options to Greenize Energy Systems. *Energy* 46:5–15.

Dincer I., Zamfirescu C. 2014. Advanced Power Generation. Elsevier: New York.

Faísca N.P., Dua V., Rustem B., Saraiva P.M., Pistikopoulos E.N. 2007. Parametric global optimization for bilevel programming. *Journal of Global Optimization* 38:609–623.

NRCAN 2010. National energy use database. Natural Resources of Canada. Internet source accessed on June 18, 2015: http://oee.nrcan.gc.ca/corporate/statistics/neud/dpa/data_e/databases.cfm?attr¼40.

Pardalos P.M., Rosen J.B. 1986. Methods for global concave minimization: a bibliographic survey. *Society of Industrial and Applied Mathematics* 28:367–379.

Vargas J.V.C., Ordonez J.C., Zamfirescu C., Campos M.C., Bejan A. 2005. Optimal ground tube length for cooling of electronics in shelters. *Heat Transfer Engineering* 26:8–20.

# 10

# Sustainability and Environmental Impact Assessment of Drying Systems

## 10.1 Introduction

It has been already stated in previous chapter that drying is an important and energy-intensive activity specific to many sectors including agriculture and food industry, manufacturing and chemical industry, and so on. Very large amounts of thermal energy, majorly derived from fossil fuel combustion, are required annually to drive drying processes in the industry. Atmospheric and other types of pollution are therefore caused by drying systems which cannot be neglected. The most significant problems include global climate change, stratospheric ozone depletion, and acid precipitation. In addition, drying requires consumption of important energy resources such as fuels which thus deplete. As a consequence of these facts, drying systems need to be assessed from sustainability and environmental points of view.

By the end of the twentieth century, science became more and more the dominant driving force influencing the development of society. Due to the science and the establishment of the information era, a global self-conscience is developed – the so-called noosphere – that made humans aware of possible catastrophic scenarios related to the unsustainable development of technology. National and international institutions were created in the last quarter of the twentieth century to promote a sustainable development of the society. As such, the World Commission on Environment and Development acted within the United Nations for a period of five years in the mid-1980s and defined formally the concept of sustainable development and showed that society development and the environment are interrelated for any foreseeable time horizon.

*Drying Phenomena: Theory and Applications*, First Edition. İbrahim Dinçer and Calin Zamfirescu.

Sustainability in the context of this chapter refers to the feature of a drying system to be sustainable, that is, the system is conceived such that it does not permanently damage the environment, it does not consume excessive natural resources, and it operates effectively for a sufficiently long period of time such that the drying needs are satisfied without jeopardizing the ability of future generations to meet their need of natural resources and clean environment. Sustainability and sustainable development are complex and thenceforth very difficult to assess. One must go beyond environmental impact assessment of sustainability and tackle aspects of political and societal assessment.

Because of its wide character spanning from environment to society, sustainability is difficult to assess based on quantifiable indicators. Some efforts to categorize the tools for sustainability assessment are due to Ness et al. (2007). In the past, attempts were made to formulate empirical indicators for sustainability assessment such as "environmental sustainability index (SI)," "well-being index," "human development index," and so on. More recently, product-related assessment methods were proposed based on product material flow analysis, life cycle assessment, and thermodynamic analysis. Among those, the sustainability and environmental assessment based on exergy analysis is remarked for furthering the goal of more efficient resource use with reduced waste rejection into the environment.

Dincer (2002) analyzes the links between energy, exergy, and environmental impact for drying systems. The paper insists on increased exergy efficiency strategies which reduce the energy requirement of drying systems, and therefore, it reduced the environmental impact associated with production, storage, distribution of primary fuels, and grid electricity. Understanding the connection between energy, exergy, and environment reveals the fundamental patterns and forces that affect and underlie changes in the environment and consequently allow for dealing better with environmental damage created by anthropogenic systems. Exergy analysis facilitates the environmental impact and sustainability assessment by clearly specifying the reference environment and system–environment interactions.

A combined analysis based on energy, exergy, environmental impact, economics, and sustainability is presented in Dincer (2011) for drying systems. The most significant impact of drying systems on the environment is due to global climate change, stratospheric ozone depletion, and acid precipitation. Global climate change due to anthropogenic activity is believed to be caused by the increasing concentration of greenhouse gases (GHGs) in the atmosphere as an effect of polluting effluents such as combustion flue gases. Higher presence of the GHG in the atmosphere entrap thermal radiation emitted by earth and consequently produce a thermal unbalance which leads to rises of the terrestrial surface temperature and sea levels. This global warming is potentially the most important environmental problem related to energy utilization.

Sustainable energy supplies are required by a sustainable development. These types of resources must be available for a reasonable long foreseeable future at an affordable cost and reasonable access without causing negative societal and environmental impacts. Therefore, better design of drying system must be developed with increased efficiency and reduced fossil fuel consumption such that (i) less primary resources are used and (ii) reduced pollutants such as GHG ($CO_2$, $CH_4$) and acid gases ($SO_2$, $NO_x$ responsible for acid precipitation) and chlorofluorocarbon (CFC, responsible of ozone layer depletion) are emitted. Considering the significance of industrial drying in the global economy, any improvement of designs toward efficient and environmentally benign drying technology will help substantially the sustainable development of society. In this respect, exergy analysis plays a major role, as shown in Kanoglu et al. (2009), because it connects sustainability with the environment.

In this chapter, the sustainability and environmental impact assessment of drying systems is discussed. There are various aspects to be analyzed in this respect. The sustainability concept and its assessment methods are introduced with a focus on exergy-based assessment. The standard environment models are presented as a basis for determining the environmental impact in conjunction with exergy. A case study is presented for a drying system sustainability and environmental impact assessment.

## 10.2   Sustainability

At the beginning of the twenty-first century, a new branch of science came out, namely, the sustainability science, having an interdisciplinary character involving the following main disciplines: physics, chemistry, biology, medicine, social and economic sciences, and engineering. One of the goals of sustainability science is to model the complex interactions between society, economy, and environment, accounting also for resource depletion. One crucial aspect is the provision of theoretical foundations and tools for sustainability assessment. Here, we review these tools and related sustainability issues with emphasis on drying operations.

### 10.2.1   Sustainability Assessment Indicators

Often in policy planning, indicators are used to assess a country or geopolitical region from various points of view: economic, social, and so on. For example, the gross domestic product (GDP) is an economic indicator which indicates the well-being of a society. Ness et al. (2007) review and categorize some indicators for sustainability assessment which are related to environmental and social assessment. The aim of sustainability assessment indicators is to confer a basis for decision-making and policy elaboration toward sustainable development considering the integrated nature–society systems for a certain temporal perspective. In general, sustainability assessment indicators quantify in an integrative manner the economic, social, environmental, and institutional development of a country or region.

The sustainability indicators (or indices) are generally extensions of the environmental assessment indicators. Therefore, they are based on a model such as the one shown in Figure 10.1 and denoted with drivers–pressures–state–impact–responses (DPSIR). In this mechanism, the drivers are the developments in society, economy, and the environment. These developments (or changes) exert pressure on the sustainability, and as a consequence, the sustainability state changes in some direction. This eventually leads on foreseeable impact on sustainable

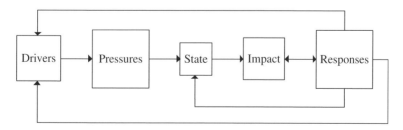

**Figure 10.1**   The DPSIR model for sustainability assessment

development: it improves or degrades the sustainability. The society can respond to this by taking actions to change the impacts directly (see the double arrow in the figure). In addition, the society can send feedback to drivers (i.e., to impose educated changes in the society, economy, environment). The responses can also act directly on the pressures and on the state of sustainability by applying adequate measures if possible. The responses must be effective, because their effectiveness affects directly the sustainable development. For sustainability assessment, this model simplifies to drivers–state–responses (DSR) or pressure–state–response (PSR).

Table 10.1 shows some significant indicators for sustainability assessment. Certainly, the indicators have some degree of subjectivity because it is not actually possible to objectively determine and fully understand the interrelationships between social change and environment and economic development. The independent formulation of environmental, economic, and social indicators is more facile.

Once independent sustainability indicators covering various aspects are determined, aggregate indicators can be constructed using weighting and scaling methods. The aggregated indicators are generally known as sustainability assessment indices. The United Nations developed more than 130 indicators for sustainability assessment of which 30% are social indicators, 17% are economic indicators, 41% are environmental indicators, and the rest of 12% represent institutional indicators. Approximately 33% of sustainability indicators are of driving force (drivers) kind, 39% are state indicators, and 18% are response-kind indicators.

Table 10.1 illustrates in fact a selection of significant indicators which can be used to elaborate aggregate SIs. In principle, the indices can be elaborated according to the purpose of the analysis. Some specific aspects on elaboration and adoption of holistic sustainability assessment methods are discussed as follows:

- Sustainability assessment should integrate ecological condition assessment of human and life system habitat with economic development, social well-being, and with equity and disparity of current population and future generations.
- Obtain a consensus on adoption of a time horizon sufficiently long to be in accordance with ecosystem timescale and anticipated society timescale.
- Sustainability indicators and indices must be able to link sustainability categories to society goals.
- Sustainability assessment should integrate the following category of models:
  ○ Economic models
  ○ Stress-response models
  ○ Multiple capital models
  ○ Social–economic–environmental models
  ○ Human-ecosystem well-being models
- Definition of a reference environment (or state) is required for determining the sustainability change.
- Monetary assessment of environmental damage should be included in sustainability assessment.

An example of development of a sustainability assessment framework is given here from Ontario Hydro, the largest power provider in Ontario. According to Hardi and Zdan (1997), there are five categories for sustainability assessment established by Ontario Hydro, namely, (i) energy and resource use efficiency, (ii) environmental integrity, (iii) renewable energy,

**Table 10.1** Categories and kinds of indicators influencing the sustainability assessment

| Category | Kind | | |
|---|---|---|---|
| | Driver indicators | State indicators | Response indicators |
| Social | • Unemployment rate<br>• Population growth rate<br>• Adult literacy rate<br>• Motor fuel consumption per capita<br>• Loss rate due to natural disasters | • Poverty gap index<br>• Income inequality index<br>• School life expectancy<br>• House price per income<br>• Floor area per person | • GDP on education<br>• Childhood immunization<br>• Infrastructure expenditure per capita<br>• Health expenditure per capita<br>• Hazardous chemicals in foods |
| Economic | • Per capita GDP<br>• Investment share of GDP<br>• Annual energy consumption<br>• Natural resource consumption<br>• Capital goods imports | • Proven mineral reserves<br>• Fossil fuel reserves<br>• Lifetime of energy reserves<br>• Share of renewable energy<br>• Manufacturing added value | • Environmental protection expenditure<br>• Funding rate on sustainable development<br>• Amount of funds on sustainability<br>• Percentage of new funds on sustainability<br>• Funding grants on technology |
| Environmental | • Water consumption per capita<br>• Generation of wastes<br>• ODP substances emission<br>• Emission of GHG, $SO_2$, $NO_x$<br>• Energy use in agriculture | • Groundwater reserves<br>• Monthly rainfall index<br>• Desertification rate<br>• Pollutant concentration<br>• Acute poisoning | • Waste water treatment expenditures<br>• Natural resource management<br>• Air pollution mitigation expenditures<br>• Waste management expenditures<br>• Number of restricted chemicals |
| Institutional | | • Scientists number per capita<br>• Engineers number per capita<br>• Internet access per capita<br>• Telephone line access<br>• Other information channels | • Policy on sustainable development<br>• Environment protection programs<br>• GDP share of R&D expenditures<br>• Number of R&D personnel per capita<br>• Ratification/implementation of agreements |

*Source:* UN (2007).

(iv) financial integrity, and (v) social integrity. For the first category of energy and resource use efficiency, the following indicators are considered in a compounded manner:

- Power consumption and transmission losses as a percentage of sales
- Fuel conversion efficiency
- Water withdrawals
- Fuel and commodity consumption
- Internal energy recovery and savings

For the environmental integrity category, the monitored emission rates are used for all activities including design, development, construction, commissioning, operation, decommissioning, and material management. Here are the considered individual sustainability indicators for this category:

- GHG emissions
- Ozone-depleting substance emissions
- Acid gas releases
- Waste management indicator
- Radioactivity levels of generated wastes
- Hazardous waste emissions
- Reportable spills
- Compliance violations
- Environmental expenditures

Regarding the specific indicators for renewable energy use, the following indicators are considered:

- Energy share generated from renewable sources
- Energy generated from wind power
- Energy generated from solar power

The consistent generation of cash flow is considered by Ontario Hydro with the help of the financial integrity category for sustainability assessment. Here, the following specific indicators are considered:

- Net income
- Interest coverage
- Debt ratio
- Total cost of the energy unit

The interaction of Ontario Hydro with the communities and with own employees determines the social integrity category for sustainability assessment. These specific indicators are selected such that innovation and greater employee involvement in sustainability are encouraged; thence, the indicators are as follows:

- Employee accident severity
- Corporate citizenship program

- Employee productivity
- Payments in lieu of taxes
- Number of public fatalities
- Aboriginal grievances
- Severity of environmental complaints and their number

The Ontario Hydro corporate developed two composite indices which assess sustainability on long-term targets and are based on the aforementioned five categories of specific indices. These are (i) resource use efficiency composite indicator focusing on inputs (e.g., fuels, water) and (ii) environmental performance indicator focusing on outputs (pollutants and waste emissions). The composite indicators are constructed such that they take positive value.

The reference value is set to zero at the level of year 1995. The value of 100 was predicted to be reached by the composite indicator in year 2000, while in year 2002, it reaches the value of 190. This case study shows how a corporate can assess sustainability internally which helps adapt its strategy to achieve its goals. In this respect, an adoption of a clear definition of the development vision is crucial.

Sustainability of production processes becomes increasingly important. In general, for industrial production processes such as drying, only certain indicators may be relevant to assess the sustainability. One question is to elucidate how to construct an index which observes the effect of drying systems on overall sustainability in a specific geopolitical context or globally. Indices that characterize the material and energy flows in the society may be very useful in this respect. This aspect is discussed subsequently and reviewed in the next section which focuses on exergy-based sustainability assessment.

Figure 10.2 shows a simplified flow diagram to be used in sustainability modeling and assessment of a drying process. There are basically two categories of inputs and two categories of outputs for a drying process (similar modeling is valid for any other production process). The inputs are the resources and the process enabled. The outputs are the actual process outputs and the process wastes. The input resources are of three kinds: moist material, drying agent, and energy. For each of the inputs, relevant indicators, quantities, and qualities can be determined. For example, the moist material is characterized by a moisture content, a specific texture, a shape and characteristic dimension, and its physicochemical properties. The energy input is required in the form of work and heat. In a heat pump dryer, the energy input is completely provided in the form of work.

The process enablers are of three kinds: the human operators, the drying system (including all machines and technical components), and the surrounding environment (characterized by certain temperature, pressure, and composition). Related to the human operators, the sustainability indicators must mainly quantify safety, health, and educational aspects. Related to the dryer system, the sustainability indicators must consider the life cycle and the specific environmental emissions and impact related to the system construction. The process output is mainly the dry material with its quantitative and qualitative characteristics. For the systems which recycle partially or totally the drying agent, there may be some other outputs considered useful: for example, the recovered moisture can be used or valorized in some ways. But the majority of system outputs besides the dry product itself are typically wasted.

Three types of wastes can be considered in drying systems: environmental pollutants, energy wastes, and material wastes. The pollutant emissions are due mainly to combustion processes directly or indirectly associated with drying. Also the emission of landfill gas due to drying

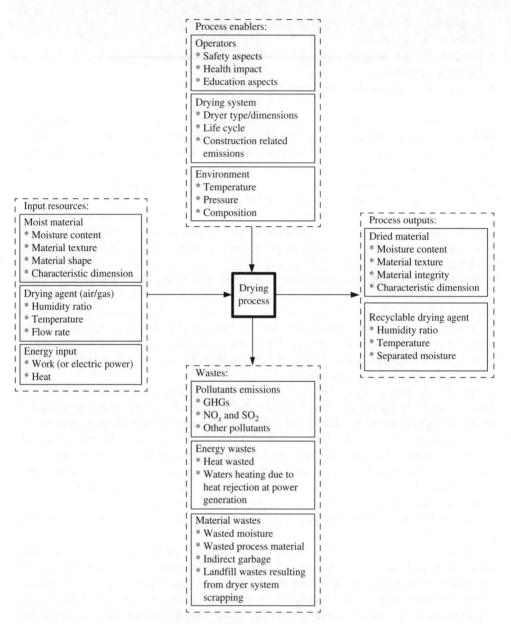

**Figure 10.2** Sustainability assessment model for a drying process, based on material and energy balances (modified from Linke et al. (2014))

system scraping at the end of the lifetime is a pollutant waste. The energy wastes can be observed in the form of heat rejected into the environment. This heat can be rejected directly by heat transfer at the system boundary. Another way of energy waste is through warm water release in lakes, rivers, or seas where power plants are installed. It is demonstrated that the cooling water for power plant condensers affects the aquatic environment by changing the local temperature which often negatively impacts the life systems. Noise is another way of energy waste with environmental impact. The material wastes include also landfill material generated at product scraping phase.

It is noteworthy that the sustainability of any production processes must be related to both the quantity and the quality of the processed material. One largely used sustainability indicator based on mass and energy balances is the eco-efficiency indicator representing the ratio between value in outputs and the environmental impact. If one denotes with EcI the eco-indicator and with DPV the drying product value and with EI the environmental impact associated to the DPV, then the eco-indicator is expressed mathematically as follows:

$$EcI = \frac{DPV}{EI} \tag{10.1}$$

Linke et al. (2014) introduce an assessment indicator for sustainability denoted as "sustainability efficiency indicator" (SEI) defined by the ratio between a selected performance parameter (PP) and quality parameter (QP) for a specified resource consumption (RC); this has the mathematical formula as follows:

$$SEI = \frac{PP}{QP \times RC} \tag{10.2}$$

The SEI can be extended for drying systems in various ways, such as in the form of specific emissions per final moisture and energy input, where the PP is the specific emissions, the QP is the final moisture content, and the RC is the energy input.

Here are briefly given some other sustainability assessment parameters as summarized in Dincer and Zamfirescu (2011) and Midilli et al. (2006):

- *Ecological footprint* (EF) analysis is an accounting tool enabling the estimation of RC and waste assimilation requirements of a defined human population or economy in terms of corresponding productive land use.
- *Sustainable process index* (SPI) is a means of measuring the sustainability of a process producing goods. The unit of measure is square meter ($m^2$) of land. It is calculated from the total land area required to provide raw materials, process energy (solar derived), infrastructure (including energy generation production facilities), and waste disposal.
- *Sectorial impact ratio* ($R_{si}$) is based on the provided financial support of public, private, and media sectors for transition to green energy-based technologies and depends on the total green energy financial budget as a reference parameter.
- *Technological impact ratio* ($R_{ti}$): this parameter quantifies the provided financial support for research and development, security, and analysis of green energy-based technologies. This parameter depends on the total green energy financial budget as a reference parameter.
- *Practical application impact ratio* ($R_{pai}$): this parameter quantifies the provided financial support for design, production, conversion, marketing, distribution, management, and consumption of green fuel from green energy sources and also depends on the total green energy financial budget.

The aggregated indicators for sustainability can be formed by various possible methods of normalization, ranking, weighting, and scaling. In order to be possible to aggregate the individual indicators, these have to be normalized first. The normalization can be done based on a reference sustainability indicator. Denote $Ind_{ref}$ an arbitrary selected sustainability indicator; this can be taken as $Ind_{ref} = \max\{Ind_i | i = 1, 2, ..., n\}$, where $n$ is the number of sustainability indicators $Ind_i$ to be aggregated. In this case, the normalized sustainability indicator can be taken as proposed in Dincer et al. (2010), namely,

$$NSI_i = \frac{\frac{1}{Ind_i}}{\frac{1}{Ind_{ref}}} \tag{10.3}$$

where $Ind_i$ must be expressed in a dimensionless format.

Other ways of indicator average are given in Singh et al. (2009). Accordingly, the normalization can be done based on the reference value as follows:

$$NSI_i = \frac{Ind_i}{Ind_{ref}} \tag{10.4}$$

where $Ind_{ref}$ can be either the maximum of the dimensionless indicators or the arithmetic mean of them.

The sustainability indicator can be also normalized based on the relative alleviation from the average, as follows:

$$NSI_i = \frac{Ind_i - \overline{Ind}}{Ind_i} \tag{10.5}$$

where $\overline{Ind}$ represents the average of dimensionless indicators.

A variation for the normalization given by Eq. (10.5) is as follows, where the weighted average is used instead of the arithmetic average as follows:

$$NSI_i = \frac{Ind_i - \overline{Ind}_\sigma}{\sigma_i} \tag{10.6}$$

where $\sigma_i$ represents the standard average of $Ind_i$.

The normalization can be also done by scaling as follows:

$$NSI_i = \frac{Ind_i - \min\{Ind_i | i \in \{1, n\}\}}{\max\{Ind_i | i \in \{1, n\}\} - \min\{Ind_i | i \in \{1, n\}\}} \tag{10.7}$$

Based on the normalized sustainability indicator, various aggregation possibilities exist. A typical approach is that of a weighted average determined as follows:

$$ASI = \frac{\sum w_i NSI_i}{\sum w_i} \tag{10.8}$$

where ASI is the aggregated sustainability index and $w_i$ are the weighting factors which can be chosen or determined based on various considerations; in principle, the weighting factors can be determined as follows:

$$w_i = \frac{\text{NSI}_i}{\sum \text{NSI}_i} \tag{10.9}$$

## 10.2.2   Exergy-Based Sustainability Assessment

Exergy analysis offers a basis for sustainability assessment because exergy is a measure of abatement of the system subjected to the analysis from the environment. Ness et al. (2007) categorize exergy as one of the emerging methods for sustainability assessment. A precursory of the exergy-based sustainability assessment is the regional and sectorial exergy analysis. Wall (1990, 1997) presented exergy analysis of Japan and the Unites States, respectively. This type of analysis has been expanded to cover environment and sustainable development by Rosen and Dincer (2001). It is shown that exergy is at the confluence of energy, environment, and sustainable development.

The relationships among exergy, energy, and environmental impact of sustainable drying systems are commented in Dincer (2002). Because sustainable development includes a component of energy security, the exergy-based assessment of sustainability becomes very relevant. Through exergy analysis and exergy-based thermodynamic optimization (see Chapter 9), increased efficiency is obtained and drying systems with reduced exergy destruction are developed. Less exergy destruction implicitly leads to reduced environmental impact. Figure 10.3 shows a bar chart comparing energy and exergy processes of two basic drying

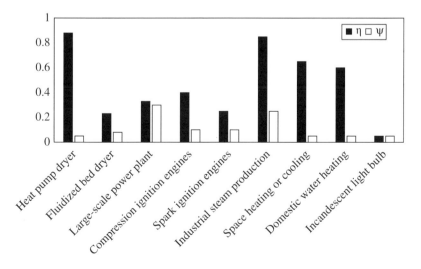

**Figure 10.3**   Energy and exergy efficiency of basic drying systems and other thermal processes (data from Dincer (2002))

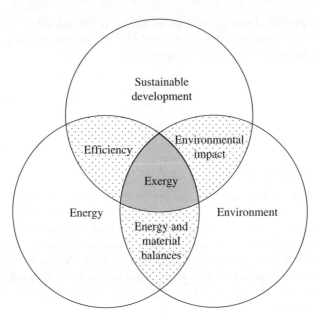

**Figure 10.4** Representation of the exergy at the confluence of energy, environment, and sustainable development (extended from Dincer (2011))

systems with other thermal processes. The two drying systems considered are heat pump dryers and combustion-driven fluidized bed dryer. The graph clearly reveals that although the energy efficiency of the heat pump drying system is high, the corresponding exergy is quite low; note that exergy efficiency of power generation is included in exergy efficiency of the heat pump dryer. It is also observed from the figure that except for large-scale power plants and industrial steam production, the exergy efficiency of thermal processes is rather low.

Sustainable drying systems demand sustainable energy sources. Let us consider a heat pump dryer of 80% energy efficiency, based on the power input. If the supplied power is generated by a hydro power plant with 80% energy efficiency, then the overall energy efficiency becomes 64%, while exergy efficiency is approximated to 12%, that is, more than double with respect to grid-connected heat pump dryer. The additional benefit of hydropowered heat pump dryer is the nonpollution feature during operation, a fact that improved the sustainability. The usage of renewable energy in drying system generally leads to better sustainability with respect to conventional systems that burn fossil fuels.

The conceptual representation of exergy at the confluence of energy, environment, and sustainable development is shown in Figure 10.4. Here, it is also suggested that the energy and material flow balance methods connect energy analysis with the environment. Indeed, in energy analysis, material and energy flow balances are inventoried as they must conserve. These balances ultimately allow to find the relation between system performance and system discharges into the environment. In addition, the energy efficiency is at the frontier between energy and sustainable environment. Better energy efficiency means less RC and less environmental impact. Furthermore, at the frontier between the environment and sustainable development, the environmental impact is interlaced, which suggests that less environmental impact

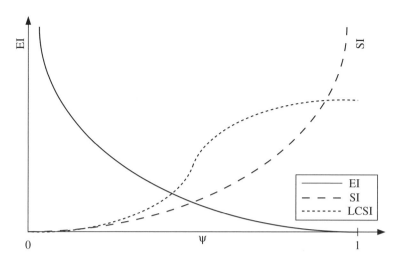

**Figure 10.5** The correlation between environmental impact index and sustainability index (modified from Dincer (2011))

leads to better system sustainability. The unique position at the center, in the figure, shows that exergy unites the three fields: energy, environment, and sustainable development. In addition, exergy is connected with efficiency, environmental impact, and energy and material inventories. As shown in Dincer (2011), exergy efficiency can be correlated with SI and with the environmental impact index (EI). It can also be correlated with the life cycle sustainability index (LCSI). Here, the difference between SI and LCSI consists of the fact that SI refers to the system utilization phase only, while LCSI considers the system life cycle including the construction and scraping phases.

These correlations are qualitatively represented in Figure 10.5. In the limiting case when exergy efficiency is significantly low, the SI approaches zero because although exergy-rich resources (fuels, ores, steam, etc.) are consumed, nothing is accomplished. Provided that exergy efficiency is very high, the environmental impact tends to become negligible since the exergy conversion approaches the ideal case with no waste emissions and negligible irreversibilities. Furthermore, the curve of LCSI should show an inflection point because any increase in exergy efficiency implies more investment in the system construction that is more RC and more environmental impact at construction phase. This component is negligible for low exergy efficiency but becomes increasingly important for high exergy efficiency.

In addition of being useful for achieving better sustainability, the exergy analysis offers the pathway toward designing systems with better cost-effectiveness under prevailing legal conditions and with regard to ecological, social, and ethical consequences. In fact, the economic component most likely limits the design and determines the optimum exergy efficiency according to the law of diminishing return. Higher efficiency involves exponentially higher system cost with neither substantial reduction of environmental impact nor important improvement of LCSI. Sustainable development requires a supply of energy resources that, in the long term, is readily and sustainably available at reasonable cost and can be utilized for all required tasks without causing negative societal impacts.

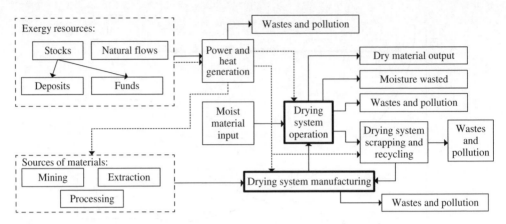

**Figure 10.6** Exergetic life cycle modeling diagram for sustainability assessment of a drying system (continuous arrows mean material flow, dashed arrows mean heat or work fluxes)

Exergy analysis of drying processes and system is presented in Chapter 4. The method is based on mass, energy, entropy, and exergy balance equation and leads to determination of exergy destructions of the system and the components. On top of exergy analysis, the exergoeconomic and exergoenvironmental analyses can be pursued as presented in Chapter 8 where some environmental impact and economic parameters were introduced for drying systems. More insight on exergy efficiency for drying systems is provided in Chapter 9 where the general mathematical expression for the reversible work of a drying process is derived. The general expressions for energy and exergy efficiency of the drying system are presented in Eqs. (9.11), (9.13), (9.17), and (9.19). Here, the analysis is to be extended to cover the connection between exergy and sustainability, that is, exergo-sustainability analysis based on a life cycle approach.

Figure 10.6 shows a process diagram for life cycle modeling of a drying system in view of sustainability assessment. The process starts by identification of exergy resources (fuels or renewable energies). Exergy resources are available in the environment in many forms. An appropriate mix of exergy resources can be considered in the analysis depending on the geopolitical region of interest. The exergy resources are categorized as stocks and natural flows where the natural flows represent renewable energies in the form of wind, solar radiation, geothermal heat, various forms of ocean energy, tidal energy, and hydraulic energy. The stocks represent storable forms of exergy.

There are two types of exergy stocks as discussed in Dincer (2002): deposits and funds. The deposits of exergy represent dead stocks. It requires an activation energy to be able to extract exergy from deposits. Once activated, the produced exergy is high and covers the activation exergy. Once it is used, the exergy deposit continuously diminishes. The deposits are basically petroleum and coal resources. The funds are living stocks that never deplete. They are basically forests and crop fields which grow while consuming solar energy and deliver biomass as a renewable resource which never depletes. The exergy resources are in nonequilibrium with the environment. Therefore, they can be converted on demand and transmitted further either in the form of work or of heat.

The generated work (power) and heat are used primarily in mining, extraction, and processing of materials required for drying system construction. Additional power and heat is consumed by the manufacturing process of the drying system. During the operational lifetime, an amount of moist material is fed into the dryer. In the same time, the dryer is supplied with the required amount of heat and work to operate. The resulting outputs are as follows: the corresponding amount of dry material produced during the operational lifetime, wasted moisture, other wastes, and polluting effluents. Another output which is available at the end of the lifetime is the drying system itself which is subjected to an operation of scrapping and recycling (this operation requires exergy input; see the figure). Thence, a part of the scrapped system is wasted with associated pollution and another part is recycled to reduce the required amounts of material input for manufacturing. The environment, as a process enabler, influences the overall process and all subprocesses. Once all exergy flows are determined, it becomes possible to assess the system sustainability in correlation with exergy.

In the model from Figure 10.6, the wastes and pollution emitted by the power and heat generation process include any indirect contributions due to power plant construction. All wastes shown in the diagram include the exergy destruction component. The analysis may start by specifying the type and characteristics of the moist material and the required reduction in moisture content. Also, the amount of material $m_p$ (dry basis) to be dried for the whole operational lifetime is specified. Then the required reversible work to dry the whole amount of material for the lifetime is

$$W_{rev} = m_p \left( \omega_0 \Delta W ex_{w,0} + \Delta ex_p \right) \tag{10.10}$$

where all terms from Eq. (10.10) are defined in Chapter 9: $\omega_0$ is the humidity ratio in the surrounding environment, $\Delta W$ is the change of moisture content (wet–dry), $ex_{w,0}$ is the specific exergy of moisture in the surrounding environment, and $\Delta ex_p$ represents the specific exergy change of the moist material on dry basis.

A detailed exergy analysis will determine the exergy efficiency of the process. Therefore, the total exergy input required to drive the process for the lifetime results from

$$Ex_{in} = \psi W_{rev} \tag{10.11}$$

The moisture extracted for the lifetime is $m_w = m_p \Delta W$ and the total wastes and pollution is proportional to the exergy destruction which is equal to exergy input since all input exergy is destroyed in the process provided that the dried product is not valorized as fuel. Otherwise, if the dried product is a fuel, then this is a useful output, and the exergy destroyed becomes $Ex_d = Ex_{in} - m_p ex_{p,out}^{ch}$, where $ex_{p,out}^{ch}$ is the chemical exergy in dry product on dry basis. The amount of environmental pollution in terms of GHG, $NO_x$, $SO_2$, CO, volatile organic compounds (VOCs), and so on can be correlated with the exergy destruction.

The analysis can also determine how much of the scrapped system is wasted (in kg curb) and out of this how much landfill gas and GHG emissions are generated; other environmental indicators as given in Table 8.4, such as eutrophication etc., can be included in the analysis. Further, the material balance for the manufacturing process of the drying system determines the required amount $m_{mat}$ of construction materials

$$m_{mat} + m_{recycled} = m_{system} \tag{10.12}$$

Once $m_{mat}$ is known, the required exergy for materials extraction and processing can be determined. The total exergy that must be provided during the lifetime to drive the drying system operation, to extract the construction material, to manufacture the system, and to scrap the system can be now calculated as follows:

$$Ex_{tot} = Ex_{system} + Ex_{extraction} + Ex_{manufacturing} + Ex_{scraping} \tag{10.13}$$

The exergy efficiency of power and heat generation $\psi_{gen}$ will consider the overall process for fuel resource extraction, production, and conversion. The total resource amount can be correlated with the total exergy of resource used for the life cycle, $Ex_{resource}$. The total exergy resource required for the lifetime is given as follows:

$$Ex_{resource} = \frac{Ex_{tot}}{\psi_{gen}} \tag{10.14}$$

The exergy destruction associated with fuel extraction and processing and power and heat generation becomes

$$Ex_{d,gen} = \left(1 - \psi_{gen}\right) Ex_{resource} \tag{10.15}$$

The pollution and wastes due to fuel extraction, processing, and power and heat generation altogether can be correlated with $Ex_{d,gen}$. Finally, the pollution amounts for overall system can be determined for the entire lifetime, by adding the pollution components of each kind due to power and heat generation, system operation, and system scraping. Thence, the total pollution in terms of GHG, $NO_x$, $SO_2$, CO, and VOCs is obtained.

Dincer (2007) suggests to group the pollution terms other than GHG in a compounded air pollution indicator. Provided that $SO_2$ emissions are negligible, the following weighting factors taken from Dincer (2007) can be used as a guideline for compounding indicator, namely, 0.017 for CO, 1 for $NO_x$, and 0.64 for VOCs. Therefore, an equivalent mass of air pollutants for the lifetime (in kilogram) is obtained as follows:

$$m_{AP} = 0.017 m_{CO} + m_{NO_x} + 0.64 m_{VOCs} \tag{10.16}$$

In order to account for the economic component of sustainability, an exergy-based capital investment effectiveness (CIEx) can be formulated in a similar manner as in Dincer (2007). This is represented by the ratio between exergy input into the drying system for operation during the lifetime and the exergy consumed for materials extraction and system manufacturing. Thence, CIEx is expressed as follows:

$$CIEx = \frac{Ex_{system}}{Ex_{extraction} + Ex_{manufacturing}} \tag{10.17}$$

An aggregated normalized sustainability indicator can be thence defined for the drying system using weighting factors for air pollution $w_{AP}$, for GHG $w_{GHG}$, and for capital investment effectiveness $w_{CIE}$. These weighting factors should be determined based on the actual data, considering a normalization method as given in Eqs. (10.3)–(10.8). The normalized sustainability indicator (dimensionless) becomes

$$\text{NSI} = w_{\text{AP}}m_{\text{AP}} + w_{\text{GHG}}m_{\text{GHG}} + w_{\text{CIE}}\text{CIEx} \tag{10.18}$$

The normalized sustainability indicator from Eq. (10.18) is based on exergy due to the fact that each of its term depends on exergy destruction of the system and its components. In Eq. (10.18), only environmental and economic terms are included. The aggregated indicator can be further expanded to include other sustainability metrics included in Table 10.1, which can be correlated with exergy destructions.

## 10.3    Environmental Impact

Environmental impact assessment is a key element of sustainability assessment. Environment must be defined in correspondence with exergy. The terrestrial environment is not at equilibrium, but for the purpose of exergy and environment analysis, it is useful to identify a surrounding environment which is at equilibrium and with respect to which a fuel resource has nonzero exergy.

The terrestrial environment is composed of four identifiable subsystems which are the atmosphere, the hydrosphere, the lithosphere, and the biosphere (within which the anthropogenic biosphere is remarked). Each of these subsystems is in obvious thermodynamic nonequilibrium with each other.

Any anthropogenic activity which leads to substantial changes in the environmental subsystems or parts of them is susceptible of being polluted, bringing negative environmental impact. Let us analyze industrial drying from the perspective of the environment and environmental impact. Figure 10.7 shows a thermodynamic model of the terrestrial surface environment. The model illustrates the interactions among four entities of the environmental ecosystems: the anthropogenic system (and its associated activity), lithosphere, hydrosphere, and atmosphere. As already mentioned earlier, in the ecosystem does exist exergy stocks and natural flows. Noteworthy, the lithosphere contains large deposits of conventional fuels which can be manipulated by humans to extract their embedded exergy.

The biomass funds are continuously replenished due to the natural ecosystem which receives solar energy, moisture, nitrogen, phosphorus, and so on from the natural cycles. Today, the humans are capable to harvest exergy from a large palette of exergy sources from the lithosphere, hydrosphere, and atmosphere. As shown in the figure, the exergy resources can be listed as follows: conventional fuels including nuclear, biomass, geothermal, hydraulic, ocean energy, solar radiation, and winds. Once used in the anthropogenic activity, exergy is destroyed; in particular, the exergy deposits are depleted, whereas the exergy associated with renewable energies is always replenished. In order to operate, any anthropogenic activity must connect an exergy source with a surrounding environment so that the exergy is destroyed while the desired action is performed.

Let us revisit the drying system discussed previously in conjunction with Figure 10.6. If the drying system is a heat pump dryer, then power is required to run the system. Also, power and heat is required to construct the drying system and to disassemble, recycle, and dispose it at the end of its lifetime. If one assumes that conventional resources (e.g., fossil fuels) are used to

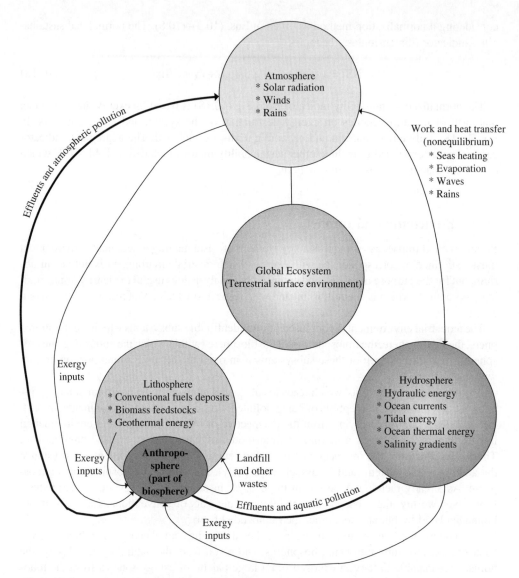

**Figure 10.7** Thermodynamic model for the terrestrial environment showing the interactions among the main subsystems (biosphere is partially represented by anthropogenic biosphere)

construct, run, and dispose the system, then the following scenario can be assumed for a drying activity, considering the whole lifespan:

- An amount of exergy is extracted from a conventional deposit of the lithosphere which is thus depleted.
- Using the extracted exergy:
  - Power and heat is generated to run all the involved subprocesses for the whole life cycle.
  - The required construction materials and mined, extracted, and processed.

- ∘ The drying system is manufactured.
- ∘ The drying system is operated.
- ∘ The drying system is dismantled, and components are partially recycled and partially scrapped.
- An amount of supplied moist material is converted to dry product which is traded for economic benefit.
- Waste substances (pollutants) and energy streams are discharged in the environment as follows:
  - ∘ Flue gases including air pollutants and GHG are released in the atmosphere due to combustion.
  - ∘ Warm water streams are discharged in the hydrosphere by power generation stations.
  - ∘ Garbage of indirect activities and the scrapped system at lifetime end is released on landfields.
  - ∘ Landfield gas pollutes the atmosphere with GHG and air pollutants.

Due to the received energy and material streams, changes occur in every component of the ecosystem, which are caused by anthropogenic activity. These changes are referred to as environmental impact. In addition, changes occur in the anthropogenic system (i.e., society) which is continuously reshaped. The sustainable development doctrine requires controlling the changes in the society and the environment such that the sustainability desiderate is maintained.

In this section, the environmental impact aspects are discussed in relation with drying system sustainability. Thenceforth, standard environmental models are introduced, and the anthropogenic environmental impact is discussed. Environmental impact data is presented as well. Finally, some exergy-based environmental impact parameters are introduced.

## 10.3.1   Reference Environment Models

Because the intensive properties (pressure, temperature, chemical potential) of the natural environment vary temporally and spatially, the natural environment is far from a thermodynamic equilibrium. Thenceforth, being departed from equilibrium, the natural environment possesses work potential; that is, it has nonzero exergy. However, in most of the cases, the environment changes slow as chemical reactions are not activated due to a reduced rate of the involved transport processes. Fossil fuel supply reserves or forests do not burn spontaneously since no activation energy is provided.

Therefore, a compromise can be made between the theoretical requirements of the reference environment and the actual behavior of the natural environment. That is, a standard reference model can be adopted which is useful for exergy and environmental impact calculations. The reference environment is in stable equilibrium, acting as an infinite sink and source for heat and materials, experiencing only internally reversible processes with unaltered intensive states.

Natural environment models attempt to simulate realistically subsystems of the natural environment. A model which considers saturated moist air and liquid water in phase equilibrium and allows for the sulfur-containing materials to be considered in the analyses is proposed in Gaggioli and Petit (1977). The temperature and pressure of this reference environment are normally taken to be 25 °C and 1 atm, respectively, and the chemical composition is taken to consist of air saturated with water vapor and the following condensed phases: water ($H_2O$), gypsum ($CaSO_4 \cdot 2H_2O$), and limestone ($CaCO_3$). The stable configurations of C, O, and N, respectively,

**Table 10.2** Atmosphere gases and their molar fractions for Gaggioli and Petit (1977) model

| Constituent | $N_2$ | $O_2$ | $H_2O$ | Ar | $CO_2$ | $H_2$ |
|---|---|---|---|---|---|---|
| Molar fraction | 0.7567 | 0.2035 | 0.0303 | 0.00911 | 0.0003 | 0.0001 |

are taken to be those of $CO_2$, $O_2$, and $N_2$ as they exist in air saturated with liquid water at $T_0$ and $P_0$ (the temperature and pressure for the reference environment); of hydrogen is taken to be in the liquid phase of water saturated with air at $T_0$ and $P_0$; and of S and Ca, respectively, are taken to be those of $CaSO_4 \cdot 2H_2O$ and $CaCO_3$ at $T_0$ and $P_0$. The atmosphere composition for this model is given in Table 10.2.

Equilibrium and constrained-equilibrium models were also formulated in which all the materials present in the atmosphere, oceans, and a layer of the crust of the earth are pooled together, and an equilibrium composition is calculated for a given temperature. The selection of the thickness of crust considered is subjective and is intended to include all materials accessible to thermal processes like drying. Thicknesses varying from 1 to 1000 m and a temperature of 25 °C were considered. Exergy values obtained using these environments are significantly dependent upon the thickness of crust considered and represent the absolute maximum amount of work obtainable from a material. Ahrendts (1980) proposed a version of equilibrium environment model in which the calculation of an equilibrium composition excludes the possibility of the formation of nitric acid ($HNO_3$) and its compounds. That is, all chemical reactions in which these substances are formed are in constrained equilibrium, and all other reactions are in unconstrained equilibrium. When a thickness of crust of 1 m and temperature of 25 °C are used, the model becomes similar to the natural environment model.

An environment model which contains only components that participate in the process being examined in a stable equilibrium composition at the temperature and pressure of the natural environment was proposed by Bosnjakovic (1963) and categorized as a process-dependent morel. This model is dependent on the process examined and is not general. Exergies evaluated for a specific process-dependent model are relevant only to the process. They cannot rationally be compared with exergies evaluated for other process-dependent models or used in environmental assessments.

Assume that the environment is a large reservoir that when a particular substance interacts with it, the following types of processes occur after a sufficiently long time: (i) mechanical equilibration (the substance pressure will equal to that of the surroundings), (ii) thermal equilibrium (meaning that the substance temperature becomes equal to that of the environment), (iii) chemical reaction equilibrium (meaning that the substance enters in a series of spontaneous chemical reactions with the environment such that it decomposes and eventually forms only chemical species present in the environment), and (iv) concentration equilibrium (the chemical species resulting from substance reaction with the environment dilute or concentrate such that they reach the concentration in the environment). The processes (i) and (ii) refer to the thermomechanical equilibrium. The departure of the substance stream from the temperature and pressure of the environment is a measure of the thermomechanical exergy. Moreover, the processes (iii) and (iv) represent chemical equilibrium which is associated with chemical exergy (see Chapter 1).

Szargut et al. (1988) proposed a reference environment model which includes chemical species abundant in the natural environment which is far from thermodynamic equilibrium.

A set of reference species are selected for the lithosphere, hydrosphere, and atmosphere as the most probable products of interaction between the analyzed substances and the environment. This model is revised and improved in Rivero and Garfias (2006) considering that the relative humidity in the atmosphere is 70%, the carbon dioxide concentration is 345 ppm in volume, and the salinity of seawater is 35‰. The list of stable components in the atmosphere and their concentration in the reference environment is given in Table 10.3. The chemical elements in the atmosphere for this model are Ar, C, $D_2$, $H_2$, He, Kr, $N_2$, Ne, $O_2$, and Xe. The molar fraction of the species in the atmosphere is related to the standard chemical exergy (Chapter 1) under the assumption of ideal gas behavior as follows:

$$y = \exp\left(-\frac{ex^{ch}}{\mathcal{R}}T_0\right) \qquad (10.19)$$

## 10.3.2 Anthropogenic Impact on the Environment

At a small scale, the impact of human activity on the environment is negligible because the environment is so vast that any action from any human or groups is immediately dumped. At a regional or global scale, the interaction between anthropogenic activity and environment leads to observable environmental impact. Two major types of anthropogenic impact on the environment can be observed in the recent era: the impact on climate change and global warming and the impact regarding environment acidification. In addition to that, various impacts on local ecosystems can be observed in a diverse manner as quantifiable using the parameters given in Table 8.4.

Here, we discuss global warming and acidification issues. The prominent impact on global climate is due to the emissions of GHGs, aerosols, and ozone-depleting substances. Figure 10.8 presents a generic power production system which consumes some fuels, generates useful work, and expels some pollutants in the environment. There may be pollutant emissions, accidents, hazards, ecosystem degradation through air and water pollution, animal poisoning, GHG emission, carbon monoxide leakages, stratospheric ozone depletion, and emission of $SO_2$, $NO_x$, VOCs, particulate matter (PM), and other aerosols.

The effluents expelled in the atmosphere by power generation systems can be categorized in two kinds: GHGs and aerosols. GHGs are those chemicals which are released in the terrestrial atmosphere and produce greenhouse effect. When released from natural and anthropogenic activities, GHGs travel through the atmosphere and reach its upper layer, the troposphere. At the troposphere level, GHGs absorb an important part of the infrared radiation emitted by the earth surface. As a matter of consequence, the earth surface temperature tends to increase, and this process is called the greenhouse effect.

On the other hand, aerosols such as VOCs, soot, PM, and so on are released continuously in the atmosphere and concentrate at its upper layers. Aerosols contribute to the earth's albedo. Due to their presence in the atmosphere, aerosols reflect and scatter back in the extraterrestrial space a part of the incident solar radiation. As a matter of consequence, the earth temperature tends to decrease. This process can be denoted as the albedo effect. The balance between greenhouse and albedo effects establishes the earth temperature and regulates the earth climate. This mechanism of climate control is a natural process. However, it is noted that since the industrial revolution, the anthropogenic impact on climate became obvious due to accentuate emission of

**Table 10.3** Reference environment described in Rivero and Garfias (2006)

| | | | | | | | | |
|---|---|---|---|---|---|---|---|---|
| **Atmosphere** E: | Ar | $CO_2$ | $D_2O$ | $H_2O$ | He | | | |
| y: | 9.13E–3 | 3.37E–4 | 3.37E–6 | 2.17E–2 | 4.89E–6 | | | |
| E: | Kr | $N_2$ | Ne | $O_2$ | Xe | | | |
| y: | 9.87E–7 | 0.7634 | 1.76E–5 | 0.2054 | 8.81E–8 | | | |
| **Hydrosphere** E: | $HAsO_4^{2-}$ | $B(OH)_3$ | $BiO^+$ | $Br^-$ | $Cl^-$ | $Cs^+$ | $IO_3^-$ | $K^+$ |
| x: | 3.87E–8 | 3.42E–4 | 9.92E–11 | 8.73E–4 | 0.5658 | 2.34E–9 | 5.23E–7 | 1.04E–2 |
| E: | $Li^+$ | $MoO_4^{2-}$ | $Na^+$ | $HPO_4^{2-}$ | $Rb^+$ | $SO_4^{2-}$ | $SeO_4^{2-}$ | $WO_4^{2-}$ |
| x: | 2.54E–5 | 1.08E–7 | 0.4739 | 4.86E–7 | 1.46E–6 | 1.24E–2 | 1.18E–9 | 5.64E–10 |
| **Lithosphere** E: | AgCl | $Al_2SiO_5$ | Au | $Be_2SiO_4$ | $CaCO_3$ | $CaF_2 \cdot 3Ca_3(PO_4)_2$ | $CdCO_3$ | $CeO_2$ |
| y: | 1E–9 | 2.07E–3 | 1.36E–9 | 4.2E–6 | 1.4E–4 | 2.24E–4 | 1.22E–8 | 1.17E–6 |
| E: | $CoFe_2O_4$ | $CuCO_3$ | $Dy(OH)_3$ | $Er(OH)_3$ | $Eu(OH)_3$ | $Fe_2O_3$ | $Ga_2O_3$ | $Gd(OH)_3$ |
| y: | 22.85E–7 | 5.89E–6 | 4.88E–8 | 4.61E–8 | 2.1E–7 | 6.78E–3 | 2.89E–7 | 9.21E–8 |
| E: | $GeO_2$ | $HfO_2$ | $HgCl_2$ | $Ho(OH)_3$ | $In_2O_3$ | $IrO_2$ | $K_2Cr_2O_7$ | $La(OH)_3$ |
| y: | 9.49E–8 | 1.15E–7 | 5.42E–10 | 1.95E–8 | 2.95E–9 | 3.59E–12 | 1.35E–6 | 5.96E–7 |
| E: | $Lu(OH)_3$ | $Mg_3Si_4O_{10}(OH)_2$ | $MnO_2$ | $Nb_2O_3$ | $Nb(OH)_3$ | NiO | $OsO_4$ | $PbCO_3$ |
| y: | 7.86E–9 | 8.67E–4 | 2.3E–5 | 1.49E–7 | 5.15E–7 | 1.76E–6 | 3.39E–13 | 1.04E–7 |
| E: | PdO | $Pr(OH)_3$ | $PtO_2$ | $PuO_2$ | $RaSO_4$ | $Re_2O_7$ | $Rh_2O_3$ | $RuO_2$ |
| y: | 6.37E–11 | 1.57E–7 | 1.76E–11 | 8.4E–20 | 2.98E–14 | 1.76E–6 | 3.29E–12 | 6.78E–13 |
| E: | $Sb_2O_5$ | $Sc_2O_3$ | $SiO_2$ | $Sm(OH)_3$ | $SnO_2$ | $SrCO_3$ | $Ta_2O_5$ | $Tb(OH)_3$ |
| y: | 1.08E–10 | 3.73E–7 | 0.407 | 1.08E–7 | 4.61E–7 | 2.91E–5 | 7.45E–9 | 1.71E–8 |
| E: | $TeO_2$ | $ThO_2$ | $TiO_2$ | $Tl_2O_4$ | $Tm(OH)_3$ | $UO_3 \cdot H_2O$ | $V_2O_5$ | $Y(OH)_3$ |
| y: | 9.48E–12 | 2.71E–7 | 1.63E–4 | 1.49E–9 | 7.59E–9 | 1.48E–8 | 1.83E–6 | 1E–6 |
| E: | $Yb(OH)_3$ | $ZnCO_3$ | $ZrSiO_4$ | | | | | |
| y: | 4.61E–8 | 7.45E–6 | 2.44E–5 | | | | | |

*Source:* Rivero and Garfias (2006).

$x$ is the mass fraction in the hydrosphere; $y$ is the molar fraction used for the atmosphere and lithosphere.

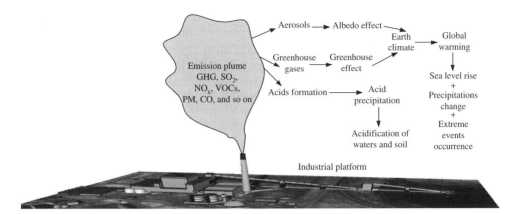

**Figure 10.8**   Anthropogenic environmental impact affecting the atmosphere and global climate

GHGs emitted massively by many activity sectors (energy, transportation, industry) which induce a tendency of global warming.

Global warming leads to changes in natural systems. There is an increased melting effect of permafrost ice, and consequently, there is a continuous tendency of sea-level increase. Moreover, there are observed changes in global precipitations (with more dry areas and more regions flooding), and there are recorded more extreme events (tsunami, extreme winds, etc.).

Another major environmental impact induced by energy systems is due to acid precipitations. Gaseous effluents expelled in the atmosphere by power generation systems can eventually form acids and with precipitations return back on earth where they contribute to acidifications of soil and seas. This effect negatively affects all life systems. Other impacts are also possible as it will be detailed subsequently. The main types of pollutant emissions due to energy systems are given in Table 10.4, and their influence is explained.

Anthropogenic activity induces an intensification of the greenhouse effect that led to global warming. With greenhouse effect, there are two possibilities:

- The anthropogenic activity results in more GHG emissions. Since these gases will be more concentrated in the lower layers of the atmosphere, where human activity occurs, the troposphere reradiates more toward the earth's surface. Therefore, the global temperature increases.
- The GHG concentration reduces in the vicinity of the earth's surface and concentrates toward the upper layers for some unknown reason. In this case, the troposphere reradiates less toward the earth. This means a decrease of earth temperature.

The relevant properties of the principal GHGs except water in the atmosphere are presented in Table 10.5. Three of the gases shown in the table, namely, carbon dioxide, methane, and nitrogen dioxide, are both part of natural cycles of carbon and nitrogen and part of anthropogenic emissions; therefore, they were present in the atmosphere prior the industrial era, for which the year 1750 is considered the reference. As seen, the concentration of these gases in year 1750 is higher than 0.

**Table 10.4** Atmospheric pollutants released by power generation systems

| Pollutant | Explanations |
|---|---|
| GHG | Greenhouse gases. These are gases that produce greenhouse effect. The main GHGs are $CO_2$, $CH_4$, and $N_2O$. Greenhouse effect is the main cause of global warming |
| CO | Carbon monoxide. Arises mostly from the incomplete combustion of fuels and poses a health and life of animals and humans at high risk upon inhalation. Intense emission of CO in open spaces may affect birds, whereas emissions in close spaces (residences, garages) may induce death of exposed persons |
| $SO_2$ | Sulfur dioxide. It is a corrosive gas which is hazardous to human health and harmful to the natural environment. It results from combustion of coal and fuel oil, smelting of nonferrous metal ores, oil refining, electricity generation, and pulp and paper manufacturing. It causes respiratory difficulties, damages green plants, and is a precursor of acid precipitation |
| $NO_x$ | Nitrogen oxides (NO and $NO_2$). It is produced by combustion of fuels at high temperature in all combustion facilities from large scale to small scale (including motor engines, furnaces, etc.). It can lead to respiratory problems of humans and animals. It can form acids in high-altitude atmosphere |
| VOCs | Volatile organic compounds. These are volatile organic particles resulting from hydrocarbon combustion in engines, have harmful effects in the atmosphere, and impede the formation of the stratospheric ozone |
| PM | Particulate matter. Particles in the air (fly ash, sea salt, dust, metals, liquid droplets, soot) come from a variety of natural and human-made sources. Particulates are emitted by factories, power plants, vehicles, and so on and are formed in the atmosphere by condensation or chemical transformation of emitted gases. PM cause of health and environmental effects including acid precipitation, damage to plant life and human structures, loss of visibility, toxic or mutagenic effects on people, and possibly nonaccidental deaths |
| CFC, HCFC, HFC | Chlorofluorocarbons. These are mainly refrigerants or propellants. They can contribute to the destruction of the stratospheric ozone layer and to greenhouse effect |

**Table 10.5** The principal greenhouse gases and their approximated concentration in the atmosphere

| Gas | Chemical formula | Spectral range $(cm^{-1})$ | Atmospheric concentration | | Atmospheric lifetime (years) |
|---|---|---|---|---|---|
| | | | Year 1750 | Current | |
| Carbon dioxide | $CO_2$ | 550–800 | 280 ppm | 387 ppm | 50–200 |
| Methane | $CH_4$ | 950–1650 | 700 ppb | 1750 ppb | 12 |
| Nitrous oxide | $N_2O$ | 1200–1350 | 270 ppb | 314 ppb | 120 |
| CFC-11 | $CFCl_3$ | 800–900 | 0 | 251 ppt | 50 |
| CFC-12 | $CF_2Cl_2$ | 875–950 | 0 | 538 ppt | 102 |

*Source*: Data from the IPCC (2007).
ppm, parts per million; ppb, parts per billion; ppt, parts per trillion.

Other two man-made substances are shown in Table 10.5, which are examples of GHGs; these are freons. The freon concentration in preindustrial era is nil. However, since their invention in the twentieth century, their concentration in the atmosphere started to increase. The table shows the spectral range, given in terms of wave number, for each gas, and also indicates the atmospheric lifetime, which is a concept detailed in subsequent paragraphs. For now, observe that the atmospheric lifetime of freons is very high, as well as that of nitrous oxide (which today is also an emission mainly from human activity). Other GHGs are a result of human activity, but they are emitted in lower quantity, and as a result, their actual concentration in the atmosphere is low. Here is a nonexhaustive list of those given in decreased order of their influence on greenhouse effect: 1,1,2-trichloro-1,2,2-trifluoroethane (CFC-113), chlorodifluoromethane (HCFC-22), CFC-141b, CFC-142b, 1,1,1-trichloroethane ($CH_3CCl_3$), carbon tetrachloride ($CCl_4$), and 1,1,1,2-tetrafluoroethane (HFC-134a).

The lifetime of a mass $m_0$ of a gas present in the atmosphere depends on the gas reactivity with other species and its circulation as a part of one of the natural biogeochemical cycles. Methane, for example, is more reactive than $CO_2$, a fact that can be seen by observing the formation enthalpies of these two substances (4.7 MJ/mol vs. 8.9 MJ/mol, respectively). The temporal decrease of the gas mass in the atmosphere, relative to the initial quantity, can be modeled as follows:

$$\frac{m}{m_0} = e^{-t/\tau} \tag{10.20}$$

where $t$ is the time measured from the initial moment and $\tau$ is known as atmospheric lifetime.

The atmospheric lifetime can be estimated for each species. Based on data from Table 10.5, the fraction of gas mass existent in the atmosphere relative to the initial moment is calculated with Eq. (10.20) and plotted as shown in Figure 10.9. The atmospheric lifetime is a useful parameter when modeling the concentration of atmospheric gases.

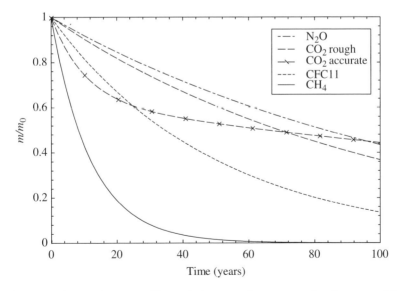

**Figure 10.9**  The residence of gaseous effluents in the atmosphere expressed as the temporal decrease of mass relative to the initial moment

The carbon dioxide lifetime is variable, because it depends on different of sources of release and uptake mechanisms specific to the carbon cycle. The prediction accuracy of carbon dioxide lifetime in the atmosphere becomes difficult due to the existence of a large variety of sources and sinks. One mentions here the following pathways through which carbon dioxide communicates with the terrestrial atmosphere:

- Ocean absorption, ~3 Gt carbon per annum
- Anthropogenic – fuel combustion and cement production, ~6 Gt carbon per annum
- Tropical deforestation, ~2 Gt carbon per annum
- Storage in the atmosphere, ~3.5 Gt carbon

The steady state of the incoming radiation crossing the tropopause in one direction 342 W/m$^2$ is equal with the outgoing radiation crossing in the other direction $107 + 229 + 6 = 342 \, \text{W}/\text{m}^2$. At this balance, the average temperature at the troposphere and the earth surface temperature are −20 and 15 °C, respectively. Assume that a reason perturbs the energy balance. If more incident radiation enters than the radiation which leaves, the climate on earth – which we define for the purpose of the present explanation through the temperature of the troposphere and at the earth surface – will also change such that the energy inventory is rebalanced. If there is a net incoming of radiation, the global temperature increases, and vice versa if there is a net outgoing energy flux, the planet cools. The following main factors can influence the net flux of radiation energy crossing the tropopause:

- A variation of the solar radiative flux – this can happen periodically because of two main reasons, namely, the sun's temperature varies with a period of ~11 years and the distance between earth and sun varies with a period of 1 year.
- The concentration of GHGs in the atmosphere changes; this can be caused by natural phenomena or more recently by anthropogenic emissions of GHGs.
- The concentration of aerosols and other PM changes in the atmosphere; this is also in great proportion an anthropogenic effect, because soot and PM form fuel combustion and airplane jet raise in the atmosphere and concentrate in higher level than in the past, increasing in this way the earth's albedo, reflecting and scattering more radiation that eventually is directed mainly toward the extraterrestrial space. Natural aerosols do also exist, for example, the water vapors and their clouds formation.
- Changes in the concentration of stratospheric and tropospheric ozone; the depletion of stratospheric ozone reduces the earth albedo and increases the net incoming flux of radiation and produces negative radiative forcing; the tropospheric $O_3$ produces positive radiative forcing.

The *radiative forcing* is defined as the net change in radiation balance at the tropopause, produced by a specified cause. By convention, the radiative forcing is positive if it induces an increase of planet temperature, and negative vice versa. The unit of measure of radiative forcing is the same as the unit of radiation energy rate per square meter of the earth surface, where the earth surface, by convention, is the area of the sphere having the average radius of the planet. The usual symbol for the radiative force is $\Delta F$. Typical values of radiative forcing are 0–2 W/m$^2$. It can be derived from what was explained earlier that change in the concentration of GHGs, aerosols, and atmospheric ozone induces radiative forcing. Depending on the concentration of the gas in the atmosphere, there are three regimes of producing radiative forcing: low, moderate, and high concentration.

Low-concentration regime regards gases present in the atmosphere in parts per billion or parts per trillion. In this case, the forcing is proportional to the concentration change, $\Delta F \sim \Delta C = C - C_0$. Freons and the tropospheric ozone fall in this category. According to the IPCC (2001), the proportionality constants are as follows: 0.25 for CFC-11, 0.32 for CFC-12, and 0.02 for tropospheric $O_3$.

In the moderate-concentration regime because the gas molecule at higher concentration absorbs much radiation where absorption band is the strongest, the absorption for the broader range of the spectrum diminishes in rate. Because of this reason, the radiative forcing is proportional with the root of the concentration, $\Delta F \sim \left(\sqrt{C} - \sqrt{C_0}\right)$. This is the case of methane and nitrous oxide molecules, of which concentration is also of ppb, but their relative effect is higher due to a more active absorption spectrum. According to the IPCC (2001), the radiative forcing of methane and $N_2O$ is given by

$$\begin{cases} \Delta F_{CH_4} = 0.036\left(\sqrt{M} - \sqrt{M_0}\right) - f(M, N_0) + f(M_0, N_0) \\ \Delta F_{N_2O} = 0.12\left(\sqrt{N} - \sqrt{N_0}\right) - f(M_0, N) + f(M_0, N_0) \end{cases} \tag{10.21}$$

where $M$ is the methane and $N$ is the nitrous oxide concentrations, respectively, and index 0 refers to the initial situation. The effect of methane and nitrous oxide on radiative forcing overlaps; this fact is accounted for by the following overlapping function given by

$$f(M, N) = 0.47 \ \ln\left[1 + 2.01 \times 10^{-5}(M \times N)^{0.75} + 5.31 \times 10^{-15}M(M \times N)^{1.52}\right] \tag{10.22}$$

The third regime – of high-concentrated gases – corresponds to carbon dioxide only, where because of the high concentration, any further change produces less effect. In this situation, the radiative forcing is proportional to the variation of the natural logarithm of the concentration as given by the following equation:

$$\Delta F_{CO_2} = 5.35 \ \ln\left(\frac{C}{C_0}\right) \tag{10.23}$$

In Figure 10.10, we show, based on the IPCC (2007), the radiative forcing produced from various causes. The total radiative forcing represented the superposition of all individual anthropogenic-type components is also indicated.

Knowing the total radiative forcing creates the opportunity to define an equivalent carbon dioxide concentration that is present alone in the atmosphere creates the same overall forcing. By equating $\Delta F_{CO_2^{eqv}} = \Delta F_{total}$ and using Eq. (10.24), one obtains

$$C_{eqv} = C_0 \ \exp\left(\frac{\Delta F_{total}}{5.35}\right) \tag{10.24}$$

where $C_0$ is the initial concentration of carbon dioxide in the atmosphere.

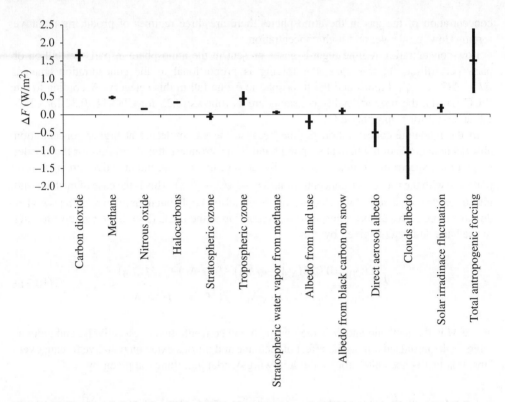

**Figure 10.10**   The radiative forcing in 2005 with respect to year 1750, due to various causes of change (data from the IPCC (2001))

Therefore, the radiative forcing can be expressed in terms of equivalent carbon dioxide concentration change, $C_{eqv}/C_0$. To show how this is useful for data interpretation, one can read from Figure 10.10 that the radiative forcing due to carbon dioxide emission in the industrial era has the average of 1.66 W/m$^2$. The total anthropogenic forcing is of 1.6. Therefore, the relative change in carbon dioxide concentration is $\exp(1.6/5.35) = 1.35$. The equivalent carbon dioxide for this growth, based on the value of $C_0 = 280$ ppm given in Table 10.5, is $C_{eqv} = 378$ ppm. Note from Table 10.5 that the carbon dioxide concentration is currently of 387 ppm, thus larger than the equivalent carbon associated to the industrial era. This situation is explained by the fact that the radiative forcing accounts for both positive and negative effects on radiation balance of the planet.

The usefulness of radiative forcing or the equivalent carbon comes from the possibility to correlate the forcing with the global temperature change. One asks the question, how much change on the planet temperature is produced by a given total radiative forcing. In this respect, a so-called climate sensitivity factor $\gamma$ can be introduced as follows:

$$\gamma = \frac{\Delta T_e}{\Delta F_{total}} \tag{10.25}$$

where $\Delta T_e$ is the corresponding temperature variation that can be caused by $\Delta F_{total}$.

Rubin (2001) compiles the results of two relevant studies published prior to 1979 that give constant approximations for the climate sensitivity factor in the range of 0.55–0.65 km$^2$/W. An average value of 0.6 km$^2$/W can be assumed for the first approximation calculations. The IPCC (2007) summarizes the result of climate change prediction based on thorough simulations that account for many relevant influencing factors. The results of the IPCC are presented in the form of temperature increase due to doubling of $CO_2$ concentration. It is estimated that the radiative forcing caused by doubling the carbon dioxide concentration with respect to the year 1992 is 4.37 W/m$^2$ and the induced global temperature increase is 2.5 K; thus, the probable value for climate sensitivity factor is 0.57 km$^2$/W. Nevertheless, there must be a time lag between the moment when a radiative forcing occur and the future moment when – potentially – the radiative balance is established again and the global temperature reaches a new equilibrium. One expects that the time to reach a new equilibrium is long because the thermal inertia of the planet surface is high (it is composed from waters in a large amount, ice, biomass, rocks, and land – all having high specific or latent heats).

Quantifying the effect on climate that a particular atmospheric gas has is a multivariable problem. The radiative forcing induced by atmospheric gas is an indication of the direction in which the presence of the respective gas can influence the climate. If, for example, the gas absorbs more in infrared spectrum, its greenhouse effect is accentuated, that is, it has associated a positive radiative forcing. However, another parameter is also important, that is, the atmospheric lifetime. It is important how long the gas is active in the atmosphere with respect to radiative balance control.

It is indicated that the radiative forcing produced by introducing $m_0 = 1$ kg of $CO_2$ in the atmosphere with $\Delta F_{CO_2}$. The net forcing produced during an infinitesimal time interval is given by the product $\Delta F_{CO_2}(t) f_{CO_2}(t) dt$. One can integrate the former quantity over a time horizon TH and obtain the total forcing produced by the respective amount of $CO_2$. If another kind of GHG except carbon dioxide is considered, its integrated forcing over the time horizon can be normalized with the integrated forcing of $CO_2$. From this reasoning results the following definition:

$$GWP = \frac{\int_0^{TH} \Delta F_{GHG}(t) f_{GHG}(t) dt}{\int_0^{TH} \Delta \Gamma_{CO_2}(t) f_{CO_2}(t) dt} \tag{10.26}$$

which is the global warming potential (GWP) of the GHG.

Table 10.6 shows the GWP of the main GHG for 3 time horizons, based on the IPCC (2007). Uncontrolled human activities since the industrial revolution have brought the planet up to a level that the amount of emissions and the magnitude of global environmental impact are indigestible. The planet has the symptoms of inadvertently catching a disease. The symptoms have of course slowly become more apparent as a slight rise in the global atmospheric temperature. There is no agreement on the cause or the consequence. Some ascribe this rise as due to an addiction to uncontrolled energy use and moreover carbon-based energy and the emissions of infrared-absorbing gases. With the so-called climate change, we are literally burning up a fever with a giant bonfire, a respiratory disease of carbon-based fuels that we are literally steadily breathing out as an added atmospheric pollution, $CO_2$ and $CH_4$ burden.

**Table 10.6** The GWP of principal greenhouse gases

| Substance | Time horizon (years) | | | Substance | Time horizon (years) | | |
|---|---|---|---|---|---|---|---|
| | 20 | 100 | 500 | | 20 | 100 | 500 |
| Carbon dioxide | 1 | 1 | 1 | HCFC-22 | 5,160 | 1,810 | 549 |
| Methane | 72 | 25 | 7.6 | Carbon tetrachloride | 2,700 | 1,400 | 465 |
| Nitrous oxide | 289 | 298 | 153 | HFC-134a | 3,830 | 1,430 | 435 |
| CFC-11 | 6,730 | 4,750 | 1,620 | Sulfur hexafluoride | 16,300 | 22,800 | 32,600 |
| CFC-12 | 11,000 | 10,900 | 5,200 | Nitrogen trifluoride | 12,300 | 17,200 | 20,700 |

*Source*: Data from the IPCC (2007).

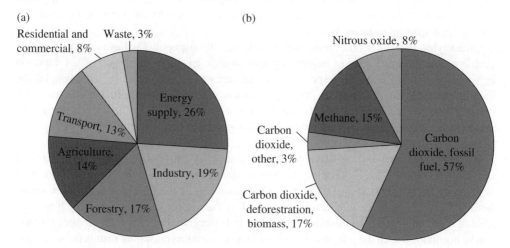

(a)

Residential and commercial, 8%   Waste, 3%

Energy supply, 26%

Transport, 13%

Agriculture, 14%

Industry, 19%

Forestry, 17%

(b)

Nitrous oxide, 8%

Methane, 15%

Carbon dioxide, other, 3%

Carbon dioxide, fossil fuel, 57%

Carbon dioxide, deforestation, biomass, 17%

**Figure 10.11** Anthropogenic GHG emissions by sectors (a) and major gas type (b) (data from the IPCC (2007))

The emissions of gases in the atmosphere specific to the industrial era affected and will affect the radiative balance of the planet and therefore influence the earth climate. Figure 10.11 presents the distribution of anthropogenic GHG emissions per sector of activity (a) and per type of major GHG (b). This distribution can be sued as a weighting factor for various emission-related ASIs.

The global value of carbon emission can be evaluated based on the value of carbon energy and ability to provide economic growth. After all, this is how improvement in modern industrial society is measured and reported. The purely economic value of the carbon emissions and power source is reflected in producing financial wealth for every country (such as the national GDP) using carbon energy. Energy is greatest in developed (rich) nations, and we observe a correlation between the growth in GDP and the growth in carbon energy use.

Table 10.7 shows the typical life cycle emissions for power generation from differing sources (g/kW h). This relationship also holds true at the global level. Hence, the global growth in GHG concentration in the atmosphere over the last 30 years (measured as ppm $CO_2$ at Mauna Loa, Hawaii, where 1 ppm $CO_2 \sim 9.1012 \, t_{CO2}$) is directly and linearly correlated to the GWP (measured in terra dollars, US $1012). To reduce the effect of the year-to-year noise

**Table 10.7** Typical life cycle emissions for power generation from differing sources (g/kW h)

| Electric energy generation source | Switzerland 2000 | Canada 2000 | IAEA 2000 | France (production only) |
|---|---|---|---|---|
| Natural gas | 605 | N/A | 696 | 500 |
| Coal | 1071 | 974 | 978 | N/A |
| Solar panels | 114–189 | N/A | 97 | N/A |
| Nuclear | 16 | 3–15 | 21 | 0 |
| Oil | 856 | 778 | 811 | 701 |
| Wind | 36 | N/A | 36 | N/A |
| Hydro | 4 | 15 | 16–23 | N/A |

*Source*: Data from EIA (2008).
IAEA, International Atomic Energy Agency.

in the atmospheric $CO_2$ concentrations, five-year averages for GWP were plotted against the change in $CO_2$ measured over those 5 years. Rather than plot ppm values of $CO_2$, the change was converted to billions of tones (Gt) of $CO_2$ released based on the 7.9 Gt of $CO_2$ required to cause a 1 ppm increase in the atmosphere accompanied by an equal release being absorbed in the oceans. So, 1 ppm was taken to be equivalent to a total of 15.8 Gt of $CO_2$ released. It is reasonable to use 1950 as the base year since the $CO_2$ buildup prior to about 1950 was small.

## 10.3.3   Exergy Destruction and Environmental Impact of Drying Systems

The environmental impact of drying can be determined based on the accounting of exergy destructions. This aspect has been extensively analyzed in Rosen and Dincer (1999) who presented the exergy analysis of waste emissions with a focus on thermal systems. Exergy analysis helps evaluating and comparing the relative environmental merits of distinctly different residual streams in an objective manner by using exergy destruction as a waste emissions accounting tool. By using exergy destruction as a measure to quantify environmental effects associated with emissions and resource depletion, the environmental impact assessment is made based on purely physical principles. An understanding of the relations between exergy and the environment may reveal the underlying fundamental patterns and forces affecting changes in the environment and help researchers to deal better with environmental damage.

For the discussion purpose, in Figure 10.12 is shown a drying system indicating the waste stream in a simplified manner. The moist material and the resources all carry exergy which is fed into the system. The outputs, that is, the dry product and the wastes, discharge exergy into the environment. As a consequence, all exergy input is destroyed (wasted in the environment) except the exergy carried by the dry product (however, in many situations, the exergy in the dry product is not valorized directly). Thence, the discharged wastes in the environment and their impact can be quantified by accounting for exergy destructions. It is known that the exergy destroyed (and wasted) by the system is of two kinds:

- Internal exergy destruction, which represents the lost opportunity to perform work; the environmental impact and rejected wastes due to all upstream processes (e.g., power generation) can be related to the internal exergy destruction.

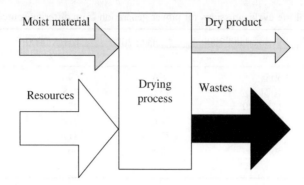

**Figure 10.12**   Thermodynamic representation of the drying process indicating the waste stream

- Lost exergy or exergy destruction at system interaction with its surroundings, which is related to the discharged wastes by the process itself. In principle, all discharged wastes by the system can be recovered to use their exergy and reduce the environmental impact; however, this action may be very expensive and generally is not undertaken in practice, except when it is justified economically or enforced by sustainability policies.

  The practical connection between wasted exergy and environmental impact can be discovered by correlating recorded emission data at the level of a region or globally with the chemical exergy of emitted pollutants. Here, some exergy destruction versus emission data correlation is presented for Ontario, as a relevant example, based on the works by Rosen and Dincer (1999) and Dincer (2007). The general aspects of environmental policy in Ontario can be described as follows:

- In Ontario, the Environmental Protection Act by the Ministry of the Environment exists giving the legislation on environmental quality and air pollution limits which are conceived such that human health and the ecosystem are not endangered.
- The potential of a substance to impact on the environment is evaluated using a set of 10 parameters:
  - Transport
  - Persistence
  - Bioaccumulation
  - Acute lethality
  - Sublethal effects on mammals
  - Sublethal effects on plants
  - Sublethal effects on nonmammalian animals
  - Teratogenicity
  - Mutagenicity/genotoxicity
  - Carcinogenicity
- An aggregated indicator is determined based on the 10 impact parameters (above), referred to as the point of impingement (PoI) which is determined based on the best known available pollution control technology.

**Table 10.8**   Estimations of exergetic cost of atmospheric pollutants for Ontario

| Pollutant | EPC ($/kg) | $M$ (kg/kmol) | MEPC ($/kmol) | $ex^{ch}$ (MJ/kmol) | $ex^{PC}$ ($/MJ) | PoI ($\mu g/m^3_{air}$) |
|---|---|---|---|---|---|---|
| $CO_2$ | 0.0402 | 44 | 1.77 | 19.60 | 0.090 | 56,764 |
| $CH_4$ | 1.2998 | 16 | 20.80 | 831.66 | 0.025 | N/A |
| $NO_x$ | 3.8944 | 38 | 148.18 | 72.4 | 2.047 | 500 |
| $SO_2$ | 3.551 | 64 | 227.26 | 310.99 | 0.731 | 830 |
| CO | 5.5074 | 28 | 154.21 | 275.00 | 0.561 | 6,000 |
| VOCs | 0.603 | 44 | 26.53 | 1,233.00 | 0.021 | N/A |
| PM | 5.5074 | 68 | 374.50 | 436.00 | 0.860 | N/A |

*Source*: Carpenter (1990), De Gouw et al. (2006), Pellizzari et al. (1999), Ontario Regulation 419/05.
*Note*: Costs are in $\$_{2013}$ based on the Canadian consumer price index.

**Table 10.9**   Approximated average VOC composition and characteristics in Ontario

| VOCs | Formula | $M$ (kg/kmol) | $y$ (kmol/kmol) | $ex^{ch}$ (MJ/kmol) | PoI ($\mu g/m^3_{air}$) |
|---|---|---|---|---|---|
| Methanol | $CH_3OH$ | 32 | 0.020 | 612 | 12,000 |
| Acetonitrile | $CH_3CN$ | 41 | 0.823 | 1,169 | 180 |
| Acetaldehyde | $CH_3CHO$ | 44 | 0.001 | 1,063 | 500 |
| Acetone | $CH_3COCH$ | 58 | 0.111 | 1,636 | 48,000 |
| Acetic acid | $CH_3COOH$ | 60 | 0.025 | 780 | 2,500 |
| Butanone | $C_2H_5COCH_3$ | 72 | 0.015 | 2,755 | 250 |
| Toluene | $C_6H_5CH_3$ | 92 | 0.005 | 3,771 | 2,000 |

*Source*: De Gouw et al. (2006), Ontario Regulation 419/05.

- The methodology denoted removal pollution costs (RPCs) is applied to correlate the exergy of the waste stream with the cost of removing pollutants from the waste stream prior to discharge into the surroundings. The costs for waste emissions are evaluated as the total fuel cost per unit fuel exergy multiplied by the chemical exergy per unit fuel exergy and divided by the exergy efficiency of the pollution removal process.
- In Canada, the environmental pollution costs (EPCs) are estimated based on qualitative and quantitative evaluations of the pollution cost to the society for compensation and correction of the environmental damage and to prevent harmful discharges into the environment. Table 10.8 gives the EPCs of Ontario pollutants.
- The average compositions of VOC emissions in Ontario are approximated as given in Table 10.9.
- The average compositions of PM emissions in Ontario are given in Table 10.10.
- The fuel costs for three types of fossil fuels, namely, coal, no. 6 fuel oil, and natural gas, are in average value of CN$ 2013 as follows:
  - Average coal: $1.411/GJ LHV
  - Average no. 6 fuel oil: $1.864/GJ LHV
  - Average natural gas: $3.899/GJ LHV
- Table 10.11 illustrates the EPC and RPC for the main fossil fuels in Ontario.

**Table 10.10** Approximated average PM composition and characteristics in Ontario

| PM | Formula | $M$ (kg/kmol) | $y$ (kmol/kmol) | $ex^{ch}$ (MJ/kmol) | PoI ($\mu g/m^3_{air}$) |
|---|---|---|---|---|---|
| Lead | Pb | 207 | 0.053 | 249.2 | 10 |
| Cadmium | Cd | 112 | 0.097 | 298.4 | 5 |
| Nickel | Ni | 59 | 0.185 | 242.6 | 5 |
| Chromium | Cr | 52 | 0.210 | 584.4 | 5 |
| Copper | Cu | 63 | 0.173 | 132.6 | 100 |
| Manganese | Mn | 55 | 3.115E–7 | 487.7 | 7.5 |
| Vanadium | V | 51 | 0.214 | 721.3 | 5 |
| Aluminum | Al | 27 | 0.006 | 795.7 | 26 |
| Calcium | Ca | 40 | 0.054 | 729.1 | 14 |
| Magnesium | Mg | 24 | 0.008 | 626.9 | 60 |

*Source*: Pellizzari et al. (1999), Ontario Regulation 419/05.

**Table 10.11** Environmental pollution costs (EPC) and removal pollution costs (RPC) for fossil fuels in Ontario

| Pollutant | EPC ($/GJ$_{fuel\ exergy}$) | | | RPC ($/GJ$_{fuel\ exergy}$) | | |
|---|---|---|---|---|---|---|
| | Coal | No. 6 fuel oil | Natural gas | Coal | No. 6 fuel oil | Natural gas |
| $CO_2$ | 5.2662 | 5.7352 | 7.7184 | 3.35 | 2.7604 | 1.7822 |
| $CH_4$ | 5.1724 | 1.1524 | 0 | 0.8174 | 0.1608 | 0 |
| $NO_x$ | 0.0469 | 0.03082 | 0.03082 | 0.938 | 0.469 | 0.3886 |
| CO | 2.1038 | 0.03082 | 0.3484 | 7.5442 | 0.0938 | 0.4556 |
| $SO_2$ | 0.4958 | 0.402 | 0 | 2.5594 | 1.5544 | 0 |

*Source*: Rosen and Dincer (1999).
*Note*: Monetary values are in CN$ 2013.

A simplified method to estimate the cost of pollutant removal from the waste stream is based on the exergy efficiency of pollutant removal. According to Rosen and Dincer (1999), this exergy efficiency is in the range of 1–5%. Therefore, once the exergy destroyed due to pollutant discharge is known, the required exergy to remove the pollutant from the waste stream can be calculated. Furthermore, in Chapter 8, it is shown that the average price of exergy can be estimated for any geopolitical region; for Canada, it is approximated $C_{ex} = 8.4 \text{ ¢/kWh} = 2.3 \text{ ¢/MJ}$. When the exergy required to remove the pollutants is multiplied with exergy price, the removal pollutant cost is obtained. Therefore, one has

$$RPC = C_{ex} \psi_{pr} Ex_{d,pw} \qquad (10.27)$$

where $C_{ex}$ is the exergy price, $\psi_{pr}$ is the exergy efficiency of pollutant removal from the waste stream, and $Ex_{d,pw}$ is the exergy destroyed due to pollutant waste in the environment.

Assuming an average exergy efficiency of pollutant removal from the waste stream of 3%, the RPC can be estimated as $0.7 \text{ GJ}^{-1}$. The RPC of power generation can be roughly estimated

**Table 10.12**   Removal pollution cost at power generation in Canada

| Primary energy | Exergy input (PJ) | Power output (PJ) | Exergy destruction (PJ) | RPC (million $) | $\psi$ (%) |
|---|---|---|---|---|---|
| Nuclear | 1073 | 339 | 734 | 514 | 32 |
| Hydro | 1698 | 1358 | 340 | 238 | 80 |
| Coal | 630 | 378 | 252 | 176 | 60 |
| Fuel oil | 62.5 | 23 | 39.5 | 28 | 37 |
| Natural gas | 254 | 112 | 142 | 99 | 44 |
| Natural gas liquids | 33.6 | 8 | 25.6 | 20 | 24 |
| Diesel fuel | 8.3 | 3.2 | 5.1 | 4 | 38 |
| Biomass | 66 | 19 | 47 | 33 | 29 |
| Secondary sources | 1629 | 543 | 1086 | 760 | 33 |
| Overall | 5454 | 2783 | 2671 | 1870 | 51 |

*Source*: Dincer and Zamfirescu (2011, Chapter 17).

**Table 10.13**   Life cycle emissions into the atmosphere for power generation technologies (kg/GJ)

| Technology | Kilogram per gigajoule fuel exergy | | | | | | | |
|---|---|---|---|---|---|---|---|---|
| | $CO_2$ | $CH_4$ | $NO_x$ | $SO_2$ | CO | VOCs | PM | |
| Coal-fired power plants | 274 | 73 | 0.180 | 0.400 | 1.374 | 0.251 | 8.463E−6 | |
| Fuel oil-fired power plants | 176 | 47 | 0.115 | 0.200 | 0.876 | 0.160 | 5.439E−6 | |
| Natural gas-fired power plants | 112 | 30 | 0.073 | 0.150 | 0.558 | 0.102 | 2.557E−6 | |
| PV power generation | 44 | 12 | 0.045 | 0.090 | 0.038 | 0.003 | 1.336E−6 | |
| Wind power generation | 33 | 9 | 0.033 | 0.070 | 0.028 | 0.003 | 1.024E−6 | |
| Hydro power | 9 | 2 | 0.006 | 0.015 | 0.044 | 0.008 | 0.268E−6 | |
| Nuclear power generation | 48 | 13 | 0.032 | 0.065 | 0.241 | 0.044 | 1.488E−6 | |
| Canada averages kilogram pollutant per megawatt hour power | 42 | 11 | 0.029 | 0.060 | 0.194 | 0.035 | 1.288E−6 | Total ($/MW h) |
| $EPC_{ex}$ ($/MW h power) | 1.7 | 14.7 | 0.1 | 0.2 | 1.1 | 0.02 | 7.1E−6 | 17.8 |

*Source*: Carpenter (1990), De Gouw et al. (2006), Pellizzari et al. (1999), and Rosen and Dincer (1999). $EPC_{ex}$, exergetic environmental pollution cost.

based on statistical data that allow for the estimation of the exergy destructions. Table 10.12 shows the rough estimate of RPC associated with Canadian power generation. The cost of pollution associated with the construction of power generation facilities, reparations, and maintenance is not included in the results shown in Table 10.12. As given, the total RPC for power generation is ~$1870 million, and the overall exergy efficiency of the power generation sector is 51%. The RPC becomes $2.5 MW$^{-1}$ h$^{-1}$ generated power.

Life cycle pollutant emissions from power generation technologies are determined in Table 10.13 based on multiple literature data sources as follows: Carpenter (1990), De Gouw

**Table 10.14**  Embodied energy, pollution, and environmental pollution cost in construction materials

| Material | EE (GJ/t) | SE (kg $CO_2$/GJ) | $EPC_{ex}$ ($\$_{2013}$/GJ) |
|---|---|---|---|
| Concrete | 1.4 | 24 | 70.5 |
| Iron | 23.5 | 11 | 29.3 |
| Steel | 34.4 | 11 | 29.3 |
| Stainless steel | 53 | 62 | 29.3 |
| Aluminum | 201.4 | 10 | 29.9 |
| Copper | 131 | 57 | 60.1 |
| Fiberglass | 13 | 62 | 66.0 |

*Source*: Rosen and Dincer (1999).
EE, embodied energy; SE, specific GHG emissions; $EPC_{ex}$, exergetic environmental pollution cost.

et al. (2006), Pellizzari et al. (1999), and Rosen and Dincer (1999). The amounts of atmospheric pollutant are given with respect to the gigajoule of source exergy. Using the data from Table 10.12, weighted average pollutant emissions are obtained for the Canadian power generation mix. The averages are given in kilogram of pollutant per megawatt hour of power generated; in order to convert from source exergy basis to generated power basis, the average Canadian exergy efficiency of power generation is used. Then the exergy-based EPC for power generation is calculated for each pollutant in $/MW h; the Canadian average of $EPC_{ex}$ is $17.8/ MW h; therefore, the cost of pollutant removal from the waste stream is much lower to the society than the cost of pollutant emission.

Materials used for system construction bring associated embodied energy and pollution. Table 10.14 demonstrates the embodied energy and pollution amount and cost with various construction materials relevant to drying systems. Concrete, copper, and fiberglass bring the highest among the EPC of the listed materials. The highest embodied energy is due to aluminum fabrication, which as it is known require an energy-intensive electrochemical process.

Using the concepts introduced here, the exergy destruction can be utilized to determine the environmental impact of a drying system in various manners as follows:

- The removal of pollution cost can be approximated by multiplying the exergy destruction with the exergy price estimated for a specific region or country.
- The rate of exergy input into the drying system can be used to determine the system physical size.
- From the system physical size, the mass amount of materials required for system construction is determined.
- The amount of each construction material correlates with the embedded energy required for its extraction and with GHG emissions and the EPC (see Table 10.14).
- The life cycle total exergy input into the system is equal to the exergy required for system construction, system operation, and system salvage.
- Based on the life cycle total exergy input and the exergy efficiency of power and heat generation system, the exergy destruction can be determined at power and heat generation.
- The emissions and EPC due to life cycle total exergy supply are determined based on the exergy destruction and exergetic EPC of the power and heat generation subsystem.

- The exergy destruction of the drying system itself allows for determination of pollutant wastes and the EPC for operation during the entire lifetime.
- If a percent of materials recycling is provided, then the wasted energy and emissions of scrapped system can be determined.
- The total pollutant emissions and EPC result from the summation of the terms associated to power and heat generation from primary sources, system manufacturing, system operation, and system scrapping.

In many cases, it is useful to compare the environmental performance of the studied system with a reference system. If the environmental impact is smaller with respect to the reference, the studied system is "greener." The concept of greenization has been introduced by Dincer and Zamfirescu (2012). When applied to a drying system, the greenization concept can be illustrated as shown in Figure 10.13.

As can be observed from the figure, the general idea is to reduce the environmental impact of drying system by relating in an increased manner to renewable energy sources and greener processes with improved exergy efficiency and reduced exergy destructions. It is noteworthy that the societal reaction to pollution can be of two kinds: adaptation and mitigation.

Adaptation involves taking measures aimed at reducing the causes leading to the society's vulnerability to various pollution effects such as climate change. In this respect, the drying system must be made more effective to reduce the polluting fuel consumption and thence to reduce the GHG emissions. Furthermore, the share of renewable must be increased gradually in the energy supply.

The mitigation refers to a set of measures aimed to eliminate the effects of environmental impact. Here also, the GHGs are the most important target of mitigation. It appears that a good

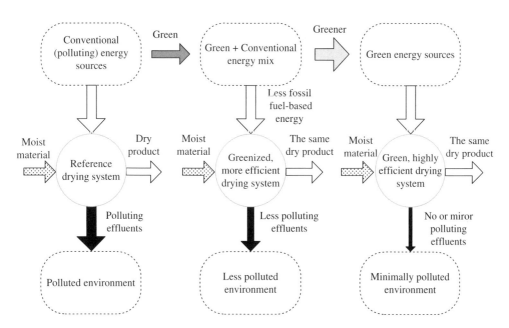

**Figure 10.13**  Illustrating the concept of greenization applied to drying systems

policy to encourage carbon mitigation entails imposing a carbon tax to encourage sustainable energy development. The carbon tax seems likely to be applied in most of the developed countries. By 2050, it is projected that developed countries will implement a carbon tax in the range of \$25–250 per ton of carbon dioxide equivalent emitted. Some other environmental impact mitigation measures of anthropogenic environmental impact are as follows:

- Applying energy efficiency and energy conservation measures
- Encouraging development of nuclear power generation systems
- Switching fuels from coal to natural gas
- Making use of renewable thermal energy from geothermal, biomass, or solar radiation
- Making use of renewable electricity from solar radiation, hydro-energy, wind, biomass, geothermal, tidal, and ocean thermal (via ocean thermal energy conversion)
- Using combined heat and power systems and multigeneration systems
- Applying carbon dioxide capture and storage
- Applying waste heat recovery and energy extraction from waste materials

In view of greenization of drying systems (and other thermal systems), a greenization factor can be developed such that the drying system is compared in relative terms with a reference system. A greenization factor of zero indicates that the system is not greenized with respect to the reference system. If the system is fully greenized, then the greenization factor is 1. Fully greenized systems depend on sustainable energy sources that have zero or minimal environmental impact during the utilization stage (although some environmental impact is associated with system construction). In greenization factor definition equation, the environmental impact factor must be specified for two cases: the reference system and the greenized system.

Depending on the specific problem analyzed, various types of environmental impact factors may be formulated. An effective and comprehensive way to define the greenization factor is through exergy destructions as follows:

$$GF = \frac{Ex_{d,ref} - Ex_d}{Ex_{d,ref}} \qquad (10.28)$$

where $Ex_d$ is the life cycle exergy destruction of the studied drying system and $Ex_{d,ref}$ is the exergy destruction of the reference system.

Both the drying system and the reference system process the same amount of moist material which is converted into dry product. But the energy, exergy, environmental, and cost parameters of the two systems differ. Observe from Eq. (10.28) that if $Ex_d \rightarrow 0$, then $GF \rightarrow 1$ meaning that the system tends to be greener. If $Ex_d = Ex_{d,ref}$, no greenization effect is observed, both the reference and the studied systems having similar environmental impact.

Noteworthy is that the greenization factor can be defined based on other environmental and sustainability parameters. The ASI (as defined earlier) can be used for determining the greenization factor. In this case, Eq. (10.28) becomes

$$GF = \frac{ASI - ASI_{ref}}{ASI} \qquad (10.29)$$

The ASI accounts for thermodynamic, technical, environmental, economic, and societal impacts of the technology. Thenceforth, this greenization factor formulation allows for a multidimensional comparison of drying systems.

## 10.4 Case Study: Exergo-Sustainability Assessment of a Heat Pump Dryer

In this case study, an industrial wood chip drying process is considered as reference system for exergy-sustainability assessment based on greenization factor and other sustainability indicators. The reference system is taken from the previous work of Coskun et al. (2009). The reference system uses combustion gases for drying. An improved system is considered in which the combustor is replaced with a heat pump which recirculates air as a drying agent, whereas the air has similar parameters as the flue gases. It is assumed that the improved system is connected to Ontario regional grid.

### 10.4.1 Reference Dryer Description

The reference industrial drying system uses a directly heated rotary kiln with an average capacity of 93 t/h moist material. The kiln is supplied with preheated wood chips, which after being dried are separated from the flue gases in cyclones. The system also includes a local gas turbine power plant with cogeneration. The core process, which is the drying process in the rotary kiln, is described in Figure 10.14.

In the reference system configuration, no humid air is recycled; therefore, all material and exergy in state 4 (Figure 10.14) are wasted. The overall reference drying system configuration is described for a life cycle exergo-sustainability assessment in Figure 10.15, whereas the state point parameters are given in Table 10.15. The following remarks are made about the reference drying system:

- State point parameters 8, 10, and 13 are taken from Coskun et al. (2009).
- State point parameters 9, 11, 12, 14, and 15 were calculated based on balance equations under the following assumptions; the calculated values are shown with italic font in the table:
  - Mixing processes are ideal (no exergy destruction).
  - There is negligible heat loss in the combustor.
  - The energy efficiency of the gas turbine is 20% (including the generator losses).

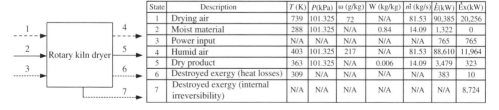

| State | Description | T (K) | P(kPa) | ω (g/kg) | W (kg/kg) | ṁ (kg/s) | Ė(kW) | Ėx(kW) |
|-------|-------------|-------|--------|----------|-----------|----------|-------|--------|
| 1 | Drying air | 739 | 101.325 | 72 | N/A | 81.53 | 90,385 | 20,256 |
| 2 | Moist material | 288 | 101.325 | N/A | 0.84 | 14.09 | 1,322 | 0 |
| 3 | Power input | N/A | N/A | N/A | N/A | N/A | 765 | 765 |
| 4 | Humid air | 403 | 101.325 | 217 | N/A | 81.53 | 88,610 | 11,964 |
| 5 | Dry product | 363 | 101.325 | N/A | 0.006 | 14.09 | 3,479 | 323 |
| 6 | Destroyed exergy (heat losses) | 309 | N/A | N/A | N/A | N/A | 383 | 10 |
| 7 | Destroyed exergy (internal irreversibility) | N/A | N/A | N/A | N/A | N/A | N/A | 8,724 |

**Figure 10.14** Wood chips drying process in an industrial rotary kiln dryer

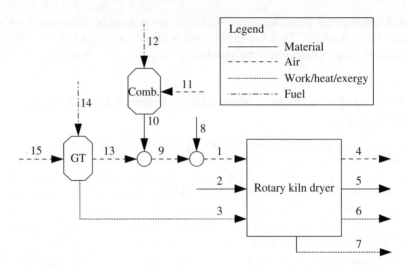

**Figure 10.15** Wood chips drying process in an industrial rotary kiln dryer

**Table 10.15** State points description and parameters for the reference drying system

| State | Description | $T$ (K) | $P$ (kPa) | $\omega$ (g/kg) | (kg/s) | $\dot{E}$ (kW) | $\dot{E}x$ (kW) |
|-------|-------------|---------|-----------|-----------------|--------|----------------|-----------------|
| 8 | Air leak in | 298 | 101.325 | 7.2 | 2.23 | 691 | 0 |
| 9 | Flue gas | 749 | 100 | 74 | 79.3 | 89,694 | 20,256 |
| 10 | Flue gas | 803 | 100 | 60.6 | 62.4 | 71,834 | 7,722 |
| 11 | Air input | 298 | 101.325 | 7.2 | 61.1 | 2,651 | 0 |
| 12 | Fuel (natural gas) | 298 | 101.325 | N/A | 1.26 | 69,183 | 65,410 |
| 13 | Flue gas | 533 | 101.325 | 123.1 | 16.9 | 17,890 | 12,534 |
| 14 | Fuel (natural gas) | 298 | 1,200 | N/A | 0.31 | 17,176 | 16,239 |
| 15 | Air input | 298 | 101.325 | 7.2 | 16.47 | 714 | 0 |

- Because no chemical reactions occur, the chemical exergy of wood is not considered; in addition, the industrial process is conducted to process wood as construction material and not firewood.
- The balance of plant requires auxiliary power of 670 kW.
- The site consumes 2 MW of power for housekeeping and other industrial uses.
- The fuel for gas turbine and combustor is natural gas.

We calculate now the energy efficiencies and the exergy destroyed by each component and the overall system. The exergy destroyed by the rotary kiln unit is the sum of the exergy wasted as humid air (stream 4); the exergy of the warm, discharged wood; the exergy wasted as heat loss (stream 6); and the internal irreversibility (stream 7). The exergy input into the kiln is the sum of exergies for drying air, moist material, and power. Therefore, from Figure 10.14, one has

$$\dot{E}x_{d,kiln} = 11,964 + 323 + 10 + 8724 = \dot{E}x_{in,kiln} = 20,256 + 0 + 765 = 21,021\,kW$$

The reversible work for the drying process is reported in Coskun et al. (2009), and it is 893 kW:

$$\psi_{kiln} = \frac{\dot{W}_{rev}}{\dot{Ex}_{in,\,kiln}} = \frac{893}{21,021} = 0.0425$$

The exergy destroyed in the combustor can be calculated based on the data from Table 10.15, and it is

$$\dot{Ex}_{d,\,comb} = \dot{Ex}_{11} + \dot{Ex}_{12} - \dot{Ex}_{10} = 57,687\,kW$$

The exergy destroyed by the gas turbine is determined as follows:

$$\dot{Ex}_{d,\,gt} = \dot{Ex}_{14} + \dot{Ex}_{15} - \dot{Ex}_{13} = 270.2\,kW$$

The total exergy destruction is

$$\dot{Ex}_d = \dot{Ex}_{d,\,gt} + \dot{Ex}_{d,\,comb} + \dot{Ex}_{d,\,kiln} = 78,979\,kW$$

The total exergy input is

$$\dot{Ex}_{in} = \dot{Ex}_{12} + \dot{Ex}_{14} = 81,649\,kW$$

The overall plant produces two useful outputs: the reversible work for drying and generated power for other process (2 MW); therefore, the exergy efficiency becomes

$$\psi = \frac{893 + 2000}{81,649} = 0.0354$$

In order to operate, the plant consumes 1.57 kg/s natural gas. The annual operation hours are assumed 260 h. The system lifetime is 25 years. The total amount of natural gas consumed is 36,738 t or 1.91 PJ of exergy.

## 10.4.2 Exergo-Sustainability Assessment for the Reference Drying System

A quick sustainability assessment of the reference system can be performed based on the exergy efficiency and according to the SI defined by Eq. (4.26). This index is calculated as follows:

$$SI = \frac{1}{1-\psi} = \frac{1}{1-0.0354} = 1.03$$

Since the SI is low (close to 1), this indicates that the system sustainability is low; there should be room for improvement. A more profound analysis must be therefore made, considering the whole life cycle of the system; this is denoted as the exergo-sustainability analysis. We will start this analysis by an attempt to estimate the sizes of system components and from here to determine the amounts of materials needed.

First, the drying time must be determined. From Table 5.2, we take the moisture diffusivity for Douglas fir of $D = 32.4\text{E}-10\,\text{m}^2/\text{s}$. Assume the wood chips having a slab form with an average half width $L = 0.0002\,\text{mm}$. Assume that the equilibrium moisture content is 10% of the final moisture content (this is a reasonable assumption for practical systems). Therefore, $W_e = 0.1 W_5 = 0.0006\,\text{kg/kg}$ (see Figure 10.14). Therefore, the final dimensionless moisture content becomes

$$\phi_f = \frac{W_6 - W_e}{W_2 - W_e} = \frac{0.006 - 0.0006}{0.84 - 0.0006} = 0.0064$$

With the use of Eq. (5.28) and under the assumption that $Bi_m > 100$, the following value is obtained for the drying time:

$$t_{dry} = -\frac{\pi L^2}{4D} \ln\left(\frac{4}{\pi}\,\Phi_f\right) = 45\,\text{s}$$

Assume that inside the kiln, the particles move at the periphery such that the particle velocity makes an angle of $\pi/6$ with air velocity. The dimensions of the kiln can be then found such that the gas flow regime is turbulent. We found that if $U_{air} = 20\,\text{m/s}$, then the required diameter to accommodate the 82 kg/s of air flow rate (Figure 10.14) is 3.3 m and the Reynolds number is 890,000. Furthermore, the wood velocity component in axial direction becomes $U_{wood} = 0.36\,\text{m/s}$. Therefore, the length of the dryer is estimated to

$$L_{kiln} = U_{wood} t_{dry} = 17\,\text{m}$$

If the kiln is made in stainless steel of 8 mm thick, then the mass of steel is $m_{ssteel} = \pi d_{kiln} L_{kiln} = 10.7\,\text{t}$. A supporting structure of carbon steel will be required, say, 80% lighter; $m_{csteel} = 3.4\,\text{t}$. A concrete platform must be casted below the kiln with an approx. size of $L_{kiln} \times d_{kiln} \times 0.3\,\text{m} \rightarrow m_{concrete} \cong 40\,\text{t}$.

The combustor generates ~62 kg/s hot gases at 803 K which is equivalent to 141 m³/s. With a combustion residence time of 5 ms (practical value) and a combustion zone of 0.5 m, the combustor can be approximated as a cylinder with 1.3 m diameter and 5 m height (length), placed vertically; therefore, the required metal mass (carbon steel) is 15 t including the supports. The concrete required for the foundation is 3 t.

The gas turbine is made of stainless steel with aluminum blades and has a connected generator that comprises mainly copper and iron. The flow rate of air at gas turbine suction is 15 m³/s. The estimated materials are 13 t stainless steel, 2 t carbon steel, 0.5 t aluminum, 2 t iron, 4 t copper for the generator, and 9 t concrete. The construction material data is summarized in Table 10.16 where the embedded energy, specific emissions, and EPC are given.

In order to perform the life cycle assessment, the model presented in Figure 10.16 is elaborated. Primary exergy resources are consumed and destroyed to generate the power required for the essential operations during lifetime. A part of consumed power is used for fuel processing (stream 17) in which the stream 22 of refined natural gas is produced and distributed to supply

**Table 10.16**  Construction materials and the associated sustainability parameters for the reference drying system

| Material | $m$ (t) | EE (GJ) | SE (kg GHG) | EPC$_{ex}$ ($\$_{2013}$) |
|---|---|---|---|---|
| Concrete | 52 | 73 | 1,752 | 5,146 |
| Iron | 2 | 47 | 517 | 1,377 |
| Carbon steel | 20 | 688 | 7,568 | 20,158 |
| Stainless steel | 23.7 | 1,256 | 77,827 | 36,800 |
| Aluminum | 0.5 | 101 | 1,010 | 3,020 |
| Copper | 4 | 524 | 29,868 | 31,492 |
| Total | | 2,689 | 118,542 | 97,993 |

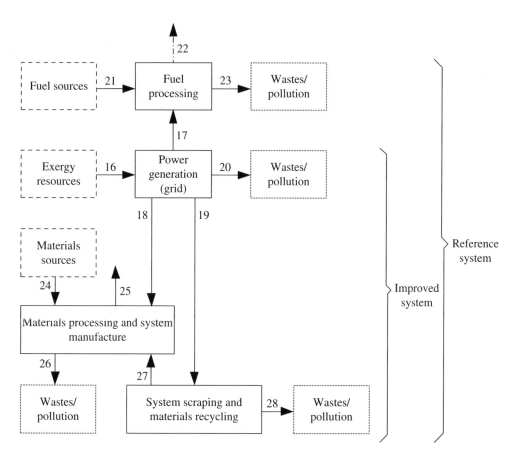

**Figure 10.16**  Model for life cycle operations for drying system manufacture, scrapping, and fuel production

the reference drying system for the lifetime. Another part of electrical grid exergy is consumed for materials processing and system manufacture, stream 18, and another part for system scrapping and materials recycling, stream 19. In all processes, wastes and pollution are emitted in the environment.

Assume that 10% of scrapped materials are recycled. Therefore, the embedded energy for materials processing and manufacture can be approximated with 90% of 2689 GJ given in Table 10.16; this is 2420 GJ. The specific emissions will be 106.7 t GHG with an EPC of $88,194. The exergy consumed to refine and distribute the natural gas represents a small fraction of the fuel exergy. Based on the data from life cycle assessment presented in Dincer et al. (2010) for the system life cycle, the exergy used to extract, refine, and distribute the fuel is estimated to 9950 GJ as given in Table 10.17.

The total exergy consumed for fuel processing, materials extraction, manufacturing, and system scrapping becomes $9950 + 2420 + 135 = 12,505$ GJ. Based on the averaged exergy efficiency of the Canadian grid, the consumed exergy resource is $12,505/0.51 = 24,520$ GJ (stream 16 in Table 10.17). Based on Table 10.13, the EPC for the Canadian grid is $17.8 \, \mathrm{MW^{-1} \, h^{-1}}$ or $4.9 \, \mathrm{GJ^{-1}}$; thence, the EPC associated to power production for materials, manufacturing, fuel processing, and system scrapping is $4.9 \times 24,520 = \$120,146$.

The amount of dried wood chips is $14.9 \mathrm{kg/s} \times (25 \times 260 \times 3600) \mathrm{s} = 3.487\mathrm{E5t}$ which are to be valorized on the market. The total expenditure in the fuel is $1.91\mathrm{E6} \, \mathrm{GJ} \, \mathrm{t} \times (3.889 \times 52/50)$ $/\mathrm{GJ} = \$3.68\mathrm{E6}$, where the fraction 52/50 is the ratio of chemical exergy to the lower heating value of natural gas (Section 10.3.2). The total exergy destruction during system operation is $78.979 \mathrm{MW} \times (25 \times 260) \mathrm{h} = 513.3 \mathrm{GWh} = 1.85\mathrm{E6} \mathrm{GJ}$. The total EPC for the lifetime is, according to data given in Table 10.17, $15.825\mathrm{E6} of which 2% (totalling $325,672) represents the pollution costs associated with materials extraction, manufacturing, system scrapping and recycling, fuel processing, and distribution; note that the exergy destruction associated with this EPC is $18.3 \, \mathrm{GW} \, \mathrm{h} = 0.07\mathrm{E6} \, \mathrm{GJ}$. Therefore, the total lifetime exergy destruction of the reference drying system is $\mathrm{Ex_{d,ref}} = 1.92\mathrm{E6} \mathrm{GJ}$. The total exergy input to sustain the lifetime operation and the system construction and scrapping is shown in Table 10.17:

**Table 10.17** Embedded exergy and environmental pollution costs for reference drying system life cycle

| Stream | Description | Ex (GJ) | EPC ($) |
|--------|-------------|---------|---------|
| 16 | Exergy extracted from primary sources for power generation | 24,520 | N/A |
| 17 | Exergy consumed for fuel processing | 9,550 | N/A |
| 18 | Exergy consumed for materials processing and system manufacture | 2,420 | N/A |
| 19 | Exergy consumed for system scrapping and materials recycling | 135 | N/A |
| 20 | Environmental waste stream for grid power generation for lifetime | N/A | 120,146 |
| 21 | Primary fuel sources extracted (natural gas) | 1.91E6 | N/A |
| 22 | Refined natural gas distributed to consumption point | 1.91E6 | 15.5E6 |
| 23 | Waste emissions of pollutants due to fuel processing | N/A | 77,332 |
| 24 | Materials extracted (ores) for system manufacturing | N/A | N/A |
| 25 | Embedded energy in the constructed system | 2,420 | N/A |
| 26 | Polluting wastes due to materials processing and system manufacture | N/A | 88,194 |
| 27 | Recycled materials | 134 | N/A |
| 28 | Wastes and pollution due to system scrapping and recycling | N/A | 40,000 |

$Ex_{lc,in} = 24,250 + 1.91E6 = 1.934E6\,GJ$. The total reversible work is $W_{lc,rev} = 873\,kW \times (25 \times 260)\,h = 20,428\,GJ$.

The sustainability of the reference drying system can be assessed based on the following parameters:

- Thermodynamic parameters – that is:
  - Total life cycle exergy destruction, $Ex_{d,ref} = 1.92E6\,GJ$.
  - Life cycle exergy efficiency is $\psi_{lc} = W_{lc,rev}/Ex_{lc,in} = 0.020428/1.934 = 0.01$.
  - Specific reversible work is $SRW = W_{lc,rev}/Ex_{d,ref} = 0.020428/1.92 = 0.011$.
- Exergoeconomic parameter – that is:
  - Specific exergetic capital investment (ExCI) defined as the ratio of exergy invested in system construction $Ex_{inv} = 9950 + 2420 + 135 = 12,505\,GJ$ and the amount of reversible work needed to dry the product $ExCI = Ex_{inv}/W_{lc,rev} = 12,505/20,428 = 0.612$
  - Exergetic investment efficiency (ExIE) defined here as the ratio between exergy expenditure for investment and exergy destroyed, $ExIE = Ex_{inv}/Ex_{d,ref} = 12,505/1.92E6 = 0.0065$
  - Capital investment effectiveness defined by Eq. (10.17), $CIEx = 1.91E6/36,490 = 52.2$
- Environmental impact parameter – that is:
  - Life cycle EPC $EPC_{lc} = \$15.825E6$
  - EPC for system construction and scrapping $EPC_{cs} = \$128,194$
  - Construction exergy expenditure to life cycle exergy destruction ratio $ExCDR = 0.0081$

The sustainability can be assessed based on an ASI. Here, we formulate the ASI based on exergy destruction using thermodynamic, economic, and environmental assessment parameters. In this respect, one notes that the social benefit of the system is proportional to the sum of the reversible work, the exergy investment in the system construction, and the exergy expenditure equivalent to EPC for system construction and scrapping. The benefit of the investment in the reversible work is in fact the dry product which has market and societal value. The benefit exergy expenditure for system construction is in fact the drying system as a good, capable for production. The benefit on exergy investment to compensate for the environmental pollution at system construction and scrapping represents a general societal good consisting of a better environment. The sum of these three terms must be compared with the life cycle exergy destruction in order to form the ASI. Therefore, the equation for ASI becomes

$$ASI - SRW + ExIE + ExCDR \qquad (10.30)$$

For the reference system, the ASI becomes $ASI_{ref} = 0.011 + 0.0065 + 0.0081 = 0.0256$. This aggregate index must be compared to that for the improved system.

## 10.4.3 Improved Dryer Description

The reference system can be improvement with respect to sustainability provided that more sustainable energy sources are used as input, and in addition, the system irreversibilities are reduced by providing a better design. In the reference system, the energy source is from natural gas. If instead of combusting natural gas, a heat pump is used, supplied from Ontario regional grid, then the associated environmental impact will be substantially reduced because in Ontario the grid power is relatively clean as in the energy mix much hydro and nuclear do exist. Also,

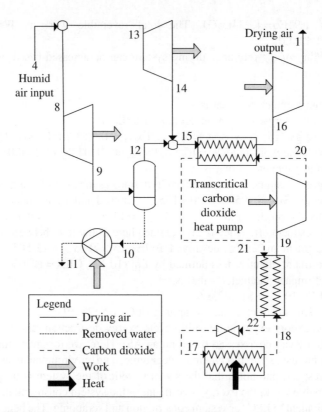

**Figure 10.17** Heat pump for improved wood chips drying system

when a heat pump is used, no drying agent is expelled to the environment; therefore, heat is regenerated internally and less energy supply is required. This explains why a heat pump dryer will have a higher exergy efficiency than the reference dryer.

We assume that the drying process as described in Figure 10.14 remains the same. Therefore, the same reversible work is required $\dot{W}_{rev} = 839 \mathrm{kW}$ for the same production rate of 14.09 kg/s dry product. The exergy input required to drive the process must be the same as for the reference case, namely, $\dot{Ex}_{in, kiln} = 21{,}021 \mathrm{kW}$.

The heat pump concept is relatively simple, and it allows for moisture extraction from humid air prior to heating. Figure 10.17 shows the heat pump system for drying air recirculation and moisture removal. In state 4, a flow rate of 81.53 kg/s of humid air with 403 K and 217 g/kg humidity ratio enters the heat pump system, and it is split in two fractions state 8 and state 13. Once expanded to a vacuum pressure, water condensates both in states 9 and 14. However, water from state 9 is gravitationally separated, whereas the moisture from state 14 continues to flow together with the air stream. Since water is separated under vacuum, a pump is used to pressurize it to atmospheric pressure in 11.

A transcritical carbon dioxide heat pump heats the drying air from state 15 to 16 where it is recompressed to atmospheric pressure and delivered to the required parameters in state 1: temperature 739 K and humidity ratio of 7.2 g/kg. The carbon dioxide heat pump uses the

**Table 10.18** State descriptions and thermodynamic parameters for the heat pump system

| State | Description | $T$ (K) | $P$ (kPa) | $\omega$ (g/kg) | $x$ (kg/kg) | $\dot{E}$ (kW) |
|---|---|---|---|---|---|---|
| 1 | Drying air (output) | 739 | 101.325 | 72 | N/A | 59,359 |
| 4 | Humid air (input) | 403 | 101.325 | 217 | N/A | 59,231 |
| 8 | Humid air | 403 | 101.325 | 217 | N/A | 39,578 |
| 9 | Saturated humid air | 321.7 | 44.32 | 217 | N/A | 33,296 |
| 10 | Liquid water (vacuum) | 321.7 | 44.32 | N/A | 1 | 30,635 |
| 11 | Liquid water | 321.7 | 101.325 | N/A | 1 | 30,636 |
| 12 | Dry air | 321.7 | 44.32 | 0 | N/A | 2,661 |
| 13 | Humid air | 403 | 101.325 | 217 | N/A | 19.653 |
| 14 | Saturated humid air | 321.7 | 44.32 | 217 | N/A | 16,533 |
| 15 | Cold drying air | 321.7 | 44.32 | 72 | N/A | 19,194 |
| 16 | Preheated drying air | 594.2 | 44.32 | 72 | N/A | 45,014 |
| 17 | Two-phase vapor–liquid $CO_2$ | 283 | 4,485 | N/A | 0.019 | −25,628 |
| 18 | Saturated $CO_2$ vapor | 283 | 4,485 | N/A | 1 | −7,731 |
| 19 | Superheated $CO_2$ vapor | 526.3 | 4,485 | N/A | N/A | 18,729 |
| 20 | Hot supercritical $CO_2$ | 614.2 | 10,133 | N/A | N/A | 26,651 |
| 21 | Cooled supercritical $CO_2$ | 381.7 | 10,133 | N/A | N/A | 831.3 |
| 22 | Subcooled $CO_2$ liquid | 297.9 | 10,133 | N/A | N/A | −25,628 |

surrounding medium (e.g., a lake) to draw heat at reference temperature $T_0 = 298\,K$ and evaporate the working fluid from state 17 (low vapor quality two-phase mixture) to state 18 (saturated vapor). The vapors are superheated to state 19 using internal heat regeneration and further are compressed so that they are able to deliver heat by heat transfer to the drying air in process 20–21. Typical mass, energy, and exergy balances are written for each component from Figure 10.17 and solved to determine the state points.

The heat pump system state points descriptions and thermodynamic parameters are given in Table 10.18. The net required work input for the heat pump system can be calculated as follows:

$$\dot{W}_{net,in} = \dot{E}_1 - \dot{E}_{16} + \dot{E}_{11} - \dot{E}_{10} + \dot{E}_{20} - \dot{E}_{18} + \dot{E}_9 - \dot{E}_8 + \dot{E}_{14} - \dot{E}_{13} = 12,866\,kW$$

The heat input for the heat pump is $\dot{Q}_{in} = \dot{E}_{18} - \dot{E}_{17} = 30,763\,kW$. The exergy of the heat input is zero because the system boundary is set at $T_0$. The heat output neglecting heat losses is $\dot{Q}_{out} = \dot{Q}_{in} + \dot{W}_{net,in} = 43,629\,kW$. The coefficient of performance becomes

$$COP = \frac{\dot{Q}_{out}}{\dot{W}_{net,in}} = \frac{43,629}{12,866} = 3.39$$

Because the heat is delivered at temperature $T_1 = 739\,K$, the exergetic COP becomes

$$COP_{ex} = \frac{\dot{Q}_{out}\left(1 - \dfrac{T_0}{T_1}\right)}{\dot{W}_{net,in}} = \frac{43,629\left(1 - \dfrac{298}{739}\right)}{12,866} = 2.0$$

The exergy balance equation allows for determination of the exergy destruction which is

$$\dot{E}x_1 + \dot{W}_{net,in} = \dot{Q}_{out}\left(1 - \frac{T_0}{T_1}\right) + \dot{E}x_{d,hp} \rightarrow \dot{E}x_{d,hp} = 3707\,kW$$

The total exergy destruction of the drying process coupled to the heat pump is equal to the sum of exergy destruction by the kiln and the exergy destruction by the heat pump

$$\dot{E}x_d = \dot{E}x_{d,hp} + \dot{E}x_{d,kiln} = 3707 + 21,021 = 24,728\,kW$$

The total power required for the drying process is equal to the power consumed by the blowers and kiln rotation system plus the power consumed by the heat pump: $\dot{W}_{act} = 765 + 12,866 = 13,631\,kW$. The exergy efficiency of the dryer becomes

$$\psi = \frac{\dot{W}_{rev}}{\dot{W}_{act}} = \frac{839}{13,631} = 0.061$$

### 10.4.4  Exergo-Sustainability Assessment for the Improved Drying System

The efficiency is improved 1.7 times with respect to the exergy efficiency of the reference system. Furthermore, the SI of the improved system becomes

$$SI = \frac{1}{1-\psi} = \frac{1}{1-0.061} = 1.065$$

Taking in account for the additional 2 MW required on-site, the total power consumed by the grid by the improved system is $\dot{W}_{in} = 15,866\,kW$. For a life cycle of 25 years and 260 h of annual operation, the total electric energy consumed from the grid becomes $W_{in} = 371,265\,GJ$. To this energy, the amount of energy required for system construction and scrapping must be added. The improved system does not have any more the combustor and the gas turbine. However, it is fair to assume for a rough estimation that the amount materials such as copper, aluminum, and stainless steel used for the gas turbine, combustor, and the electric power generator are now used to construct the heat pump which itself include similar components as for the reference system: electrical motors, compressors, and turbines.

Therefore, Table 10.16 shows data that is kept unchanged for the improved system. However, in Table 10.17, the streams 17, 21, 22, and 23 do not exist because only electrical power is demanded by the improved system. The life cycle operations of the improved system are described as shown in Figure 10.16. Table 10.19 demonstrates the embedded exergy and EPCs for improved drying system.

The sustainability assessment parameters are calculated for the improved system in a similar manner as for the reference system. The sustainability for the two systems is compared as shown by the results given in Table 10.20. The total exergy destruction for the lifetime is given by the sum of the exergy destruction by the system itself and the exergy destruction at power

**Table 10.19** Embedded exergy and environmental pollution costs for the improved drying system life cycle

| Stream | Description | Ex (GJ) | EPC ($) |
|--------|-------------|---------|---------|
| 16 | Exergy extracted from primary sources for power generation | 373,820 | N/A |
| 17 | Electric energy consumed by the system for operation (for lifetime) | 371,265 | N/A |
| 18 | Exergy consumed for materials processing and system manufacture | 2,420 | N/A |
| 19 | Exergy consumed for system scrapping and materials recycling | 135 | N/A |
| 20 | Environmental waste stream for grid power generation for lifetime | N/A | 1.848E6 |
| 24 | Materials extracted (ores) for system manufacturing | N/A | N/A |
| 25 | Embedded energy in the system constructed and delivered | 2,420 | N/A |
| 26 | Polluting wastes due to materials processing and system manufacture | N/A | 88,194 |
| 27 | Recycled materials | 134 | N/A |
| 28 | Wastes and pollution due to system scrapping and recycling | N/A | 40,000 |

**Table 10.20** Comparison of the reference and improved drying systems

| Parameter type | Parameter | Drying system | |
|----------------|-----------|---------------|---------|
| | | Reference | Improved |
| Thermodynamic | $Ex_{d,ref}$ | 1.92E6 GJ | 0.21E6 |
| | $\psi_{lc}$ | 0.01 | 0.055 |
| | SRW | 0.011 | 0.098 |
| Exergoeconomic | ExCI | 0.612 | 0.612 |
| | ExIE | 0.0065 | 0.060 |
| | CIEx | 52.2 | 145 |
| Environmental | $EPC_{lt}$ | $15.825E6 | $1.98E6 |
| | $EPC_{cs}$ | $128,194 | $128,194 |
| | ExCDR | 0.0081 | 0.065 |
| Sustainability | ASI | 0.0256 | 0.223 |
| | $GF_x$ | — | 0.89 |
| | $GF_{ASI}$ | — | 0.88 |

generation, which is $373,828(1-0.51)+24,728=207,904\,GJ$, where 0.51 is the exergy efficiency of the grid. The exergy input for the life cycle is given in Table 10.19: $Ex_{lc,in}=373,820\,GJ$. The total reversible work amount is previously calculated $W_{lc,rev}=20,428\,GJ$. Thence, the life cycle exergy efficiency becomes $\psi_{lc}=0.055$ showing five times improvement with respect to the reference case.

The specific reversible work becomes $SRW=W_{lc,rev}/Ex_{d,ref}=20,428/207,904=0.098$. The specific ExCI remains unchanged because the investment and the reversible work are not affected by the system change, since the same investment has been assumed for both systems. Also, the EPC for system construction does not change. However, the ExIE changes to $ExIE=Ex_{inv}/Ex_{d,ref}=12,505/207,904=0.060$. Finally, the ASI for the improved system is determined by Eq. (10.30) and becomes $ASI=0.098+0.060+0.065=0.223$.

Further comparison of the improved and reference system is done using the greenization factor. Using Eq. (10.17), the exergy destruction-based GF becomes

$$GF_{ex} = \frac{Ex_{d,ref} - Ex_d}{Ex_{d,ref}} = \frac{1.92 - 0.21}{1.92} = 0.89$$

If the ASI is used for the greenization factor, then from Eq. (10.18) one gets

$$GF_{ASI} = \frac{ASI - ASI_{ref}}{ASI} = \frac{0.223 - 0.0256}{0.223} = 0.88$$

### 10.4.5   Concluding Remarks

In this case study, the exergo-sustainability assessment of a drying system is demonstrated. Although, the analysis is approximated, it clearly demonstrates the main steps to follow in order to assess the system. The scope of the analysis is extended to the entire lifetime which considers three phases: materials extraction and system construction (including fuel processing), system operation, and system scrapping and materials recycling. For all phases, the embedded exergy, the EPC, and exergoeconomic costs can be determined such that eventually the overall sustainability can be assessed by an aggregated index. Furthermore, if the system is improved or other system is comparatively assessed for sustainability, the greenization factor can be used, which is a simple method to quantify the system improvement toward greenization. The example shows that application of heat pumps in conditions when clean grid power is available locally is beneficial.

## 10.5   Conclusions

In this chapter, sustainability and environmental impact aspects are presented with a focus on drying systems. Sustainability assessment is relatively subjective, but there are several well-established criteria that assess sustainability from various points of view. A generic sustainability system for a drying system is shown in which the notion of process enablers is introduced as incorporating the effect of the environment, operators, and drying apparatus on the system performance. Also, many relevant sustainability assessment parameters are introduced together with a procedure to compound these indices in an aggregated sustainability index.

Exergy plays a major role in sustainability assessment and defining the reference environment. Various reference environment models are presented and the anthropogenic impact on the environment explained. Many useful environmental impact data are compiled in the chapter with a detailed view for Canada and Ontario. A case study illustrates the application of exergo-sustainability assessment for a reference drying system and an improved drying system.

## 10.6   Study Problems

10.1   Among the sustainability indicators presented herein, name three that represent better sustainability of drying systems.

10.2　What is the role of process enablers in sustainability modeling of a drying system?

10.3　Explain the difference between the eco-indicator and the sustainability efficiency indicator.

10.4　Explain the two possible methods for compounding sustainability indices in an aggregate index.

10.5　What is the role of exergy in sustainability assessment of drying systems?

10.6　What is the reason why a reference environment is assumed at thermodynamic equilibrium although the natural environment is not at equilibrium?

10.7　What is the meaning of radiative forcing?

10.8　Explain the definition of the global warming potential.

10.9　What is the role of exergy destruction in quantifying the environmental impact?

10.10　Explain the difference between "environmental pollution cost" and "removal pollution cost."

10.11　What is the meaning of greenization factor?

10.12　Rework case study 10.4 for a different data set, modified from the presented one.

# References

Ahrendts J. 1980. Reference states. *Energy* 5:667–668.

Bosnjakovic F. 1963. Reference state for exergy in reaction systems. *Research in Engineering* 20:151–152.

Carpenter S. 1990. The environmental cost of energy in Canada. In: Sustainable Energy Choices for the 90s. Proceedings of the 16th Annual Conference of the Solar Energy Society of Canada, Halifax, Nova Scotia, 337–342.

Coskun C., Bayraktar M., Oktay Z., Dincer I. 2009. Energy and exergy analyses of an industrial wood chips drying process. *International Journal of Low-Carbon Technologies* 4:224–229.

De Gouw J.A., Warneke C., Stohl A., Wollny A.G., Brock C.A., Cooper O.R., Holloway J.S., Trainer M., Fehsenfeld F. C., Atlas E.L., Donnelly S.G., Stroud V., Lueb A. 2006. Volatile organic compounds composition of merged and aged forest fire plumes from Alaska and western Canada. *Journal of Geophysical Research* 111:D10303.

Dincer I. 2002. On energetic, exergetic and environmental aspects of drying systems. *International Journal of Energy Research* 26:717–727.

Dincer I. 2007. Environmental and sustainability aspects of hydrogen and fuel cell systems. *International Journal of Energy Research* 31:29–55.

Dincer I. 2011. Exergy as a tool for sustainable drying systems. *Sustainable Cities and Societies* 1:91–96.

Dincer I., Zamfirescu C. 2011. Sustainable Energy Systems and Applications. Springer: New York.

Dincer I., Zamfirescu C., 2012. Potential options to greenize energy systems. *Energy* 46:5–15.

Dincer I., Rosen M.A., Zamfirescu C. 2010. Economic and environmental comparison of conventional and alternative vehicle options. In: Electric and Hybrid Vehicles. G. Pistoia Eds., Elsevier: New York.

EIA 2008. Energy Information Administration, US Government accessed on June 18th: http://www.eia.doe.gov/.

Gaggioli R.A., Petit P.J. 1977. Use the second law first. *Chemtech* 7:496–506.

Hardi P., Zdan T.J. 1997. Assessing Sustainable Development: Principles in Practice. The International Institute for Sustainable Development: Winnipeg, Manitoba.

IPCC 2001. Climate Change 2001: The Scientific Basis. Intergovernmental Panel on Climate Change. J.T. Houghton et al. Eds., Cambridge University Press: Cambridge, UK.

IPCC 2007. Climate change 2007: synthesis report. Intergovernmental panel on climate change. IPCC Plenary XXVII, Valencia, Spain.

Kanoglu M., Dincer I., Cengel Y.A. 2009. Exergy for better environment and sustainability. *Environment, Development and Sustainability* 11:971–988.

Linke B., Das J., Lam M., Ly C. 2014. Sustainability indicators for finishing operations based on process performance and part quality. *Procedia CIRP* 14:564–569.

Midilli A., Dincer I., Ay M. 2006. Green energy strategies for sustainable development. *Energy Policy* 34:3623–3633.

Ness B., Urbel-Piirsalu E., Anderberg S., Olsson L. 2007. Categorising tools for sustainability assessment. *Ecological Economics* 60:498–508.

Pellizzari E.D., Clayton C.A., Rodes C.E., Mason R.E., Piper L.L., Forth B., Pfeifer G., Lynam D. 1999. Particulate matter and manganese exposures in Toronto, Canada. *Atmospheric Environment* 33:721–734.

Rivero R., Garfias M. 2006. Standard chemical exergy of the elements updated. *Energy* 31:3310–3326.

Rosen M.A., Dincer I. 1999. Exergy analysis of waste emissions. *International Journal of Energy Research* 23:1153–1163.

Rosen M.A., Dincer I. 2001. Exergy as the confluence of energy, environment and sustainable development. *Exergy, An International Journal* 1:3–13.

Rubin E.S. 2001. Introduction to Engineering and the Environment. McGraw-Hill Higher Education: New York.

Singh R.K., Murty H.R., Gupta S.K., Dikshit A.K. 2009. An overview of sustainability assessment methodologies. *Ecological Indicators* 9:189–212.

Szargut J., Morris D.R., Steward F.R. 1988. Exergy Analysis of Thermal, Chemical, and Metallurgical Processes. Hemisphere: New York.

UN 2007. Indicators of Sustainable Development: Guidelines and Methodology. 3rd ed. United Nations: New York.

Wall G. 1990. Exergy conversion in Japanese society. *Energy* 15:435–444.

Wall G. 1997. Exergy use in the Swedish society 1994. In: TAIES '97 Thermodynamic Analysis and Improvement of Energy Systems Conference; Beijing, China, June 10–13.

# 11

# Novel Drying Systems and Applications

## 11.1 Introduction

The fact that the energy demand for drying processes represents 16% worldwide average of the energy consumption by the industrial sector (Raghavan et al., 2005) created an impressive market pull toward progress and innovation in drying technology. Starting from this observation, Mujumdar (2004) remarked that a constant preoccupation is observed for development of novel drying systems with a nearly exponential R&D growth in the last four decades.

As inferred in Dincer (2011), the development goal is toward dryers with higher efficiency, reduced environmental impact, improved sustainability, improved safety, and improved product quality (i.e., reduced shrinkage, avoid texture degradation by cracking and brittleness, improved nutrient retention, better product color). Moreover, dying applications are expanded in new areas much different than traditional ones such as drying of nanomaterials (Mujumdar and Huang, 2007), preparation of enzymes and catalysts, and other special materials.

One clear tendency toward sustainability is the use of renewable energy to conduct drying processes (Dincer, 2011). Present trends in passive (or with natural circulation) and active (or with forced convection) solar drying systems are reviewed in Sharma et al. (2009). Additionally, some novel solar drying systems and research and development trends are reviewed by Fudholi et al. (2010) including agricultural and marine products.

Often, the development of new processes or new products put a demand of innovation in drying technology. Several novel and emerging drying methods are reviewed in Chou and Chua (2001). Among the named systems that emerge, one notes infrared drying, microwave drying, radio-frequency drying, pressure regulating drying, and multistage dryers. When a drying process is conducted in multistage, the external conditions (heat input, temperature and humidity ratio of drying agent, pressure) are adjusted to optimal values corresponding to the drying phase (constant rate phase, variable rate phase, etc.). Thenceforth, the energy

*Drying Phenomena: Theory and Applications*, First Edition. İbrahim Dinçer and Calin Zamfirescu.
© 2016 John Wiley & Sons, Ltd. Published 2016 by John Wiley & Sons, Ltd.

and drying agent consumption is highly reduced. Besides the reduction of energy consumption, there must be a constant preoccupation of dryers' designers to find novel solutions to reduce the investment cost for the system and equipment. Chua and Chou (2003) show that fluidized bed dryers, spouted bed dryers, infrared dryers, solar dryers, and agricultural convective dryers present the most potential for a reduced investment cost.

The use of superheated steam as drying agent is one of the most promising methods with an expanded use in current drying technology. As indicated in Mujumdar (1996), although the use of superheated steam as a drying agent is proposed since the last part of the eighteenth century, its practical implementation in the industry can be observed only after 1950. As it has been reported for many cases, superheated steam increases product quality.

One of the very promising pathways to enhance the exergy efficiency of drying and in the same time to stimulate an increased use of renewable energy is represented by the heat pump dryers. Heat pump dryers can be directly supplied with electricity which can be derived from renewable sources (solar-PV, wind, hydro). Daghigh et al. (2010) review the solar-assisted heat pump drying systems for agricultural and marine products.

The heat pump dryers use advanced heat recovery technology, recycle the drying agent, and extract the moisture in liquid form as a separate stream. A heat pump-driven drying tunnel allows, for example, not only for a fine regulation of drying agent temperature and humidity, but also it can regulate the working pressure. Historical development and newer trends in heat pump dryer development is reviewed in Colak and Hepbasli (2009a,b). Additional review on the topic is presented in Chua et al. (2002).

It is shown that the coefficient of performance (COP) of the heat pump integrated with a drying system can reach very high values such as 8 for two-stage, engine-driven vapor compression heat pumps with subcooler, while the COP of single-stage heat pumps is around 4. Closed cycle, air source, and ground source heat pump dryers were developed. Among heat pump dryers, the application of chemical heat pumps to drying process presents some peculiarities. In principle, a chemical heat pump can be used to supply heat to the moist material and in the same time to absorb humidity from the humidified drying agent. Ogura and Mujumdar (2000) estimated the potential of chemical heat pump application to industrial drying processes and proposed the name of chemical heat pump dryers (CHPD) to refer drying systems that integrate chemical heat pumps. This CHPD is indeed an emerging technology.

Some emerging drying methods with fast commercialization potential were identified by Jangam (2011) as follows: pulse-combustion spray drying, pulsed and ultrasound-assisted osmotic dehydration, impinging streams and pulsed fluidized beds, and intermittent and hybrid drying. In addition, Mujumdar (1991) identified the emerging Carver-Greenfield process as one of the most promising for sludge drying.

Frei et al. (2004) proposed a novel drying process applicable to permeable biomass materials which when subjected to a flow of dry air induce slow exothermic reactions due to thermophilic activity of aerobic microorganisms. This process removes the moisture from sludge–wood waste biomass while reducing the moisture content and enhancing the heating value.

A new method of drying for small-scale application is the corona wind drying as described in Goodenough et al. (2007). In this method, forced convection is induced by injection of ionic species which are accelerated in an electric field created between two electrodes. Although the method is demonstrated at very small scale and the efficiency of a corona wind blower is below 2%, the demonstrated overall efficiency is comparably high due to the ability of corona wind to be properly directed toward the surface of the moist material.

The ultrasonic method is generally known in unit operation as a mean of enhancement of process rate. Schösler et al. (2012) proposed a new freeze-drying method which is assisted by contact ultrasound to reduce the drying time. In this case, the moist product is placed on screen trays which are vibrated using ultrasonic frequencies. Heating effects may be induced by the ultrasonic energy dissipation. The ultrasonic process is performed with intermittency with about 90% recovery time which assures that the boundary layer is always broken. De la Fuente-Blanco et al. (2006) developed an ultrasonic dryer prototype of 100 W scale which helped in studying the mechanical and thermal effects during foodstuff drying. It is argued that the ultrasonic drying technology has high industrial relevance.

A novel direct contact drying method is the refractance window dehydration technology as reviewed in Nindo and Tang (2007). This method reduces drastically the drying times of slurries, purees, or juices due to a high rate of moisture removal. The liquid or semiliquid (puree) material to be dried is spread in a thin layer on a conveyor belt heated from below by direct contact with a hot water pool. The product removal from the belt is done with the help of a scraping blade. Alternatively, the refractance window dehydration technology can be implemented on drum dryers.

Drying as a nonlinear time-dependent process possesses special problems to system control in a dynamic mode. Model-based controlling of drying systems has been reviewed in Thyagarajan et al. (1998) with a focus on using artificial neural network (ANN) and other conventional controlling methods such as conventional feedback. Some future directions in controlling of drying processes indicate that fuzzy logic and genetic algorithms together with ANN provide good promises toward development of new control algorithm with smoother output which enhance the lifetime of the actuators.

El-Dessouky et al. (2000) proposed a new application of drying, namely, membrane drying of humid air as a first step for air conditioning. The application of humid air drying extends the operating range of the evaporative cooler for air conditioning and leads to 50% or more reduction of energy consumption with respect to conventional systems.

A novel application of spray drying for aqueous $CuCl_2$ solutions is developed at Clean Energy Research Laboratory (CERL) at University of Ontario Institute of Technology (UOIT), as reviewed in Dincer and Naterer (2014) and Naterer et al. (2013), Section 6.7.1. A spray drying apparatus is developed using a pneumatic atomization nozzle, and experiments were performed to assess the dryer performance. The dried powder is separated from the drying air using a cyclone separator. A method is devised to determine the drying coefficient at room temperature. This process is essential within a newly emerging hybrid copper–chlorine water-splitting cycle for hydrogen production.

Spray drying has been used in the past to microencapsulate food ingredients as reviewed in Gharsallaoui et al. (2007). Novel applications of drying occurred in recent years in different fields starting from pharmaceutical drug preparation to nanomaterials and nanopowder drying methods. Ré (2006) reviews the recent improvements on spray drying method applied to drug delivery systems in the form of microspheres and microcapsules. In general, these methods are used for spray-dried powders, spray-dried silica gels, and spray-dried aqueous solutions. He et al. (1999) developed a novel spray drying method and its pharmaceutical application to chitosan (a glucosamine-based drug component which facilitates drug delivery through the skin; it is also used in agriculture, viticulture, and other industries). The microencapsulation is done by emulsion spray drying which is a new method that differs from the conventional spray drying. Another spray drying adaptation for nanoparticle processing in "pharmaceutical nanotechnology" is developed in Lee et al. (2011) with application to protein nanotherapeutics for cancer, diabetes, and asthma. The main problem encountered by the conventional spray drying when applied to nanopowders is due to the collection of dried material which is not possible for particles with size below 2000 nm. Therefore, integration of particle collection methods such as electrostatic collectors into the spray dryer is necessary.

Controlled crystallization during freeze-drying has been developed by De Waard et al. (2008) for production of drug nanocrystals. This method increases the dissolution behavior of lipophilic drugs. The use of nanocrystals in medicine expanded from drug processing and delivery to creation of advanced composite materials for prosthetics. For example, hydrox-yapatite HP mineral in the form of nanocrystals blended with gelatin and starch produces bio-compatible materials for bone repair and regeneration, providing enhanced mechanical properties. Sundaram et al. (2008) developed a novel microwave drying method to coagulate HP nanocrystals with gelatin–starch for prosthetics materials.

One can briefly conclude that enough innovation exists in drying technology today to create a technology push such that – as mentioned in Mujumdar and Huang (2007) – the research, development, and practical implementation of novel drying systems and application is highly stimulated. In this chapter, some relevant developments on drying systems and application are presented.

## 11.2   Drying with Superheated Steam

Among many other emergent drying technologies, the systems which use superheated steam as a drying agent appear the most promising ones at industrial-scale level. Superheated steam dryers (SSDs) use in the industry is relatively new – since last 30 years, although some indus-trial implementation appeared since 1950s (Mujumdar, 1996). The main applications of super-heated steam drying are in paper industry, biomass, peat and coal drying, beet pulp drying, and textile drying including silk cocoon drying. Of course, superheated steam drying is not appli-cable for cases when high-temperature heating of moist material is not permitted. An additional drawback is that the investment cost is higher for SSDs than for conventional systems. On the other hand, it is easy to understand the advantages of superheated steam as drying agent, which are listed as follows:

- Impedes material oxidation, preserving thus its structure and often its color.
- Impedes combustion of sensitive materials subjected to drying such as biomass or coal.
- Allows a reduced energy consumption provided that steam is recycled; the heat input is around 400 J/g of removed moisture as compared to the latent heat of water evaporation of ~2260 J/g.
- Allows for an optimized dryer control in accordance with the drying regime.
- The operation is closed loop; therefore, it can recover toxic effluents present in the moisture in some cases.

The general description of drying systems with superheated steam is presented in the dia-gram shown in Figure 11.1. The system has a number of main components as described sub-sequently. The main component is the drying chamber which is of a sealed construction with typical operation in a semibatch mode (although continuous mode of operation is possible for

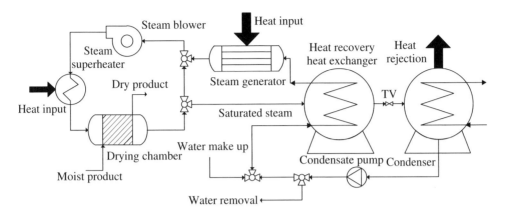

**Figure 11.1** Schematic of a drying system with superheated steam

some designs). Other components are the steam superheater, the steam generator (which produces saturated steam), the heat recovery heat exchanger, the condenser, the condensate pump, and a number of valves among which that indicated with TV is a throttling valve. Overall, the system extracts the moisture from the moist material, and it rejects it out as a condensate stream.

In fact, the system must compensate also for the heat losses and the required heat addition needed to superheat the steam and to preheat the material. Therefore, more steam is needed to compensate for the losses. Consequently, saturated steam is continuously added to the loop from the steam generator, and more heat is added to the superheated steam within the loop. In the same time, the moisture removed from the loop increases the flow rate of steam. Thenceforth, some saturated steam is extracted from the loop such that the mass balance is maintained. The extracted steam is cooled with heat recovery and further condensed to remove water. The valves with automatic actuation regulate continuously the stream such that the dryer operation is maintained at the desired parameters.

One technical problem with the SSDs is due to the requirement to supply the moist material already preheated at the process temperature. If this is not done, then steam will condense initially at the product surface. Thereafter, once sufficiently long time passes, all surface moisture evaporates and the product is heated up at the process temperature and the drying starts. Of course, in such a case, the drying process is much delayed. In any of the encountered cases, for the constant drying rate period, the mass balance requires that

$$j_{\mathrm{m}} = \frac{h_{\mathrm{tr}}\left(T_{\mathrm{steam}} - T_{\mathrm{sat}}\right)}{\Delta h_{\mathrm{lv}}} \tag{11.1}$$

where $j_{\mathrm{m}}$ represents the mass flux (kg/m$^2$s), $h_{\mathrm{tr}}$ is the heat transfer coefficient, $T_{\mathrm{steam}}$ is the temperature of the superheated steam, $T_{\mathrm{sat}}$ is the saturation temperature corresponding to the operation pressure (this is equal to the surface temperature), and $\Delta h_{\mathrm{lv}}$ represents the latent heat of moisture evaporation.

Due to a better Prandtl number of steam, the heat transfer coefficient is higher in SSD than in conventional dryers using air as drying agent ($Pr \cong 1$ for superheated steam at 120 °C, vs. ~0.7 for air in the same conditions). This is why for sufficient superheating temperature the rate of

**Table 11.1**   Energy demand comparison of conventional and superheated steam dryers

| Energy (kJ/kg moisture) | Conventional dryer | Superheated steam dryer |
| --- | --- | --- |
| Moisture evaporation | 2,594 | 2,594 |
| Material heating | 50 | 100 |
| Structural heat losses | 100 | 150 |
| Exhaust losses | 700 | 0 |
| Gross energy demand | 3,444 | 28,441 |
| Recycled energy | 0 | 2,170 |
| Net energy demand | 3,444 | 674 (or 20% of 3,444) |

*Source*: Mujumdar and Huang (2007).

drying is better for superheated steam system than for conventional systems. Typically, the steam is superheated with 50 K, meaning that for operation at 1 atm the steam temperature is 150 °C. Furthermore, because in a conventional dryer the material is maintained at wet-bulb temperature until water evaporates, the critical moisture is higher than that which can be obtained in an SSD that operates under similar conditions. Table 11.1 shows a comparison of energy requirements in conventional and SSD dryers.

Of special implementation in steam drying is the low-pressure superheated steam drying (LPSSD) case in which the drying chamber is maintained in deep vacuum of the order of 10 kPa. Some foodstuff products can be then dried because lower temperature of steam is possible in these conditions. Applications to chitosan, banana, and potato chips drying, carrots, and onions prove that LPSSD is successful while bringing the additional advantage of destroying undesired organisms and microbes (Chou and Chua, 2001); also, the method can be applied to paper drying as reported in Mujumdar (1991) or drying of dairy sludge (Raghavan et al., 2005).

One alternative loop design of an SSD system uses a steam compressor to mechanically recompress the steam after its exit out of the drying chamber. A mechanical compressor such as centrifugal or screw type can be used, or instead, steam ejectors can be used – especially for LPSSD (Mujumdar, 1991). Thenceforth, the specific steam enthalpy is increased and steam is superheated. Also in some cases, there is a local need of steam; therefore, the steam resulted from moisture extraction can be directly consumed as a useful by-product on-site. Furthermore, SSD can be intrinsically coupled with other processes such as sterilization, pasteurization, deodorization, and so on.

The drying system is not necessarily restricted to batch tray driers when SSD is implemented. The method can be applied to various dryer configurations including flash drying, impinging jet drying. In addition, SSD can be applied to fluidized bed dryers or recently developed methods of pulsed fluidization which may lead to 10 times reduction of energy consumption with respect to conventional convective dryers with air as drying agent (Mujumdar and Huang, 2007).

## 11.3   Chemical Heat Pump Dryers

Heat pump dryers and their advantages – such as the ability to control the humidity, pressure, and temperature of the drying air – were discussed in detail in this book. In many applications, heat pumps take advantage of heat sources from the surrounding environment in lithosphere (ground source heat pumps), atmosphere (air source heat pumps), or hydrosphere (lake/river source heat pump) and rise the temperature level while consuming work.

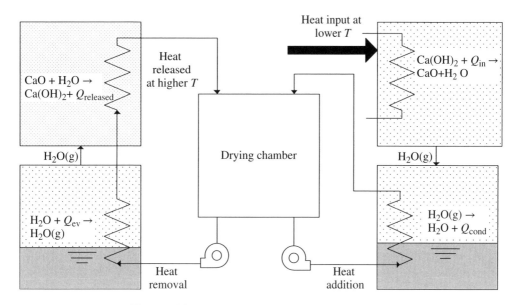

**Figure 11.2**  Example of a chemical heat pump dryer

Here, an emerging technology is presented, namely, the chemical heat dryer which brings some additional advantages with respect to dryer systems assisted by vapor compression heat pumps. In principle, a chemical heat pump involves absorption–desorption reaction which are exothermic/endothermic. It absorbs heat due to desorption process at low temperature by recovering heat typically from the exhausted drying air. In a subsequent reverse reaction, the chemical heat pump rejects heat at a higher temperature, thus heating up the drying agent.

In some cases, as reviewed in Jangam (2011), chemical heat pump-assisted dryers are coupled with solar energy as input source. The solar-assisted chemical heat pump dryer systems are reviewed in Daghigh et al. (2010). One of the common reversible reactions for chemical heat pump systems is the hydration/dehydration of calcium oxide/calcium hydroxide.

Figure 11.2 shows the operation principle of a heat pump dryer driven with the help of a chemical heat pump that operates based on the $CaO/H_2O/Ca(OH)_2$ system. The heat source is at low grade; for example, solar flat-type solar collectors. The heat sink is the drying chamber itself. Heat input is provided to the endothermic reaction of calcium hydroxide dehydration as follows:

$$Ca(OH)_2 + Q_{in} \rightarrow CaO + H_2O(g) \qquad (11.2)$$

Once steam is released, this is collected and condensed in a separate chamber. The condensation temperature is higher than the drying temperature because the steam release is at higher pressure. Therefore, the condensation process adds heat to the drying chamber. In parallel, saturated steam is generated in another chamber at a temperature below the drying temperature and further absorbed by the reverse exothermic reaction which spontaneously occurs in a separate reactor, as follows:

$$CaO + H_2O(g) \rightarrow Ca(OH)_2 + Q_{release} \qquad (11.3)$$

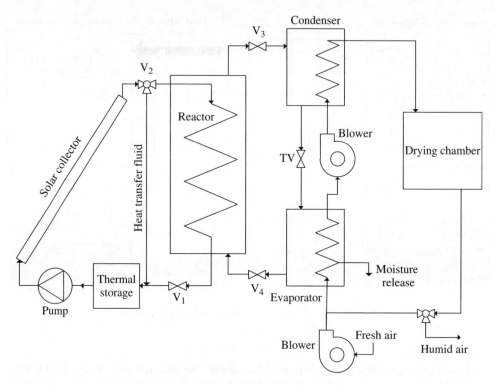

**Figure 11.3**  Solar chemical heat pump dryer with the CaCl$_2$/NH$_3$ pair or MgCl$_2$/NH$_3$ pair

The released heat is transferred to the drying air which is heated at a temperature higher than the drying chamber temperature. It is in principle possible that the humid air is put in direct contact with the exothermic reaction (11.3) in packed bed reactor; consequently, some humidity will be absorbed by the bed, and the humidity ratio of the drying air will be reduced.

Another known chemical heat pump system operates with the CaCl$_2$/NH$_3$ pair. In Figure 11.3, a solar-driven dryer system integrated with the chemical heat pump is shown. The spontaneous exothermic reaction of ammonia absorption goes as follows:

$$CaCl_2 \cdot 2NH_3 + 6NH_3 \rightarrow CaCl_2 \cdot 8NH_3 + Q_{release} \qquad (11.4)$$

The system may also operate in a similar manner with MgCl$_2$/NH$_3$ pair. As shown in the figure, the low-grade heat input from a flat plate solar collector is transferred to the heat absorption reaction, namely, the reverse of reaction (11.4). When this phase occurs, the valves $V_1$, $V_2$, $V_3$ are opened and TV and $V_4$ closed. Ammonia is released in a confined space, and the pressure increases until the pressure value corresponding to the condensation temperature is reached. While ammonia condenses, liquid accumulates at high pressure at the bottom of the condenser. This heat is transferred to the drying air which is heated to sufficient temperature level prior its return to the drying chamber.

In the second phase, $V_1$ is closed off and $V_2$ is shifted to bypass the reactor. Therefore, all the absorbed thermal radiation is accumulated as thermal energy in the thermal storage tank (see the Figure 11.3). This accumulated heat is used again in the next desorption phase to preheat the heat transfer fluid prior the thermal collector. In fact, the thermal storage unit can be connected in parallel with the solar collector as an option. At the beginning of the ammonia absorption phase – reaction (11.4) – the packed bed inside the reactor is depleted of ammonia and the evaporator is at low pressure. The throttling valve and $V_4$ are then opened, and the liquid ammonia from the condenser is throttled, and the cooling effect occurs in the evaporator. Ammonia is continuously withdrawn from the evaporator in gaseous phase and absorbed in the reactor which warms up operating at low pressure. The drying air is cooled in the evaporator and its moisture removed. Further, the drying air with reduced humidity ratio is heated in the condenser, where, this time, a subcooling process of ammonia occurs. Once ammonia is fully absorbed, the valves TV and V4 are closed to preserve the low pressure in the evaporator. Then $V_1$ is opened and $V_2$ shifted toward the reactor which is already preheated and charged with ammonia. The cycle repeats.

In essence, this heat pump system allows for drying air heating and dehumidification. As reviewed in Daghigh et al. (2010), the COP of these types of heat pump is superior to 1.4 for common drying operations and can reach values as high as 2.

## 11.4 Advances on Spray Drying Systems

Remarkable attention is received by the spray drying technology where impressive innovation can be observed together with an increase of the number of application. Today, spray drying is extensively applicable in pharmaceutical processes and nanomaterials. Here, some novel developments are presented.

### 11.4.1 Spray Drying of $CuCl_2(aq)$

A novel spray drying application has been developed at CERL at UOIT to dry an aqueous solution of $CuCl_2$. This process is required within a copper–chlorine hybrid thermochemical cycle for water splitting and hydrogen production as described in Dincer and Naterer (2014). The process may be relevant for other applications, for example, in copper industry.

The copper dichloride is soluble in water with a solubility limit of 0.1075 mol $CuCl_2$ per mole of water at 303 K and 0.1333 mol $CuCl_2$ per mole of water at 353 K. In the aqueous solution, copper dichloride hydrates and it forms the complex $CuCl_2 \cdot 2H_2O$ in the form of crystals which appear as an agglomeration of planar square cells with copper atoms surrounded by two chlorine atoms and two water molecules placed approximately in the corners of a square. In order to fully dry the solution and fully eliminate water to form anhydrous cupric chlorine, sufficient heating must be applied. Two possibilities to apply a spray drying process and extract the anhydrous $CuCl_2$ exist:

- Spray drying of aqueous solution – the droplets of aqueous $CuCl_2(aq)$ are sprayed in hot air maintained in a large vessel at 353 K. An evaporative heat and mass transfer process occurs which removes water.

- Spray draying of cupric chloride slurry obtained by concentrating the solution above the solubility limit – the slurry containing about 55% solids per volume with a ratio of 3.5 mol of water per mole of cupric chloride is sprayed to induce an evaporative heat and mass transfer process to remove water molecules.

A low-temperature spray drying unit has been set up at CERL based on Yamato D41 spray dryer which was modified to avoid corrosion caused by the corrosive cupric chloride solution. A titanium two-fluid nozzle was used, and stainless steel parts downstream of the nozzle were coated with a thin layer of Viton.

The spray drying test system is shown in Figures 11.4 (schematics) and 11.5 (photograph) and comprises the following elements:

- Cylindrical glass drying chamber of 0.45 m diameter with 1.0 m cylindrical height.
- Glass-made cyclone and particle collection vessel.
- 1 two-fluid atomization nozzle of 0.7 mm diameter made of titanium.
- Air is cleaned via a scrubber.
- Peristaltic pump for solution to spray.
- Flow meter for measurement of the flow rate of air for atomization.
- Evaporation capacity of 3 l/h.
- Temperature sensors mounted in the drying air at the inlet and outlet.
- Digital instruments for barometric pressure and ambient air temperature and humidity.

In the experiments, a flow rate of 0.58 kg/h of aqueous cupric solution with 10% solid fraction at 293 K has been atomized using air at 4 bar and a flow rate of 3.84 kg/h. The drying air has been provided at 393 K with 72 kg/h, while the air exit temperature has been 328 K.

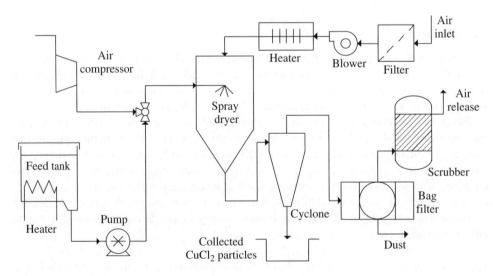

**Figure 11.4** Setup for spray drying of aqueous cupric chloride at UOIT (modified from Daggupati et al. (2011))

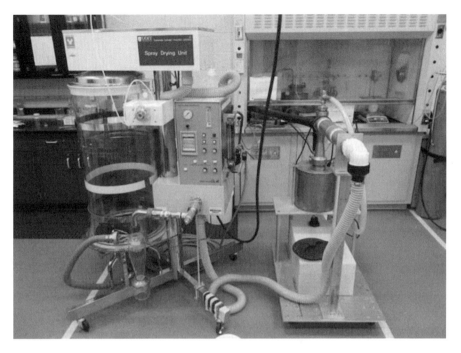

**Figure 11.5**   Photograph of the low-temperature spray drying system for $CuCl_2$(aq) dehydration (taken at Clean Energy Research Laboratory (UOIT))

Using the experimental data, the Hausner ratio which indicates the flow behavior of powders from cohesive nonflowing to free-flowing regimes has been determined. The Hausner ratio represents the ratio of tapped density ($\rho_{tap}$) to aerated bulk density ($\rho_{bulk}$) of a powder, defined as follows:

$$HR = \frac{\rho_{tap}}{\rho_{bulk}} \qquad (11.5)$$

The tap density and Hausner ratio were determined with the Electrolab ETD-1020 microcontroller-based tap density tester. The particle diameter is determined via light scattering using the Microtrac S3500 particle analyzer which gives the particle volume mean diameter, number mean diameter, and area mean diameter and also the particle size distribution. Thenceforth, the Sauter mean diameter of the solid particle is estimated as follows:

$$d_p = \left(\frac{\sum d^3}{\sum d^2}\right) \qquad (11.6)$$

For a distribution analysis of particle size, the span is calculated using the following diameters: $d_{10\%}$, the diameter for which 10% of the sample is smaller; $d_{50\%}$, the diameter for which

50% of the sample is smaller; and $d_{90\%}$, the diameter for which 90% of the sample is smaller. The span is calculated as follows:

$$SP = \frac{d_{90\%} - d_{10\%}}{d_{50\%}} \tag{11.7}$$

The experimental results are reported in detail in Daggupati et al. (2011). The determined moisture content is 20.88 by weight, with tap density of 652 kg/m$^3$ and bulk density of 411 kg/m$^3$, with a Sauter diameter of the particles of 21.56 μm, with the span 1.58 and Hausner ratio of 1.58. The product recovery rate for these runs was above 80%.

Powder characteristics and properties can be controlled and maintained uniformly throughout a drying operation. A sample scan electron microscopy (SEM) image of CuCl$_2$ particles obtained by spray drying is shown in Figure 11.6. The particle morphology was investigated with a JEOL JSM-6400 scanning electron microscope. Since the SEM requires a vacuum, the procedure required the complete dehydration of the sample before placing it into the SEM vacuum chamber. Therefore, the crystal structure observed under the SEM was the anhydrous cupric chloride.

In order to investigate the effect of temperature on the spray drying process and the form of solid particles, a second set of experiments is performed at higher temperature, as reported in Daggupati et al. (2011). The SEM image of the high-temperature spray drying is shown in Figure 11.6b. As the drying temperature increases, the solid becomes amorphous, because there was less time for crystals to grow before contacting another crystal growing in the droplet. Because of the increased number of crystals, there is less precipitate per crystal, and the crystals are much smaller in size.

For the high-temperature experiments, the drying air has been provided at 443 and 473 K (for two runs) with a mass flow rate of 82 kg/h. The output air temperature is around 378 K. The

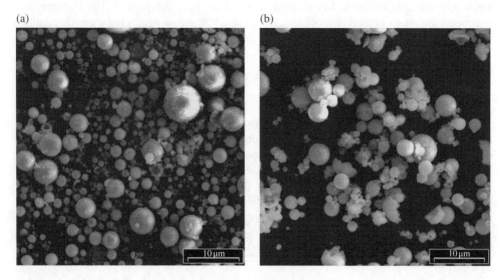

(a)                                                                (b)

**Figure 11.6**  SEM images of CuCl$_2$ particles formed by spray drying at UOIT: (a) low-temperature experiments, (b) high-temperature experiments (Naterer et al. (2013))

atomization air for the two experiments was at 7.3 kg/h and 1.5 bar and at 10.4 kg/h and 2 bar, respectively. The atomization liquid is $CuCl_2$(aq) with 44.85% solid concentration and flow rates of 0.864 and 0.464 kg/h, respectively. A Niro spray drying unit is used for the high-temperature experiments, having the following characteristics:

- Cylindrical drying chamber: 0.8 m diameter, 0.6 m height, and a 60° conical bottom.
- Two-fluid nozzle of 0.8 mm diameter.
- Drying gas is introduced from the top.
- Cyclone separator is used to separate air and powder.
- It has an air cleaning unit.

In the process of spray drying, the aqueous cupric chloride and volatile compounds (water and hydrochloric acid) evaporate from the droplets generated by the atomization nozzle. Because they are evaporated at the surface of the droplet, nucleation points occur first on the surface. This causes crystals to form and grow internally into the droplet. Based on the SEM images, it was observed that a higher temperature increases the number of nucleation points. Ongoing research is examining the crystal structure of the $CuCl_2$ product, particularly to determine if there are interstitial water molecules inside a crystal lattice of $CuCl_2$. Other issues under investigation include whether there are water molecules bonded to the edges of the $CuCl_2$ crystal, whether there are interstitial Cu and Cl atoms inside a lattice of the dihydrate, and gaining a deeper understanding of the precise mechanism of drying. The molecular structures of both the dihydrate and anhydrous forms are planar. Bonds are formed between adjacent planes at specific angles. In the hydrate form, removal of water leads to collapse of the crystals.

## 11.4.2 Spray Drying of Nanoparticles

Nano spray drying or spray drying of nanoparticles represents an emerging drying technology. There is a high market pull for this technology with the most promising applications in drug delivery for nanotherapeutics of various diseases such as asthma, diabetes, and cancers. The nanotherapeutics involves drug administration through peroral, nasal, pulmonary, and transdermal delivery. In this case, the drug must be formulated in the form of a dry nanopowder. Recent modifications of conventional spray drying technology led to the development of new methods permitting nanopowder drying for applications in nanotherapeutics.

Conventional spray drying methods are described in Chapter 3 of this book and involve liquid atomization followed by particle drying through intense evaporation followed by particle separation from the solid–gas mixture. Conventional spray drying methods are difficult to apply for micrometer-size particles and impossible for nanoparticles. Therefore, modifications must be done – especially in atomizers – to reduce the allowable particle size. Nanotherapeutics is much more effective than conventional drug delivery in some cases because the right quantities of drug – which are very small quantities – are delivered right at the targeted spots. This process leads to economy of expensive materials, while the drug assimilation is better. Moreover, when spray dryers are used for drug formulation, important savings of energy and time are obtained since the drying and particle separation process is done in one single step. The atomizers of the conventional sprays need to be essentially modified in order to handle nanoparticles. In addition, special adaptations of particle separators must be implemented.

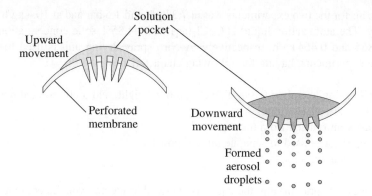

**Figure 11.7**   Principle of operation of the vibrating membrane atomizer (modified from Lee et al. (2011))

A vibrating mesh atomizer is described in Lee et al. (2011) which uses a piezoelectric actuator and a perforated membrane with an array of holes each having few micron diameters. The mesh is vibrated in one direction as shown in Figure 11.7 in the ultrasound spectrum (~60,000 Hz). In this system, fine aerosol droplets are formed which eventually are dried to form nanoparticles collected at the bottom side of the spray dryer.

The preferred flow regime is laminar while the droplets flow downward in cocurrent with the drying air which can provide better quality product due to slower heating rather than in turbulent flow when the process is more abrupt and can negatively affect the product quality. An electrostatic particle separator is placed at the bottom side of the spray dryer which creates a strong electric field between the dryer axis and the periphery. Figure 11.8 shows the electrostatic separator principle. When the particles are negatively charged, they move toward the periphery and agglomerate at the anode (the reverse movement happens with positively charged particles is any).

### 11.4.3   Microencapsulation through Spray Drying

Besides the emerging field of nanopowder dryers, remarkable progress is encountered in spray drying of microparticles and forming microcapsules. Various applications are envisaged in this respect starting with drug delivery systems as microspheres or microcapsules to microencapsulation of food ingredients such as lipids or flavors. Microencapsulation emerged as method of confining finer particles or droplets by application of a coating or by embedding the active components in a heterogeneous matrix. The microencapsulation has several benefits, among the major one is the placement of inert barrier between the active elements inside the capsule and the exterior environment. Therefore, any reaction with the environment is retarded or completely impeded. Furthermore, the microcapsule can be used in a controlled manner, when, for example, placed in a solution where the coating can be dissolved.

As mentioned in Gharsallaoui et al. (2007), the inside active material is often referred as the core, fill, or internal phase, whereas the coating is sometimes called shell or membrane. Figure 11.9 shows the main types of microcapsules. The spray drying technique can be conveniently adapted to manufacture encapsulated microcapsules. Although this encapsulation

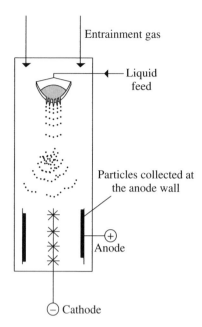

**Figure 11.8** Principle of operation of the electrostatic separator for nanoparticles (modified from Lee et al. (2011))

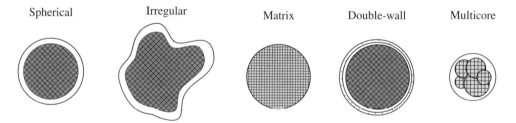

**Figure 11.9** Configuration of the main types of microcapsules

technique is known since nearly 80 years (milk fat encapsulation in lactose is one of the oldest applications), the progress toward smaller sizes, toward 1 μm, is remarked only recently.

The following steps are involved in a microencapsulation process using spray dryers:

- The emulsion or dispersed phase or free-flowing slurry is prepared.
- The dispersed phase is homogenized.
- The feed emulsion is atomized inside the spray dryer.
- The atomized droplets are dehydrated and thence dried.

The typical setup for spray dryer microencapsulation is shown in Figure 11.10. There is no much difference with respect to a regular spray drying setup. The most important aspect relates

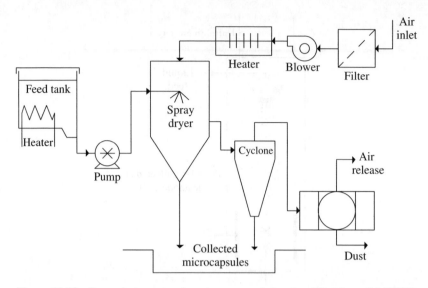

**Figure 11.10**  Spray drying setup for microencapsulation (modified from Ré (2006))

to setting the right/optimum operating conditions and devising separation devices in concordance with the particle size. In many cases, the microcapsules are formulated using chitosan or hydrophilic polymers or biodegradable organic materials or silica gels. A detailed review of new trends in spray drying microencapsulation is given in Ré (2006).

## 11.5  Membrane Air Drying for Enhanced Evaporative Cooling

Here, a novel drying application is presented consisting of a membrane-based air drying system integrated with an evaporative cooling process. This system has been proposed by El-Dessouky et al. (2000). The idea of the process is to use a membrane air dryer as a first-stage process, which absorbs the humidity of air at constant temperature.

Figure 11.11 shows the configuration of the novel system. Basically, the humid air is dried using a membrane system specially selected. Alternatively, a desiccant which absorbs air humidity can be used instead. After the first stage of isothermal membrane drying, a sensible cooling process is applied in an extended surface heat exchanger. Further, evaporative cooling is applied to drop the air temperature even more while humidifying.

The process can be observed in the Mollier diagram from where its benefits can be understood as shown in Figure 11.12. Assume, for example, that air during summer is at 35 °C with 30 °C wet-bulb temperature. While passing over the membrane system, air rejects humidity at constant temperature according to process 1–2. Further, air is slightly cooled at constant humidity ratio in the heat exchanger according to the process 2–3. Next, air is humidified in the cooling tower according to process 3–4 until it reaches the wet-bulb temperature corresponding to state 3. Therefore, for the illustrated example, the final air temperature becomes 20 °C. If one compares the process 1–4 with a common evaporative cooling, that is, process 1–5, then one observes that the final air temperature is 30 °C, that is, the wet-bulb temperature corresponding

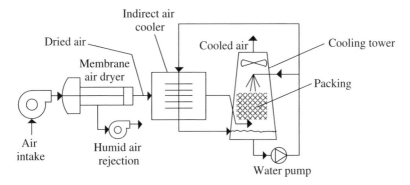

**Figure 11.11**   Emerging air-conditioning system with membrane air dryer integrated with evaporative cooling (modified from El-Dessouky et al. (2000))

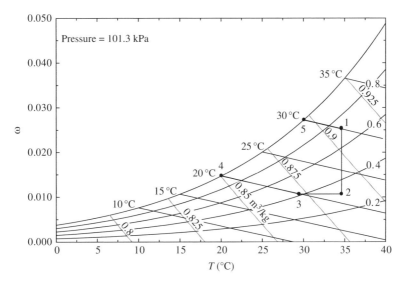

**Figure 11.12**   Cooling process representation in Mollier diagram for the membrane air dryer integrated with evaporative cooling

to state 1. Thenceforth, when membrane dehumidification is applied, the cooling becomes possible until 20 °C, that is, 10 °C less as with direct evaporative cooling.

## 11.6   Ultrasound-Assisted Drying

When a moist material is exposed to a sufficiently intense ultrasound field, high-energy dissipation occurs due to fast contraction and expansion of the material. This dissipation warms up the material within its all volume. Moreover, a cavitation effect is observed in the micropores because of their repeated expansion. This leads to local liquid evaporation. Furthermore, the

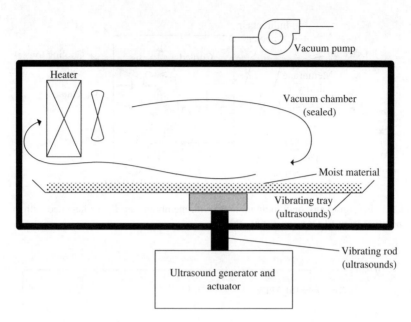

**Figure 11.13**   Ultrasound-assisted vacuum drying system

forces induced within the material by the ultrasound effects can be higher than the superficial tension within the capillaries; if this happens, moisture is released to flow toward surface in a liquid phase. The overall effect is an increase of moisture transport from the material core to the periphery, where moisture diffusion into the drying agent is even better facilitated by the vibrating surface which ruptures the boundary layer.

Figure 11.13 shows the configuration of an ultrasound-assisted vacuum drying system. Here, an ultrasonic generator and actuator system is used to vibrate a tray at ultrasonic frequency. The moist material is placed on the tray. In the chamber, vacuum pressure is maintained and heat provided at low temperature.

The experiments reported in De la Fuente-Blanco et al. (2006) confirm that the effect of ultrasounds enhances the drying rate (or reduces the drying time). A lab-scale freeze-drying setup has been equipped with an ultrasonic system to vibrate the tray with sound powers of up to 100 W. As observed from the results shown in Figure 11.14, the beneficial effect of ultrasound is confirmed. The relative moisture removal is expressed with respect to the initial mass of moist material. When ultrasound power is set to 25 W then the moisture removal after 90 min doubles with respect to the baseline case with no ultrasound applied. With 50 W ultrasound power, the relative moisture removal is 47%, whereas when 100 W ultrasound is applied, the moisture removed amounts 73% of the initial mass of moist material.

The effect of moist material temperature increase when exposed to ultrasound field is investigated by Schösler et al. (2012). For a lab-scale experiment, it is shown that for a 2 h run of a freeze-drying process, the sample temperature increased from −15 to 60 °C when ultrasounds were applied, whereas in the same conditions, the sample reached 10 °C when no ultrasounds were applied. This helps obviously in increasing the drying rate, but in addition, the equilibrium moisture content is reduced which allows for more moisture extraction from the material.

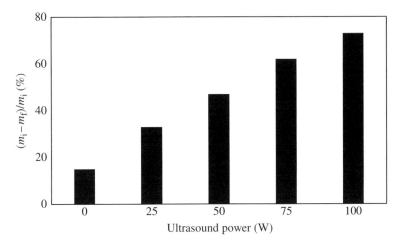

**Figure 11.14**   Relative moisture removal after 90 min of vacuum drying under ultrasound exposure at various intensities (data from De la Fuente-Blanco et al. (2006))

## 11.7   Conclusions

In this chapter, some novel drying systems and applications are reviewed based on literature sources. There is an obvious market pull toward progress and innovation in drying due to an increased need of more and diversified product, desirably obtained in a better way, with high quality, reduced energy consumption and reduced environmental impact.

One of the most promising pathways for drying system improvement today and in very near future is the use of superheated steam as drying agent because of its ability to increase the product quality while reducing the operational costs, although the investment cost is higher than with the conventional air dried systems.

Heat pump dryers and special CHPD bring some good promise as advanced methods for drying, with better exergy efficiency and better capability of using renewable energy sources. The $CaO/Ca(OH)_2$ pair within a heat pump can bring a potential advantage of application of direct heat and mass transfer with the absorption bed for humidity removal from the drying air. This definitely increases the processes efficiency and has the potential to reduce both the operational and the capital costs.

Spray drying systems show impressive progress and a continuous expansion of their application domain. One remarks the newly emerging application of nanoparticle production with spray drying. Other methods such as ultrasound-assisted drying, membrane drying, and so on are briefly presented. The field of drying is very dynamic with a high demand of innovation, research, and development.

## 11.8   Study Problems

11.1   In your opinion, is drying innovation stimulated by market pull or rather by technology push?

11.2   What are the remarkable novelties in heat pump-assisted drying?

11.3    Explain why superheated steam drying is beneficial.

11.4    What are the most important development goals for drying systems?

11.5    Explain why spray drying is important in nanomaterials.

11.6    How can nanoparticles be atomized?

11.7    How can nanoparticles be separated?

11.8    What is the benefit of microencapsulations?

11.9    What is the benefit of membrane drying of humid air in air-conditioning applications?

11.10   Explain the mechanisms through which ultrasound application improves the drying process.

# References

Chou S.K., Chua K.J. 2001. New hybrid drying technologies for heat sensitive foodstuffs. *Trends in Food Science and Technology* 12:359–369.

Chua K.J., Chou S.K. 2003. Low-cost drying methods for developing countries. *Trends in Food Science and Technology* 14:519–528.

Chua K.J., Chou S.K., Ho J.C., Hawlader M.N.A. 2002. Heat pump drying: recent developments and future trends. *Drying Technology* 20:1579–1610.

Colak N., Hepbasli A. 2009a. A review of heat pump drying: Part 1 – Systems, models and studies. *Energy Conversion and Management* 50:2180–2186.

Colak N., Hepbasli A. 2009b. A review of heat pump drying (HPD): Part 2 – Applications and performance assessment. *Energy Conversion and Management* 50:2187–2199.

Daggupati V.N., Naterer G.F., Gabriel K.S., Gravelsins R.J., Wang Z.L. 2011. Effects of atomization conditions and flow rates on spray drying for cupric chloride particle formation. *International Journal of Hydrogen Energy* 36:11353–11359.

Daghigh R., Ruslan M.H., Sulaiman M.Y., Sopian K. 2010. Review of solar assisted heat pump drying systems for agricultural and marine products. *Renewable and Sustainable Energy Reviews* 14:2564–2579.

De la Fuente-Blanco S., Riera-Franco de Sarabia E., Acosta-Aparicio V.M., Blanco-Blanco A., Gallego-Juárez J.A. 2006. Food drying process by power ultrasound. *Ultrasonics* 44:523–527.

De Waard H., Hinrichs W.L.J., Frijlink H.W. 2008. A novel bottom-up process to produce drug nanocrystals: controlled crystallization during freeze-drying. *Journal of Controlled Release* 128:179–183.

Dincer I. 2011. Exergy as a tool for sustainable drying systems. *Sustainable Cities and Societies* 1:91–96.

Dincer I., Naterer G.F. 2014. Overview of hydrogen production research in the Clean Energy Research Laboratory (CERL) at UOIT. *International Journal of Hydrogen Energy* 39:20592–20613.

El-Dessouky H.T., Ettouney H.M., Bouhamra W. 2000. A novel air conditioning system. Membrane air drying and evaporative cooling. *Transaction of the Institution of Chemical Engineers* 78A:999–1009.

Frei K.M., Cameron D., Stuart P.R. 2004. Novel drying process using forced aeration through a porous biomass matrix. *Drying Technology* 22:1191–1215.

Fudholi A., Sopian K., Ruslan M.H., Alghoul M.A., Sulaiman M.Y. 2010. Review of solar dryers for agricultural and marine products. *Renewable and Sustainable Energy Reviews* 14:1–30.

Gharsallaoui A., Roudaut G., Chambin O., Voilley A., Saurel R. 2007. Applications of spray-drying in microencapsulation of food ingredients: an overview. *Food Research International* 40:1107–1121.

Goodenough T.I.J., Goodenough P.W., Goodenough S.M. 2007. The efficiency of corona wind drying and its application to food industry. *Journal of Food Engineering* 80:1233–1238.

He P., Davis S.S., Illum L. 1999. Sustained release of chitosan microspheres prepared by novel spray drying methods. *Journal of Microencapsulation* 16:343–355.

Jangam S.V. 2011. An overview of recent developments and some R&D challenges related to drying of foods. *Drying Technology* 29:1343–1357.

Lee S.H., Heng D., Ng W.K., Cham H.-K., Tan R.B.H. 2011. Nano spray drying: a novel method for preparing protein nanoparticles for protein therapy. *International Journal of Pharmaceutics* 403:192–200.

Mujumdar A.S. 1991. Drying technologies of the future. *Drying Technology* 9:325–347.

Mujumdar A.S. 1996. Innovation in drying. *Drying Technology* 14:1459–1475.

Mujumdar A.S. 2004. Research and development in drying: recent trends and future prospects. *Drying Technology* 22:1–26.

Mujumdar A.S., Huang L.X. 2007. Global R&D needs in drying. *Drying Technology* 25:647–658.

Naterer G.F., Dincer I., Zamfirescu C. 2013. Hydrogen Production from Nuclear Energy. Springer: New York.

Nindo C.I., Tang J. 2007. Refractance window dehydration technology: a novel contact drying method. *Drying Technology* 25:37–48.

Ogura H., Mujumdar A.S. 2000. Proposal for a novel chemical heat pump dryer. *Drying Technology* 18:1033–1053.

Raghavan G.S.V., Rennie T.J., Sunjka P.S., Orsat V., Phaphuangwittayakul W., Terdtoon P. 2005. Overview of new techniques for drying biological materials with emphasis on energy aspects. *Brazilian Journal of Chemical Engineering* 22:195–201.

Ré M.-I. 2006. Formulating drug delivery systems by spray drying. *Drying Technology* 24:433–446.

Schösler K., Jäger H., Knorr D. 2012. Novel contact ultrasound system for the accelerated freeze-drying of vegetables. *Innovative Food Science and Emerging Technologies* 16:113–120.

Sharma A., Chen C.R., Vu Lan N. 2009. Solar-energy drying systems: a review. *Renewable and Sustainable Energy Reviews* 13:1185–1210.

Sundaram J., Durance T.D., Wang R. 2008. Porous scaffold of gelatin–starch with nanohydroxyapatite composite processed via novel microwave vacuum drying. *Acta Biomaterialia* 4:932–942.

Thyagarajan T., Shanmugam J., Panda R.C., Ponnavaikko M., Rao P.G. 1998. Artificial neural networks: principle and application to model based control of drying systems. *Drying Technology* 16:931–966.

# Appendix A

## Conversion Factors

| Quantity | SI to English | English to SI |
|---|---|---|
| Area | $1\ m^2 = 10.764\ ft^2$ $= 1{,}550.0\ in.^2$ | $1\ ft^2 = 0.00929\ m^2$ $1\ in.^2 = 6.452 \times 10^{-4}\ m^2$ |
| Density | $1\ kg/m^3 = 0.06243\ lb_m/ft^3$ | $1\ lb_m/ft^3 = 16.018\ kg/m^3$ $1\ slug/ft^3 = 515.379\ kg/m^3$ |
| Energy | $1\ J = 9.4787 \times 10^{-4}\ Btu$ | $1\ Btu = 1{,}055.056\ J$ $1\ cal = 4.1868\ J$ $1\ lb_f\ ft = 1.3558\ J$ $1\ hp\ h = 2.685 \times 10^6\ J$ |
| Mass specific energy | $1\ J/kg = 4.2995 \times 10^{-4}\ Btu/lb_m$ | $1\ Btu/lb_m = 2{,}326\ J/kg$ |
| Force | $1\ N = 0.22481\ lb_f$ | $1\ lb_f = 4.448\ N$ $1\ pdl = 0.1382\ N$ |
| Heat flux | $1\ W/m^2 = 0.3171\ Btu/h\ ft^2$ | $1\ Btu/h\ ft^2 = 3.1525\ W/m^2$ $1\ kcal/h\ m^2 = 1.163\ W/m^2$ $1\ cal/s\ cm^2 = 41{,}870.0\ W/m^2$ |
| Heat generation (volumetric) | $1\ W/m^3 = 0.09665\ Btu/h\ ft^3$ | $1\ Btu/h\ ft^3 = 10.343\ W/m^3$ |
| Heat transfer coefficient | $1\ W/m^2\ K = 0.1761\ Btu/h\ ft^2\ °F$ | $1\ Btu/h\ ft^2\ °F = 5.678\ W/m^2\ K$ $1\ kcal/h\ m^2\ °C = 1.163\ W/m^2\ K$ $1\ cal/s\ m^2\ °C = 41{,}870.0\ W/m^2\ K$ |
| Heat transfer rate | $1\ W = 3.4123\ Btu/h$ | $1\ Btu/h = 0.2931\ W$ |
| Length | $1\ m = 3.2808\ ft$ $= 39.370\ in.$ $1\ km = 0.621371\ mi$ | $1\ ft = 0.3048\ m$ $1\ in. = 2.54\ cm = 0.0254\ m$ $1\ mi = 1.609344\ km$ $1\ yd = 0.9144\ m$ |

*(continued overleaf)*

*Drying Phenomena: Theory and Applications*, First Edition. İbrahim Dinçer and Calin Zamfirescu.
© 2016 John Wiley & Sons, Ltd. Published 2016 by John Wiley & Sons, Ltd.

*(continued)*

| Quantity | SI to English | English to SI |
|---|---|---|
| Mass | $1 \text{ kg} = 2.2046 \text{ lb}_m$ <br> $1 \text{ t (metric)} = 1{,}000 \text{ kg}$ <br> $1 \text{ grain} = 6.47989 \times 10^{-5} \text{ kg}$ | $1 \text{ lb}_m = 0.4536 \text{ kg}$ <br> $1 \text{ slug} = 14.594 \text{ kg}$ |
| Mass flow rate | $1 \text{ kg/s} = 7{,}936.6 \text{ lb}_m/\text{h}$ <br> $\quad = 2.2046 \text{ lb}_m/\text{s}$ | $1 \text{ lb}_m/\text{h} = 0.000126 \text{ kg/s}$ <br> $1 \text{ lb}_m/\text{s} = 0.4536 \text{ kg/s}$ |
| Power | $1 \text{ W} = 1 \text{ J/s} = 3.4123 \text{ Btu/h}$ <br> $\quad = 0.737562 \text{ Ib}_f \text{ ft/s}$ <br> $1 \text{ hp (metric)} = 0.735499 \text{ kW}$ <br> $1 \text{ t of refrig.} = 3.51685 \text{ kW}$ | $1 \text{ Btu/h} = 0.2931 \text{ W}$ <br> $1 \text{ Btu/s} = 1{,}055.1 \text{ W}$ <br> $1 \text{ lb}_f \text{ ft/s} = 1.3558 \text{ W}$ <br> $1 \text{ hp}^{UK} = 745.7 \text{ W}$ |
| Pressure | $1 \text{ Pa} = 0.020886 \text{ lb}_f/\text{ft}^2$ <br> $\quad = 1.4504 \times 10^{-4} \text{ lb}_f/\text{in.}^2$ <br> $\quad = 4.015 \times 10^{-3} \text{ in water}$ <br> $\quad = 2.953 \times 10^{-4} \text{ in Hg}$ | $1 \text{ lb}_f/\text{ft}^2 = 47.88 \text{ Pa}$ <br> $1 \text{ lb}_f/\text{in.}^2 = 1 \text{ psi} = 6{,}894.8 \text{ Pa}$ <br> $1 \text{ stand. atm} = 1.0133 \times 10^5 \text{ Pa}$ <br> $1 \text{ bar} = 1 \times 10^5 \text{ Pa}$ |
| Specific heat | $1 \text{ J/kg K} = 2.3886 \times 10^{-4} \text{ Btu/lb}_m \, {}^\circ\text{F}$ | $1 \text{ Btu/lb}_m \, {}^\circ\text{F} = 4{,}187 \text{ J/kg K}$ |
| Surface tension | $1 \text{ N/m} = 0.06852 \text{ lb}_f/\text{ft}$ | $1 \text{ lb}_f/\text{ft} = 14.594 \text{ N/m}$ <br> $1 \text{ dyn/cm} = 1 \times 10^{-3} \text{ N/m}$ |
| Temperature | $T \, (\text{K}) = T \, (^\circ\text{C}) + 273.15$ <br> $\quad = T \, (^\circ\text{R})/1.8$ <br> $\quad = [T \, (^\circ\text{F}) + 459.67]/1.8$ <br> $T \, (^\circ\text{C}) = [T \, (^\circ\text{F}) - 32]/1.8$ | $T \, (^\circ\text{R}) = 1.8 T \, (\text{K})$ <br> $\quad = T \, (^\circ\text{F}) + 459.67$ <br> $\quad = 1.8 T \, (^\circ\text{C}) + 32$ <br> $\quad = 1.8[T \, (\text{K}) - 273.15] + 32$ |
| Temperature difference | $1 \text{ K} = 1 \, ^\circ\text{C} = 1.8 \, ^\circ\text{R} = 1.8 \, ^\circ\text{F}$ | $1 \, ^\circ\text{R} = 1 \, ^\circ\text{F} = 1 \text{ K}/1.8 = 1 \, ^\circ\text{C}/1.8$ |
| Thermal conductivity | $1 \text{ W/m K} = 0.57782 \text{ Btu/h ft} \, ^\circ\text{F}$ | $1 \text{ Btu/h ft} \, ^\circ\text{F} = 1.731 \text{ W/m K}$ <br> $1 \text{ kcal/h m} \, ^\circ\text{C} = 1.163 \text{ W/m K}$ <br> $1 \text{ cal/s cm} \, ^\circ\text{C} = 418.7 \text{ W/m K}$ |
| Diffusivity | $1 \text{ m}^2/\text{s} = 10.7639 \text{ ft}^2/\text{s}$ | $1 \text{ ft}^2/\text{s} = 0.0929 \text{ m}^2/\text{s}$ <br> $1 \text{ ft}^2/\text{h} = 2.581 \times 10^{-5} \text{ m}^2/\text{s}$ |
| Thermal resistance | $1 \text{ K/W} = 0.52750 \, ^\circ\text{F h/Btu}$ | $1 \, ^\circ\text{F h/Btu} = 1.8958 \text{ K/W}$ |
| Velocity | $1 \text{ m/s} = 3.2808 \text{ ft/s}$ <br> $1 \text{ km/s} = 0.62137 \text{ mi/h}$ | $1 \text{ ft/s} = 0.3048 \text{ m/s}$ <br> $1 \text{ ft/min} = 5.08 \times 10^{-3} \text{ m/s}$ |
| Viscosity (dynamic) <br> $(\text{kg/m s} = \text{N s/m}^2)$ | $1 \text{ kg/m s} = 0.672 \text{ lb}_m/\text{ft s}$ <br> $\quad = 2{,}419.1 \text{ lb}_m/\text{fh h}$ | $1 \text{ lb}_m/\text{ft s} = 1.4881 \text{ kg/m s}$ <br> $1 \text{ lb}_m/\text{ft h} = 4.133 \times 10^{-4} \text{ kg/m s}$ <br> $1 \text{ centipoise (cP)} = 10^{-2} \text{ P}$ <br> $\quad = 1 \times 10^{-3} \text{ kg/m s}$ |
| Viscosity (kinematic) | $1 \text{ m}^2/\text{s} = 10.7639 \text{ ft}^2/\text{s}$ <br> $\quad = 1 \times 10^4 \text{ St}$ | $1 \text{ ft}^2/\text{s} = 0.0929 \text{ m}^2/\text{s}$ <br> $1 \text{ ft}^2/\text{h} = 2.581 \times 10^{-5} \text{ m}^2/\text{s}$ <br> $1 \text{ St} = 1 \text{ cm}^2/\text{s}$ |
| Volume | $1 \text{ m}^3 = 35.3134 \text{ ft}^3$ <br> $1 \text{ l} = 1 \text{ dm}^3 = 0.001 \text{ m}^3$ | $1 \text{ ft}^3 = 0.02832 \text{ m}^3$ <br> $1 \text{ in.}^3 = 1.6387 \times 10^{-5} \text{ m}^3$ <br> $1 \text{ gal}^{US} = 0.003785 \text{ m}^3$ <br> $1 \text{ gal}^{UK} = 0.004546 \text{ m}^3$ |
| Volumetric flow rate | $1 \text{ m}^3/\text{s} = 35.3134 \text{ ft}^3/\text{s}$ <br> $\quad = 1.2713 \times 10^5 \text{ ft}^3/\text{h}$ | $1 \text{ ft}^3/\text{s} = 2.8317 \times 10^{-2} \text{ m}^3/\text{s}$ <br> $1 \text{ ft}^3/\text{min} = 4.72 \times 10^{-4} \text{ m}^3/\text{s}$ <br> $1 \text{ ft}^3/\text{h} = 7.8658 \times 10^{-6} \text{ m}^3/\text{s}$ <br> $1 \text{ gal}^{US}/\text{min} = 6.309 \times 10^{-5} \text{ m}^3/\text{s}$ |

# Appendix B

## Thermophysical Properties of Water

**Table B.1** Thermophysical properties of pure water at atmospheric pressure

| $T$ (°C) | $\rho$ (kg/m³) | $\mu$ (Pa s) | $C_p$ (J/kg K) | $k$ (W/m K) | $Pr$ | $\beta$ (K⁻¹) | $c$ (m/s) | $\sigma$ (N/m) |
|---|---|---|---|---|---|---|---|---|
| 5 | 1000 | 0.001519 | 4200 | 0.5576 | 11.44 | 0.00001135 | 1426 | 0.07494 |
| 10 | 999.7 | 0.001307 | 4188 | 0.5674 | 9.642 | 0.00008743 | 1448 | 0.07422 |
| 15 | 999.1 | 0.001138 | 4184 | 0.5769 | 8.253 | 0.0001523 | 1467 | 0.07348 |
| 20 | 998.2 | 0.001002 | 4183 | 0.5861 | 7.152 | 0.000209 | 1483 | 0.07273 |
| 25 | 997.1 | 0.0008905 | 4183 | 0.5948 | 6.263 | 0.0002594 | 1497 | 0.07197 |
| 30 | 995.7 | 0.0007977 | 4183 | 0.603 | 5.534 | 0.0003051 | 1509 | 0.07119 |
| 35 | 994 | 0.0007196 | 4183 | 0.6107 | 4.929 | 0.000347 | 1520 | 0.0704 |
| 40 | 992.2 | 0.0006533 | 4182 | 0.6178 | 4.422 | 0.0003859 | 1528 | 0.06959 |
| 45 | 990.2 | 0.0005963 | 4182 | 0.6244 | 3.994 | 0.0004225 | 1534 | 0.06877 |
| 50 | 988 | 0.0005471 | 4181 | 0.6305 | 3.628 | 0.0004572 | 1537 | 0.06794 |
| 55 | 985.7 | 0.0005042 | 4182 | 0.636 | 3.315 | 0.0004903 | 1538 | 0.0671 |
| 60 | 983.2 | 0.0004666 | 4183 | 0.641 | 3.045 | 0.0005221 | 1537 | 0.06624 |
| 65 | 980.6 | 0.0004334 | 4184 | 0.6455 | 2.81 | 0.0005528 | 1534 | 0.06536 |
| 70 | 977.8 | 0.000404 | 4187 | 0.6495 | 2.605 | 0.0005827 | 1529 | 0.06448 |
| 75 | 974.9 | 0.0003779 | 4190 | 0.653 | 2.425 | 0.0006118 | 1523 | 0.06358 |
| 80 | 971.8 | 0.0003545 | 4194 | 0.6562 | 2.266 | 0.0006402 | 1514 | 0.06267 |
| 85 | 968.6 | 0.0003335 | 4199 | 0.6589 | 2.125 | 0.0006682 | 1504 | 0.06175 |
| 90 | 965.3 | 0.0003145 | 4204 | 0.6613 | 2 | 0.0006958 | 1491 | 0.06081 |
| 95 | 961.9 | 0.0002974 | 4210 | 0.6634 | 1.888 | 0.000723 | 1475 | 0.05987 |
| 100 | 0.5896 | 0.00001227 | 2042 | 0.02506 | 0.9996 | 0.002881 | 472.8 | 0.05891 |

*Source*: Data obtained with EES software – Engineering Equation Solver (S.A. Klein, version 9.698).
$T$ = temperature, $\rho$ = density, $\mu$ = dynamic viscosity, $c_p$ = specific heat, $k$ = thermal conductivity,
$Pr$ = Prandtl number, $\beta$ = volume expansion coefficient, $c$ = sound speed, $\sigma$ = superficial
tension, $\omega = 0.3443$ = accentric factor.
Critical parameters: $T_c = 373.984°C, P_c = 220.64\,bar, v_c = 3.106\,dm^3/kg$.
Triple point parameters: $T_t = 0.01°C, P_t = 611.732\,Pa$.

*Drying Phenomena: Theory and Applications*, First Edition. İbrahim Dinçer and Calin Zamfirescu.
© 2016 John Wiley & Sons, Ltd. Published 2016 by John Wiley & Sons, Ltd.

**Table B.2** Thermophysical properties of water at saturation

| T (°C) | P (bar) | σ (mN/m) | ρL (kg/m³) | Cpl (kJ/kg K) | kl (mW/m K) | μl (μPa s) | Prl | βl (m/K) | vv (m³/kg) | Cpv (kJ/kg K) | kv (mW/m K) | μv (μPa s) | Prv | βv (m/K) | γ |
|---|---|---|---|---|---|---|---|---|---|---|---|---|---|---|---|
| 0.01 | 0.00612 | 75.64 | 1000 | 4.23 | 547.5 | 1792 | 13.84 | N/A | 206.0 | 1.87 | 17.07 | 9.216 | 1.008 | 3.672 | 1.33 |
| 10 | 0.0123 | 74.22 | 999.7 | 4.19 | 567.4 | 1307 | 9.645 | 0.0872 | 106.3 | 1.87 | 17.62 | 9.461 | 1.006 | 3.548 | 1.33 |
| 20 | 0.0234 | 72.73 | 998.2 | 4.18 | 586 | 1002 | 7.154 | 0.209 | 57.78 | 1.88 | 18.22 | 9.727 | 1.004 | 3.435 | 1.33 |
| 30 | 0.0425 | 71.19 | 995.6 | 4.18 | 602.9 | 797.7 | 5.535 | 0.305 | 32.9 | 1.89 | 18.88 | 10.01 | 1.003 | 3.332 | 1.33 |
| 40 | 0.0738 | 69.59 | 992.2 | 4.18 | 617.8 | 653.3 | 4.423 | 0.386 | 19.53 | 1.90 | 19.59 | 10.31 | 1.002 | 3.239 | 1.33 |
| 50 | 0.1234 | 67.94 | 988 | 4.18 | 630.4 | 547.1 | 3.629 | 0.457 | 12.04 | 1.92 | 20.36 | 10.62 | 1.001 | 3.156 | 1.33 |
| 60 | 0.1993 | 66.24 | 983.2 | 4.18 | 640.9 | 466.6 | 3.045 | 0.522 | 7.674 | 1.94 | 21.18 | 10.93 | 1 | 3.082 | 1.33 |
| 70 | 0.3118 | 64.48 | 977.7 | 4.19 | 649.5 | 404 | 2.605 | 0.583 | 5.045 | 1.96 | 22.06 | 11.26 | 0.9995 | 3.017 | 1.33 |
| 80 | 0.4737 | 62.67 | 971.8 | 4.19 | 656.2 | 354.5 | 2.266 | 0.64 | 3.409 | 1.98 | 23 | 11.59 | 0.9993 | 2.962 | 1.33 |
| 90 | 0.7012 | 60.81 | 965.3 | 4.20 | 661.3 | 314.5 | 2 | 0.696 | 2.362 | 2.01 | 24 | 11.93 | 0.9994 | 2.917 | 1.34 |
| 100 | 1.013 | 58.91 | 958.4 | 4.22 | 665.1 | 281.9 | 1.787 | 0.75 | 1.674 | 2.04 | 25.08 | 12.27 | 1 | 2.882 | 1.34 |
| 110 | 1.432 | 56.96 | 951 | 4.23 | 667.6 | 254.8 | 1.615 | 0.804 | 1.211 | 2.08 | 26.22 | 12.61 | 1.001 | 2.857 | 1.34 |
| 120 | 1.985 | 54.96 | 943.2 | 4.25 | 669.1 | 232.1 | 1.474 | 0.858 | 0.892 | 2.12 | 27.44 | 12.96 | 1.004 | 2.843 | 1.35 |
| 130 | 2.7 | 52.93 | 934.9 | 4.27 | 669.7 | 213 | 1.357 | 0.912 | 0.669 | 2.17 | 28.73 | 13.3 | 1.007 | 2.84 | 1.36 |
| 140 | 3.612 | 50.85 | 926.2 | 4.29 | 669.4 | 196.6 | 1.259 | 0.968 | 0.509 | 2.23 | 30.09 | 13.65 | 1.012 | 2.849 | 1.36 |
| 150 | 4.757 | 48.74 | 917.1 | 4.31 | 668.3 | 182.5 | 1.178 | 1.026 | 0.393 | 2.30 | 31.54 | 13.99 | 1.02 | 2.872 | 1.37 |
| 160 | 6.177 | 46.59 | 907.5 | 4.34 | 666.4 | 170.3 | 1.109 | 1.087 | 0.3071 | 2.37 | 33.06 | 14.34 | 1.029 | 2.908 | 1.39 |
| 170 | 7.915 | 44.4 | 897.5 | 4.37 | 663.7 | 159.6 | 1.051 | 1.152 | 0.243 | 2.46 | 34.66 | 14.68 | 1.042 | 2.959 | 1.40 |
| 180 | 10.02 | 42.19 | 887.1 | 4.40 | 660.2 | 150.2 | 1.002 | 1.221 | 0.194 | 2.56 | 36.34 | 15.03 | 1.057 | 3.027 | 1.42 |
| 190 | 12.54 | 39.94 | 876.1 | 4.44 | 655.9 | 141.8 | 0.9607 | 1.296 | 0.156 | 2.67 | 38.09 | 15.37 | 1.077 | 3.114 | 1.44 |
| 200 | 15.54 | 37.67 | 864.7 | 4.49 | 650.7 | 134.4 | 0.9269 | 1.377 | 0.127 | 2.80 | 39.93 | 15.71 | 1.1 | 3.221 | 1.46 |
| 210 | 19.06 | 35.38 | 852.8 | 4.54 | 644.7 | 127.7 | 0.8994 | 1.467 | 0.104 | 2.94 | 41.85 | 16.06 | 1.129 | 3.351 | 1.48 |
| 220 | 23.18 | 33.06 | 840.3 | 4.60 | 637.6 | 121.6 | 0.8777 | 1.567 | 0.086 | 3.11 | 43.86 | 16.41 | 1.162 | 3.508 | 1.52 |
| 230 | 27.95 | 30.73 | 827.3 | 4.67 | 629.5 | 116 | 0.8615 | 1.679 | 0.071 | 3.30 | 45.96 | 16.76 | 1.203 | 3.697 | 1.55 |
| 240 | 33.45 | 28.39 | 813.5 | 4.76 | 620.3 | 110.9 | 0.8508 | 1.807 | 0.06 | 3.51 | 48.15 | 17.12 | 1.25 | 3.923 | 1.59 |

(continued overleaf)

Table B.2  (continued)

| T (°C) | P (bar) | σ (mN/m) | ρ_L (kg/m³) | C_pl (kJ/kg K) | k_l (mW/m K) | μ_l (μPa s) | Pr_l | β_l (m/K) | v_v (m³/kg) | C_pv (kJ/kg K) | k_v (mW/m K) | μ_v (μPa s) | Pr_v | β_v (m/K) | γ |
|---|---|---|---|---|---|---|---|---|---|---|---|---|---|---|---|
| 250 | 39.74 | 26.04 | 799.1 | 4.86 | 609.8 | 106.2 | 0.8455 | 1.954 | 0.050 | 3.77 | 50.46 | 17.49 | 1.306 | 4.195 | 1.64 |
| 260 | 46.89 | 23.68 | 783.8 | 4.97 | 598 | 101.7 | 0.846 | 2.126 | 0.042 | 4.06 | 52.9 | 17.88 | 1.373 | 4.523 | 1.70 |
| 270 | 55 | 21.33 | 767.7 | 5.11 | 584.5 | 97.55 | 0.853 | 2.33 | 0.036 | 4.41 | 55.49 | 18.28 | 1.453 | 4.923 | 1.77 |
| 280 | 64.13 | 18.99 | 750.5 | 5.28 | 569.4 | 93.56 | 0.8673 | 2.576 | 0.031 | 4.83 | 58.28 | 18.7 | 1.549 | 5.415 | 1.86 |
| 290 | 74.38 | 16.66 | 732.2 | 5.48 | 552.3 | 89.71 | 0.8908 | 2.88 | 0.026 | 5.33 | 61.31 | 19.15 | 1.666 | 6.032 | 1.97 |
| 300 | 85.84 | 14.35 | 712.4 | 5.74 | 533.1 | 85.95 | 0.9262 | 3.266 | 0.022 | 5.97 | 64.67 | 19.65 | 1.813 | 6.822 | 2.11 |
| 310 | 98.61 | 12.08 | 691 | 6.08 | 511.3 | 82.22 | 0.9782 | 3.773 | 0.018 | 6.78 | 68.45 | 20.21 | 2.001 | 7.861 | 2.29 |
| 320 | 112.8 | 9.858 | 667.4 | 6.54 | 486.8 | 78.46 | 1.054 | 4.47 | 0.015 | 7.87 | 72.84 | 20.84 | 2.251 | 9.278 | 2.54 |
| 330 | 128.5 | 7.697 | 641 | 7.2 | 458.8 | 74.57 | 1.17 | 5.489 | 0.013 | 9.41 | 78.1 | 21.6 | 2.603 | 11.31 | 2.91 |
| 340 | 145.9 | 5.62 | 610.8 | 8.24 | 426.9 | 70.45 | 1.359 | 7.116 | 0.011 | 11.8 | 84.68 | 22.55 | 3.137 | 14.43 | 3.47 |
| 350 | 165.2 | 3.66 | 574.7 | 10.1 | 389.7 | 65.88 | 1.711 | 10.1 | 0.0088 | 15.9 | 93.48 | 23.81 | 4.058 | 19.77 | 4.47 |
| 360 | 186.6 | 1.872 | 528.1 | 14.7 | 344.4 | 60.39 | 2.574 | 17.11 | 0.007 | 25.2 | 106.6 | 25.71 | 6.085 | 30.79 | 6.74 |
| 370 | 210.3 | 0.3846 | 453.1 | 41.7 | 280.7 | 52.25 | 7.765 | 45.88 | 0.0050 | 70.4 | 132.6 | 29.57 | 15.7 | 64.53 | 18.0 |
| 373.9 | 220.4 | 0.0004 | 349.4 | N/A | 210.8 | 41.95 | 277.1 | 87.51 | 0.0034 | 340.1 | 180 | 37 | 69.91 | 104.5 | 84.5 |

*Source:* Data obtained with EES software – Engineering Equation Solver (S.A. Klein, version 9.698).

$T$ = temperature, $P$ = pressure, $\sigma$ = surface tension, $\rho$ = density, $c_p$ = specific heat, $k$ = thermal conductivity, $\mu$ = dynamic viscosity, $Pr$ = Prandtl number, $\beta$ = volumetric expansion coefficient, $v$ = specific volume, $\gamma$ = isentropic expansion coefficient of saturated vapor, indices: l = liquid; v = vapor.

# Appendix C

## Thermophysical Properties of Some Foods and Solid Materials

**Table C.1**  Thermophysical properties of some solid materials

| Material | $T$ (K) | $\rho$ (kg/m$^3$) | $k$ (W/m K) | $C_p$ (J/kg K) |
|---|---|---|---|---|
| Asphalt | 300 | 2115 | 0.0662 | 920 |
| Bakelite | 300 | 1300 | 1.4 | 1465 |
| Chrome brick | 473 | 3010 | 2.3 | 835 |
| Diatomaceous silica, fired | 478 | — | 0.25 | — |
| Fire clay brick | 478 | 2645 | 1.0 | 960 |
| | 922 | — | 1.5 | — |
| Clay | 300 | 1460 | 1.3 | 880 |
| Coal (anthracite) | 300 | 1350 | 0.26 | 1260 |
| Concrete (stone mix) | 300 | 2300 | 1.4 | 880 |
| Leather (sole) | 300 | 998 | 0.159 | — |
| Paper | 300 | 930 | 0.18 | 1340 |
| Paraffin | 300 | 900 | 0.24 | 2890 |
| Sand | 300 | 1515 | 0.27 | 800 |
| Soil | 300 | 2050 | 0.52 | 1840 |
| Teflon | 300 | 2200 | 0.35 | — |
| Animal skin | 300 | — | 0.37 | — |
| Animal fat layer (adipose) | 300 | — | 0.2 | — |
| Animal muscle | 300 | — | 0.41 | — |
| Fir wood | 300 | 415 | 0.11 | 2720 |
| Oak wood | 300 | 545 | 0.17 | 2385 |
| Yellow pine wood | 300 | 640 | 0.15 | 2805 |
| White pine wood | 300 | 435 | 0.11 | — |

*Source*: Dincer (1997) and Incropera and DeWitt (1998).

*Drying Phenomena: Theory and Applications*, First Edition. İbrahim Dinçer and Calin Zamfirescu.
© 2016 John Wiley & Sons, Ltd. Published 2016 by John Wiley & Sons, Ltd.

**Table C.2**  Average water content moisture diffusivity of selected foodstuff at room temperature

| Food | $W$ (kg/kg d.b.) | $D$ (m$^2$/s) | Food | $W$ (kg/kg d.b.) | $D$ (m$^2$/s) |
|------|------|------|------|------|------|
| Alfalfa stem | 3.7 | 1.3E–9 | Garlic | 0.9 | 1E–10 |
| Apple, whole | 85 | 1.4E–7 | Grape, jelly | 42 | 1.2E–7 |
| Apple, dried | 42 | 9.6E–8 | Ham | 72 | 1.4E–7 |
| Apple, sauce | 37 | 1.1E–7 | Ham, smoked | 64 | 1.2E–7 |
| Apricots, dried | 44 | 1.1E–7 | Mango, dried | N/A | 8E–10 |
| Avocado | 11 | 2.2E–10 | Milk, foam | 0.2 | 1.1E–9 |
| Banana, flesh | 76 | 1.3E–7 | Milk, skim | 0.52 | 1.3E–10 |
| Beef, chuck | 66 | 1.2E–7 | Muffin | 0.35 | 7.6E–8 |
| Beef, steak | 37 | 1.1E–7 | Onion | 9.5 | 2.75E–9 |
| Beef, tongue | 68 | 1.3E–7 | Peach | N/A | 1.4E–7 |
| Beet | N/A | 1.5E–9 | Peach, dried | 43 | 1.2E–6 |
| Biscuit | 0.37 | 5E–8 | Pepper, green | 8.15 | 4.9E–9 |
| Bread | 0.4 | 2.7E–7 | Pepperoni | 0.19 | 5.2E–11 |
| Carrot | 6.0 | 2.1E–9 | Potato | 3.5 | 1.7E–9 |
| Cheese | N/A | 6.25E–10 | Potato, mashed | 78 | 1.35E–7 |
| Cherries, flesh | N/A | 1.3E–7 | Prune | 43 | 1.2E–7 |
| Chocolate | 32 | 1.2E–7 | Raisins | 32 | 1.1E–7 |
| Coconut, albumen | 0.4 | 6.9E–10 | Salami | 36 | 1.3E–7 |
| Coffee, extract | 0.6 | 1.4E–10 | Semolina, pasta | 0.13 | 7.6E–11 |
| Corn | 0.14 | 5E–11 | Sorghum, whole | N/A | 2E–11 |
| Corn, pasta | 0.45 | 6.7E–11 | Starch, granular | 0.35 | 1.7E–9 |
| Dates | 35 | 1E–7 | Strawberry, flesh | 92 | 1.3E–7 |
| Figs | 40 | 9.6E–8 | Strawberry, jam | 41 | 1.2E–7 |
| Fish, cod | 81 | 1.2E–7 | Wheat | 0.21 | 1.4E–10 |
| Fish, halibut | 76 | 1.5E–7 | Wheat, pasta | 0.25 | 5E–9 |
| Fish, muscle | 0.17 | 2.1E–10 | Yogurt, concentr. | 1.3 | 1.5E–9 |

*Source*: Marinos-Kouris and Maroulis (2006), ASHRAE Handbook of Refrigeration (1998), Doulia et al. (2014).

$W$, moisture content; $D$, moisture diffusivity.

# References

Dincer I. 1997. Heat Transfer in Food Cooling Applications. Taylor & Francis: Washington, DC.

Incropera F.P., DeWitt D.P. 1998. Fundamentals of Heat and Mass Transfer. Wiley: New York.

Marinos-Kouris D., Maroulis Z.B. 2006. Transport properties in the drying of solids. In: Handbook of Industrial Drying, Chapter 4. 3rd ed. CRC Press, Taylor and Francis Group: Boca Raton, FL.

American Society of Heating, Refrigeration and Air Conditioning. 1998. ASHRAE Handbook of Refrigeration. American Society of Heating, Refrigeration and Air Conditioning: New York.

Doulia D., Tzia K., Gekas V. 2014. A knowledge base for the apparent mass diffusion coefficient of foods. Database of Physical Properties of Food accessed on January 12, 2015. http://www.nelfood.com. Document 328.

# Appendix D

## Psychometric Properties of Humid Air

*Drying Phenomena: Theory and Applications*, First Edition. İbrahim Dinçer and Calin Zamfirescu.
© 2016 John Wiley & Sons, Ltd. Published 2016 by John Wiley & Sons, Ltd.

**Table D.1**  Properties of air at standard atmospheric pressure

| $T$ (°C) | $\rho$ (g/dm$^3$) | $\mu$ (µPa s) | $C_p$ (J/kg K) | $k$ (mW/m K) | $Pr$ | $\beta$ (K$^{-1}$) | $c$ (m/s) | $\gamma$ |
|---|---|---|---|---|---|---|---|---|
| −200 | 47.61 | 5.201 | 1002 | 6.667 | 0.7821 | 0.01367 | 171.6 | 1.401 |
| −170 | 33.77 | 7.311 | 1002 | 9.668 | 0.7581 | 0.009695 | 203.7 | 1.401 |
| −140 | 26.16 | 9.296 | 1002 | 12.53 | 0.7435 | 0.00751 | 231.4 | 1.401 |
| −110 | 21.35 | 11.16 | 1003 | 15.26 | 0.7332 | 0.006129 | 256.2 | 1.401 |
| −80 | 18.03 | 12.93 | 1003 | 17.87 | 0.7251 | 0.005177 | 278.8 | 1.401 |
| −50 | 15.61 | 14.6 | 1003 | 20.37 | 0.7186 | 0.004481 | 299.6 | 1.401 |
| −20 | 13.76 | 16.19 | 1003 | 22.77 | 0.7131 | 0.00395 | 319.1 | 1.401 |
| 10 | 12.3 | 17.7 | 1004 | 25.09 | 0.7086 | 0.003532 | 337.4 | 1.4 |
| 40 | 11.12 | 19.15 | 1006 | 27.32 | 0.7049 | 0.003193 | 354.7 | 1.4 |
| 70 | 10.15 | 20.54 | 1008 | 29.49 | 0.702 | 0.002914 | 371.2 | 1.398 |
| 100 | 9.334 | 21.88 | 1011 | 31.6 | 0.6999 | 0.00268 | 386.8 | 1.397 |
| 130 | 8.639 | 23.18 | 1014 | 33.64 | 0.6985 | 0.00248 | 401.8 | 1.395 |
| 160 | 8.041 | 24.43 | 1018 | 35.64 | 0.6978 | 0.002309 | 416.2 | 1.393 |
| 190 | 7.52 | 25.64 | 1023 | 37.59 | 0.6977 | 0.002159 | 429.9 | 1.39 |
| 220 | 7.063 | 26.82 | 1028 | 39.5 | 0.6982 | 0.002028 | 443.2 | 1.387 |
| 250 | 6.658 | 27.96 | 1034 | 41.37 | 0.6991 | 0.001912 | 456 | 1.384 |
| 280 | 6.296 | 29.07 | 1041 | 43.2 | 0.7004 | 0.001808 | 468.3 | 1.381 |
| 310 | 5.973 | 30.16 | 1047 | 45.0 | 0.7019 | 0.001715 | 480.3 | 1.378 |
| 340 | 5.68 | 31.22 | 1054 | 46.77 | 0.7038 | 0.001631 | 491.9 | 1.374 |
| 370 | 5.415 | 32.26 | 1061 | 48.51 | 0.7058 | 0.001555 | 503.1 | 1.371 |
| 400 | 5.174 | 33.28 | 1069 | 50.23 | 0.7079 | 0.001486 | 514.1 | 1.367 |
| 430 | 4.953 | 34.27 | 1076 | 51.92 | 0.7101 | 0.001422 | 524.7 | 1.364 |
| 760 | 3.371 | 44.21 | 1147 | 69.34 | 0.7314 | 0.0009679 | 629 | 1.334 |
| 790 | 3.276 | 45.04 | 1153 | 70.84 | 0.7328 | 0.0009406 | 637.5 | 1.332 |
| 820 | 3.186 | 45.86 | 1158 | 72.33 | 0.7341 | 0.0009148 | 646 | 1.33 |
| 850 | 3.101 | 46.67 | 1163 | 73.82 | 0.7353 | 0.0008904 | 654.3 | 1.328 |
| 880 | 3.02 | 47.48 | 1168 | 75.29 | 0.7364 | 0.0008672 | 662.6 | 1.326 |
| 910 | 2.944 | 48.28 | 1172 | 76.75 | 0.7374 | 0.0008452 | 670.7 | 1.324 |
| 940 | 2.871 | 49.07 | 1177 | 78.21 | 0.7383 | 0.0008243 | 678.8 | 1.323 |
| 970 | 2.802 | 49.85 | 1181 | 79.65 | 0.7391 | 0.0008044 | 686.7 | 1.321 |
| 1000 | 2.736 | 50.63 | 1185 | 81.09 | 0.7398 | 0.0007855 | 694.5 | 1.32 |

*Source*: Table produced with Engineering Equation Solver (S.A. Klein 2014, version 9.698).
$T$ = temperature, $\rho$ = density, $\mu$ = dynamic viscosity, $c_p$ = specific heat, $k$ = thermal conductivity, $Pr$ = Prandtl number, $\beta$ = volume expansion coefficient, $c$ = sound speed, $\gamma$ = adiabatic expansion coefficient, $\omega = -0.009278$ = accentric factor. Critical parameters: $T_c = 132.531$ K, $P_c = 37.86$ bar, $\nu_c = 2.917$ dm$^3$/kg.

**Table D.2** Properties of humid air at standard atmospheric pressure

| T (°C) | φ | $T_{dp}$ (°C) | $T_{wb}$ (°C) | ω (kg/kg) | ρ (kg/m³) | $C_p$ (J/kg K) | k (W/mK) | μ (Pa s) | h (kJ/kg) | s (kJ/kg K) |
|---|---|---|---|---|---|---|---|---|---|---|
| 20 | 0.1 | −11.18 | 7.604 | 0.001439 | 1.201 | 1.009 | 0.02514 | 0.00001825 | 23.78 | 5.693 |
| | 0.325 | 3.039 | 11.24 | 0.004701 | 1.195 | 1.015 | 0.02516 | 0.00001824 | 32.06 | 5.724 |
| | 0.55 | 10.69 | 14.47 | 0.007998 | 1.189 | 1.022 | 0.02517 | 0.00001823 | 40.43 | 5.754 |
| | 0.775 | 15.95 | 17.38 | 0.01133 | 1.183 | 1.028 | 0.02519 | 0.00001823 | 48.88 | 5.783 |
| | 1 | 20 | 20 | 0.01468 | 1.176 | 1.034 | 0.0252 | 0.00001822 | 57.43 | 5.812 |
| 30 | 0.1 | −4.352 | 13.24 | 0.002617 | 1.16 | 1.012 | 0.02589 | 0.00001871 | 36.89 | 5.738 |
| | 0.325 | 11.75 | 18.51 | 0.008588 | 1.149 | 1.023 | 0.02592 | 0.0000187 | 52.15 | 5.793 |
| | 0.55 | 19.98 | 22.93 | 0.01467 | 1.138 | 1.034 | 0.02595 | 0.00001869 | 67.71 | 5.847 |
| | 0.775 | 25.63 | 26.7 | 0.02088 | 1.127 | 1.046 | 0.02597 | 0.00001867 | 83.58 | 5.901 |
| | 1 | 30 | 30 | 0.02718 | 1.116 | 1.057 | 0.026 | 0.00001866 | 99.75 | 5.954 |
| 40 | 0.1 | 2.629 | 18.57 | 0.004566 | 1.119 | 1.015 | 0.02664 | 0.00001917 | 52.02 | 5.789 |
| | 0.325 | 20.42 | 25.81 | 0.01509 | 1.101 | 1.035 | 0.0267 | 0.00001915 | 79.12 | 5.884 |
| | 0.55 | 29.23 | 31.46 | 0.02597 | 1.082 | 1.055 | 0.02674 | 0.00001912 | 107.1 | 5.978 |
| | 0.775 | 35.3 | 36.08 | 0.03723 | 1.064 | 1.076 | 0.02679 | 0.0000191 | 136.1 | 6.073 |
| | 1 | 40 | 40 | 0.04883 | 1.045 | 1.098 | 0.02683 | 0.00001907 | 166.2 | 6.17 |
| 50 | 0.1 | 10.08 | 23.78 | 0.007674 | 1.079 | 1.021 | 0.0274 | 0.00001962 | 70.24 | 5.85 |
| | 0.325 | 29.02 | 33.23 | 0.02565 | 1.049 | 1.055 | 0.02749 | 0.00001958 | 116.9 | 6.009 |
| | 0.55 | 38.44 | 40.1 | 0.04469 | 1.019 | 1.091 | 0.02757 | 0.00001954 | 166.3 | 6.17 |
| | 0.775 | 44.96 | 45.52 | 0.06488 | 0.9892 | 1.128 | 0.02765 | 0.00001949 | 218.6 | 6.337 |
| | 1 | 50 | 50 | 0.08623 | 0.9593 | 1.168 | 0.02772 | 0.00001943 | 274.3 | 6.51 |
| 60 | 0.1 | 17.45 | 29 | 0.01249 | 1.039 | 1.031 | 0.02816 | 0.00002006 | 93.03 | 5.925 |
| | 0.325 | 37.58 | 40.8 | 0.04251 | 0.9918 | 1.087 | 0.02831 | 0.00001999 | 171.5 | 6.185 |
| | 0.55 | 47.63 | 48.84 | 0.07551 | 0.9449 | 1.149 | 0.02844 | 0.00001991 | 257.7 | 6.458 |
| | 0.775 | 54.6 | 54.99 | 0.112 | 0.898 | 1.217 | 0.02855 | 0.00001982 | 352.9 | 6.753 |
| | 1 | 60 | 60 | 0.1522 | 0.851 | 1.293 | 0.02864 | 0.0000197 | 458.7 | 7.074 |
| 70 | 0.1 | 24.74 | 34.32 | 0.01976 | 0.9971 | 1.045 | 0.02893 | 0.00002049 | 122.5 | 6.02 |
| | 0.325 | 46.08 | 48.52 | 0.06916 | 0.9258 | 1.137 | 0.02917 | 0.00002038 | 252.5 | 6.44 |
| | 0.55 | 56.79 | 57.66 | 0.1268 | 0.8545 | 1.246 | 0.02936 | 0.00002023 | 404.2 | 6.909 |
| | 0.775 | 64.22 | 64.5 | 0.1949 | 0.7833 | 1.373 | 0.02949 | 0.00002003 | 583.5 | 7.45 |

*(continued overleaf)*

**Table D.2** (continued)

| T (°C) | φ | $T_{dp}$ (°C) | $T_{wb}$ (°C) | ω (kg/kg) | ρ (kg/m³) | $C_p$ (J/kg K) | k (W/mK) | μ (Pa s) | h (kJ/kg) | s (kJ/kg K) |
|---|---|---|---|---|---|---|---|---|---|---|
| 80 | 1 | 70 | 70 | 0.2763 | 0.712 | 1.526 | 0.02954 | 0.00001978 | 798.7 | 8.085 |
| | 0.1 | 31.94 | 39.79 | 0.03053 | 0.9528 | 1.065 | 0.02972 | 0.0000209 | 161.5 | 6.145 |
| | 0.325 | 54.53 | 56.35 | 0.1115 | 0.8476 | 1.218 | 0.03008 | 0.00002072 | 376.2 | 6.821 |
| | 0.55 | 65.91 | 66.53 | 0.2155 | 0.7424 | 1.413 | 0.0303 | 0.00002042 | 651.8 | 7.653 |
| | 0.775 | 73.83 | 74.01 | 0.3539 | 0.6371 | 1.673 | 0.03034 | 0.00002 | 1018 | 8.731 |
| | 1 | 80 | 80 | 0.5459 | 0.5319 | 2.034 | 0.03016 | 0.00001941 | 1530 | 10.2 |
| 90 | 0.1 | 39.05 | 45.43 | 0.04629 | 0.9047 | 1.096 | 0.03055 | 0.00002131 | 214.2 | 6.311 |
| | 0.325 | 62.93 | 64.26 | 0.1807 | 0.7533 | 1.349 | 0.03104 | 0.00002097 | 572.9 | 7.415 |
| | 0.55 | 75 | 75.42 | 0.3827 | 0.6018 | 1.729 | 0.03115 | 0.00002036 | 1112 | 9.003 |
| | 0.775 | 83.43 | 83.54 | 0.7208 | 0.4503 | 2.365 | 0.03074 | 0.00001937 | 2015 | 11.59 |
| | 1 | 90 | 90 | 1.397 | 0.2988 | 3.638 | 0.02958 | 0.00001786 | 3832 | 16.68 |
| 100 | 0.1 | 46.09 | 51.2 | 0.06911 | 0.8514 | 1.14 | 0.03141 | 0.00002169 | 286.5 | 6.535 |
| | 0.325 | 71.28 | 72.2 | 0.2995 | 0.6386 | 1.575 | 0.03197 | 0.00002106 | 905.7 | 8.397 |
| | 0.55 | 84.07 | 84.29 | 0.7601 | 0.4257 | 2.446 | 0.03154 | 0.00001971 | 2144 | 11.96 |
| | 0.775 | 93.01 | 93.03 | 2.142 | 0.2129 | 5.057 | 0.02941 | 0.00001719 | 5859 | 22.37 |
| | 0.9 | 97.07 | 97.06 | 5.596 | 0.09463 | 11.59 | 0.02707 | 0.000015 | 15144 | 48.03 |

*Source*: Table produced with Engineering Equation Solver (S.A. Klein 2014, version 9.698).

$T$ = temperature, $\phi$ = relative humidity, $T_{dp}$ = dew point temperature, $T_{wb}$ = wet bulb temperature, ω = humidity ratio, ρ = density, $C_p$ = specific heat, k = thermal conductivity, μ = dynamic viscosity, h = specific enthalpy, s = specific entropy.

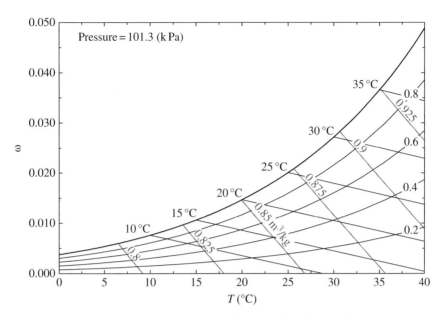

**Figure D.1**   Psychometric chart of humid air (produced with Engineering Equation Solver, developed by S.A. Klein 2014, version 9.698)

# Index

Note: Page numbers in *italics* refer to Figures; those in **bold** to Tables.

*Drying Phenomena: Theory and Applications*, First Edition. İbrahim Dinçer and Calin Zamfirescu.
© 2016 John Wiley & Sons, Ltd. Published 2016 by John Wiley & Sons, Ltd.